Unit Groups of Classical Rings

Unit Groups of Classical Rings

Gregory Karpilovsky

Department of Mathematics,
University of the Witwatersrand,
Johannesburg,
South Africa

CLARENDON PRESS · OXFORD
1988

Oxford University Press, Walton Street, Oxford OX2 6DP
Oxford New York Toronto
Delhi Bombay Calcutta Madras Karachi
Petaling Jaya Singapore Hong Kong Tokyo
Nairobi Dar es Salaam Cape Town
Melbourne Auckland
and associated companies in
Beirut Berlin Ibadan Nicosia

Oxford is a trade mark of Oxford University Press

Published in the United States
by Oxford University Press, New York

British Library Cataloguing in Publication Data
Karpilovsky, Gregory,
Unit groups of classical rings.
1. algebra. Rings
I. title
512'.4
ISBN 0-19-853557-0

Library of Congress Cataloging in Publication Data
Karpilovsky, Gregory, 1940–
Unit groups of classical rings/Gregory Karpilovsky.
Bibliography:
Includes indexes.
1. Fields, Algebraic—Units. 2. Rings (Algebra) 3. Groups,
Theory of. 4. Representations of groups. I. Title.
QA247.K325   1988   512'.4—dc 19   88-6972
ISBN 0-19-853557-0

Typeset and printed by The Universities Press (Belfast) Ltd

Dedicated to the memory
of my teacher
S. D. Berman,
who was and will remain my
source of inspiration

Preface

The subject matter of this book lies at the crossroads of four areas: ring theory, group theory, group representation theory, and algebraic number theory. Like so many interdisciplinary studies, it has its fascinations and attractions, but also its inherent dilemmas. A multitude of concepts and devices pertaining to the structure of groups and rings, algebraic K-theory, algebraic and analytic number theory, valuation theory, and group representation theory are used in open or disguised form to investigate the structure of the unit group of a particular ring. This provides for a vivid interplay among, and is a source of enrichment for, the mentioned disciplines. The *leitmotif* in our discussion is created by the following two related problems.

Problem A. Given a ring R, determine the isomorphism class of the unit group $U(R)$ of R in terms of natural invariants associated with R.
Problem B. Given a ring R, find an effective method for the construction of units of R.

Historically, the study of units was initiated by Gauss (1832), who dealt with $\mathbb{Z}[i]$ and whose work culminated in the celebrated Dirichlet Unit Theorem proved in 1840. The latter result provides a complete solution of Problem A when R is the ring of integers of an algebraic number field. Note, however, that Problem B is still far from being solved, even for such R. For example, the question of finding explicit formulae for the fundamental units of $\mathbb{Z}[\varepsilon_n]$ is still an open and challenging problem, even after more than 135 years of investigation pioneered by Kummer. In fact, with few exceptions, explicit formulae for fundamental units of $\mathbb{Z}[\varepsilon_n]$ are known only for the case where n is a prime power and $\phi(n) \leqslant 66$. The advance of our knowledge of unit groups of various rings has gone hand in hand with progress in algebraic number theory and ring theory. It has therefore been found convenient to relegate certain number-theoretic and ring-theoretic results to separate sections so as not to interrupt our discussion of unit groups at an awkward stage.

The purpose of this book is to give, in as self-contained a manner as possible, an up-to-date account of the structure of unit groups of classical rings. This exposition is not intended to be encyclopaedic in nature, nor is it a historical listing of the entire theory. Instead, it concentrates on what seem to be the most important and fruitful results. The guiding principle has been to tie together various threads of the development in an effort to convey a comprehensive picture of the current state of the subject. Enough examples have been included to help the research worker who needs to compute explicitly unit groups of certain rings. I have tried to avoid making the discussion too technical. With this view in mind, maximum generality has not been achieved in those places where this would entail a loss of clarity or a lot of technicalities.

The present monograph is written on the assumption that the reader has had the equivalent of a standard first-year graduate algebra course. Thus a familiarity with basic ring-theoretic, number-theoretic and group-theoretic concepts is assumed, as is an understanding of elementary properties of modules, tensor products, and fields. There is a fairly large bibliography of works which are either directly relevant to the text or offer supplementary material of interest.

A word about notation: as is customary, Theorem 3.4.2 denotes the second result of section 4 of Chapter 3; however, for simplicity, all references to this result within Chapter 3 itself are designated as Theorem 4.2.

The following is a brief description of the content of the book. In Chapter 1 algebraic preliminaries are established, and those basic results which will be assumed without proof are listed explicitly.

Chapter 2 is confined to the study of algebraic units. Among other results, the celebrated Dirichlet Unit Theorem is proved and the circumstances discovered under which there exists a complete independent system of units $u_1, u_2, \ldots, u_n$ of one of the following types: (a) the units $u_1, \ldots, u_n$ are real; (b) the units $u_1, \ldots, u_n$ are conjugate. The final two sections are devoted to a detailed discussion of unit groups of quadratic and pure cubic fields. The chapter culminates in the proof of a beautiful result known as the Delaunay–Nagell Theorem.

The cyclotomic fields $\mathbb{Q}(\varepsilon_n)$, $n \geq 2$, and their rings of integers $\mathbb{Z}[\varepsilon_n]$ play an important role in algebraic number theory. It is therefore appropriate to investigate the unit group of $\mathbb{Z}[\varepsilon_n]$ in detail. This constitutes the content of Chapter 3, where a powerful result due to Bass, pertaining to the independence of certain units is also proved.

In Chapter 4 our attention is confined to the study of multiplicative groups of fields. After proving some preliminary results, the isomorphism class of F^*, where F is a distinguished field such as local or global, is investigated. Concentrating on field extensions E/F, it is then proved that if F is infinite and $E \neq F$, then E^*/F^* is not finitely generated. The rest of the chapter, apart from Kneser's theorem, is based on works of May, who made fundamental contributions to the subject. Numerous examples are given to illustrate that the results obtained are the best possible.

The aim of Chapter 5 is to investigate the multiplicative structure of division rings. Among other results, it is shown that every subnormal solvable subgroup of the multiplicative group of a division ring is necessarily central. In particular, for any division ring D, all non-abelian subnormal subgroups of D^* are not solvable. Special attention is drawn to the following conjecture: a non-central subnormal subgroup of the multiplicative group of a division ring contains a non-cyclic free subgroup. Among other properties, we prove Lichtman's theorem which asserts that if G is a subnormal (originally stated normal) subgroup of D^*, and G has a non-abelian nilpotent-by-finite subgroup, then G contains a noncyclic free subgroup.

In Chapter 6 rings with cyclic unit groups are examined. The programme is carried out in two steps. First finite commutative rings with cyclic unit group are investigated. As a second step, using information obtained in the preceding step, the problem is solved in the general case.

The aim of Chapter 7 is to investigate circumstances under which the unit group of a ring is finitely generated. In case R is a finitely generated commutative ring, it is shown that $U(R)$ is finitely generated if and only if the additive group of the nilradical of R is finitely generated. This is derived as a consequence of a general result pertaining to unit groups of Krull domains. In the second part, some important facts of algebraic K-theory are introduced and it is shown that the knowledge of $K_1(R)$ is fundamental in the study of the finite generation of $GL_n(R)$. As an application, it is finally proved that if $R = \mathbb{Z}$ or $\mathbb{F}_q[X]$, then

(i) $GL_n(R)$ is finitely generated for all $n \geq 3$, and
(ii) $GL_n(R[X_1, \ldots, X_d])$ is finitely generated for all $n \geq d + 3$.

Chapter 8, the final chapter, provides a detailed analysis of the structure of unit groups of group rings.

No author sees the fruition of his labours without enlisting the assistance of many authors. I would like to express my gratitude to my wonderful wife for the tremendous help and constant encouragement which she has given me in the preparation of this book. My grateful thanks to my daughter, Suzanne, and my son, Elliott, for bearing my great despair, forgiving my inattentions, and denying me pity. For answering specific queries on topics contained in the text I am indebted to J. W. S. Cassels, W. May and J.-P. Serre. Finally, my thanks go to Arlene Harris and Lucy Rich for their excellent typing.

Johannesburg,   G. Karpilovsky
June 1987

Contents

Notation

Number systems

$\mathbb{N}$	the natural numbers	37
$\mathbb{Z}$	the integers	2
$\mathbb{Q}$	the rational numbers	6
$\mathbb{Q}_p$	the rational p-adic numbers	108
$\mathbb{R}$	the real numbers	7
$\mathbb{C}$	the complex numbers	25
$\mathbb{F}_q$	the field of q elements	102

Set theory

$|X|$	the cardinality of the set X	93
$Y - X$	the complement of X in Y	1
$\aleph_0$	the cardinal of $\mathbb{N}$	92
$\subset$	proper inclusion	1
$\subseteq$	inclusion	3
$g \circ f$	the composite of g and f	1

Number theory

$a \mid b$	a divides b	93
$a \nmid b$	a does not divide b	47
(a, b)	the greatest common divisor of a and b	78
$\left(\dfrac{n}{p}\right)$	the quadratic symbol	174

Group theory

$Z(G)$	the centre of G	179
S_n	the symmetric group of degree n	207
$\mathbb{Z}_n$	the cyclic group of order n	1
Q_8	the quaternion group of order 8	11
DC_{12}	the dicyclic group of order 12	11
BT_{24}	the binary tetrahedral group of order 24	11
$G' = [G, G]$	the derived group of G	312

$N \lhd G$	N is a normal subgroup of G	185
$(G:H)$	the index of H in G	27
$\langle X \rangle$	the subgroup generated by X	1
$[x, y]$	the commutator of x and y	179
$[X, Y]$	the subgroup generated by all commutators . $[x, y]$, $x \in X$, $y \in Y$	225
F^*	the multiplicative group of a field F	1
$C_G(X)$	the centralizer of X in G	191
$N_G(X)$	the normalizer of X in G	267
$\prod_{i \in I} G_i$	direct product of the family $\{G_i \mid i \in I\}$	97
$\bigoplus_{i \in I} G_i$	direct sum of the family $\{G_i \mid i \in I\}$	27
$\hat{G}$	$= \mathrm{Hom}(G, \mathbb{C}^*)$	74
χ_0	the principal character	74
$E_n(R)$	the elementary subgroup of $GL_n(R)$	234
$GL(R)$	the infinite general linear group	235
$K_1(R)$	the Whitehead group of R	237
$SK_1(R)$	the special Whitehead group of R	238
$\coprod_{i \in I} G_i$	the restricted direct product	12
χ^*	the Dirichlet character modulo n	76
A^n	$= \{a^n \mid a \in A\}$	2
$A[n]$	$= \{a \in A \mid a^n = 1\}$	96
$t(A)$ or A_0	the torsion subgroup of A	1
$Z(p^\infty)$	the group of the p^nth complex roots of unity with n running over $\mathbb{N}$	95
A_p	the p-component of A	1
Aut G	the automorphism group of G	11
$r_p(G)$	the p-rank of G	97
supp G	the set of all primes p for which $G_p \neq 1$	317
A^λ	the direct product of λ copies of A	92
$S(A)$	the socle of A	96
$Cl(G, x)$	the group generated by the conjugacy class of x	191
$\mathrm{core}_G\, H$	the core of H in G	206

Rings and modules

$A \otimes B$	the tensor product	336
char R	the characteristic of R	2
$\bigoplus_{I \in I} R_i$	the direct sum	219
$\prod_{i \in I} R_i$	the direct product of rings	255
$Z(R)$	the centre of R	2

Fields and Dedekind domains

1
Introduction

The main purpose of this chaper is to give the reader some idea of the mathematical background we are assuming, as well as to fix conventions and notation for the rest of the book. Because we presuppose a familiarity with various elementary group-theoretic and ring-theoretic terms, only a brief description of them is presented. Many readers may wish to glance briefly at the contents of this chapter, referring back to the relevant sections when they are needed later.

1.1 Notation and terminology

It is important to establish at the outset various conventions that we shall use throughout the book.

All rings in this book are *associative* with $1 \neq 0$ and subrings of a ring R are assumed to have the same identity element as R. Each ring homomorphism will be assumed to preserve identity elements. All modules are *left* and *unital*. We shall write $A - B$ for the complement of the subset B in the set A, while $A \subset B$ will mean that A is proper subset of B. The symbol for a map will be written before the element affected, and consequently, when the composition $f \circ g$ of maps is indicated, g is the first to be carried out.

Unless explicitly stated otherwise, all groups are assumed to be *multiplicative*. We use pointed brackets to denote 'subgroup generated by'. The cyclic group of order n is denoted by $\mathbb{Z}_n$, and the multiplicative group of a field F by F^*.

Let G be an abelian group. If every element of G is of finite order, G is called a *torsion group*, while G is *torsion-free* if all its elements, except for 1, are of infinite order, The set G_0 (or $t(G)$) of elements of finite order in G s a subgroup of G such that G/G_0 is torsion-free. We call G_0 the *maximal torsion subgroup* of G. It will also be called the *torsion subgroup* of G if no confusion can arise. We say that G is *free modulo torsion* if G/G_0 is free. Thus G is free modulo torsion if and only if $G = G_0 \times F$, where F is a free subgroup of G. Let G_p consist of all $g \in G$ whose order is a power of a prime p. Then G_p is a subgroup of G, called the *p-component* of G. Note that if G is a torsion group, then G is a (restricted) direct product of all such G_p. We say that G *splits* over its subgroup H if H is a direct factor of G. A group G is *bounded* if $G^n = \{g^n \mid g \in G\} = 1$ for some $n \geqslant 1$. In case $G^n = G$ for all $n \geqslant 1$, we refer to G as a *divisible group*. It is easy to verify that the product G_d of all divisible subgroups of G is a divisible group; the group G_d is called the *maximal divisible subgroup* of G. A group G is *reduced* if $G_d = 1$.

Let R be a ring. The mapping $\mathbb{Z} \to R$, $n \mapsto n \cdot 1$, is a ring homomorphism whose image is called the *prime subring* of R; its kernel is an ideal $m\mathbb{Z}$ for a unique $m \geqslant 0$, called the *characteristic of R* and denoted by char R.

An element u of a ring R is said to be a *unit* if $uv = vu = 1$ for some $v \in R$. The set $U(R)$ of all units in R constitute a group, called the *unit group* of R.

Let R be a ring. An element $x \in R$ is *nilpotent* if $x^n = 0$ for some positive integer n. An ideal J of R is *nil* if every element of J is nilpotent, while J is *nilpotent* if there is a positive integer n such that $J^n = 0$, where J^n is the product of J with itself n times. An element $x \in R$ is *idempotent* if $x^2 = x$. An idempotent is *trivial* if it is 0 or 1. Two idempotents u, v are *orthogonal* if $uv = vu = 0$. A non-zero idempotent is *primitive* if it cannot be written as a sum of two non-zero orthogonal idempotents. The centre of a ring R will be denoted by $Z(R)$.

The *Jacobson radical* $J(R)$ of a ring R is the intersection of maximal left ideals of R; equivalently $J(R)$ consists of all $x \in R$ such that for all $y, z \in R$, $1 - yxz$ is a unit. In particular, $J(R)$ contains any nil ideal of R. A ring R is *semisimple* if $J(R) = 0$.

Let R be a commutative ring. The set $N(R)$ of all nilpotent elements of R is an ideal called the *nilradical* of R. The ring R is *reduced if $N(R) = 0$*. Let R be a subring of a commutative ring S. For any subset A of S, denote by $R[A]$ the smallest subring of S containing R and A (for $A = \{a_1, \ldots, a_n\}$, we write $R[a_1, \ldots, a_n]$ instead of $R[A]$). We say that S is *finitely generated over R* (or simply *finitely generated* if R is the prime subring of S) if there is a finite subset A of S such that $S = R[A]$.

1.2 Assumed results

In this section we list explicitly those basic results which will be assumed without proof. Details may be found in standard texts covering these topics; for example, Bourbaki (1959, 1965), P. M. Cohn (1977), Janusz (1973) and Lang (1965, 1970). For convenience, we divide the relevant information into subsections.

1.2A Commutative rings

Throughout this subsection, all rings are assumed to be commutative. Given a ring R, we denote by $R[X_1, \ldots, X_n]$ the polynomial ring in the indeterminates X_i over R.

1.2.1 Proposition *Let F be a field and let $f(X) \in F[X]$ be a non-constant polynomial. If $f(X) = f_1(X)^{e_1} \ldots f_m(X)^{e_m}$ is a factorization of $f(X)$ into product of powers of distinct irreducible polynomials, then*

$$F[X]/(f(X)) \cong F[X]/(f_1(X)^{e_1}) \times \ldots \times F[X]/(f_m(X)^{e_m}).$$

1.2.2 Proposition (Hilbert's basis theorem) *If R is a noetherian ring, then so is $R[X_1, \ldots, X_n]$.*

1.2.3 Corollary *Any ring which is finitely generated over its noetherian subring is noetherian. In particular, a ring is noetherian if it is finitely generated.*

Let R be a subring of a ring S. An element $\alpha \in S$ is *integral* over R if α is a root of a monic polynomial in $R[X]$.

1.2.4 Proposition *Let R be a subring of S and let $\alpha \in S$. Then the following conditions are equivalent*:

  (i)   *α is integral over R;*
 (ii)   *the ring $R[\alpha]$ is a finitely generated R-module;*
(iii)   *there exists a subring R_1 of S with $\alpha \in R_1$, $R_1 \supseteq R$ and such that R_1 is a finitely generated R-module.*

An integral domain R is called a *Dedekind domain* if it is noetherian, integrally closed and every non-zero prime ideal of R is maximal. By a *fractional ideal* of R we understand any non-zero R-submodule of the quotient field F of R for which there exists a non-zero $d \in R$ such that $d \cdot I \subseteq R$. The product $I \cdot J$ of two fractional ideals of R is the set of all finite sums $\sum x_i y_i$ with $x_i \in I$, $y_i \in J$. It is clear that $I \cdot J$ is a fractional ideal. A *principal fractional ideal* is an ideal of the form Rx with $0 \neq x \in F$. The principal fractional ideals form a group which we denote by $P(R)$.

1.2.5 Proposition *Let R be a Dedekind domain. Then the set $I(R)$ of all fractional ideals of R is a free abelian group with the collection of non-zero prime ideals of R as free generators.*

For any Dedekind domain R, we write $C(R) = I(R)/P(R)$ and refer to $C(R)$ as the *class group* of R.

1.2.6 Proposition *Let R be a Dedekind domain and F the quotient field of R. Let E/F be a finite separable field extension and let S be the integral closure of R in E. Then S is again a Dedekind domain. Moreover, if for* *every non-zero ideal I of R, the factor ring R/I is finite, then the analogous property is true for S.*

1.2.7 Proposition   *Let A and B be non-zero ideals of a Dedekind domain R.*

(i) $A \subseteq B$ *if and only if* $A = BC$ *for some ideal C of R.*
(ii) *If* $A = P_1^{a_1} \ldots P_n^{a_n}$, $B = P_1^{b_1} \ldots P_n^{b_n}$, $a_i, b_i \in \mathbb{Z}$, $a_i, b_i \geq 0$, *where the* $\{P_i\}$ *are distinct prime ideals and* P_i^0 *is taken to be R, then*

$$A + B = \prod_{i=1}^{n} P_i^{\min(a_i, b_i)}, \qquad A \cap B = \prod_{i=1}^{n} P_i^{\max(a_i, b_i)}.$$

Let R be a Dedekind domain such that R/I is finite for any non-zero ideal I of R. Then the number of elements in R/I is called the *norm* of I and is denoted by $N(I)$.

1.2.8 Proposition   *Let R be the ring of integers of an algebraic number field K of degree n. Then, for any rational prime p, $N(pR) = p^n$ and pR is a product of, at most, n proper ideals of R.*

Let R be a Dedekind domain of characteristic 0, F the quotient field of R, E/F a field extension of degree n, and S the integral closure of R in E. Then S is a Dedekind domain and so, given a non-zero prime ideal P of R, we can write $SP = P_1^{e_1} \ldots P_r^{e_r}$, where the P_i's are distinct prime ideals of S and the e_i's are positive integers. The exponent e_i is called the *ramification index* of P_i over R. The prime ideal P is said to *ramify* in S (or in E) if $e_i > 1$ for some $i \in \{1, \ldots, r\}$. We may identify R/P with the subring of S/P_i for all $i \in \{1, \ldots, r\}$. Both R/P and S/P_i are fields, and S/P_i is a finite extension of R/P. We shall write f_i for the dimension of S/P_i over R/P, and call f_i the *residue degree* of P_i over R.

1.2.9 Proposition   *With the above notation, $\sum_{i=1}^{r} e_i f_i = (S/SP : R/P)$, and if E/F is normal, then $e_1 = e_2 = \ldots = e_r$ and $f_1 = f_2 = \ldots = f_r$.*

Let F and E be algebraic number fields with $F \subseteq E$, and let R and S be the rings of integers of F and E, respectively. The *discriminant ideal* of S over R, written $D_{S/R}$, is the ideal of R generated by the discriminants of bases of E over F which are contained in S.

1.2.10 Proposition   *Let $F \subseteq E$ be algebraic number fields, let $R \subseteq S$ be their rings of integers and let P be a non-zero prime ideal of R. Then:*

(i)   *P ramifies in S if and only if $P \supseteq D_{S/R}$;*
(ii) *there are only finitely many prime ideals of R which ramify in S.*

In what follows, $F \subseteq E$ are algebraic number fields and $R \subseteq S$ their rings of integers. The *norm homomorphism* $N_{E/F} : I(S) \to I(R)$ is defined as follows. Let $Q \neq 0$ be a prime ideal of S with residue degree f, and let $P \neq 0$ be the prime ideal in R which lies below Q, i.e. $P = Q \cap R$. We define $N_{E/F}(Q) = P^f$ and extend it to a unique homomorphism from $I(S)$ to $I(R)$.

1.2.11 Proposition

(i)   *If I is a non-zero ideal of S and $F = \mathbb{Q}$, then $N_{E/F}(I)$ is the principal ideal of $\mathbb{Z}$ generated by $N(I)$.*

(ii) *If I is a non-zero ideal of S, then $N_{E/F}(I) \subseteq R$.*

(iii) *If $I \subseteq J$, then $N_{E/F}(I) \subseteq N_{E/F}(J)$.*

(iv) *If $I \neq 0$, $J \neq 0$ are ideals of S such that I and $N_{E/F}(J)S$ are relatively prime, then $N_{E/F}(I)$ and $N_{E/F}(J)$ are also relatively prime.*

(v)   *For every I in $I(R)$, $N_{E/F}(IS) = I^n$, where $n = (E:F)$.*

(vi) *For any chain of fields $F \subseteq E \subseteq L$, $N_{E/F}N_{L/E} = N_{L/F}$.*

1.2.12 Proposition *Let M be the minimal normal extension of F containing E, let $A \in I(S)$ and let R' be the integral closure of R in M. Then*

$$N_{E/F}(A)R' = \prod_{\phi \in T} \phi(AR')$$

where T is a transversal in $\mathrm{Gal}(M/F)$ with respect to the subgroup acting trivially on E.

1.2.13 Proposition *If A is a principal fractional ideal generated by an element a in E, then $N_{E/F}(A)$ is the principal fraction ideal generated by $N_{E/F}(a)$.*

1.2.14 Proposition *If I is a fractional ideal of S, then $N_{E/F}(I)$ is the smallest fractional ideal of R which contains all norms $N_{E/F}(a)$ with $a \in I$.*

For the proof of Propositions 2.10–2.14 we refer to Narkiewicz (1974).

1.2B Fields and their valuations

Let E/F be a field extension. If S is a subset of E, we write $F(S)$ for the smallest subfield of E containing S and F. In case $S = \{\alpha_1, \ldots, \alpha_n\}$, we also write $F(\alpha_1, \alpha_2, \ldots, \alpha_n)$ instead of $F(S)$. A field extension E/F is *finitely generated* (or E is *finitely generated over* F) if $E = F(S)$ for some finite subset S of E. A field is *finitely generated* if it is finitely generated over its prime subfield.

Let F be any field and G be a group of automorphisms of F. The set of all elements in F fixed by G is a subfield called the *fixed field* of G. We say that E/F is a *Galois extension* if E is algebraic over F and E/F is normal and separable.

1.2.15 Proposition   *Let E/F be a finite field extension.*

(i)   *E/F is Galois if and only if F is the fixed field of $\mathrm{Gal}(E/F)$.*
(ii) *E/F is separable if and only if there exists exactly $(E:F)$ F-homomorphisms of E into a normal closure of E/F. Moreover, if E/F is separable, then $E = F(\alpha)$ for some $\alpha \in E$.*

Given a field extension E/F, we write $N_{E/F}$ and $Tr_{E/F}$ for the norm and trace maps, respectively. By a *discriminant* of $\lambda_1, \ldots, \lambda_n \in E$, written $D_{E/F}(\lambda, \ldots, \lambda_n)$ we understand $\det(Tr_{E/F}(\lambda_i\lambda_j))$, $1 \le i, j \le n = (E:F)$.

1.2.16 Proposition   *Let E/F be a finite separable field extension of degree n, let $\sigma_1, \sigma_2, \ldots, \sigma_n$ be all distinct F-homomorphisms of E into the algebraic closure of F and let $\lambda_1, \lambda_2, \ldots, \lambda_n$ be an F-basis of E. Then*

$$D_{E/F}(\lambda_1, \lambda_2, \ldots, \lambda_n) = (\det(\sigma_i(\lambda_j)))^2 \ne 0.$$

1.2.17 Proposition   *Let F be a field, n a positive integer prime to char F and assume that there is a primitive nth root of unity in F.*

(i)   *If E/F is a cyclic extension of degree n, then $E = F(\alpha)$ for some $\alpha \in E$ such that α satisfies an equation $X^n - a$ for some $a \in F$.*
(ii) *If $a \in F$ and α is a root of $X^n - a$, then $F(\alpha)$ is cyclic of degree d such that $d \mid n$ and $\alpha^d \in F$.*

1.2.18 Proposition   *Let E/F be a finite Galois extension with Galois group G and assume that G can be written as a direct product $G = G_1 \times \ldots \times G_n$. Let E_i be the fixed field of $G_1 \times \ldots \times G_{i-1} \times 1 \times G_{i+1} \times \ldots \times G_n$. Then E_i/F is Galois, $E_{i+1} \cap (E_1 \ldots E_i) = F$ and $E = E_1 \ldots E_n$.*

1.2.19 Proposition   *Let R be the ring of integers of an algebraic number field K.*

(i)   *R is a free $\mathbb{Z}$-module with $(R:\mathbb{Z}) = (K:\mathbb{Q})$.*
(ii) *An element $r \in R$ is a unit if and only if r is of norm ± 1.*

Let K be an algebraic number field and let R be the ring of integers of K. Then R is a free $\mathbb{Z}$-module, and the discriminants of the bases of the $\mathbb{Z}$-module R differ by a square of a unit of $\mathbb{Z}$ and hence coincide. Thus the discriminant of the $\mathbb{Z}$-module R is a well-defined element of $\mathbb{Z}$. It is also called the *discriminant* of K and is denoted by $d(K)$.

1.2.20 Proposition   *Let ε_n be a primitive nth root of 1 over $\mathbb{Q}$ and let R be the ring of integers of $\mathbb{Q}(\varepsilon_n)$. Then*

$$R = \mathbb{Z}[\varepsilon_n] \quad and \quad (R:\mathbb{Z}) = (\mathbb{Q}(\varepsilon_n):\mathbb{Q}) = \phi(n),$$

where ϕ is the Euler function.

Let F be an arbitrary field. A *valuation* v on F is any mapping $v:F \rightarrow \mathbb{R}$ such that:

(i) $v(x) \geq 0$ for all $x \in F$, and $v(x) = 0$ if and only if $x = 0$;
(ii) $v(xy) = v(x)v(y)$ for all $x, y \in F$;
(iii) $v(x + y) \leq v(x) + v(y)$ for all $x, y \in F$.

If, instead of (iii), v satisfies the stronger condition

$$v(x + y) \leq \max\{v(x), v(y)\}$$

then v is called a *non-archimedean valuation*. All the remaining valuations are called *archimedean*. Two valuations v_1, v_2 are said to be *equivalent* if there exists a postitive $r \in \mathbb{R}$ such that $v_2(x) = v_1(x)^r$ for all $x \in F$. By a *prime* of F, we understand the equivalence class of a valuation on F. The image $v(F^*)$ of v is called the *value group* of v. In case $v(F^*) = \{1\}$, v is called *trivial*, while if $v(F^*)$ is an infinite cyclic group, v is called *discrete*.

Assume now that F is the quotient field of a Dedekind domain R which is not a field. Then for any non-zero $x \in F$, we may factor Rx into a product of powers of prime ideals. Let $v_P(x)$ denote the exponent to which P occurs in this factorization. If P does not occur, set $v_P(x) = 0$ and put $v_P(0) = \infty$. Then, for any $a > 1$, the map $v_P:F \rightarrow \mathbb{R}$ given by $v_P(x) = a^{-v_P(x)}$ is a discrete valuation of F, called the *P-adic valuation associated with P*. If R/P is finite and $a = N(P)$, then v_P is called *normalized*.

1.2.21 Proposition   *With the above notation, the following holds:*

(i)   *R_P is the valuation ring of v_P (i.e. $R_P = \{x \in F \mid v_P(x) \leq 1\}$).*
(ii) *$R/P^n \cong R_P/P^n R_P$ for all $n \geq 1$.*
(iii) *The intersection of all R_P taken over all non-zero prime ideals P of R equals R.*
(iv) *$U(R) = \{x \in F \mid v_P(x) = 1$ for all non-zero prime ideals P of $R\}$.*

1.2.22 Proposition   *Let $v_1, \ldots, v_n$ be non-trivial non-equivalent valuations on a field F. Then, for any $\alpha_1, \ldots, \alpha_n \in F$ and any $\varepsilon > 0$, there exists $\alpha \in F$ such that $v_i(\alpha - \alpha_i) < \varepsilon$ for all $i \in \{1, \ldots, n\}$.*

The traditional formulation of the archimedean axiom runs as follows:

($*$) If x and y are non-zero elements of F, then there exists a positive integer n such that $v(nx) > v(y)$.

1.2.23 Proposition *Let v be a valuation of F. Then the following conditions are equivalent*:

(i) v *is non-archimedean*;
(ii) $v(n) \leqslant 1$ *for any* $n \in \mathbb{N}$;
(iii) *there exists* $c \in \mathbb{R}$ *such that* $v(n) \leqslant c$ *for all* $n \in \mathbb{N}$;
(iv) v *does not satisfy* ($*$).

Consider the set $\mathbb{R}_\infty$ obtained by adjoining to $\mathbb{R}$ symbol ∞. Then the linear order in $\mathbb{R}$ can be extended to that in $\mathbb{R}_\infty$ by putting $\alpha < \infty$ for all $\alpha \in \mathbb{R}$. Furthermore, $\mathbb{R}_\infty$ can be endowed with a commutative monoid structure by setting $\infty + \infty = \infty$, $\alpha + \infty = \infty + \alpha = \infty$, $\alpha \in \mathbb{R}$. We now define an *exponential* valuation v on F as any map $v : F \to \mathbb{R}_\infty$ such that:

(i) $v(x) = \infty$ if and only if $x = 0$;
(ii) $v(xy) = v(x) + v(y)$ for all $x, y \in F$;
(iii) $v(x + y) \geqslant \min\{v(x), v(y)\}$ for all $x, y \in F$.

The image $v(F^*)$ of F^* is called the value group of v. If the value group of v is $a\mathbb{Z}$ for some $a > 0$ in $\mathbb{R}$, we say that v is a *principal* valuation. Two exponential valuations v_1 and v_2 are called equivalent if there exists $r > 0$ in $\mathbb{R}$ such that $v_1(x) = rv_2(x)$ for all $x \in F$. For any $a > 1$ in $\mathbb{R}$, put $a^{-\infty} = 0$ and $-\log_a 0 = \infty$. For any exponential valuation v, define $v_v : F \to \mathbb{R}$ by $v_v(x) = a^{-v(x)}$. Then the map $v \mapsto v_v$ is an equivalence preserving bijective correspondence between the exponential and non-archimedean valuations of F.

A field F is said to be *complete* with respect to a valuation v if every Cauchy sequence relative to v converges to an element of F. By a *completion* of F at v, we mean a pair $(\bar{F}, \bar{v})$ consisting of a field $\bar{F}$ and a valuation $\bar{v}$ on $\bar{F}$ with the following properties:

(i) $\bar{F}$ is complete with respect to $\bar{v}$;
(ii) F is a subfield of $\bar{F}$ with $v(x) = \bar{v}(x)$ for all $x \in F$;
(iii) F is dense in $\bar{F}$ (i.e. every element of $\bar{F}$ is a limit of a convergent sequence of elements of F).

Let v_i be a valuation on a field F_i, $i = 1, 2$. A map $f : F_1 \to F_2$ is said to be an *isometric isomorphism* if f is an isomorphism of fields such that $v_2(f(x)) = v_1(x)$ for all $x \in F_1$.

Given Cauchy sequences $\{x_n\}$ and $\{y_n\}$, we write $\{x_n\} \sim \{y_n\}$ to mean $\lim_{n \to \infty} v(x_n - y_n) = 0$. Then ' $\sim$ ' is obviously an equivalence relation on the set of Cauchy sequences of F. We denote by $\{x_n\}^*$ the equivalence class of $\{x_n\}$ and by $\bar{F}$ the set of all such $\{x_n\}^*$. For each $x \in F$, the sequence $\{x_n\}$ defined by $x_1 = x_2 = \ldots = x$ is obviously a Cauchy sequence; let us denote this $\{x_n\}^*$ by x^*. We define $\{x_n\}^* + \{y_n\}^* = \{x_n + y_n\}^*$, $\{x_n\}^*\{y_n\}^* = \{x_n y_n\}^*$ and so $\bar{F}$ becomes a field.

1.2.24 Proposition *Let v be a valuation on F, let $\bar{F}$ be defined as above and let $\bar{v} : \bar{F} \to \mathbb{R}$ be defined by $\bar{v}(\{x_n\}^*) = \lim_{n \to \infty} v(x_n)$.*

(i)   *$(\bar{F}, \bar{v})$ is a completion of F at v (we regard F as embedded in $\bar{F}$ by means of $x \mapsto x^*$). In particular, by Proposition 2.23, v is archimedean if and only if so is $\bar{v}$.*

(ii) *$\bar{F}$ is uniquely determined by F and v, up to isometric isomorphism which is the identity map on F.*

(iii) *Assume that v is non-archimedean and let R_v, $R_{\bar{v}}$ and P_v, $P_{\bar{v}}$ be the valuation rings and maximal ideals of v and $\bar{v}$, respectively. Then:*

   (a)   *the value groups of v and $\bar{v}$ are the same (in particular, v is discrete if and only if so is $\bar{v}$);*

   (b)   *R_v is dense in $R_{\bar{v}}$, P_v is dense in $P_{\bar{v}}$ and the map $R_v/P_v \to R_{\bar{v}}/P_{\bar{v}}$ induced by the embedding $F \to \bar{F}$ is an isomorphism.*

1.2.25 Proposition (Hensel's Lemma) *Let F be a complete field with respect to a non-archimedean valuation v, let R be the valuation ring of v and let $-: R[X] \to (R/I)[X]$ be the natural homomorphism, where I is the maximal ideal of R. Let $f \in R[X]$ be such that $\bar{f} \neq 0$ and let $g_0, h_0 \in R[X]$ be such that $\bar{g}_0$ and $\bar{h}_0$ are relatively prime and $\bar{f} = \bar{g}_0 \bar{h}_0$. Then there exist $g, h \in R[X]$ with*

$$f = gh, \qquad \bar{g} = \bar{g}_0, \qquad \bar{h} = \bar{h}_0 \qquad and \qquad \deg g = \deg g_0.$$

1.2.26 Proposition *Let F be a complete field with respect to a valuation v and let E be an algebraic extension of F. Then the map $w : E \to \mathbb{R}$ given by*

$$w(\lambda) = [v(N_{F(\lambda)/F}(\lambda))]^{1/n}$$

where $n = (F(\lambda):F)$, is a unique extension of v to a valuation on E. Furthermore, if E/F is a finite extension of degree m, then E is complete and

$$w(\lambda) = [v(N_{E/F}(\lambda))]^{1/m} \qquad for\ all \qquad \lambda \in E.$$

   Let E/P be a finite field extension and let v be a valuation on F. Let $(\bar{F}, \bar{v})$ be the completion of F at v and let (K, ϕ, ψ) be a composite of $\bar{F}$ and E over F. Since $(E:F) = m < \infty$, we have $(K:\phi(\bar{F})) \leqslant m < \infty$. The valuation $\bar{v}$ on $\bar{F}$ can be transferred to $\phi(\bar{F})$ by defining $\bar{v}_\phi(\phi(x)) = \bar{v}(x)$ for all $x \in \bar{F}$. Clearly, $\bar{v}_\phi$ coincides with v on F. Since $\bar{F}$ is complete relative to $\bar{v}$, it is clear that $\phi(\bar{F})$ is complete relative to $\bar{v}_\phi$. Since K is a finite extension of $\phi(\bar{F})$, $\bar{v}_\phi$ has a unique extension to a valuation $\bar{w}_\phi$ on K. Hence the map $w_\phi : E \to \mathbb{R}$ defined by $w_\phi(x) = \bar{w}_\phi(\psi(x))$ is a valuation which extends v. We shall refer to w_ϕ as the *valuation on E associated with the composite* (K, ϕ, ψ).

1.2.27 Proposition *Let E/F be a finite field extension, let v be a valuation on F and let $(\bar{F}, \bar{v})$ be the completion of F at v. Let (K_i, ϕ_i, ψ_i), $1 \leqslant i \leqslant n$, be all non-equivalent composites of $\bar{F}$ and E over F and, for each i let v_i be the valuation on E associated with (K_i, ϕ_i, ψ_i). Then $v_1, \ldots, v_n$ are all distinct valuations on E extending v (in fact, all the v_i are non-equivalent). Moreover, if E_i is the completion of E with respect to v_i, then $(E:F) \geqslant \sum_{i=1}^n (E_i : F)$ with equality if E/F is separable.*

1.2.28 Proposition *Let E/F be a finite field extension and let v be a valuation of E such that the restriction w of v to F is non-trivial. Then v is discrete if and only if so is w.*

Let $F = \mathbb{Q}(\theta)$ be an algebraic number field and let $f(x) \in \mathbb{Q}[X]$ be the minimal polynomial of θ. Denote by $\alpha_1, \alpha_2, \ldots, \alpha_{r_1}$ (respectively, $\{\beta_1, \bar{\beta}_{r_1}\}, \ldots, \{\beta_{r_2}, \bar{\beta}_{r_2}\}$) all real roots (respectively, all pairs of conjugate complex roots) of $f(X)$. Then $f(X)$ splits into $r_1 + r_2$ irreducible factors over $\mathbb{R}$ and

$$(F : \mathbb{Q}) = r_1 + 2r_2.$$

The pair $[r_1, r_2]$ is called the *signature* of F. Those fields F for which $r_2 = 0$ are called *totally real*, and those with $r_1 = 0$ are called *totally complex*. Let $\lambda_1, \lambda_2, \ldots, \lambda_{r_1}$ be the real embeddings $F \to \mathbb{R}$ defined by $\lambda_i(\theta) = \alpha_i$, $1 \leqslant i \leqslant r_1$, and let $\mu_1, \mu_2, \ldots, \mu_{r_2}$ be the complex embeddings $F \to \mathbb{C}$ defined by $\mu_j(\theta) = \beta_j$, $1 \leqslant j \leqslant r_2$. Let R be the ring of integers of F and, for each non-zero prime ideal P of R, let v_P be the P-adic valuation associated with P.

1.2.29 Proposition *Further to the notation above, let Ω be the set of all non-zero prime ideals of R and let $v_i : F \to \mathbb{R}$ and $w_j : F \to \mathbb{R}$ be defined by $v_i(x) = |\lambda_i(x)|$, $w_j(x) = |\mu_j(x)|$, $1 \leqslant i \leqslant r_1$, $1 \leqslant j \leqslant r_2$. Then*

$$\{v_i \mid 1 \leqslant i \leqslant r_1\} \cup \{w_j \mid 1 \leqslant j \leqslant r_2\} \cup \{v_P \mid P \in \Omega\}$$

is a full set of non-equivalent non-trivial valuations on F. In particular, F has precisely $r_1 + r_2$ archimedean primes and all other primes are discrete.

The valuations of F exhibited in Proposition 2.29, where each v_P is normalized, are called *canonical*. The valuations v_i $(1 \leqslant i \leqslant r_1)$ and w_j $(1 \leqslant j \leqslant r_2)$ are called *real* and *complex*, respectively. As is customary, we adopt the convention of calling each w_j^2 a valuation, although they need not satisfy the triangular inequality. This convention allows us to define the set of all *normalized valuations* on F by

$$\{v_i \mid 1 \leqslant i \leqslant r_1\} \cup \{w_j^2 \mid 1 \leqslant j \leqslant r_2\} \cup \{v_P \mid P \in \Omega\}$$

where Ω is the set of all non-zero prime ideals of R and each v_P is normalized.

Let K be an arbitrary field, let $x \in K$ and let p be a prime of K. If $v(x) = 1$ for some v in p, then $v(x) = 1$ for all v in p. We may therefore write unambiguously $p(x) = 1$ to mean that $v(x) = 1$ for some (hence all) v in p.

1.2.30   Corollary   *Let R be the ring of integers of an algebraic number field F and let $\{p_i \mid i \in I\}$ be the set of all non-archimedean primes of F. Then*

$$U(R) = \{x \in F \mid p_i(x) = 1 \quad \text{for all } i \in I\}.$$

1.2C   Torsion-free rings

A ring R is said to be *torsion-free* if the additive group R^+ of R is torsion-free. Given an abelian group A, we write $\mathrm{End}(A)$ and $\mathrm{Aut}(A)$ for the endomorphism ring and the automorphism group of A, respectively. For the proof of the following two results we refer to Fuchs (1973).

1.2.31   Proposition   *Let R be a torsion-free ring whose additive group is reduced and of finite rank. Then there exists a torsion-free abelian group A which is reduced, of finite rank and such that $R \cong \mathrm{End}(A)$.*

1.2.32   Proposition   *If the finite group G is the automorphism group of a torsion-free abelian group A, then G is ismorphic to a subgroup of finite direct product of groups of the following types:*

  (i)   *cyclic groups of orders 2, 4 or 6;*
 (ii)   *the quaternion group $Q_8 = \langle \alpha, \beta \mid \alpha^2 = \beta^2 = (\alpha\beta)^2 \rangle$ of order 8;*
(iii)   *the dicyclic group $DC_{12} = \langle \alpha, \beta \mid \alpha^3 = \beta^2 = (\alpha\beta)^2 \rangle$ of order 12;*
 (iv)   *the binary tetrahedral group $BT_{24} = \langle \alpha, \beta \mid \alpha^3 = \beta^3 = (\alpha\beta)^2 \rangle$ of order 24.*

Conversely, these six groups and all finite direct products of them occur as automorphism groups of torsion-free abelian groups.

1.2.33   Corollary   *Let G be a finite group. The following conditions are equivalent:*

 (i)   *There exists a torsion-free abelian group A such that $G \cong \mathrm{Aut}(A)$.*
(ii)   *There exists a torsion-free ring R such that $G \cong U(R)$.*

Proof.

(i) $\Rightarrow$ (ii):   Since Aut(A) is the unit group of the ring $R = \text{End}(A)$, we have $G \cong U(R)$. Furthermore, since A is torsion-free, the additive group of R is also torsion-free, as required.

(ii) $\Rightarrow$ (i):   Let R be torsion-free ring such that $G \cong U(R)$. Denote by S the subring of R generated by $U(R)$. Then $U(S) = U(R)$ and the elements of S are obviously $\mathbb{Z}$-linear combinations of the elements of $U(R)$. Since $U(R)$ is finite and S^+ is torsion-free, it follows that S^+ is free of finite rank. But then S^+ is obviously reduced and hence, by Proposition 2.31, there exists a torsion-free abelian group A such that $R \cong \text{End}(A)$. Thus $U(R) \cong \text{Aut}(A)$ and the result follows.   ∎

1.2.34 Proposition *If the finite group G is the unit group of a torsion-free ring, then G is isomorphic to a subgroup of a finite direct product of groups of the following types: (i) cyclic groups of orders 2, 4 or 6; (ii) Q_8; (iii) DC_{12}; (iv) BT_{24}. Conversely, these six groups and all finite direct products of them occur as unit groups of torsion-free rings.*

Proof.   By Corollary 2.33, there exists a torsion-free abelian group A such that $G \cong \text{Aut}(A)$. The desired conclusion is therefore a consequence of Proposition 2.32.   ∎

Let $(G_i)_{i \in I}$ be a family of groups. The subgroup $\coprod_{i \in I} G_i$ of $\prod_{i \in I} G_i$ consisting of all (g_i) with finitely many $g_k (k \in I)$ distinct from 1 is called the *restricted direct product* of $(G_i)_{i \in I}$. In the future, unless explicitly stated otherwise, 'direct product' *will always mean restricted direct product.*

2
Algebraic units

Let R be the ring of integers of an algebraic number field F. In this chapter we shall provide some general information on the structure of the unit group of R. After establishing finiteness of the class group of R, we determine the isomorphism class of the S-unit group of F. As a consequence, we prove the celebrated Dirichlet's Unit Theorem which asserts that $U(R)$ is finitely generated of torsion-free rank $n = r_1 + r_2 - 1$, where $[r_1, r_2]$ is the signature of F. We then discover circumstances under which there exists an independent system of units $\varepsilon_1, \varepsilon_2, \ldots, \varepsilon_n$ of R of one of the following types: (a) The units $\varepsilon_1, \ldots, \varepsilon_n$ are real; (b) The units $\varepsilon_1, \ldots, \varepsilon_n$ are conjugate. The final two sections are devoted to a detailed discussion of unit groups of quadratic and pure cubic fields. Among other properties, we prove a beautiful result known as Delaunay–Nagell theorem, which asserts that if $F = \mathbb{Q}(\sqrt[3]{d})$, where $d > 1$ is a cube-free rational integer, then any unit $u \neq 1$ of R which is of norm 1 and of the form $x + y\sqrt[3]{d}$, x, $y \in \mathbb{Z}$, $y \neq 0$, is either the canonical fundamental unit or its square.

2.1 Finiteness of the class group

Throughout this section, R denotes the ring of integers of an algebraic number field F, n the degree of F, $d(F)$ the discriminant of F, and $\sigma_1, \ldots, \sigma_n$ all distinct $\mathbb{Q}$-homomorphisms $F \to \mathbb{C}$.

Since R is a Dedekind domain, the group $I(R)$ of all fractional ideals of R is freely generated by the non-zero prime ideals of R. Recall that the class group $C(R)$ of R is defined by $C(R) = I(R)/P(R)$, where $P(R)$ is the subgroup of $I(R)$ consisting of principal fractional ideals.

This section is devoted to the proof of the remarkable fact that $C(R)$ is actually a finite group. The order of this group is called the *class number* of R, or, by abuse of language, of F.

We begin by assembling some facts which will be useful later. Let I be a non-zero ideal of R. Since the additive group R/I is finite and R is $\mathbb{Z}$-free of rank n, I must also be $\mathbb{Z}$-free of rank n. Thus there exists a $\mathbb{Z}$-basis $x_1, \ldots, x_n$ of R, and integers $l_i > 0$ such that $l_1 x_1, \ldots, l_n x_n$ is a $\mathbb{Z}$-basis of I. Furthermore, we obviously have

$$N(I) = l_1 l_2, \ldots, l_n.$$

Applying this information, we now prove

2.1.1 Lemma   *Let I be a non-zero ideal of R and let $y_1, \ldots, y_n$ be a $\mathbb{Z}$-basis of I. Then $D_{F/\mathbb{Q}}(y_1, \ldots, y_n) = N(I)^2 \cdot d(F)$.*

Proof.   Let $x_1, \ldots, x_n$ be a $\mathbb{Z}$-basis of R such that $l_1 x_1, \ldots, l_n x_n$ is a $\mathbb{Z}$-basis of I for some integers $l_i > 0$. Then

$$D_{F/\mathbb{Q}}(l_1 x_1, \ldots, l_n x_n) = \left(\prod_{i=1}^{n} l_i \right)^2 D_{F/\mathbb{Q}}(x_1, \ldots, x_n) = N(I)^2 \cdot d(F)$$

and

$$D_{F/\mathbb{Q}}(l_1 x_1, \ldots, l_n x_n) = D_{F/\mathbb{Q}}(y_1, \ldots, y_n),$$

as asserted.   ∎

Let I be a fractional ideal of R, say $I = \prod_{i=1}^{s} P_i^{n_i}$ for some prime ideals P_i of R and some $n_i \in \mathbb{Z}$. We define the *norm* of I, written $N(I)$, by

$$N(I) = \prod_{i=1}^{s} N(P_i)^{n_i}.$$

Then it is clear that the map

$$N : I(R) \to \mathbb{Q}^*$$

is a homomorphism. We need a simple result connecting the norm of an element of F with the norm of the fractional ideal generated by it.

2.1.2 Lemma   *Let x be a non-zero element of F and $I = Rx$ the fractional ideal generated by x. Then $N(I) = |N_{F/\mathbb{Q}}(x)|$.*

Proof.   As both sides of the asserted equality are multiplicative in x, we may harmlessly assume that $x \in R$. If $x_1, \ldots, x_n$ is a $\mathbb{Z}$-basis of R, then $x_1 x, \ldots, x_n x$ is a $\mathbb{Z}$-basis of I. Hence, by Lemma 1.1,

$$N(I)^2 = D_{F/\mathbb{Q}}(x_1 x, \ldots, x_n x) d(F)^{-1}.$$

On the other hand, by Proposition 1.2.16,

$$\begin{aligned} D_{F/\mathbb{Q}}(x_1 x, \ldots, x_n x) &= \det((\sigma_i(x_j x))^2 \\ &= \det(\sigma_i \cdot (\delta_{ij} x))^2 \cdot \det(\sigma_i(x_j))^2 \\ &= (N_{F/\mathbb{Q}}(x))^2 \cdot d(F). \end{aligned}$$

Therefore $N(I)^2 = (N_{F/\mathbb{Q}}(x))^2$,   and   since $N(I) > 0$ then $N(I) = |N_{F/\mathbb{Q}}(x)|$.   ∎

The next observation will enable us to take full advantage of the previous results.

2.1.3 *Lemma*

(i)   *If I is a non-zero ideal of R and $m = N(I)$, then $mR \subseteq I$.*

(ii) *For any integer $m > 0$, there are only finitely many ideals of R with norm equal to m.*

(iii) *There exists an integer $\mu > 0$ such that for any non-zero ideal I of R there exists an element $a \in I$, $a \neq 0$, such that*

$$|N_{F/\mathbb{Q}}(a)| \leq N(I)\mu.$$

Proof

(i)   Since the additive group R/I is of order m, we have $m(R/I) = 0$, and thus $mR \subseteq I$.

(ii) Let I be an ideal of R with $N(I) = m$. By (i), $mR \subseteq I$ and so, by Proposition 1.2.7, I divides mR. Since mR has only finitely many divisors, the required assertion follows.

(iii) Given a $\mathbb{Z}$-basis $x_1, \ldots, x_n$ of R, choose $\mu \in \mathbb{Z}$ such that

$$\mu \geq \prod_{j=1}^{n} \left(\sum_{i=1}^{n} |\sigma_j(x_i)| \right).$$

To prove that μ has the required property, fix a non-zero ideal I of R and choose $k \geq 1$ such that $k^n \leq N(I) < (k+1)^n$. Consider the set S of elements $\sum_{i=1}^{n} d_i x_i$, $d_i \in \mathbb{Z}$, $0 \leq d_i \leq k$. Because $|S| = (k+1)^n > |R/I|$, there exist b, $c \in S$, $b \neq c$, and $a_i \in \mathbb{Z}$ with $|a_i| \leq k$ such that

$$b - c = \sum_{i=1}^{n} a_i x_i \in I.$$

Setting $a = b - c$, it follows that

$$|N_{F/\mathbb{Q}}(a)| = \prod_{j=1}^{n} \left| \sum_{i=1}^{n} a_i \sigma_j(x_i) \right| \leq \prod_{j=1}^{n} k \left(\sum_{i=1}^{n} \sigma_j(x_i) \right)$$

$$= k^n \prod_{j=1}^{n} \left(\sum_{i=1}^{n} |\sigma_j(x_i)| \right) \leq N(I)\mu$$

as required. ∎

We have now come to the demonstration for which this section has been developed.

2.1.4 *Theorem*   (Dedekind 1871, Kronecker 1882)   *Let R be the ring of integers of an algebraic number field. Then the class group $C(R)$ of R is finite.*

Proof.   Choose an integer μ as in Lemma 1.3(iii). Then, by Lemma 1.3(ii), there are only finitely many non-zero ideals of R, say

$I_1, I_2, \ldots, I_k$, such that $N(I_i) \leqslant \mu$. Let I be any fractional ideal of R. Then there exists $c \in R$, $c \neq 0$, such that cI^{-1} is an ideal of R. By Lemmas 1.2 and 1.3(iii), there exists an element $b \in cI^{-1}$, $b \neq 0$, such that

$$N(Rb) \leqslant N(cI^{-1}) \cdot \mu.$$

The latter implies that

$$N(Ibc^{-1}) \cdot N(Rc) = N(Ibc^{-1} \cdot Rc) = N(Rb)N(I)$$
$$\leqslant N(cI^{-1})N(I)\mu = N(Rc)\mu,$$

whence $N(Ibc^{-1}) \leqslant \mu$. But $Ibc^{-1} \subseteq R$, so $Ibc^{-1} = I_i$ for some $i \in \{1, 2, \ldots, k\}$. Therefore the images of I and I_i in $C(R)$ coincide, proving that $|C(R)| \leqslant k$. ∎

2.2   The Dirichlet–Chevalley–Hasse Unit Theorem

Throughout this section, R denotes the ring of integers of an algebraic number field F of degree r, $[r_1, r_2]$ the signature of F, $t(F^*)$ the torsion subgroup of F^* and S a finite set of primes of F which contains all archimedean primes. Our aim is to prove the celebrated Dirichlet's Unit Theorem which asserts that $U(R)$ is a finitely generated group of torsion-free rank $r_1 + r_2 - 1$. This result was established by Dirichlet in 1840, and a more general version, due to Chevalley and Hasse, was proved a century later. The following considerations serve to motivate the Chevalley–Hasse approach which we adopt.

An element $x \neq 0$ in F is called an *S-unit* if $p(x) = 1$ for all primes p of F not contained in S. One can easily verify that the S-units form a subgroup of the multiplicative group F^* of F. We shall refer to this subgroup as the *S-unit group* of F and denote it by F_S^*. Thus, if S consists of all archimedean primes, then

$$|S| = r_1 + r_2 \qquad \text{and} \qquad U(R) = F_S^*$$

by Proposition 1.2.29 and Corollary 1.2.30, respectively. Another interesting example is the case where $F = \mathbb{Q}$ and S consists of the ordinary absolute value together with the p_i-adic valuations, $1 \leqslant i \leqslant |S| - 1$. In this case we do have

$$F_S^* = \{ \pm p_1^{n_1} p_2^{n_2} \ldots p_{s-1}^{n_{s-1}} \mid n_i \in \mathbb{Z} \} \cong \mathbb{Z}_2 \times \mathbb{Z}^{s-1},$$

which again shows that F_S^* is a finitely generated group of torsion-free rank $|S| - 1$.

The above examples make it plausible to expect that F_S^* is always a finitely generated group of torsion-free rank $|S| - 1$. That this is indeed the case was proved by Chevalley and Hasse, and the rest of this section is aimed to demonstrate this remarkable fact.

We begin with two subsidiary results which are key ingredients for the proof of the main theorem.

2.2.1 Lemma   (The Product Formula)   *Let V be the set of all normalized valuations of F and let x be a non-zero element of F. Then $v(x) = 1$ for all but a finite number of $v \in V$ and*

$$\prod_{v \in V} v(x) = 1.$$

Proof.   Let $Rx = P_1^{n_1} P_2^{n_2} \dots P_k^{n_k}$ be the factorization into non-zero powers of distinct prime ideals of R. For any non-zero prime ideal P of R, let v_P be the normalized P-adic valuation of F. Then $v_P(x) = 1$ for all $P \notin \{P_1, \dots, P_k\}$. This proves the first statement since V contains only finitely many, namely $r_1 + r_2$, non-discrete valuations.

To prove the required formula, we first note that

$$\prod_{i=1}^{k} v_{P_i}(x) = \prod_{i=1}^{k} N(P_i)^{-n_i} = N(Rx)^{-1}.$$

On the other hand, the product $\prod v(x)$ over all normalized archimedean valuations v equals $|N_{F/\mathbb{Q}}(x)|$. The desired conclusion is therefore a consequence of Lemma 1.2.   ∎

2.2.2 Lemma   *Let c be a positive real number. Then there exists only finitely many $x \in R$ such that $|x_i| \leqslant c$ for all conjugates x_i of x.*

Proof.   Let $\sigma_1, \sigma_2, \dots, \sigma_r$ be all distinct $\mathbb{Q}$-homomorphisms $F \to \mathbb{C}$, let $x_i = \sigma_i(x)$, $1 \leqslant i \leqslant r$, and let

$$A = \{x \in R \mid |\sigma_i(x)| \leqslant c \quad \text{for all } i \in \{1, 2, \dots, r\}\}.$$

To prove that A is finite, it suffices to verify that the set of polynomials

$$\left\{ \prod_{i=1}^{r} (X - x_i) \mid x \in A \right\}$$

is finite. If $s_1, s_2, \dots, s_r$ are the elementary symmetric functions in r variables, then for all $x \in A$,

$$|s_k(x_1, \dots, x_r)| \leqslant l \quad (1 \leqslant k \leqslant r)$$

where

$$l = \max \left\{ rc, \binom{r}{2} c, \dots, \binom{r}{k} c^k, \dots, c^r \right\}.$$

Since $s_k(x_1, \dots, x_r) \in \mathbb{Z}$, $1 \leqslant k \leqslant r$, the result follows.   ∎

Turning our attention to the torsion subgroup $t(F^*)$ of F^*, we next prove

2.2.3 Lemma
(i) $t(F^*)$ is a finite cyclic group of even order.
(ii) $x \in t(F^*)$ if and only if $x \in R$ and all conjugates of x have absolute value 1.

Proof.

(i) Since $(F : \mathbb{Q}) = r < \infty$, the orders of elements of $t(F^*)$ are bounded. Let $\varepsilon \in t(F^*)$ be an element of maximal order m. If $x \in t(F^*)$, then $x^m = 1$ by our choice of m. Hence $x \in \langle \varepsilon \rangle$, and therefore $t(F^*) = \langle \varepsilon \rangle$. Furthermore, $-1 \in t(F^*)$ so that $t(F^*)$ is of even order.

(ii) If $x \in t(F^*)$, then $x \in R$ and all conjugates of x are in $t(F^*)$, and hence have absolute value 1. Conversely, assume that $x \in R$ is such that all conjugates of x have absolute value 1. Since $x, x^2, \ldots, x^n \ldots$ all share this property, it follows from Lemma 2.2 that $x^t = x^s$ for some integers t, s with $t < s$. Hence $x^{s-t} = 1$ and therefore $x \in t(F^*)$. ∎

Our next observation will be needed for the proof of a rather technical result, namely Lemma 2.5 below.

2.2.4 Lemma
If c is a positive real number, then the number of ideals of R with their norm bounded by c is finite.

Proof. Since the norm is multiplicative and its values are natural numbers, it suffices to prove that only finitely many prime ideals of R have their norm bounded by c. Assume by way of contradiction that there exists an infinite family $\{P_i \mid i \in I\}$ of prime ideals of R such that $|N(P_i)| \leq c$ for all $i \in I$. Choose a positive integer m with $c \leq m$ and consider any set of $m + 1$ distinct elements $x_1, \ldots, x_{m+1}$ of R. Then for any $P \in \{P_i \mid i \in I\}$, there exist x_i, x_j with $0 \neq x_i - x_j \in P$. But the set of differences $x_k - x_s$ is finite, hence there are infinitely many such P, a contradiction. ∎

The proof of the following result relies heavily on the equality $|N_{F/\mathbb{Q}}(a)| = N(Ra)$, established in Lemma 1.2.

2.2.5 Lemma
There exists a constant $c \in \mathbb{R}$ with the following property: if $a \in F$ and

$$1/2 < |N_{F/\mathbb{Q}}(a)| < 1$$

then there exists $\varepsilon \in U(R)$ such that

$$v(\varepsilon a) \leqslant c \qquad \text{for all normalized archimedean valuations } v.$$

Proof. Let $w_1, w_2, \ldots, w_r$ be a $\mathbb{Z}$-basis of R, let V be the set of all normalized archimedean valuations of F and let

$$X = \left\{ \sum_{i=1}^{r} c_i w_i \,\middle|\, |c_i| \leqslant 1, \, c_i \in \mathbb{Q} \right\}.$$

Denote by d an upper bound of $\{v(x) \mid v \in V, x \in X\}$. We first claim that if $m \in \mathbb{N}$, $a \in F$ and $|N_{F/\mathbb{Q}}(a)| \leqslant m^r$, then for a suitable non-zero $b \in R$, we have either $v(ab) \leqslant dm$ or $v(ab) \leqslant dm^2$, according as to whether v corresponds to a real or complex embedding of F. Indeed, if $a \in R$ then the sums

$$c_1 w_1 + \ldots + c_r w_r \qquad (0 \leqslant c_i \leqslant m, \, c_i \in \mathbb{Q})$$

consider all be distinct (mod Ra) for there are

$$(1 + m)^r > |N_{F/\mathbb{Q}}(a)| = N(Ra)$$

of them. Hence, by taking an appropriate difference of two such sums, we obtain

$$ab = c_1 w_1 + \ldots + c_r w_r$$

for some $0 \neq b \in R$ and some $c_i \in \mathbb{Q}$ such that $|c_i| \leqslant m$. Now $v(m^{-1}ab) \leqslant d$ since $m^{-1}ab \in X$, so $v(ab) \leqslant dm$ or $v(ab) \leqslant dm^2$ according as to whether v corresponds to a real or complex embedding of F. If $a \notin R$, then write $a = a_0/t$ with $a_0 \in R$ and $t \in \mathbb{N}$. Obviously

$$|N_{F/\mathbb{Q}}(a_0)| = |N_{F/\mathbb{Q}}(a)| \, |N_{F/\mathbb{Q}}(t)| \leqslant (mt)^r$$

and so, by the preceding argument, $v(a_0 b) \leqslant dmt$ or $v(a_0 b) \leqslant d(mt)^2$ for some non-zero b in R. This substantiates our claim by substituting $a_0 = at$.

By the foregoing, given $a \in F$ with $1/2 < |N_{F/\mathbb{Q}}(a)| < 1$, there exists non-zero $b = b(a)$ in R such that $v(ab) \leqslant d$ for all $v \in V$. This implies $|N_{F/\mathbb{Q}}(ab)| \leqslant d^{r_1 + r_2}$, which in turn shows that $|N_{F/\mathbb{Q}}(b)| \leqslant 2d^{r_1 + r_2}$ since $|N_{F/\mathbb{Q}}(a)| > 1/2$. Because $|N_{F/\mathbb{Q}}(b)| = N(Rb)$, Lemma 2.4 shows that the numbers $b = b(a)$ generate only a finite number of principal ideals, say $Rb_1, Rb_2, \ldots, Rb_k$. Hence, for every $a \in F$ satisfying our condition, we have $b(a) = \varepsilon b_i$ for some $i \in \{1, 2, \ldots, k\}$ and some unit ε of R. Thus we can see that

$$v(\varepsilon a) = v(ab)v(\varepsilon b^{-1}) = v(ab)v(b_i^{-1}) \leqslant dv(b_i^{-1}) \qquad \text{for all } v \in V.$$

Setting $c = \max\{dv(b_i^{-1}) \mid 1 \leqslant i \leqslant k\}$, the result follows.   ∎

The following result is needed in order to establish independence of certain units.

2.2.6 Lemma *Let* $(a_{ij}) \in M_n(\mathbb{R})$ *be such that* $a_{ij} < 0$ *for* $i \neq j$ *and* $a_{i1} + \ldots + a_{in} > 0$, $1 \leq i \leq n$. *Then* $\det(a_{ij})$ *is non-zero.*

Proof. Deny the statement. Then we can find a system $x_1, \ldots, x_n$ of numbers, at least one of which is positive and such that

$$\sum_{j=1}^{n} a_{ij}x_j = 0 \qquad (i = 1, 2, \ldots, n).$$

If $x_t = \max\{x_1, \ldots, x_n\}$ then $x_t > 0$ and

$$0 = \sum_{j=1}^{n} a_{tj}x_j = a_{tt}x_t + \sum_{j \neq t} a_{tj}x_j \geq x_t \sum_{j=1}^{n} a_{tj} > 0,$$

a contradiction. ∎

Let $\varepsilon_1, \varepsilon_2, \ldots, \varepsilon_n$ be a system of S-units of F. We say that this system is *independent* if

$$\varepsilon_1^{l_1} \varepsilon_2^{l_2} \ldots \varepsilon_n^{l_n} = 1 \qquad (l_i \in \mathbb{Z})$$

implies $l_1 = l_2 = \ldots = l_n = 0$. Expressed otherwise, $\varepsilon_1, \varepsilon_2, \ldots, \varepsilon_n$ is an independent system if the group $\langle \varepsilon_1, \varepsilon_2, \ldots, \varepsilon_n \rangle$ generated by the ε_i is free abelian of rank n. Of course, the above definition applies to the units of R, by taking S to be the set of all archimedean primes of F. The following easy consequence of Lemma 2.6 is extremely useful.

2.2.7 Corollary *Let* $v_1, \ldots, v_s$, $s = r_1 + r_2$, *be all normalized archimedean valuations of* F *and let* $\varepsilon_1, \varepsilon_2, \ldots, \varepsilon_{s-1}$ *be units of* R *such that*

$$v_j(\varepsilon_i) < 1 \qquad for\ j \neq i,\ 1 \leq i \leq s-1,\ 1 \leq j \leq s.$$

Then the units $\varepsilon_1, \varepsilon_2, \ldots, \varepsilon_{s-1}$ *are independent.*

Proof. Consider the homomorphism $U(R) \xrightarrow{f} \mathbb{R}^s$ given by

$$f(r) = (\log v_1(r), \log v_2(r), \ldots, \log v_s(r)).$$

It suffices to show that $f(\varepsilon_1), f(\varepsilon_2), \ldots, f(\varepsilon_{s-1})$ are $\mathbb{R}$-linearly independent or, equivalently, that $\det(\log v_j(\varepsilon_i)) \neq 0$, $i, j \in \{1, 2, \ldots, s-1\}$. Since $v_j(\varepsilon_i) < 1$ for $i \neq j$, we have $\log v_j(\varepsilon_i) < 0$ for $i \neq j$. On the other hand, by Lemma 2.1, we have

$$\sum_{j=1}^{s-1} v_j(\varepsilon_i) = -\log v_s(\varepsilon_i) > 0 \qquad (1 \leq i \leq s-1).$$

The required assertion now follows by virtue of Lemma 2.6. ∎

We next examine some topological properties of $\mathbb{R}^n$. A subset X of $\mathbb{R}^n$

is said to be *bounded* if there is a real number c such that for each $(x_1, \ldots, x_n) \in X$, $|x_i| \leq c$ for all $i \in \{1, 2, \ldots, n\}$. The following two properties of $\mathbb{R}^n$ are standard (see Bourbaki 1966):

(a) A subset X of $\mathbb{R}^n$ is compact if and only if X is closed and bounded.
(b) A subgroup G of $\mathbb{R}^n$ is discrete if and only if, for any compact subset X of $\mathbb{R}$, the intersection $G \cap X$ is finite.

A typical example of a discrete subgroup of $\mathbb{R}^n$ is $\mathbb{Z}^n$. The following lemma surveys all discrete subgroups of $\mathbb{R}^n$.

2.2.8 Lemma *Let G be a discrete subgroup of $\mathbb{R}^n$. Then G is generated as a $\mathbb{Z}$-module by m vectors which are $\mathbb{R}$-linearly independent (so $m \leq n$).*

Proof. Let $\{x_1, x_2, \ldots, x_m\}$ be a maximal set of $\mathbb{R}$-linearly independent elements of G and let

$$X = \left\{ x \in \mathbb{R}^n \;\middle|\; x = \sum_{i=1}^{m} \lambda_i x_i, \; 0 \leq \lambda_i \leq 1 \right\}.$$

Clearly X is closed and bounded, hence compact, and therefore $G \cap X$ is finite. Given $g \in G$, it follows from the maximality of $\{x_1, x_2, \ldots, x_m\}$ that $g = \sum_{i=1}^{m} \mu_i x_i$ for some $\mu_i \in \mathbb{R}$. For each $j \in \mathbb{Z}$, put

$$g_j = jg - \sum_{i=1}^{m} [j\mu_i] x_i,$$

where $[\lambda]$ denotes the largest integer less than or equal to $\lambda \in \mathbb{R}$. Then $g_i \in G \cap X$ and $g = g_1 + \sum_{i=1}^{m} [\mu_i] x_i$. Since each x_i lies in $G \cap X$, we conclude that the $\mathbb{Z}$-module G is generated by the finite set $G \cap X$. Furthermore, since $\mathbb{Z}$ is infinite, there exist distinct integers j and k such that $g_j = g_k$. Hence, by the definition of g_j,

$$(j - k)\mu_i = [j\mu_i] - [k\mu_i],$$

which implies that the μ_i's are rational. Thus the $\mathbb{Z}$-module G is generated by a finite number of elements which are $\mathbb{Q}$-linear combinations of the x_i's. Hence $dG \subseteq \bigoplus_{i=1}^{m} \mathbb{Z}x_i \subseteq G$ for some non-zero $d \in \mathbb{Z}$. It follows that G is a free group of rank m. Choose a $\mathbb{Z}$-basis $\{y_1, \ldots, y_m\}$ of G such that $\{d_1 y_1, \ldots, d_m y_m\}$ is a $\mathbb{Z}$-basis of $\bigoplus_{i=1}^{m} \mathbb{Z}x_i$ for some $d_i \in \mathbb{Z}$. Then the elements $d_1 y_1, \ldots, d_m y_m$ are $\mathbb{R}$-linearly independent and hence so are $y_1, \ldots, y_m$. ■

We have now come to the demonstration for which this section has been developed.

2.2.9 Theorem (Chevalley–Hasse; Chevalley 1940) *Let S be a finite set of primes of F which contains all archimedean primes. Then F_S^* is a direct product of $t(F^*)$ and a free abelian group of rank $|S| - 1$.*

Proof. For any $p \in S$, choose the normalized valuation v in p and denote by V the set of all such v. Then $|S| = |V| = s$, say. Consider the homomorphism

$$\begin{cases} F_S^* \xrightarrow{f} \mathbb{R}^s, \\ x \mapsto [\log v(x)]_{v \in V}. \end{cases}$$

The idea of the proof is to show that $\operatorname{Ker} f = t(F^*)$ and that $\operatorname{Im} f$ is free of rank $s - 1$. Since $t(F^*)$ is finite by Lemma 2.3(i), it will follow that $F_S^* \cong t(F^*) \times f(F_S^*)$, as required. The fact that $\operatorname{Ker} f = t(F^*)$ is almost obvious. Indeed, $t(F^*) \subseteq \operatorname{Ker} f$ since for any valuation w of F and any $x \in t(F^*)$, $w(x) = 1$. The opposite containment is a consequence of Lemma 2.3(ii) and Corollary 1.2.30. We are therefore left to verify that $f(F_S^*)$ is free of rank $s - 1$. For the sake of clarity, we divide the proof into three steps.

Step 1: here we show that $f(F_S^*)$ is free of rank $\leq s - 1$. By Lemma 2.1, $f(F_S^*)$ is contained in the hyperplane defined by the equation $X_1 + \ldots + X_s = 0$. Hence $f(F_S^*)$ contains at most $(s - 1)$ $\mathbb{R}$-linearly independent vectors. Let A be a bounded subset of $\mathbb{R}^n$ and let $B = \{x \in F_S^* \mid f(x) \in A\}$. By Lemma 2.8, it suffices to show that B is a finite set. Since A is bounded, there exists a real number $\alpha > 1$ such that for all $x \in B$

$$\alpha^{-1} < v(x) < \alpha \qquad \text{for all } v \in V. \tag{1}$$

Let $P_1, P_2, \ldots, P_k$ be prime ideals of R corresponding to the P_i-adic valuations in V. Then, for any $x \in F_S^*$, $Rx = P_1^{n_1} \ldots P_k^{n_k}$ for some $n_i \in \mathbb{Z}$. It follows from (1) that if $x \in B$, then $-\alpha < N(P_i)^{-n_i} < \alpha$, which shows that each n_i takes only finitely many values. Thus there are only finitely many principal fractional ideals of the form Rx, $x \in B$, say $Rx_1, \ldots, Rx_n$. Hence, if $x \in B$, then $x = rx_i$ for some $r \in U(R)$ and some $i \in \{1, \ldots, n\}$. But then (1) implies that $\alpha^{-1} v(r) < v(r) v(x_i) = v(x) < \alpha$, and hence $v(r) < \alpha^2$ for all $v \in V$. Since V includes all archimedean valuations, the required assertion follows by virtue of Lemma 2.2.

Step 2: the aim of this step is to verify that there exist $s - 1$ units in $U(R)$, $s = r_1 + r_2 > 1$, whose images in $\mathbb{R}^s$ are $\mathbb{R}$-linearly independent. Let $v_1, v_2, \ldots, v_s$ be all normalized archimedean valuations of F ordered so that $v_1, \ldots, v_{r_1}$ correspond to the real embeddings of F into $\mathbb{C}$. By the proof of Corollary 2.7, it suffices to exhibit units $\varepsilon_1, \varepsilon_2, \ldots, \varepsilon_{s-1}$ of R such that

$$v_j(\varepsilon_i) < 1 \qquad \text{for } j \neq i, \ 1 \leq i \leq s - 1, \ 1 \leq j \leq s. \tag{2}$$

To this end, let $0 < a_i < b_i (1 \leq i \leq s)$ be given real numbers. We claim that it is possible to find an element a in F such that

$$a_i < v_i(a) < b_i \qquad 1 \leqslant i \leqslant s. \tag{3}$$

Indeed, define

$$A_i = \begin{cases} a_i & \text{for } 1 = 1, 2, \ldots, r_1, \\ a_i^{1/2} & \text{for } i = r_1 + 1, \ldots, s; \end{cases}$$

and

$$B_i = \begin{cases} b_i & \text{for } i = 1, 2, \ldots, r_1, \\ b_i^{1/2} & \text{for } i = r_1 + 1, \ldots, s. \end{cases}$$

Then (3) can be rewritten in the form

$$A_i < |\sigma_i(a)| < B_i \qquad 1 \leqslant i \leqslant s, \tag{4}$$

where σ_i is the embedding corresponding to v_i. Now take an arbitrary $\mathbb{Z}$-basis of R, say $w_1, w_2, \ldots, w_r$, and consider the system of linear equations

$$x_1 \sigma_i(w_1) + \ldots + x_r \sigma_i(w_r) = h_i \qquad (i = 1, 2, \ldots, r),$$

where for $i = s + 1, s + 2, \ldots, r$ the embeddings σ_i are defined by $\sigma_i(x) = \sigma_{i-r_2}(x)$, and h_i are real numbers with $A_i < h_i < B_i$, $1 \leqslant i \leqslant s$, and $h_i = h_{i-r_2}$ for $i > s$. By Cramer's rule this system has a real solution. If we now choose rational numbers $y_1, \ldots, y_r$ with the differences $|x_i - y_i|$ small enough to satisfy the inequalities

$$A_i < |y_1 \sigma_i(w_1) + \ldots + y_r \sigma_i(w_r)| < B_i \qquad (i = 1, 2, \ldots, r)$$

(where for $i = s + 1, \ldots, r$ we define $A_i = A_{i-r_2}$, $B_i = B_{i-r_2}$), then the number $a = y_1 w_1 + \ldots + y_r w_r$ satisfies (4) and hence (3). Let c be the constant of Lemma 2.5. By (3), we may find $a_1, a_2, \ldots, a_s$ in F satisfying the inequalities

$$\begin{aligned} c < v_i(a_j) < dc & \qquad (1 \leqslant i, j \leqslant s, i \neq j) \\ 1/(2c^{s-1}) < v_j(a_j) < 1/(dc)^{s-1} & \qquad (1 \leqslant j \leqslant s), \end{aligned} \tag{5}$$

where the constant $d > 1$ is chosen to satisfy $d^{s-1} < 2$. Then $1/2 < |N_{F/\mathbb{Q}}(a_i)| < 1$ and so, by Lemma 2.5, for a certain $\varepsilon_i \in U(R)$, we obtain

$$v_j(\varepsilon_i a_i) < c \qquad (1 \leqslant j \leqslant s).$$

This proves (2) by applying (5).

Step 3: completion of the proof. We first observe that $U(R)$ is a subgroup of F_S^* and, by Steps 1 and 2, $F_S^*/U(R)$ has torsion-free rank at most $(s - 1) - (r_1 + r_2 - 1) = s - r_1 - r_2 = m$, say. Thus it suffices to exhibit elements $a_1, a_2, \ldots, a_m$ in F_S^* which are independent modulo $U(R)$. Let $v_1, \ldots, v_m$ be the discrete valuations in V and let $P_1, P_2, \ldots, P_m$ be the corresponding prime ideals of R. By Theorem 1.4, the class number $h(F)$ of F is finite. Setting $k = h(F)$ it follows that $P_i^k = a_i R$, $1 \leqslant i \leqslant m$, where the a_i are some S-units. Assume that

$$a = a_1^{n_1} \ldots a_m^{n_m} \in U(R) \qquad \text{for some } n_i \in \mathbb{Z}.$$

Then $Ra = (P_1^{n_1} \ldots P_m^{n_m})^k = R$, and thus by the uniqueness of factorization we obtain $n_1 = n_2 = \ldots = n_m = 0$. This completes the proof of the theorem. ∎

We now derive a number of consequences of interest.

2.2.10 Theorem (Dirichlet's Unit Theorem; Dirichlet 1840) *Let R be the ring of integers of an algebraic number field F and let $[r_1, r_2]$ be the signature of F. Then $U(R)$ is a direct product of $t(F^*)$ and a free abelian group of rank $r_1 + r_2 - 1$.*

Proof. Apply Theorem 2.9 to the special case where S consists of all archimedean primes of F. ∎

Given an ideal I of R, define the following subgroups of $U(R)$:

$$U_I(R) = \{r \in U(R) \mid r \equiv 1 \ (\text{mod } I)\},$$

$$t_I(F^*) = \{x \in t(F^*) \mid x \equiv 1 \ (\text{mod } I)\}.$$

2.2.11 Corollary *Let R be the ring of integers of an algebraic number field and let $[r_1, r_2]$ be the signature of F. Then, for any non-zero ideal I of R, $U_I(R)$ is a direct product of $t_I(F^*)$ and a free abelian group of rank $r_1 + r_2 - 1$.*

Proof. The natural homomorphism $R \to R/I$ induces a group homomorphism $U(R) \to U(R/I)$ whose kernel is $U_I(R)$. Since R/I is a finite ring, $U_I(R)$ is of finite index in $U(R)$. Hence, by Theorem 2.10, the torsion-free rank of $U_I(R)$ is $r_1 + r_2 - 1$. Since the torsion subgroup of $U_I(R)$ is $t_I(F^*)$, the result follows. ∎

2.2.12 Corollary *Let ε_n be a primitive nth root of 1 over $\mathbb{Q}$, where $n > 2$. Then*

$$U(\mathbb{Z}[\varepsilon_n]) = \begin{cases} \langle \varepsilon_n \rangle \times A & \text{if } n \text{ is even,} \\ \langle -\varepsilon_n \rangle \times A & \text{if } n \text{ is odd,} \end{cases}$$

where A is a free abelian group of rank $\phi(n)/2 - 1$.

Proof. By Proposition 1.2.20, $\mathbb{Z}[\varepsilon_n]$ is the ring of integers of $\mathbb{Q}(\varepsilon_n)$ and $\phi(n)$ is the degree of $\mathbb{Q}(\varepsilon_n)$. Since $n > 2$, no primitive nth root of 1 in $\mathbb{Q}(\varepsilon_n)$ is real. Thus $r_1 = 0$, $r_2 = \phi(n)/2$ so that $r_1 + r_2 - 1 = \phi(n)/2 - 1$.

Furthermore, the torsion subgroup of $\mathbb{Q}(\varepsilon_n)^*$ is $\langle \varepsilon_n \rangle$ or $\langle -\varepsilon_n \rangle$, according as to whether n is even or odd. The desired conclusion is therefore a consequence of Theorem 2.10. ∎

It is a consequence of Theorem 2.9 and Lemma 2.3(i) that there exist

$$n = |S| - 1$$

S-units $u_1, u_2, \ldots, u_n$ of F and a primitive $|t(F^*)|$-th root of unity ε such that any S-unit u may be uniquely expressed in the form

$$u = \varepsilon^k u_1^{\lambda_1} u_2^{\lambda_2} \ldots u_n^{\lambda_n} \qquad (\lambda_i \in \mathbb{Z}, 0 \leq k < |t(F^*)|).$$

Every such system $u_1, u_2, \ldots, u_n$ of S-units is called a *fundamental system of S-units*, while each u_i is called a *fundamental S-unit*. In the case where S consists of archimedean primes only, we speak simply about *fundamental system of units* and *fundamental units* of R. Of course, any fundamental system of S-units is necessarily independent, but the converse need not hold.

The determination of a fundamental system of units of R requires deep results and delicate computations. In fact, even in the case $R = \mathbb{Z}[\varepsilon_n]$ no specific fundamental system of units is known, with the exception of some particular values of n. We shall return to the problem of constructing a fundamental system of units in our subsequent investigations.

We close this section by providing some criteria for independence of units and introducing the notion of a regulator. It will be convenient to fix the following notation for the rest of this section:

$s = r_1 + r_2, \qquad n = r_1 + r_2 - 1;$

$v_1, \ldots, v_s$ are all normalized archimedean valuations of F;

σ_i is the embedding $F \to \mathbb{C}$ corresponding to v_i, $1 \leq i \leq s$;

$\lambda_i = 1$ or 2 according as to whether σ_i is real or complex, $1 \leq i \leq s$;

$x^{(i)} = \sigma_i(x), \qquad x \in F, 1 \leq i \leq s.$

2.2.13 Proposition *Let* $u_1, u_2, \ldots, u_k$ *be units of* R *and let* $x_1, x_2, \ldots, x_k$ *and* $y_1, y_2, \ldots, y_k \in \mathbb{R}^n$ *be defined by*

$$x_i = (\log |u_i^{(1)}|, \ldots, \log |u_i^{(n)}|) \qquad (1 \leq i \leq k),$$

$$y_i = (\log v_1(u_i), \ldots, \log v_n(u_i)) \qquad (1 \leq i \leq k).$$

Then the following conditions are equivalent:

(i) $u_1, u_2, \ldots, u_k$ *are independent;*
(ii) $x_1, x_2, \ldots, x_k$ *are* $\mathbb{Z}$-*linearly independent;*
(iii) $x_1, x_2, \ldots, x_k$ *are* $\mathbb{R}$-*linearly independent;*
(iv) $y_1, y_2, \ldots, y_k$ *are* $\mathbb{Z}$-*linearly independent;*
(v) $y_1, y_2, \ldots, y_k$ *are* $\mathbb{R}$-*linearly independent.*

Proof.   Define the homomorphism $f : U(R) \to \mathbb{R}^n$ *and* $g : U(R) \to \mathbb{R}^n$ by

$$f(u) = (\log |u^{(1)}|, \ldots, \log |u^{(n)}|),$$

$$g(u) = (\log v_1(u), \ldots, \log v_n(u)).$$

Then $x_i = f(u_i)$, $y_i = g(u_i)$ and thus (ii) $\Rightarrow$ (i), (iii) $\Rightarrow$ (i), (iv) $\Rightarrow$ (i), and (v) $\Rightarrow$ (i). By the definition of v_i, we have $v_i(u) = |u^{(i)}|$ or $|u^{(i)}|^2$ according as to whether σ_i is real or complex. Hence, by Lemmas 2.1 and 2.3(ii), $\operatorname{Ker} f = \operatorname{Ker} g = t(F^*)$. Thus (i) $\Rightarrow$ (ii) and (i) $\Rightarrow$ (iv). We are therefore left to verify that (i) $\Rightarrow$ (iii) and (i) $\Rightarrow$ (v).

To this end, note that by Theorem 2.10, $\operatorname{Im} f$ and $\operatorname{Im} g$ are free abelian groups of rank n. Furthermore, by Lemmas 2.1 and 2.2 any bounded region of $\mathbb{R}^n$ has only finitely many elements of $\operatorname{Im} f$ and $\operatorname{Im} g$. Hence both $\operatorname{Im} f$ and $\operatorname{Im} g$ are discrete subgroups of $\mathbb{R}^n$. It follows, from Lemma 2.8, that the $\mathbb{R}$-linear spans of $\operatorname{Im} f$ and $\operatorname{Im} g$ coincide with $\mathbb{R}^n$.

Assume that $u_1, u_2, \ldots, u_k$ are independent so that $A = \langle u_1, \ldots, u_k \rangle$ is free of rank k. By Theorem 2.10, we may choose free generators $a_1 t(F^*), \ldots, a_n t(F^*)$ of $U(R)/t(F^*)$ such that $a_1^{n_1} t(F^*)$, $a_2^{n_2} t(F^*), \ldots,$ $a_k^{n_k} t(F^*)$ are free generators of $A \cdot t(F^*)/t(F^*)$ for some $0 \neq n_i \in \mathbb{Z}$. By the foregoing, $f(a_1), \ldots, f(a_n)$ are $\mathbb{R}$-linearly independent and hence so are $f(a_1^{n_1}), \ldots, f(a_k^{n_k})$. But the $\mathbb{R}$-linear span of $f(A)$ is the same as the $\mathbb{R}$-linear span of $f(a_1^{n_1}), \ldots, f(a_k^{n_k})$. This implies that $x_1, x_2, \ldots, x_k$ are $\mathbb{R}$-linearly independent. A similar argument applied to g shows that $y_1, y_2, \ldots, y_k$ are $\mathbb{R}$-linearly independent, thus completing the proof. ∎

We now come to the definition of a regulator. Let $\varepsilon_1, \varepsilon_2, \ldots, \varepsilon_n$ be a fundamental system of units of R. By the *regulator of F*, written $\operatorname{reg}(F)$, we understand the absolute value of the determinant of the matrix $[\log v(\varepsilon_i)]$, with v running over all normalized archimedean valuations of F except one, which may be chosen arbitrarily. It is routine to verify that $\operatorname{reg}(F)$ depends neither on the choice of the fundamental system nor on the deleted valuation. Thus, in terms of the notation introduced above,

$$\operatorname{reg}(F) = |\det[\log v_j(\varepsilon_i)]| \qquad (1 \leq i, j \leq n)$$

or, to more explicitly, $\operatorname{reg}(F)$ is the absolute value of the following determinant:

$$\begin{vmatrix} \lambda_1 \log |\varepsilon_1^{(1)}| & \lambda_1 \log |\varepsilon_2^{(1)}| \ldots \lambda_1 \log |\varepsilon_n^{(1)}| \\ \lambda_2 \log |\varepsilon_1^{(2)}| & \lambda_2 \log |\varepsilon_2^{(2)}| \ldots \lambda_2 \log |\varepsilon_n^{(2)}| \\ \vdots & \vdots \quad\quad \vdots \\ \lambda_n \log |\varepsilon_1^{(n)}| & \lambda_n \log |\varepsilon_2^{(n)}| \ldots \lambda_n \log |\varepsilon_n^{(n)}| \end{vmatrix}$$

Finally note that, by Proposition 2.13, $\operatorname{reg}(F)$ is a *positive* real number.

Now assume that $\varepsilon_1, \ldots, \varepsilon_n$ is an arbitrary system of n units of R. We define the *regulator* of this system, written $\text{reg}(\varepsilon_1, \varepsilon_2, \ldots, \varepsilon_n)$ by the same formula as in the case $\varepsilon_1, \ldots, \varepsilon_n$ is a fundamental system. Thus $\text{reg}(\varepsilon_1, \varepsilon_2, \ldots, \varepsilon_n)$ is the absolute value of $\det(\lambda_i \log |\varepsilon_j^{(i)}|)$, $1 \leqslant i, j \leqslant n$. Again, one immediately verifies $\text{reg}(\varepsilon_1, \varepsilon_2, \ldots, \varepsilon_n)$ does not depend on the choice of the deleted valuation. Note also that, by Proposition 2.13, $\text{reg}(\varepsilon_1, \varepsilon_2, \ldots, \varepsilon_n)$ is non-zero (or, equivalently, positive) if and only if $\varepsilon_1, \ldots, \varepsilon_n$ are independent.

For future use, we finally record the following result and its consequence.

2.2.14 Proposition *Let $\varepsilon_1, \varepsilon_2, \ldots, \varepsilon_n$ and $\eta_1, \eta_2, \ldots, \eta_n$ be two independent systems of units of R. Let $A = t(F^*)\langle \varepsilon_1, \varepsilon_2, \ldots, \varepsilon_n \rangle$ and $B = t(F^*)\langle \eta_1, \eta_2, \ldots, \eta_n \rangle$. If $A \subseteq B$, then*

$$(B : A) = \frac{\text{reg}(\varepsilon_1, \varepsilon_2, \ldots, \varepsilon_n)}{\text{reg}(\eta_1, \eta_2, \ldots, \eta_n)}.$$

Proof. We may write

$$\varepsilon_i = \left(\prod_k \eta_k^{a_{ik}} \right)(\text{root of unity}) \qquad (a_{ik} \in \mathbb{Z}).$$

Therefore

$$\lambda_j \log |\varepsilon_i^{(j)}| = \sum_k a_{ik} \lambda_j \log |r_k^{(j)}|.$$

and thus

$$\text{reg}(\varepsilon_1, \ldots, \varepsilon_n) = |\det(a_{ik})|.$$

By the theory of elementary divisors, there exist integer matrices X and Y of determinant ± 1 such that $X(a_{ik})Y = \text{diag}(d_1, \ldots, d_n)$; so $\det(a_{ik}) = \pm \prod_{i=1}^n d_i$. But X and Y correspond to changing bases of $A/t(F^*)$ and $B/t(F^*)$, so we have bases $x_1, \ldots, x_n$ of $A/t(F^*)$ and $y_1, \ldots, y_n$ of $B/t(F^*)$ with $x_i = d_i y_i$. Therefore

$$B/A \cong B/t(F^*)/A/t(F^*) \cong \bigoplus_{i=1}^n \mathbb{Z}/d_i\mathbb{Z}$$

and

$$(B : A) = \left| \prod_{i=1}^n d_i \right| = |\det(a_{ik})|$$

as asserted. ∎

2.2.15 Corollary *Let $\varepsilon_1, \varepsilon_2, \ldots, \varepsilon_n$ be an independent system of units of R. Then*

$$(U(R) : t(F^*)\langle \varepsilon_1, \varepsilon_2, \ldots, \varepsilon_n \rangle) = \frac{\text{reg}(\varepsilon_1, \varepsilon_2, \ldots, \varepsilon_n)}{\text{reg}(F)}.$$

Proof. This is a particular case of Proposition 2.14 in which $\eta_1, \eta_2, \ldots, \eta_n$ is a fundamental system of units of R. ∎

2.3 Existence of real and conjugate independent units

In what follows, R is the ring of integers of an algebraic number field F of degree r, $t(F^*)$ the torsion subgroup of F^*, $[r_1, r_2]$ the signature of F and $n = r_1 + r_2 - 1$.

The problem that motivates this section is to discover circumstances under which there exists an independent system of units $\varepsilon_1, \varepsilon_2, \ldots, \varepsilon_n$ of R of one of the following types:

(a) the units $\varepsilon_1, \varepsilon_2, \ldots, \varepsilon_n$ are real;
(b) the units $\varepsilon_1, \varepsilon_2, \ldots, \varepsilon_n$ are conjugate.

Of course, it would be desirable to solve the corresponding problem for fundamental systems. However, this is a more formidable task, as one would expect. Nevertheless, our first result tackles case (a) for fundamental systems. To treat case (b), it is natural to assume that $F/\mathbb{Q}$ is normal in order to ensure that a conjugate of a unit of R is again a unit of R. It turns out that, under this restriction, it is always possible to find an independent system $\varepsilon_1, \varepsilon_2, \ldots, \varepsilon_n$ of units of R satisfying (b). Unfortunately, in this case our proof does not shed any light on the existence of sets of conjugate fundamental units.

When is it possible to find a fundamental system of units consisting entirely of real units? Obviously this happens if and only if for every $\varepsilon \in U(R)$ there exists $x_\varepsilon \in t(F^*)$ such that

$$\varepsilon x_\varepsilon \in \mathbb{R}.$$

It will be shown below that the latter always holds provided that $F/\mathbb{Q}$ is *cyclic*, i.e. $F/\mathbb{Q}$ is normal and the Galois group $\mathrm{Gal}(F/\mathbb{Q})$ is cyclic. The special case of the following result, in which $F = \mathbb{Q}(\varepsilon_p)$, p is an odd prime and ε_p a primitive pth root of 1, was established by Kummer (1870).

2.3.1 Theorem (Latimer 1934) *If $F/\mathbb{Q}$ is cyclic, then one can find in R a fundamental system of real units.*

Proof. Since $F/\mathbb{Q}$ is normal, if $r_1 \neq 0$ then $F \subseteq \mathbb{R}$ and there is nothing to prove. We may therefore assume that $r_1 = 0$, in which case $n = r_2 - 1 = r/2 - 1$. Let $\varepsilon_1, \varepsilon_2, \ldots, \varepsilon_n$ be a fundamental system of units of R, let σ be the generator of $\mathrm{Gal}(F/\mathbb{Q})$ and let ε be the generator of $t(F^*)$ of order m.

Consider the matrix $A = (a_{ij}) \in M_n(\mathbb{Z})$ whose entries a_{ij} are uniquely defined by

$$\sigma(\varepsilon_i) = \varepsilon^{a_i} \prod_{j=1}^{n} \varepsilon_j^{a_{ij}} \qquad (1 \leqslant i \leqslant n, \, a_i, \, a_{ij} \in \mathbb{Z}).$$

The main step in the proof is to show that

$$A^n + A^{n-1} + \ldots + A + I = 0, \qquad (1)$$

where I is the identity matrix.

Simple induction shows that if, for natural k, we put

$$\sigma^k(\varepsilon_i) = \varepsilon^{a_i(k)} \prod_{j=1}^{n} \varepsilon_j^{a_{ij}^{(k)}} \qquad (1 \leqslant i \leqslant n)$$

then $(a_{ij}^{(k)}) = A^k$, which in turn implies that

$$\varepsilon_i \sigma(\varepsilon_i) \ldots \sigma^n(\varepsilon_i) = \varepsilon^{b_i} \prod_{j=1}^{n} \varepsilon_j^{b_{ij}} \qquad (1 \leqslant i \leqslant n)$$

with $(b_{ij}) = A^n + A^{n-1} + \ldots + A + I$. Put $\lambda_i = \varepsilon_i \sigma(\varepsilon_i) \ldots \sigma^n(\varepsilon_i)$, $1 \leqslant i \leqslant n$. Since $\varepsilon_1, \varepsilon_2, \ldots, \varepsilon_n$ is a fundamental system of units, (1) will follow provided that we show that

$$\lambda_i \in t(F^*) \qquad \text{for all } i \in \{1, 2, \ldots, n\}. \qquad (2)$$

Because $\sigma^{n+1} = \sigma^{r/2}$ is the only element of order 2 in $\mathrm{Gal}(F/\mathbb{Q})$, it coincides with complex conjugation. Hence

$$\sigma^{r/2}(\varepsilon_i)\sigma^{r/2+1}(\varepsilon_i) \ldots \sigma^{r-1}(\varepsilon_i) = \sigma^{r/2}[\varepsilon_i \sigma(\varepsilon_i) \ldots \sigma^{r/2-1}(\varepsilon_i)]$$

$$= \sigma^{r/2}(\lambda_i)$$

$$= \bar{\lambda}_i \qquad (1 \leqslant i \leqslant n),$$

which shows that

$$N_{F/\mathbb{Q}}(\varepsilon_i) = \varepsilon_i \sigma(\varepsilon_i) \ldots \sigma^{r-1}(\varepsilon_i) = \lambda_i \bar{\lambda}_i = |\lambda_i|^2 \qquad (1 \leqslant i \leqslant n).$$

But ε_i is a unit, so $N_{F/\mathbb{Q}}(\varepsilon_i) = \pm 1$, and hence $|\lambda_i| = 1$ for all $i \in \{1, 2, \ldots, n\}$. Moreover, we obtain $|\sigma^k(\lambda_i)| = 1$ for $k \in \{1, 2, \ldots, r-1\}$. Indeed, for $k = 1$ we have

$$\sigma(\lambda_i) = \sigma(\varepsilon_i) \ldots \sigma^{n+1}(\varepsilon_i) = \lambda_i \varepsilon_i^{-1} \bar{\varepsilon}_i$$

whence $|\sigma(\lambda_i)| = 1$, and if $|\sigma^j(\lambda_i)| = 1$ holds for $j \in \{1, 2, \ldots k-1\}$, then the equalities

$$\sigma^k(\lambda_i) = \sigma^{k-1}(\sigma(\lambda_i)) = \sigma^{k-1}(\lambda_i \varepsilon_i^{-1} \bar{\varepsilon}_i)$$

$$= \sigma^{k-1}(\lambda_i)\sigma^{k-1}(\varepsilon_i^{-1})\overline{\sigma^{k-1}(\varepsilon_i)}$$

$$= \sigma^{k-1}(\lambda_i)\overline{\sigma^{k-1}(\varepsilon_i)}/\sigma^{k-1}(\varepsilon_i)$$

imply that $|\sigma^k(\lambda_i)| = 1$. Thus each conjugate of λ_i has absolute value 1 and therefore, by Lemma 2.3(ii), $\lambda_i \in t(F^*)$. This proves (2) and hence (1).

Multiplying both sides of (1) by A and taking into account that $n + 1 = r/2$, we derive $A^{r/2} = I$. Hence

$$\sigma^{r/2}(\varepsilon_i) = \varepsilon^{t_i}\varepsilon_i \qquad (1 \leq i \leq n), \tag{3}$$

where

$$t_i = \sum_{j=1}^{n} a_j(\delta_{ij}\beta^n + a_{ij}^{(1)}\beta^{n-1} + a_{ij}^{(2)}\beta^{n-2} + \ldots + a_{ij}^{(n)}) \qquad (1 \leq i \leq n) \tag{4}$$

and β is defined by $1 \leq \beta \leq m - 1$, $\sigma(\varepsilon) = \varepsilon^\beta$, and δ_{ij} is the Kronecker symbol.

Now, $\varepsilon^{-1} = \bar{\varepsilon} = \sigma^{r/2}(\varepsilon) = \varepsilon^{\beta^{r/2}}$ which implies that $\beta^{r/2} + 1 \equiv 0 \pmod{m}$. SInce m is even (Lemma 2.3(i)), we deduce that β is odd. Because $A^k = (a_{ij}^{(k)})$, $1 \leq k \leq n$, and β is odd it follows from (1) that the expressions in parenthesis in (4) are all even, and hence the same holds for the t_i's. Put $t_i = 2l_i$ and $\eta_i = \varepsilon^{l_i}\varepsilon_i$ for $i = 1, 2, \ldots, n$, so that $\eta_1, \eta_2, \ldots, \eta_n$ is a fundamental system of units. Then, applying (3), we have

$$\bar{\eta}_i = \sigma^{r/2}(\eta_i) = \varepsilon^{-l_i}\sigma^{r/2}(\varepsilon_i) = \varepsilon^{-l_i}\varepsilon^{t_i}\varepsilon_i$$
$$= \varepsilon^{l_i}\varepsilon_i = \eta_i,$$

proving that $\eta_i \in \mathbb{R}$ for all $i \in \{1, 2, \ldots, n\}$. This establishes the theorem. ∎

Next we prove the following classical result.

2.3.2 Theorem (Minkowski 1900) *If $F/\mathbb{Q}$ is a normal extension, then there exists a system of $r_1 + r_2 - 1$ independent units of R which are all conjugate.*

Proof. Let $\sigma_1, \sigma_2, \ldots, \sigma_r$ be all distinct $\mathbb{Q}$-homomorphisms $F \to \mathbb{C}$ and let $G = \mathrm{Gal}(F/\mathbb{Q})$. Since $F/\mathbb{Q}$ is normal, $G = \{\sigma_1, \sigma_2, \ldots, \sigma_r\}$ and either $r_1 = 0$ or $r_2 = 0$. Define $\sigma \in G$ by $\sigma = 1$ if $r_2 = 0$ and σ is the complex conjugation if $r_1 = 0$.

Observe that every canonical archimedean valuation of F is of the form v_g, $g \in G$, where

$$v_g(x) = v_1(g^{-1}(x)) \quad \text{and} \quad v_1(x) = |x| \qquad (x \in F).$$

Furthermore, if T is a left transversal for $H = \{1, \sigma\}$ in G then $\{v_t \mid t \in T\}$ is the set of all distinct canonical archimedean valuations of F. Now, if w_g is the normalized valuation corresponding to v_g, then for all $x \in F$, $w_g(x) < 1$ if and only if $v_g(x) < 1$. Hence, by Step 2 of the proof of

Theorem 2.9 we may find a unit ε of R such that

$$v_g(\varepsilon) < 1 \qquad \text{for all } v_g \neq v_1. \tag{5}$$

Now define $\eta = \varepsilon$ if $r_2 = 0$ and $\eta = \varepsilon\bar{\varepsilon}$ if $r_1 = 0$. If $r_2 = 0$, then $v_g(\eta) < 1$ for all $v_g \neq v_1$, by virtue of (5). If $r_1 = 0$, then by (5)

$$v_g(\eta) = v_1(g^{-1}(\eta))$$
$$= v_1(g^{-1}(\varepsilon))v_1((g^{-1}\sigma)(\varepsilon)) = v_g(\varepsilon)v_{\sigma g}(\varepsilon) < 1 \qquad \text{for all } v_g \neq v_1$$

since $v_g \neq v_1$ implies $v_{\sigma g} \neq v_1$. Thus

$$v_g(\eta) < 1 \qquad \text{for all } v_g \neq v_1. \tag{6}$$

Now let T' be obtained from T by deletion of one of the elements of T. Since $|T| = r_1 + r_2$, we have $|T'| = r_1 + r_2 - 1$. For each $t \in T'$, define $\eta_t = t(\eta)$. Then the units η_t, $t \in T'$ of R are all conjugate. Furthermore, by (6), for all $g \in T$ and $t \in T'$

$$v_g(\eta_t) = v_1(g^{-1}t(\eta)) = v_{tg^{-1}}(\eta) < 1 \qquad \text{if } g \neq t, \tag{7}$$

since $g \neq t$ implies $v_{tg^{-1}} \neq v_1$. The desired conclusion now follows by virtue of (7) and Corollary 2.7. ∎

2.4   Units in quadratic fields

The aim of this section is to investigate the group $U(R)$, where R is the ring of integers of a quadratic field.

An integer $d \in \mathbb{Z}$ is said to be *square-free* if $d \neq 1$ and d is not divisible by any square except 1. Thus d is square-free if and only if $d = -1$ or $|d|$ is a product of distinct primes. Every integer n which is not a square can be uniquely written in the form $n = m^2 s$, where s is square-free. We refer to s as the *square-free part* of n.

For convenience, we assume that all quadratic fields are embedded in $\mathbb{C}$. In what follows by $\sqrt{x}$, $x \in \mathbb{Q}$, we mean one of the two elements of $\mathbb{C}$ whose square is x. Finally, from Lemma 4.7 onwards, d *always denotes a positive square-free integer*.

2.4.1 Lemma
(i)   *Every quadratic field is of the form* $\mathbb{Q}(\sqrt{d})$, *where d is a square-free integer.*
(ii) *If $n, m \in \mathbb{Z}$ are non-squares, then* $\mathbb{Q}(\sqrt{n}) = \mathbb{Q}(\sqrt{m})$ *if and only if the square-free parts of n and m are the same.*

Proof
(i)   Let F be a quadratic field. Then any element $x \in F - \mathbb{Q}$ is of degree 2 over $\mathbb{Q}$. Let $f(X) = X^2 + bX + c$ $(b, c \in \mathbb{Q})$ be the minimal polynomial of

such an element $x \in F$. Solving the quadratic equation $x^2 + bx + c = 0$ gives

$$2x = -b \pm \sqrt{b^2 - 4c}$$

and thus $F = \mathbb{Q}(\sqrt{b^2 - 4c})$. Now, $b^2 - 4c \in \mathbb{Q}$ so that $b^2 - 4c = m/n = mn/n^2$, with $m, n \in \mathbb{Z}$, $n \neq 0$. Hence $F = \mathbb{Q}(\sqrt{mn})$ with $m, n \in \mathbb{Z}$. Let d be the square-free part of mn, say $mn = da^2$ for some $a \in \mathbb{Z}$. Then $F = \mathbb{Q}(\sqrt{d})$ as required.

(ii) We may harmlessly assume that both n and m are square-free, in which case we need only show that $\mathbb{Q}(\sqrt{n}) = \mathbb{Q}(\sqrt{m})$ implies $n = m$. If $\mathbb{Q}(\sqrt{n}) = \mathbb{Q}(\sqrt{m})$, then

$$\sqrt{m} = x + y\sqrt{n} \qquad \text{for some } x, y \in \mathbb{Q},$$

whence

$$m = x^2 + ny^2 + 2xy\sqrt{n}$$

and thus

$$m = x^2 + ny^2, \qquad 2xy = 0.$$

If $y = 0$, then $m = x^2$, which is impossible. If $x = 0$, then $m = ny^2$ and hence $m = n$, as asserted. ∎

2.4.2 Lemma *Let $F = \mathbb{Q}(\sqrt{d})$ be a quadratic field with $d \in \mathbb{Z}$ square-free (so $d \not\equiv 0 \pmod 4$) and let R be the ring of integers of F.*

(i) *If $d \equiv 2 \pmod 4$, or $d \equiv 3 \pmod 4$, then $R = \{a + b\sqrt{d} \mid a, b \in \mathbb{Z}\}$.*
(ii) *If $d \equiv 1 \pmod 4$, then $R = \{(1/2)(a + b\sqrt{d}) \mid a, b \in \mathbb{Z} \text{ of the same parity}\}$.*

Proof. Let σ be the $\mathbb{Q}$-automorphism of F which sends $\sqrt{d}$ to $-\sqrt{d}$. Given $x = a + b\sqrt{d} \in R$ with $a, b \in \mathbb{Q}$, we have

$$x + \sigma(x) = 2a \in \mathbb{Q} \cap R = \mathbb{Z}$$

and

$$x\sigma(x) = a^2 - db^2 \in \mathbb{Q} \cap R = \mathbb{Z},$$

so that

$$2a \in \mathbb{Z} \qquad \text{and} \qquad a^2 - db^2 \in \mathbb{Z}. \tag{1}$$

Conversely, if $a, b \in \mathbb{Q}$ satisfy (1), then $x = a + b\sqrt{d} \in R$. This is so since x is a root of $X^2 - 2aX + (a^2 - db^2) = 0$.

We now claim that (1) implies $2b \in \mathbb{Z}$. Indeed, by (1), $(2a)^2 - d(2b)^2 \in \mathbb{Z}$. Since $2a \in \mathbb{Z}$, we have $d(2b)^2 \in \mathbb{Z}$ too. On the other hand, d is square-free, so if $2b$ were not an integer, its denominator would have to include a prime factor p. But this prime factor would have to appear as p^2 in the denominator of $(2b)^2$ and thus $d(2b)^2 \notin \mathbb{Z}$, a contradiction.

We now take $a = u/2$, $b = v/2$ with $u, v \in \mathbb{Z}$, so that condition (1) becomes

$$u^2 - dv^2 \in 4\mathbb{Z}. \tag{2}$$

If v is even, (2) shows that u is even too. In this case $a, b \in \mathbb{Z}$. If v is odd, then $v^2 \equiv 1 \pmod 4$. Hence $u^2 \equiv 0$ or $1 \pmod 4$. Since d is square-free, $d \not\equiv 0 \pmod 4$ and so $u^2 \equiv 1 \pmod 4$ (hence u is odd) and $d \equiv 1 \pmod 4$. ∎

2.4.3 Corollary *Further to the notation of Lemma 4.2, let D be the discriminant of $\mathbb{Q}(\sqrt d)$.*

(i) *If $d \equiv 2 \pmod 4$ or $d \equiv 3 \pmod 4$, then $\{1, \sqrt d\}$ is a $\mathbb{Z}$-basis of R and $D = 4d$.*

(ii) *If $d \equiv 1 \pmod 4$, then $\{1, (1/2)(1 + \sqrt d)\}$ is a $\mathbb{Z}$-basis of R and $D = d$.*

Proof

(i) That $\{1, \sqrt d\}$ is a $\mathbb{Z}$-basis of R is a direct consequence of Lemma 4.2(i). The field $\mathbb{Q}(\sqrt d)$ has precisely two $\mathbb{Q}$-homomorphisms; $\sigma_i : \mathbb{Q}(\sqrt d) \to \mathbb{C}$, $i = 1, 2$, given by $\sigma_1(a + b\sqrt d) = a + b\sqrt d$, and $\sigma_2(a + b\sqrt d) = a - b\sqrt d$. Setting $l_1 = 1$, $l_2 = \sqrt d$, it follows from Proposition 1.2.16 that

$$D = (\det(\sigma_i(l_j)))^2 = \left(\det\begin{pmatrix} 1 & \sqrt d \\ 1 & -\sqrt d \end{pmatrix} \right)^2 = 4d.$$

(ii) To show that $(1/2)(u + v\sqrt d)$ (with $u, v \in \mathbb{Z}$ of the same parity) is a $\mathbb{Z}$-linear combination of 1 and $(1/2)(1 + \sqrt d)$, one may, by subtracting $(1/2)(1 + \sqrt d)$, reduce the problem to the case where u and v are even. Since

$$(1/2)(u + v\sqrt d) = (1/2)(u - v) \cdot 1 + v(1/2)(1 + \sqrt d),$$

we see that $\{1, (1/2)(1 + \sqrt d)\}$ is a $\mathbb{Z}$-basis of R. Setting $l_1 = 1$ and $l_2 = (1/2)(1 + \sqrt d)$, we have

$$D = (\det(\sigma_i(l_j)))^2 = \left(\det\begin{pmatrix} 1 & (1/2)(1 + \sqrt d) \\ 1 & (1/2)(1 - \sqrt d) \end{pmatrix} \right)^2 = d,$$

as required. ∎

2.4.4 Corollary *Let R be the ring of integers of $\mathbb{Q}(\sqrt d)$ where $d \in \mathbb{Z}$ is square-free.*

(i) *If $d \equiv 2 \pmod 4$ or $d \equiv 3 \pmod 4$, then*

$$U(R) = \{a + b\sqrt d \mid a, b \in \mathbb{Z}, a^2 - db^2 = \pm 1\}.$$

(ii)   *If $d \equiv 1 (\mathrm{mod}\ 4)$, then*

$$U(R) = \{(1/2)(a + b\sqrt{d}) \mid a,b \in \mathbb{Z}, a^2 - db^2 = \pm 4\}.$$

Proof.   If $x = a + b\sqrt{d}$, $a,b \in \mathbb{Q}$, then for $F = \mathbb{Q}(\sqrt{d})$

$$N_{F/\mathbb{Q}}(x) = (a + b\sqrt{d})(a - b\sqrt{d}) = a^2 - db^2.$$

Note also that if $a^2 - db^2 = \pm 4$ with $a,b \in \mathbb{Z}$, then since d is square-free, a and b are of the same parity. The desired conclusion is therefore a consequence of Lemma 4.2 and Proposition 1.2.19(ii).   ∎

2.4.5   Remark   *The formula*

$$U(R) = \{(1/2)(a + b\sqrt{d}) \mid a,b \in \mathbb{Z}, a^2 - db^2 = \pm 4\}$$

is true for any square-free d.

Indeed, if $a \equiv b \equiv 1 (\mathrm{mod}\ 2)$, then $a^2 \equiv b^2 \equiv 1 (\mathrm{mod}\ 4)$ and hence $a^2 - db^2 = \pm 4$ implies $d \equiv 1 (\mathrm{mod}\ 4)$. Now apply Corollary 4.4.   ∎

If $d > 0$, $\mathbb{Q}(\sqrt{d})$ is called a *real quadratic field*, while if $d < 0$ then $\mathbb{Q}(\sqrt{d})$ is said to be an *imaginary quadratic field*. In what follows $[r_1, r_2]$ denotes the signature of $\mathbb{Q}(\sqrt{d})$. Thus, if $d > 0$, then $r_1 = 0$ and $r_2 = 1$, while if $d > 0$, then $r_1 = 2$ and $r_2 = 0$. In the former case we have $r_1 + r_2 - 1 = 0$ so that $U(R)$ is a finite cyclic group (Theorem 2.10 and Lemma 2.3). With a little calculation we shall prove this result directly and make it more precise.

2.4.6   Proposition   *Let R be the ring of integers of an imaginary quadratic field F. Then $U(R) = \{-1, 1\}$, except in the following two cases;*

(i)   *If $F = \mathbb{Q}(i)$, $i^2 = -1$, then $U(R) = \langle i \rangle$.*
(ii)   *If $F = \mathbb{Q}(\sqrt{-3})$, then $U(R) = \langle \varepsilon \rangle$ where ε is a primitive sixth root of 1.*

Proof.   Write $F = \mathbb{Q}(\sqrt{-m})$, where $m = 1$ or m is a square-free positive integer.

(i)   If $m \equiv 1$ or $2 (\mathrm{mod}\ 4)$, then we may apply Corollary 4.4(i). For $x = a + b\sqrt{-m}$, $a,b \in \mathbb{Z}$, to be a unit of R we must have $a^2 + mb^2 = 1$. If $m \geq 2$, this implies $b = 0$ and $a = \pm 1$, so $x = \pm 1$. If $m = 1$, besides the solution $x = \pm 1$, there are the solutions $a = 0$, $b = \pm 1$, i.e. $x = \pm i$ with $i^2 = -1$.
(ii)   If $m \equiv 3 (\mathrm{mod}\ 4)$, then we may apply Corollary 4.4(ii). For $x = (1/2)(a + b\sqrt{-m})$, $a,b \in \mathbb{Z}$, to be a unit of R we must have $a^2 + mb^2 = 4$. If $m \geq 7$, this implies $b = 0$ so $a = \pm 2$ and $x = \pm 1$. If

$m = 3$ and $b \neq 0$, then the relations $a = \pm 1$, $b = \pm 1$ entail the additional solutions $x = (1/2)(\pm 1 \pm \sqrt{-3})$. Therefore the proposition is true. ∎

From now on, we concentrate on the more interesting and challenging case where F is a real quadratic field. As a preliminary observation, we first prove

2.4.7 Lemma   *Let R be the ring of integers of a real quadratic field F.*

(i)   *There is a unique unit $\varepsilon > 1$ in R such that $U(R) = \{\pm 1\} \times \langle \varepsilon \rangle$. In particular,*

$$U(R) \cong \mathbb{Z}_2 \times \mathbb{Z}$$

and

$$\pm \varepsilon, \pm \varepsilon^{-1}$$

are all fundamental units of R.

(ii)   *If $F = \mathbb{Q}(\sqrt{d})$, where $d \geq 2$ is a square-free integer and $u > 1$ is a unit of R, then $u = a + b\sqrt{d}$ for some $a > 0$, $b > 0$ in $\mathbb{Q}$.*

Proof

(i)   By hypothesis, $r_1 = 2$ and $r_2 = 0$, so that $r_1 + r_2 - 1 = 1$. On the other hand, since $F \subseteq \mathbb{R}$, $t(F^*) = \{\pm 1\}$. Now apply Theorem 2.10.

(ii)   Let $u = a + b\sqrt{d}$ $(a, b \in \mathbb{Q})$ be a unit of R. The numbers u, u^{-1}, $-u$, and u^{-1} are units of R and, since $N(u) = (a + b\sqrt{d})(a - b\sqrt{d}) = \pm 1$, these four numbers are $\pm a \pm b\sqrt{d}$. For $u \neq \pm 1$ only one of the four numbers u, u^{-1}, $-u$, $-u^{-1}$ is greater than one, and it is the largest of the four. Thus $u > 1$ implies $a > 0$, $b > 0$ as required. ∎

In what follows we refer to the unit ε in Lemma 4.7(i) as the *canonical fundamental unit* of R. Thus a knowledge of $U(R)$ is equivalent to a knowledge of the canonical fundamental unit ε, and therefore we henceforth restrict ourselves to the computation of ε.

For the rest of this section, we fix the following notation:

$d \geq 2$ is a square-free integer

R is the ring of integers of $\mathbb{Q}(\sqrt{d})$

ε_d is the canonical fundamental unit of R

$N(\varepsilon_d)$ is the norm of ε_d

Occasionally, by abuse of language, we say 'unit of $\mathbb{Q}(\sqrt{d})$' to mean 'unit of R'.

In view of Corollary 4.4, it is important to investigate the Diophantine equation

$$x^2 - dy^2 = m \tag{3}$$

in the special case where $m = \pm 1$, ± 4. The equation (3), commonly known as *Pell's equation*, was actually never considered by Pell; it was because of a mistake on Euler's part that his name has been attached to it. If $x = a$ and $y = b$ are integers which satisfy (3) we say, for simplicity, that the number

$$a + b\sqrt{d}$$

is a *solution* of eqn (3). It will be called *positive* if a and b are positive integers. The positive solutions will be ordered by the size of $a + b\sqrt{d}$. If $a_1 + b_1\sqrt{d}$ and $a_2 + b_2\sqrt{d}$ are two solutions of (3), then $a_1^2 - a_2^2 = d(b_1^2 - b_2^2)$. Hence, if both $a_1 + b_1\sqrt{d}$ and $a_2 + b_2\sqrt{d}$ are positive solutions of (3), then

$$a_1 + b_1\sqrt{d} > a_2 + b_2\sqrt{d} \Leftrightarrow a_1 > a_2 \Leftrightarrow b_1 > b_2 \tag{4}$$

The reader is warned that if $a_1 + b_1\sqrt{d}$ and $a_2 + b_2\sqrt{d}$ are positive solutions of the double equation $x^2 - dy^2 = \pm m$, then $a_1 + b_1\sqrt{d} > a_2 + b_2\sqrt{d}$ need not imply $b_1 > b_2$. Indeed, consider the double equation $x^2 - 5y^2 = \pm 4$. Then $3 + \sqrt{5}$ and $1 + \sqrt{5}$ are positive solutions with $3 + \sqrt{5} > 1 + \sqrt{5}$, but $b_1 = b_2 = 1$.

The following result provides a useful characterization of ε_d. Although it is possible to treat the cases $d \equiv 2$ or $3 \pmod 4$ and $d \equiv 1 \pmod 4$ simultaneously by applying Remark 1.5, we consider them separably for practical purposes.

2.4.8   Proposition

(i) If $d \equiv 2$ or $3 \pmod 4$, then $\varepsilon_d = a + b\sqrt{d}$ where $a + b\sqrt{d}$ is the minimal positive solution of $x^2 - dy^2 = \pm 1$.

(ii) If $d \equiv 1 \pmod 4$, then $\varepsilon_d = (1/2)(a + b\sqrt{d})$ where $a + b\sqrt{d}$ is the minimal positive solution of $x^2 - dy^2 = \pm 4$.

Proof

(i) By Corollary 4.4(i), we may write $\varepsilon_d = a + b\sqrt{d}$, with $a, b \in \mathbb{Z}$ satisfying $a^2 - db^2 = \pm 1$. Because $\varepsilon_d > 1$, Lemma 4.7(ii) implies that $a > 0$ and $b > 0$. Thus $a + b\sqrt{d}$ is a positive solution of $x^2 - dy^2 = \pm 1$. Let $a_1 + b_1\sqrt{d}$ be any positive solution of $x^2 - dy^2 = \pm 1$. Then $a_1 + b_1\sqrt{d} > 1$ and, by Corollary 4.4(i), $a + b\sqrt{d} \in U(R)$. Hence, by Lemma 4.7(i), $a_1 + b_1\sqrt{d} = \varepsilon_d^n$ for some $n \geq 1$. Since $\varepsilon_d > 1$, this shows that $a_1 + b_1\sqrt{d} \geq a + b\sqrt{d}$, as required.

(ii) Apply arguments of (i) with Corollary 4.4(i) replaced by Corollary 4.4(ii).   ∎

Note that if $N(\varepsilon_d) = 1$, then $N(u) = 1$ for all units u of R. Hence, if $d \equiv 2$ or $3 \pmod 4$ and $\varepsilon_d = a + b\sqrt{d}$ then, by Proposition 4.8(i), the following is true

(a) $a + b\sqrt{d}$ is the minimal positive solution of $x^2 - dy^2 = 1$ or $x^2 - dy^2 = -1$ according to as whether $N(\varepsilon_d) = 1$ or $N(\varepsilon_d) = -1$. In particular, $N(\varepsilon_d) = 1$ if and only if the equation $x^2 - dy^2 = -1$ has no solution in $\mathbb{Z}$.

Similarly, if $d \equiv 1 \pmod 4$ and $\varepsilon_d = (1/2)(a + b\sqrt{d})$, then by Proposition 4.8(ii) we have:

(b) $a + b\sqrt{d}$ is the minimal positive solution of $x^2 - dy^2 = 4$ or $x^2 - dy^2 = -4$ according to as whether $N(\varepsilon_d) = 1$ or $N(\varepsilon_d) = -1$. In particular, $N(\varepsilon_d) = 1$ if and only if the equation $x^2 - dy^2 = -4$ has no solution in $\mathbb{Z}$.

2.4.9 Proposition

(i) *Assume that $d \equiv 2$ or $3 \pmod 4$ and let $\varepsilon_d = a_1 + b_1\sqrt{d}$ with $a_1, b_1 \in \mathbb{N}$. Then the sequence*

$$a_n + b_n\sqrt{d} = (a_1 + b_1\sqrt{d})^n \qquad (n \geqslant 1)$$

gives all positive solutions of $x^2 - dy^2 = \pm 1$.

(ii) *Assume that $d \equiv 1 \pmod 4$ and let $\varepsilon_d = (1/2)(a_1 + b_1\sqrt{d})$ with $a_1, b_1 \in \mathbb{N}$. Then the sequence*

$$a_n + b_n\sqrt{d} = 2^{1-n}(a_1 + b_1\sqrt{d})^n \qquad (n \geqslant 1)$$

gives all positive solutions of $x^2 - dy^2 = \pm 4$.

Proof

(i) Since ε_d is of infinite order, all the $a_n + b_n\sqrt{d}$, $n \geqslant 1$, are distinct. Furthermore, by Corollary 4.4(i) and Lemma 4.7(ii), they are positive solutions of $x^2 - dy^2 = \pm 1$. If $z = a + b\sqrt{d}$ is any positive solution of $x^2 - dy^2 = \pm 1$, then $z \in U(R)$ and $z > 1$ so that, by Lemma 4.7(i), $z = \varepsilon_d^n = a_n + b_n\sqrt{d}$ for some $n \geqslant 1$.

(ii) Apply arguments of (i) with Corollary 4.4(i) replaced by Corollary 4.4(ii). ∎

We now apply Proposition 4.9 to provide a rather crude method for calculating the canonical fundamental unit ε_d of R. It will be convenient to treat the cases $d \equiv 2$ or $3 \pmod 4$ and $d \equiv 1 \pmod 4$ separately.

2.4.10 Proposition *Assume that $d \equiv 2$ or $3 \pmod 4$. In order to calculate*

$$\varepsilon_d = a_1 + b_1\sqrt{d} \qquad (a_1, b_1 \in \mathbb{N})$$

it suffices to write down the sequence db^2, $b = 1, 2, \ldots,$ and to stop at the first number db_1^2 of this sequence which differs from ± 1 by a square a_1^2.

Proof. Let $a_n + b_n\sqrt{d}$ be the sequence defined in Proposition 4.9(i). It suffices to show that

$$b_{n+1} > b_n \qquad \text{for all } n \geq 1.$$

Indeed, in this case ε_d can be characterized as the unique positive solution $a + b\sqrt{d}$ of $x^2 - dy^2 = \pm 1$ with smallest possible b. Since

$$a_{n+1} + b_{n+1}\sqrt{d}$$
$$= (a_1 + b_1\sqrt{d})(a_n + b_n\sqrt{d}) = (a_1a_n + b_1b_nd) + (b_1a_n + a_1b_n)\sqrt{d}$$

we have

$$b_{n+1} = b_1a_n + a_1b_n > b_n$$

as required. ∎

To illustrate the case $d \equiv 2$ or $3(\mathrm{mod}\ 4)$, let $d \in \{7, 6, 2\}$. If $d = 7$, the sequence db^2, $b = 1, 2, \ldots,$ is $7, 28, 7 \cdot 3^2 = 8^2 - 1$, so taking $b_1 = 3$ and $a_1 = 8$, we see that $8 + 3\sqrt{7}$ is the canonical fundamental unit of $\mathbb{Q}(\sqrt{7})$. A similar argument shows that $5 + 2\sqrt{6}$ is the canonical fundamental unit of $\mathbb{Q}(\sqrt{6})$. Note that both units $8 + 3\sqrt{7}$ and $5 + 2\sqrt{6}$ have norm 1. By taking $d = 2$, and hence $\varepsilon_d = 1 + \sqrt{2}$, we obtain an example of a canonical fundamental unit of norm -1.

Turning to the case $d \equiv 1(\mathrm{mod}\ 4)$, note that it is not true in general that $b_{n+1} > b_n$ for all $n > 1$. Indeed, it will be shown below that $(1/2)(1 + \sqrt{5})$ is the canonical fundamental unit of $\mathbb{Q}(\sqrt{5})$. Hence, if $a_n + b_n\sqrt{5} = 2^{1-n}(1 + \sqrt{5})^n$, then $b_1 = b_2 = 1$. This remark explains why we have to modify slightly the method of Proposition 4.10 in order to treat the case $d \equiv 1(\mathrm{mod}\ 4)$.

2.4.11 Proposition *Assume that $d \equiv 1(\mathrm{mod}\ 4)$. In order to calculate*

$$\varepsilon_d = (1/2)(a_1 + b_1\sqrt{d}) \qquad (a_1, b_1 \in \mathbb{N})$$

it suffices to write down the sequence db^2, $b = 1, 2, \ldots,$ and to stop at the first term db_1^2 of this sequence which differs from ± 4 by a square a_1^2 (if there are two such squares, then as a_1^2 choose the minimal one).

Proof. Let $a + b\sqrt{d}$ be the positive solution of $x^2 - dy^2 = \pm 4$ with smallest possible b and such that if there are two positive solutions of the form $c_1 + b\sqrt{d}$ and $d_1 + b\sqrt{d}$ then $a = \min\{c_1, d_1\}$. The assertion of the proposition amounts to saying that

$$\varepsilon_d = (1/2)(a + b\sqrt{d}).$$

Put $a_n + b_n = 2^{1-n}(a_1 + b_1\sqrt{d})^n$ so that by Proposition 4.9(ii), the $a_n + b_n\sqrt{d}$, $n \geqslant 1$, are all positive solutions of $x^2 - dy^2 = \pm 4$. We distinguish two cases.

Suppose first that $N(\varepsilon_d) = 1$. Then $x^2 - dy^2 = \pm 1$ implies $x^2 - dy^2 = 4$. Since

$$a_{n+1} + b_{n+1}\sqrt{d} = 2\varepsilon_d^{n+1} > 2\varepsilon_d^n = a_n + b_n\sqrt{d}$$

it follows from (4) that $b_{n+1} > b_n$ for all n. Hence $\varepsilon_d = a + b\sqrt{d}$.

Now assume that $N(\varepsilon_d) = -1$. Then

$$a_1 + b_1\sqrt{d} < a_3 + b_3\sqrt{d} < a_5 + b_5\sqrt{d} < \ldots$$

are solutions of $x^2 - dy^2 = -4$, while

$$a_2 + b_2\sqrt{d} < a_4 + b_4\sqrt{d} < a_6 + b_6\sqrt{d} < \ldots$$

are solutions of $x^2 - dy^2 = 4$. Hence, by (4),

$$b_1 < b_3 < b_5 < \ldots \qquad \text{and} \qquad b_2 < b_4 < b_6 < \ldots$$

This shows that $a + b\sqrt{d} = a_1 + b_1\sqrt{d}$ or $a_2 + b_2\sqrt{d}$. If $b_1 = b_2$, then $a = \min\{a_1, a_2\} = a_1$ so that $a + b\sqrt{d} = a_1 + b_1\sqrt{d}$. We are therefore left to verify that $b_1 \leqslant b_2$. But

$$b_{n+1} = (1/2)(a_1 b_n + b_1 a_n) \qquad (n \geqslant 1)$$

and so $b_2 = a_1 b_1 \geqslant b_1$, as required. ∎

2.4.12 Remark *Assume that* $d \equiv 1 \pmod{4}$ *and that* $d \neq 5$. *If* $\varepsilon_d = (1/2)(a_1 + b_1\sqrt{d})$ *then* $a_1 + b_1\sqrt{d}$ *is the unique positive solution of* $x^2 - dy^2 = \pm 4$ *with smallest possible* b_1.

Indeed, if $a + b\sqrt{d}$ is a positive solution of $x^2 - dy^2 = \pm 4$ with smallest possible b then, by the above, $a + b\sqrt{d} = a_1 + b_1\sqrt{d}$ or $a + b\sqrt{d} = a_2 + b_2\sqrt{d}$. Therefore, if $a + b\sqrt{d} = a_2 + b_2\sqrt{d}$, then $b_1 = b_2 = a_1 b_1$ in which case $a_1 = 1$. Hence $1 - db_1^2 = -4$ and so $db_1^2 = 5$. Since $d \geqslant 2$ we deduce that $d = 5$, as asserted. ∎

The above remark, together with Proposition 4.10, shows that *for any square-free integer $d \geqslant 2$ distinct from 5, $\varepsilon_d = (1/2)(a + b\sqrt{d})$, where $a + b\sqrt{d}$ is a unique positive solution of $x^2 - dy^2 = \pm 4$ with smallest possible b.*

To illustrate the case $d \equiv 1 \pmod 4$, let $d \in \{5, 13, 17, 69\}$. If $d = 5$, the first term of the sequence db^2, $b = 1, 2, \ldots$, is $5 = 1^2 + 4 = 3^2 - 4$. Hence $b_1 = 1$ and $a_1 = \min\{1, 3\} = 1$. Thus $(1/2)(1 + \sqrt{5})$ is the canonical fundamental unit of $\mathbb{Q}(\sqrt{5})$.

If $d = 13$, then the first term of the sequence db^2, $b = 1, 2, \ldots$, is $13 = 3^2 + 4$. Hence $b_1 = 1$ and $a_1 = 3$ so that $(1/2)(3 + \sqrt{13})$ is the canonical fundamental unit of $\mathbb{Q}(\sqrt{13})$. If $d = 17$, then the sequence db^2, $b = 1, 2, \ldots$, is 17, $68 = 8^2 + 4$ and hence $(1/2)(8 + 2\sqrt{17}) = 4 + \sqrt{17}$ is the canonical fundamental unit of $\mathbb{Q}\sqrt{17}$). Note that all the above three units have norm -1. If $d = 69$, the sequence db^2, $b = 1, 2, \ldots$, is

$$69, 276, 621 = 25^2 - 4$$

so that $a_1 = 25$ and $b_1 = 3$. Thus $(1/2)(25 + 3\sqrt{69})$ is the canonical fundamental unit of $\mathbb{Q}(\sqrt{69})$ with norm 1.

The above calculations become very awkward for large d. However, using the theory of continuous fractions, one can find other, more rapid, procedures for calculating the canonical fundamental unit.

The rest of this section will be devoted to the investigation of the sign of $N(\varepsilon_d)$.

Let a and $m > 1$ be coprime integers. We say that a is a *quadratic residue modulo* m if $x^2 \equiv a(\mathrm{mod}\,m)$ for some $x \in \mathbb{Z}$. Otherwise, we say that a is a *non-quadratic residue modulo* m. For each $n \in \mathbb{Z}$, let $\bar{n}$ be its image in $\mathbb{Z}/m\mathbb{Z}$ and let $U(m)$ be the unit group of $\mathbb{Z}/m\mathbb{Z}$. Then a is a quadratic residue modulo m if and only if $\bar{a}$ is a square in $U(m)$. Of course, if m is a power of an odd prime, the $U(m)$ is cyclic as is well known.

Let p be an odd prime and let a be an integer coprime to p. We define the *Legendre symbol* $(a \mid p)$ of a, relative to p, as follows:

$$(a \mid p) = \begin{cases} 1 & \text{if } a \text{ is a quadratic residue modulo } p, \\ -1 & \text{if } a \text{ is a non-quadratic residue modulo } p. \end{cases}$$

A straightforward verification shows that:

(i) If $a \equiv b(\mathrm{mod}\,p)$, then $(a \mid p) = (b \mid p)$.
(ii) $(ab \mid p) = (a \mid p)(b \mid p)$ (in particular, $(a^2 \mid p) = 1$).

2.4.13 Lemma *Let p be an odd prime and let a be an integer coprime to p. Then*

$$(a \mid p) \equiv a^{(p-1)/2}(\mathrm{mod}\,p)$$

and, in particular,

$$(-1 \mid p) = (-a^2 \mid p) = (-1)^{(p-1)/2}.$$

Proof. Let $a \equiv w^t(\mathrm{mod}\,p)$, where w is a primitive root modulo p and $0 \le t < p - 1$. Since $\bar{w}$ is not a square in $U(p)$ and $w^{(p-1)/2} \equiv -1(\mathrm{mod}\,p)$ we have

$$(a \mid p) = (w^t \mid p) = (w \mid p)^t = (-1)^t = (w^{(p-1)/2})^t$$
$$= (w^t)^{(p-1)/2} \equiv a^{(p-1)/2}(\mathrm{mod}\,p)$$

as asserted. ■

The following application of the above lemma is very useful.

2.4.14 Lemma *If $a, b \in \mathbb{N}$ satisfy $a^2 - db^2 = -4$, then all odd prime factors of d and b are congruent to 1 modulo 4 and $b \not\equiv 0 \pmod 4$.*

Proof. Since $a^2 - db^2 = -4$, we have $a^2 \equiv -4 \pmod{db^2}$. Hence, for any odd prime factor p of db^2, we have $a^2 \equiv -4 \pmod{p}$. Therefore, by Lemma 4.13,

$$1 = (-4 \mid p) = (-1)^{(p-1)/2}$$

which implies $p \equiv 1 \pmod 4$.

If $b \equiv 0 \pmod 4$, then $a^2 - db^2 = -4$ implies $a \equiv 0 \pmod 2$, and hence we may put $b = 4b_0$, $a = 2a_0$, in which case $a_0^2 - 4db_0^2 = -1$. Therefore $a_0^2 \equiv -1 \pmod 4$, which is a contradiction. ∎

It is now an easy matter to provide a necessary condition for $N(\varepsilon_d) = -1$.

2.4.15 Lemma *If $N(\varepsilon_d) = -1$, then all odd prime factors of d are congruent to 1 modulo 4.*

Proof. By Remark 4.5, $\varepsilon_d = (1/2)(a + b\sqrt{d})$, where $a, b \in \mathbb{N}$ satisfy $a^2 - db^2 = -4$. Now apply Lemma 4.14. ∎

The converse of the above result is decidedly false. Indeed, let $d = 34$ so that the only odd prime factor of 34 is $17 \equiv 1 \pmod 4$. Applying Proposition 4.10, we see that $\varepsilon_d = 35 + 6\sqrt{34}$ and thus $N(\varepsilon_d) = 1$.

The following property is a partial converse of Proposition 4.15.

2.4.16 Proposition *If d is a prime congruent to 1 modulo 4, then $N(\varepsilon_d) = -1$.*

Proof. It suffices to exhibit at least one unit of norm -1. The latter will hold provided we show that the equation $x^2 - dy^2 = -1$ is solvable in $\mathbb{Z}$.

Let $x_1 + y_1\sqrt{d}$ be the minimal positive solution of $x^2 - dy^2 = 1$. Then

$$x_1^2 - 1 = dy_1^2. \tag{5}$$

Hence x_1 cannot be even, for in this case we should have $-1 \equiv d \pmod 4$. Therefore x_1 is odd, in which case the numbers $x_1 - 1$ and $x_1 + 1$ have the greatest common divisor 2. It follows from (5) that

$$x_1 \pm 1 = 2a^2, \qquad x_1 \mp 1 = 2db^2,$$

where a and b are natural numbers and $y_1 = 2ab$. By elimination of x_1, we obtain

$$\pm 1 = a^2 - db^2.$$

Since $b < y_1$ we cannot have the upper sign. Thus the lower sign must be taken and the result follows. ∎

In order to investigate those $\mathbb{Q}(\sqrt{d})$ for which $N(\varepsilon_d) = -1$, we first give a generating function for all such fields. All the remaining results of this section are based on a work of Yokoi (1968).

2.4.17 Lemma *A unit ε of R has norm -1 if and only if*

$$\varepsilon = (t \pm \sqrt{t^2 + 4})/2$$

for some $t \in \mathbb{Z}$ with $\mathbb{Q}(\sqrt{d}) = \mathbb{Q}(\sqrt{t^2 + 4})$. In particular, $N(\varepsilon_d) = -1$ if and only if $\mathbb{Q}(\sqrt{d}) = \mathbb{Q}(\sqrt{n^2 + 4})$ for some $n \in \mathbb{N}$ or, equivalently, if and only if d is the square-free part of $n^2 + 4$ for some $n \in \mathbb{N}$.

Proof. If ε is a unit of R with norm -1, then by Remark 4.5, $\varepsilon = (t + u\sqrt{d})/2$ where $t, u \in \mathbb{Z}$ satisfy

$$t^2 - du^2 = -4. \tag{6}$$

By (6), we have $t^2 + 4 = du^2$, so that $\mathbb{Q}(\sqrt{d}) = \mathbb{Q}(\sqrt{t^2 + 4})$ and $u\sqrt{d} = \pm\sqrt{t^2 + 4}$. Thus $\varepsilon = (t \pm \sqrt{t^2 + 4})/2$.

Conversely, assume that $\varepsilon = (t \pm \sqrt{t^2 + 4})/2$, where $t \in \mathbb{Z}$ is such that $\mathbb{Q}(\sqrt{d}) = \mathbb{Q}(\sqrt{t^2 + 4})$. Then, by Lemma 4.1(ii), $t^2 + 4 = du^2$ for some $u \in \mathbb{N}$ and thus $\varepsilon = (t \pm u\sqrt{d})/2$. Since $t^2 - du^2 = -4$, we conclude by Remark 4.5, that ε is a unit of R with norm -1. ∎

It is now possible to provide a global picture of canonical fundamental units of norm -1.

2.4.18 Proposition *The sequence*

$$(n + \sqrt{n^2 + 4})/2$$

with $n \in \mathbb{N}$ such that $n < m$ whenever $n \neq m$ and $n^2 + 4$, $m^2 + 4$ have the same square-free part, gives all canonical fundamental units of norm -1. In fact, for any such n,

$$(n + \sqrt{n^2 + 4})/2 = \varepsilon_d$$

where d is the square-free part of $n^2 + 4$.

Proof. Fix $n \in \mathbb{N}$ such that $n < m$ whenever $n \neq m$ and $n^2 + 4$, $m^2 + 4$ have the same square-free part. We claim that $\varepsilon_d = (n + \sqrt{n^2 + 4})/2$ and $N(\varepsilon_d) = -1$, where d is the square-free part of $n^2 + 4$. Indeed, by Lemma 4.17, $(n + \sqrt{n^2 + 4})/2$ is a unit of R with norm -1. Hence $N(\varepsilon_d) = -1$ and so, by Lemma 4.17,

$$\varepsilon_d = (m + \sqrt{m^2 + 4})/2$$

for some $m \in \mathbb{N}$, such that d is the square-free part of $m^2 + 4$. If $n \neq m$, then $n < m$ by hypothesis and so $(n + \sqrt{n^2 + 4})/2 < \varepsilon_d$, a contradiction. Thus $m = n$ and $\varepsilon_d = (n + \sqrt{n^2 + 4})/2$.

Conversely, assume that $N(\varepsilon_d) = -1$. Then, by Lemma 4.17, there exists $n \in \mathbb{N}$ such that d is the square-free part of $n^2 + 4$ and $\varepsilon_d = (n + \sqrt{n^2 + 4})/2$. If $n \neq m$ and d is the square-free part of $m^2 + 4$, then $\varepsilon = (m + \sqrt{m^2 + 4})/2$ is a unit of R, again by Lemma 4.17. If $n > m$, then $\varepsilon_d > \varepsilon$, which is impossible. Thus $n < m$ and the result follows. ∎

As an easy application of the above, we prove

2.4.19 Proposition

(i) If $d = n^2 + 4$ for some $n \in \mathbb{N}$, then

$$\varepsilon_d = (n + \sqrt{n^2 + 4})/2 \qquad and \qquad N(\varepsilon_d) = -1.$$

(ii) If $d = m^2 + 1$ with $2 \neq m \in \mathbb{N}$, then

$$\varepsilon_d = m + \sqrt{m^2 + 1} \qquad and \qquad N(\varepsilon_d) = -1.$$

Proof

(i) Assume that $m \in \mathbb{N}$ is such that the square-free parts of $m^2 + 4$ and $n^2 + 4$ are the same. Since d is square-free, if $m \neq n$, then $n^2 + 4 < m^2 + 4$ and hence $n < m$. Now apply Proposition 4.18.

(ii) Put $\varepsilon = m + \sqrt{m^2 + 1}$, so that $\varepsilon = (1/2)(2m + \sqrt{(2m)^2 + 4})$. By Proposition 4.18, it suffices to prove that if $k \neq 2m$ and d is the square-free part of $k^2 + 4$, then $2m < k$. Now

$$4m^2 + 4 = 2^2 d \qquad and \qquad k^2 + 4 = a^2 d$$

for $a \geq 1$. Hence we need only verify that $a \neq 1$. Assume by way of contradiction that $k^2 + 4 = d$. Then $k^2 + 4 = m^2 + 1$, which implies $(k - m)(k + m) = -3$. Since the latter is possible only in the case $m = 2$, $k = 1$, we derive a desired contradiction. ∎

Finally, applying Proposition 4.19, we derive

2.4.20 Corollary If $\varepsilon_d = (1/2)(t + u\sqrt{d})$, then the following conditions are equivalent:

(i) $N(\varepsilon_d) = -1$ and $u \in \{1, 2\}$;
(ii) $d = n^2 + 4$ or $d = n^2 + 1$ for some $n \in \mathbb{N}$.

Proof

(i) $\Rightarrow$ (ii) Since $N(\varepsilon_d) = -1$, $t^2 - du^2 = -4$. Taking into account that $u = 1$ or 2 we obtain, respectively, $d = t^2 + 4$ or $d = t_0^2 + 1$ for $t_0 = t/2$.

(ii) $\Rightarrow$ (i) If $d = n^2 + 4$, then by Proposition 4.19, $\varepsilon_d = (n + \sqrt{d})/2$ (so

$u = 1$) and $N(\varepsilon_d) = -1$. If $d = n^2 + 1$ with $n \neq 2$, then by Proposition 4.19, $\varepsilon_d = n + \sqrt{n^2 + 1}$ (so $u = 2$) and $N(\varepsilon_d) = -1$. Finally, if $d = n^2 + 1$ and $n = 2$, then $d = 1^2 + 4$, and we may apply the first case. ■

We close this section by remarking that the following problem is still outstanding:

Problem   *Given a square-free integer $d \geq 2$, determine ε_d exclusively in terms of d.*

In this vein we quote the following result due to Richaud (1866) and Degert (1958), as formulated by Yokoi (1968).

2.4.21   Theorem   *Let $d \geq 2$ be a square-free integer and write $d = n^2 + r$, $-n < r \leq n$. If r divides $4n$, then*

$$\varepsilon_d = \begin{cases} n + \sqrt{d} \text{ with } N(\varepsilon_d) = -\operatorname{sgn} r \text{ for } |r| = 1 \\ \qquad\qquad\qquad\qquad\qquad (\text{except for } d = 5, \ n = 2, \ r = 1); \\ (n + \sqrt{d})/2 \text{ with } N(\varepsilon_d) = -\operatorname{sgn} r \text{ for } |r| = 4; \\ [(2n^2 + r) + 2n\sqrt{d}]/r \text{ with } N(\varepsilon_d) = 1 \quad \text{ for } |r| \neq 1, 4. \end{cases}$$

2.5   Units in pure cubic fields

2.5A   Prologue

Let R be the ring of integers of $\mathbb{Q}(\sqrt[3]{d})$, where $d > 1$ is a cube-free rational integer. Then one can show (see Proposition 5.3) that

$$U(R) = \{\pm 1\} \times \langle \varepsilon_d \rangle$$

for a unique unit ε_d of R with $0 < \varepsilon_d < 1$. Let us refer to ε_d as the canonical fundamental unit. Thus effective knowledge of $U(R)$ ultimately depends upon our ability to express ε_d exclusively in terms of d. To solve this problem, however, seems to pose almost insurmountable difficulties. Indeed, write $d = ab^2$ where ab is square-free. Then if we could express ε_d in terms of d, it would also be possible to provide all solutions of the Diophantine equation

$$x^3 + ab^2y^3 + a^2bz^3 - 3abxyz = 1.$$

The latter, needless to say, is an exceptionally difficult task. It is nevertheless possible, for a given numerical value of d, to calculate ε_d by means of algorithms. For example, with the aid of the computer, Wada (1970) calculated

$$\varepsilon_d \qquad \text{for all } 2 \leq d < 250.$$

Another possibility is to put a certain constraint on d and then try to find a formula for ε_d. It is in this direction that most of the successful results have been achieved. The following two cases are among the significant ones:

(i) $d = D^3 + k$, where D, k are rational integers with $D > 0$, $|k| > 1$, and $k \mid 3D^2$;

(ii) $d = (n^3 + 1)(n^3 + 2)$, where $n \in \mathbb{N}$.

In case (i), extending a result of Stender (1969), it was shown by Rudman (1973) that

$$\varepsilon_d = (\sqrt[3]{d} - D)^3 / k$$

except for $(D, k) = (2, -6)$, $(1, 3)$, $(2, 2)$, $(3, 1)$, $(5, -25)$ and $(2, -4)$. In case (ii), a noteworthy result of Bernstein (1975a) (established by ingenious calculations) asserts that

$$\varepsilon_d = 1 - 3n(n^3 + 1)\sqrt[3]{d} + 3n^2\sqrt[3]{d^2}.$$

The proofs of the aforementioned facts are very computational in nature, and therefore are not suitable to be presented here. Instead, we decided, after discussing some general facts, to prove a beautiful result known as the Delaunay–Nagell theorem, which asserts that any unit $u \neq 1$ of R which is of norm 1 and of the form

$$x + y\sqrt[3]{d} \qquad (x, y \in \mathbb{Z}, \, y \neq 0)$$

is either the canonical fundamental unit or its square. As an application, we disclose ε_d for infinitely many d. Namely, we prove that for

$$d = n(u^6 n^2 + 3u^3 n + 3) \qquad (n, u \in \mathbb{N})$$

where $nu \equiv 0 \pmod 3$ or $n \equiv u \pmod 3$,

$$\varepsilon_d = (1 + u^3 n) - u\sqrt[3]{d}.$$

Applying a similar argument, we also show that if

$$d = n(u^6 n^2 - 3u^3 n + 3) \qquad (n, u \in \mathbb{N})$$

where $nu \equiv 0 \pmod 3$ or $n + u \equiv 0 \pmod 3$, then

$$\varepsilon_d = (1 - u^3 n) + u\sqrt[3]{d}.$$

Finally, we provide a number of numerical examples pertaining to the above results.

2.5B   Preliminary results

A rational integer $d > 1$ is said to be *cube-free* if d is not divisible by a cube of a prime. The field $F = \mathbb{Q}(\sqrt[3]{d})$, in which $d > 1$ is a cube-free rational integer and $\sqrt[3]{d}$ is real, is called a *pure cubic field*. Throughout R denotes the ring of integers of the pure cubic field $F = \mathbb{Q}(\sqrt[3]{d})$. In this subsection we determine a $\mathbb{Z}$-basis for R and note certain other properties.

Since d is cube-free, we can write

$$d = ab^2$$

where ab is square-free. Taking into account that $\sqrt[3]{d^2} = b\sqrt[3]{a^2b}$, we see that

$$\{1, \sqrt[3]{ab^2}, \sqrt[3]{a^2b}\}$$

is a $\mathbb{Q}$-basis for F. Following Dedekind, we say that F is of the *first* or *second kind*, according as 9 does not, or does, divide $a^2 - b^2$. The reason for the distinction is made clear in the following classical result.

2.5.1 Proposition (Dedekind 1900) *A $\mathbb{Z}$-basis for R is given by*

$$\{1, \sqrt[3]{ab^2}, \sqrt[3]{a^2b}\}$$

if F is of the first kind and by

$$\{(1/3)(1 + a\sqrt[3]{ab^2} + b\sqrt[3]{a^2b}, \sqrt[3]{ab^2}, \sqrt[3]{a^2b}\}$$

if F is of the second kind.

Proof. Given $r \in R$, we may write

$$r = x_1 + x_2\sqrt[3]{ab^2} + x_3\sqrt[3]{a^2b}$$

for some $x_i \in \mathbb{Q}$, $1 \leq i \leq 3$. Then the conjugates of r are

$$r' = x_1 + \rho x_2\sqrt[3]{ab^2} + \rho^2 x_3\sqrt[3]{a^2b},$$

$$r'' = x_1 + \rho^2 x_2\sqrt[3]{ab^2} + \rho x_3\sqrt[3]{a^2b}$$

where ρ is a primitive cube root of unity. It follows that

$$r + r' + r'' = 3x_1 \in R \cap \mathbb{Q} = \mathbb{Z},$$

$$\sqrt[3]{a^2b}(r + \rho^2 r' + \rho r'') = 3abx_2 \in R \cap \mathbb{Q} = \mathbb{Z},$$

$$\sqrt[3]{ab^2}(r + \rho r' + \rho^2 r'') = 3abx_3 \in R \cap \mathbb{Q} = \mathbb{Z},$$

and so, for any $r \in R$, there are y_1, y_2, y_3 in $\mathbb{Z}$ such that

$$3abr = y_1 + y_2\sqrt[3]{ab^2} + y_3\sqrt[3]{a^2b}. \tag{1}$$

We now show that ab is a divisor of y_1, y_2 and y_3, and so can be omitted in (1).

To this end, let p be a rational prime dividing a, and let P be a prime ideal of R which divides (p). It was supposed that ab is square-free; *a fortiori*, $(a, b) = 1$, and $P \nmid (b)$. If we put

$$\alpha = \sqrt[3]{ab^2}, \qquad \beta = \sqrt[3]{a^2b},$$

then $P \mid (\alpha)^3$, so $P^3 \mid (\alpha)^3$. Because F is of degree 3, it follows from Proposition 1.2.8 that $(p) = P^3$. Hence $P \parallel (\alpha)$ and $P^2 \parallel (\beta)$.

Now assume, in accordance with (1), that

$$y_1 + y_2\alpha + y_3\beta \equiv 0 (\text{mod } 3ab).$$

Then

$$\left.\begin{aligned}
y_1 + y_2\alpha + y_3\beta &\equiv 0 (\text{mod } P^3), \\
y_1 &\equiv 0 (\text{mod } P), \\
y_1 &\equiv 0 (\text{mod } p), \\
y_1 &\equiv 0 (\text{mod } P^3);
\end{aligned}\right\} \tag{2}$$

$$\left.\begin{aligned}
y_2\alpha + y_3\beta &\equiv 0 (\text{mod } P^3); \\
y_2\alpha &\equiv 0 (\text{mod } P^2); \\
y_2 &\equiv 0 (\text{mod } p);
\end{aligned}\right\} \tag{3}$$

$$\left.\begin{aligned}
y_3\beta &\equiv 0 (\text{mod } P^3), \\
y_3 &\equiv 0 (\text{mod } p).
\end{aligned}\right\} \tag{4}$$

Invoking (2), (3), and (4), and that p was an arbitrary prime divisor of a, we conclude that a divides y_1, y_2, and y_3. Similarly, b divides y_1, y_2, and y_3. Thus there are z_1, z_2, z_3 in $\mathbb{Z}$ such that

$$3r + z_1 + z_2\alpha + z_3\beta. \tag{5}$$

Let the defining equation of r be

$$x^3 + c_1x^2 + c_2x + c_3 = 0 \qquad (c_1, c_2, c_3 \in \mathbb{Z}).$$

Then, by (5) and the analogous equations for $3r'$ and $3r''$, we have

$$c_1 = -(r + r' + r'') = -z_1,$$

$$c_2 = rr' + rr'' + r'r'' + r'r'' = (1/3)(z_1^2 - abz_2z_3); \tag{6}$$

$$c_3 = -rr'r'' = (-1/27)(z_1^3 + ab^2z_2^3 + a^2bz_3^3 - 3abz_1z_2z_3). \tag{7}$$

Assume that 3 divides a; then 3 does not divide b, and F is of the first kind. Because $c_2 \in \mathbb{Z}$, 3 divides z_1. Since $c_3 \in \mathbb{Z}$,

$$0 \equiv -27c_3 \equiv 3(a/3)b^2z_2^3(\text{mod } 9),$$

whence 3 divides z_2, and by (7), 3 divides z_3. In this case, then $\{1, \sqrt[3]{ab^2}, \sqrt[3]{a^2b}\}$ is a $\mathbb{Z}$-basis for R. A similar argument applies in the case 3 divides b.

Suppose now that $3 \nmid ab$, so that

$$a^2 \equiv b^2 \equiv 1 (\text{mod } 3). \tag{8}$$

If $3 \mid z_1$, then by (6), $3 \mid z_2 z_3$; if $3 \mid z_2$, say, then it follows from (7) that also $3 \mid z_3$. Similarly, if $3 \mid z_2$, then also $3 \mid z_1$ and $3 \mid z_3$. The conclusion is that 3 divides all or none of z_1, z_2, z_3; in the first case r is of the form specified in the proposition.

We next examine the possibility that r in (6) is an integer, but $3 \nmid z_1 z_2 z_3$. Then by (7), (8) and Fermat's theorem,

$$z_1^3 + ab^2 z_2^3 + a^2 b z_3^3 \equiv 0 \pmod{3},$$

$$z_1 + az_2 + bz_3 \equiv 0 \pmod{3},$$

$$z_1 \equiv az_2 \equiv bz_3 \pmod{3},$$

$$z_2 \equiv az_1, \qquad z_3 \equiv bz_1 \pmod{3},$$

$$z_2 = az_1 + 3t_2, \qquad z_3 = bz_1 + 3t_3.$$

Substituting these expressions for z_2 and z_3 into (7), we derive

$$
\begin{aligned}
-27 &= z_1^3 + ab^2 (az_1 + 3t_2)^3 + a^2 b (bz_1 + 3t_3)^3 \\
&\quad -3abz_1 (az_1 + 3t_2)(bz_1 + 3t_3) \\
&= z_1^3 (1 + a^4 b^2 + a^2 b^4 - 3a^2 b^2) \\
&\quad + 9z_1^2 (a^3 b^2 t_2 + a^2 b^3 t_3 - a^2 bt_3 - ab^2 t_2) \\
&\quad + 27z_1 (a^2 b^2 t_2^2 + a^2 b^2 t_3^2 - abt_2 t_3) + 27(ab^2 t_2^3 + a^2 bt_3^3) \\
&\equiv z_1^3 (1 + a^4 b^2 + a^2 b^4 - 3a^2 b^2) \\
&\quad + 9z_1^2 (ab^2 t(a^2 - 1) + a^2 bt_3 (b^2 - 1)) \pmod{27}.
\end{aligned}
$$

However, by (8),

$$0 \equiv -27c_3 \equiv z_1^3 (1 + a^4 b^2 + a^2 b^4 - 3a^2 b^2) \pmod{27}$$

and therefore

$$1 + a^4 b^2 + a^2 b^4 - 3a^2 b^2 \equiv 0 \pmod{27}. \tag{9}$$

Applying (8), we can put

$$b^2 = a^2 + 3m + 9n,$$

where m and n are in $\mathbb{Z}$ and $0 \leqslant m \leqslant 2$. Then the congruence (9) reduces to

$$\phi(m, a) = 2a^6 + (9m - 3)a^4 + 9(m^2 - m)a^2 + 1 \equiv 0 \pmod{27}.$$

For $m = 0$, this becomes

$$(a^2 - 1)^2 (2a^2 + 1) \equiv 0 \pmod{27},$$

which is true for every a not divisible by 3, since

$$2a^2 + 1 \equiv a^2 - 1 \equiv 0 \pmod{3}.$$

Furthermore, for every a such that $3 \nmid a$,

$$\phi(1, a) \equiv \phi(0, a) + 9a^4 \equiv 9a^4 \not\equiv 0 (\bmod 27),$$

$$\phi(2, a) \equiv \phi(0, a) + 18a^4 + 18a^2 \equiv 18a^2(a^2 + 1) \not\equiv 0 (\bmod 27).$$

We have therefore shown that, if $3 \nmid z_1 z_2 z_3$, then $c_3 \in \mathbb{Z}$ if and only if

$$a^2 \equiv b^2 \not\equiv 0 (\bmod 9)$$

(i.e. if and only if F is of the second kind), and $az_2 \equiv bz_3 (\bmod 3)$. If this is the case, then c_1 and c_2 are also in $\mathbb{Z}$ and

$$r = (1/3)(z_1 + (az_1 + 3t_2)\alpha + (bz_1 + 3t_3)\beta)$$
$$= z_1(1/3)(1 + a\alpha + b\beta) + t_2\alpha + t_3\beta$$

is an element of R. This completes the proof. ■

Remark 1:   The second basis represents every integer represented by the first. This is so since

$$3z_1(1 + a\sqrt[3]{ab^2} + b\sqrt[3]{a^2b})/3 + (z_2 - az_1)\sqrt[3]{ab^2} + (z_3 - bz_1)\sqrt[3]{a^2b}$$
$$= z_1 + z_2\sqrt[3]{ab^2} + z_3\sqrt[3]{a^2b} \qquad (z_i \in \mathbb{Z}).$$

Remark 2:   In the course of the proof, it appeared that if

$$(1/3)(z_1 + z_2\sqrt[3]{ab^2} + z_3\sqrt[3]{a^2b}) \in R \qquad (z_i \in \mathbb{Z})$$

then 3 divides all or none of the z_1, z_2, z_3.

Remark 3:   It can be easily verified that $9 \mid a^2 - b^2$ if and only if $ab^2 \equiv \pm 1 (\bmod 9)$. Thus $\mathbb{Q}(\sqrt[3]{d})$ is of the first or second kind, according as $d \not\equiv \pm 1 (\bmod 9)$ or $d \equiv \pm 1 (\bmod 9)$.

As an application of Proposition 5.1, we now prove

2.5.2 Proposition *Let R be the ring of integers of $\mathbb{Q}(\sqrt[3]{d})$, where $1 \neq d \in \mathbb{N}$ is cube-free, and let $d = ab^2$, where ab is square-free.*

(i)   *If $\mathbb{Q}(\sqrt[3]{d})$ is of the first kind, then $U(R)$ consists precisely of those elements*

$$x + y\sqrt[3]{ab^2} + z\sqrt[3]{a^2b} \qquad (x, y, z \in \mathbb{Z}),$$

for which

$$x^3 + ab^2y^3 + a^2bz^3 - 3abxyz = \pm 1. \tag{10}$$

(ii)   *If $\mathbb{Q}(\sqrt[3]{d})$ is of the second kind, then $U(R)$ consists precisely of those elements*

$$(u/3)(1 + a\sqrt[3]{ab^2} + b\sqrt[3]{a^2b}) + v\sqrt[3]{ab^2} + w\sqrt[3]{a^2b} \qquad (u, v, w \in \mathbb{Z}),$$

for which

$$x^3 + ab^2y^3 + a^2bz^3 - 2abxyz = \pm 27, \tag{11}$$

where $u = x$, $au + 3v = y$, and $bu + 3w = z$.

Proof

(i)   Let $w = x + y\sqrt[3]{ab^2} + z\sqrt[3]{a^2b}$ and let ρ be a primitive cube root of unity. Then $N(w) = ww'w''$, where

$$w' = x + \rho y\sqrt[3]{ab^2} + \rho^2 z\sqrt[3]{a^2b} \quad \text{and} \quad w'' = x + \rho^2 y\sqrt[3]{ab^2} + \rho z\sqrt[3]{a^2b}.$$

A straightforward calculation shows that

$$N(w) = x^3 + ab^2y^3 + a^2bz^3 - 3abxyz \tag{12}$$

and the required assertion follows by Proposition 5.1.

(ii)   Apply the same argument as in (i).   ∎

2.5.3   Proposition   *There is a unique unit ε_d of R with $0 < \varepsilon_d < 1$ such that*

$$U(R) = \{\pm 1\} \times \langle \varepsilon_d \rangle.$$

Furthermore:

(i)   *for any w of R, $N(w) = 1$ if and only if $w > 0$;*

(ii)   *if $w = x + y\sqrt[3]{ab^2}$ or $w = x + y\sqrt[3]{a^2b}$, $x \in \mathbb{Z}$, $0 \neq y \in \mathbb{Z}$, and w is a positive unit of R, then $w < 1$. In particular, $w = \varepsilon_d^n$ for some $n > 0$.*

Proof.   The field F has the property that each of its elements is either rational or of degree 3: for if there were an element of degree 2, F would be an extension of the field generated by that element, and so would be of even degree. It follows that ± 1 are the only roots of unity in F.

Since $\sqrt[3]{d}$ has two complex conjugates, we have $r_1 = r_2 = 1$ and hence $r_1 + r_2 - 1 = 1$. Applying Theorem 2.10, we see that $U(R) = \{\pm 1\} \times \langle \delta \rangle$ for some $\delta > 0$ in R. If $\delta' \neq \delta$ is any other positive fundamental unit, then $\delta' = \delta^{-1}$. Hence $\varepsilon_d = \min\{\delta, \delta^{-1}\}$ is a unique unit of R, with $0 < \varepsilon_d < 1$ and $U(R) = \{\pm 1\} \times \langle \varepsilon_d \rangle$.

If w', w'' are conjugates of w, then $N(w) = ww'w''$. But w', w'' are complex conjugates, so $w > 0$ if and only if $N(w) > 0$. In particular, for $w \in U(R)$, $N(w) = 1$ if and only if $w > 0$.

Assume that $w = x + y\sqrt[3]{ab^2}$, $x, y \in \mathbb{Z}$, is a positive unit of R and $y \neq 0$. Then $N(w) = 1$ and, by (12),

$$x^3 + ab^2y^3 = 1,$$

which shows that $xy < 0$. It therefore follows that, for $\alpha = \sqrt[3]{ab^2}$,

$$w^{-1} = w'w'' = x^2 - xy\alpha + y^2\alpha^2 \geq 1 + \alpha + \alpha^2 > 1,$$

and hence that $w < 1$. A similar argument can be applied to the case $w = x + y\sqrt[3]{a^2b}$.   ∎

From now on, we refer to the unit ε_d in Proposition 5.3 as the *canonical fundamental unit* of R.

We close this subsection by proving two technical lemmas to be applied later. For simplicity in notation, we define the binomial coefficient $\binom{m}{k}$ to be zero for $k > m$.

2.5.4 Lemma *Let m be a positive integer. Then*

$$\binom{m}{0} + \binom{m}{3} + \binom{m}{6} + \ldots \not\equiv 0 \pmod 3.$$

Proof. Put

$$s_0 = \binom{m}{0} + \binom{m}{3} + \binom{m}{6} + \ldots,$$

$$s_1 = \binom{m}{1} + \binom{m}{4} + \binom{m}{7} + \ldots,$$

$$s_2 = \binom{m}{2} + \binom{m}{5} + \binom{m}{8} + \ldots.$$

Then

$$s_0 + s_1 + s_2 = 2^m \equiv (-1)^m \pmod 3$$

and

$$s_2 = \binom{m}{1}(m-1)/2 + \binom{m}{4}(m-4)/5 + \ldots \equiv -ms_1 + s_1 \pmod 3,$$

$$s_1 = \binom{m}{0}(m/1) + \binom{m}{3}(m-3)/4 + \ldots \equiv ms_0 \pmod 3.$$

Thus

$$(1 + 2m - m^2)s_0 \equiv (-1)^m \pmod 3$$

and the result follows. ∎

2.5.5 Lemma *Suppose that x and y are integers such that $(x, dy) = 1$ and let*

$$(x + y\sqrt[3]{d})^n = X + Y\sqrt[3]{d} + Z(\sqrt[3]{d})^2,$$

where X, Y, and Z are rational and $n > 1$. Then $XYZ \neq 0$ except in the following cases:

$$(\sqrt[3]{10} - 1)^5 = 99 - 45\sqrt[3]{10};$$
$$(\sqrt[3]{4} - 1)^4 = -15 + 12\sqrt[3]{2}.$$

Proof. We first note that, since $(x, d) = 1$, we have $X \neq 0$. Assume that $Z = 0$, so that

$$\binom{n}{2}x^{n-2}y^2 + \binom{n}{5}x^{n-5}y^5d + \binom{n}{8}x^{n-8}y^8d^2 + \ldots = 0. \tag{13}$$

Dividing by $\binom{n}{2}y^2$, we derive

$$-x^{n-2} = \sum_{k \geqslant 1} \binom{n-2}{3k} \frac{2x^{n-3k-2}y^{3k}d^k}{(3k+1)(3k+2)}. \tag{14}$$

Let p be a prime divisor of y. Then since $p^{3k} \geqslant 2^{3k} > 3k + 2$ for $k \geqslant 1$, each term in the last sum is divisible by p, which is impossible since $(x, y) = 1$. Thus $y = \pm 1$.

When $n \equiv 0 \pmod 3$, eqn (14) can be written in the form

$$-y^{n-3}d^{(1/3)(n-3)} = \sum_{k \geqslant 1} \binom{n-1}{3k} \frac{x^{3k}y^{n-3k-3}d^{(1/3)(n-3)-k}}{3k+1};$$

when $n \equiv 1 \pmod 3$,

$$-y^{n-4}d^{(1/3)(n-4)} = \sum_{k \geqslant 1} \binom{n-2}{3k} \frac{2x^{3k}y^{n-3k-4}d^{(1/3)(n-4)-k}}{(3k+1)(3k+2)};$$

and when $n \equiv 2 \pmod 3$,

$$-y^{n-2}d^{(1/3)(n-2)} = \sum_{k \geqslant 1} \binom{n}{3k} x^{3k}y^{n-3k-2}d^{(1/3)(n-2)-k}.$$

A similar argument now shows that $x = \pm 1$, and since it is clear from (13) that $xy < 0$, we have $x = -y$.

Now assume that p is a prime divisor of d with $p^\alpha \mid d$ but $p^{\alpha+1} \nmid d$. If $p^\alpha > 5$, then $p^{\alpha k} > 5^k \geqslant 3k + 2$ for $k \geqslant 1$, so each term in the sum in (14) is divisible by p, which is impossible since $(x, d) = 1$. If $p = 3$, then because $3 \nmid (3k+1)(3k+2)$ we reach the same contradiction. Thus $p^\alpha = 2$ or 5, and $d \in \{2, 5, 10\}$.

The information obtained so far shows that

$$1 - \binom{n-2}{3} \frac{2d}{45} + \binom{n-2}{6} \frac{2d^2}{7 \cdot 8} - \ldots = 0. \tag{15}$$

If $d = 10$, this becomes

$$\binom{n-2}{3} - 1 = \tfrac{1}{6}(n-5)(n^2 - 4n + 6)$$

$$= \sum_{k \geqslant 2} (-1)^k \binom{n-2}{3k} \frac{2 \cdot 10^k}{(3k+1)(3k+2)}.$$

This equation is true for $n = 5$, and leads to the first of the exceptions mentioned in the lemma. For other values of n, we may divide through by $(n - 5)/6$ and obtain

$$n^2 - 4n + 6$$
$$= -\sum_{k \geqslant 1} (-1)^k \binom{n - 6}{3k - 1} \frac{(n - 2)(n - 3)(n - 4) \cdot 12 \cdot 10^{k+1}}{3k(3k + 1)(3k + 2)(3k + 3)(3k + 4)(3k + 5)}.$$

The highest power of 5 which divides the denominator of a term in the sum is obviously at most $5(3k + 5)$, and because $5^{k+1} > 5(3k + 5)$ for $k \geqslant 2$, we have

$$n^2 - 4n + 6 = (n - 2)^2 + 2$$
$$\equiv \binom{n - 6}{2} \frac{(n - 2)(n - 3)(n - 4) \cdot 12 \cdot 10^2}{3 \cdot 4 \cdot 5 \cdot 6 \cdot 7 \cdot 8}$$
$$\equiv 0 \pmod 5,$$

which is impossible since -2 is a quadratic non-residue of 5.

When $d = 2$ or 5, eqn (15) implies the congruence

$$1 + \binom{n - 2}{3} + \binom{n - 2}{6} + \ldots \equiv 0 \pmod 3,$$

which is again impossible, by Lemma 5.4.

We are therefore left to examine the possibility that $Y = 0$. The proof that this happens only in the case of the second exception mentioned in the lemma is the same as what has just been done for the case $Z = 0$ (the only variation lies in the fact that d may not have the sole prime divisor 2, so that $d = 2$ or 4). Thus the lemma is proved. ∎

2.5C   The Delaunay–Nagell theorem

Throughout this subsection, $d > 1$ is a cube-free integer and R the ring of integers of $\mathbb{Q}(\sqrt[3]{d})$. As before, we write

$$d = ab^2, \qquad \alpha = \sqrt[3]{ab^2} \qquad \text{and} \qquad \beta = \sqrt[3]{a^2 b},$$

where ab is square-free. Then $u = x + y\alpha + z\beta$ $(x, y, z \in \mathbb{Z})$ is a unit of norm 1 if and only if

$$x^3 + ab^2 y^3 + a^2 b z^3 - 3abxyz = 1. \tag{16}$$

Finding such u supplies infinitely many solutions of (16), by taking all powers of u; finding one non-trivial solution of (16) means having a unit $u \neq 1$, which again supplies infinitely many solutions of (16). Of course, if we know a fundamental unit u satisfying (16), then we have all solutions of (16). Thus there is an intimate connection between units of R and diophantine equation (16). It is probably a hopeless task to try to solve (16) for any a, b. Instead, it is reasonable to specify a certain value for x, y, z, and a natural choice would fall on $z = 0$. Thus we are confronted with the following equation:

$$x^3 + dy^3 = 1 \qquad (x, y \in \mathbb{Z}). \qquad (17)$$

This equation was first considered by Delaunay [for a detailed account of this work refer to Delaunay and Faddeev (1964)]. His method was later refined by Nagell, and it is the purpose of this section to prove Nagell's result. In the context of units Nagell's theorem asserts that any unit $u \neq 1$ of R which is of norm 1 and of the form

$$x + y\sqrt[3]{d} \qquad (x, y \in \mathbb{Z},\ y \neq 0)$$

is either the canonical fundamental unit ε_d or ε_d^2. Whether for a given d eqn (17) has a non-trivial solution is a very delicate problem. Some partial results can be found in Bernstein (1971b) and J. H. E. Cohn (1967). Here we only mention that for values $d \leqslant 100$ such solutions exist if $d \in \{2, 7, 9, 17, 19, 20, 26, 28, 37, 43, 63, 65, 91\}$ (see J. H. E. Cohn 1967).

The initial step in the proof of our main result is the following.

2.5.6 Lemma   *The only solutions in $\mathbb{Z}$ of the equation*

$$x^2 + 2 = y^3 \qquad (18)$$

are $x = \pm 5$, $y = 3$.

Proof.   By Lemma 4.2, the integers of the field $\mathbb{Q}(\sqrt{-2})$ are of the form $a + b\sqrt{-2}$, where a, $b \in \mathbb{Z}$. A routine calculation shows that they form a Euclidean domain: given $a, b, c, k \in \mathbb{Z}$, with $ck \neq 0$, there are $e, f, g, h \in \mathbb{Z}$, such that

$$a + b\sqrt{-2} = (c + k\sqrt{-2})(e + f\sqrt{-2}) + (g + h\sqrt{-2}),$$
$$g^2 + 2h^2 < c^2 + 2k^2.$$

It follows that $\mathbb{Z}[\sqrt{-2}]$ is a unique factorization domain.

We now prove that if x and y satisfy (18), then $x + \sqrt{-2}$ and $x - \sqrt{-2}$ are relatively prime. It is obvious that

$$(x + \sqrt{-2},\ x - \sqrt{-2}) \mid -2\sqrt{-2},$$

and because $-2\sqrt{-2} = (\sqrt{-2})^3$ and $\sqrt{-2}$ is prime in $\mathbb{Z}[\sqrt{-2}]$ (by Proposition 1.2.8), we must have

$$(x + \sqrt{-2},\ x - \sqrt{-2}) = (\sqrt{-2})^m \qquad (0 \leqslant m \leqslant 3).$$

But if $x + \sqrt{-2} = (a + b\sqrt{-2})\sqrt{-2}$, then $x = -2b$ whence, by (18),

$$4b^2 + 2 = y^3,$$

$$y^3 \equiv 2 \pmod{4},$$

which is impossible.

By Proposition 4.6, the only units of $\mathbb{Z}[\sqrt{-2}]$ are ± 1. It follows from (18) that

$$x + \sqrt{-2} = (a + b\sqrt{-2})^3$$

for some $a, b \in \mathbb{Z}$, and equating real and imaginary parts gives

$$a^3 - 6ab^2 = x, \tag{19}$$

$$3a^2b - 2b^3 = 1. \tag{20}$$

From (20) it follows that $b = \pm 1$ and so $3a^2 - 2 = \pm 1$, or $a = \pm 1$. From eqn (19), we therefore derive $x = \pm 1 \mp 6 = \pm 5$, thus completing the proof. ∎

The next four lemmas carry most of the burden of the proof of Nagell's theorem.

2.5.7 Lemma *The square of a positive irrational unit of R of the form*

$$\lambda = x + y\alpha + z\beta \qquad (x, y, z \in \mathbb{Z})$$

is itself of the form $X + Y\alpha$ only if $d = 20$ and

$$\lambda = 1 + \sqrt[3]{20} - \sqrt[3]{50}.$$

Proof. Let $\lambda = x + y\alpha + z\beta$ be a positive unit of R so that, by (12) and Proposition 5.3(i),

$$x^3 + ab^2y^3 + a^2bz^3 - 3abxyz = 1 \tag{21}$$

and

$$\lambda^2 = (x^2 + 2abyz) + (2xy + az^2)\alpha + (2xz + by^2)\beta.$$

If $2xz + by^2 = 0$, then

$$z = -by^2/2x,$$

and substituting this into (21) we have

$$x^3 + dy^3 - d^2 \frac{y^6}{8x^3} + 3d \frac{y^3}{2} = 1,$$

or

$$d^2y^6 - 20x^3dy^3 - 8(x^6 - x^3) = 0,$$

whence
$$dy^3 = 10x^3 \pm 2x\sqrt{27x^4 - 2x}. \tag{22}$$

Hence the number $27x^4 - 2x$ must be a square:
$$(27x^3 - 2)x = t^2. \tag{23}$$

If x is even, then $(27x^3 - 2, x) = 2$, so that
$$27x^3 - 2 = \pm 2u^2, \qquad x = \pm 2v^2.$$

Because -1 is a quadratic non-residue of 3, we must choose the lower sign, and eliminating x we derive
$$108v^6 + 1 = u^2,$$
$$(u - 1)(u + 1) = 108v^6.$$

Because $(u - 1, u + 1) = 2$, this implies that
$$u \pm 1 = 54r^6, \qquad u \mp 1 = 2s^6,$$
whence
$$27r^6 - s^6 = (3r^2)^3 - (s^2)^3 = \pm 1.$$

From the truth of Fermat's conjecture for $n = 3$, we deduce that $r = 0$, which gives $v = 0$ and $x = 0$. It follows also that $y = 0$, by (22), which is impossible since $z\beta$ is not a unit.

Now assume that x is odd. Then (23) gives
$$27x^3 - 2 = \pm u^2, \qquad x = \pm v^2.$$

Here the upper sign must be chosen, and we obtain
$$(3x)^3 = u^2 + 2$$

which by Lemma 5.6 has the sole solution $x = 1$, $y = \pm 5$. By (22), $dy^3 = 10 \pm 10$, so that $d = 20$, $y = 1$ (if $y = 0$, then $z = 0$, and λ is rational). The sole solution is therefore
$$(1 + \sqrt[3]{20} - \sqrt[3]{50})^2 = -19 + 7\sqrt[3]{20}$$

as asserted. ∎

2.5.8 Lemma   *If the square of a positive unit of R of the form*
$$\lambda = (1/3)(x + y\alpha + z\beta), \qquad 3 \nmid xyz \qquad (x, y, z \in \mathbb{Z})$$

is itself of the form $X + Y\alpha$, then either $x = 2$, $y = 2$, $z = -1$ and $d = 19$, or x is odd and $x = -v^2$ for some $v \in \mathbb{Z}$.

Proof.   By (11), we have
$$x^3 + ab^2y^3 + a^2bz^3 - 3abxyz = 27 \tag{24}$$

and

$$9\lambda^2 = (x^2 + 2abyz) + (2xy + az^2)\alpha + (2xz + by^2)\beta. \tag{25}$$

If the coefficient of β in the expression for λ^2 is 0, we have

$$z = -by^2/2x.$$

Substituting this into (24), it follows that

$$dy^3 = 10x^3 \pm 6x\sqrt{3x^4 - 6x} \tag{26}$$

so that

$$3x^4 - 6x = t^2. \tag{27}$$

If x is even, the fact that $3 \nmid x$ implies that

$$x^3 - 2 = \pm 6u^2, \qquad x = \pm 2v^2$$

whence

$$\pm 4v^6 - 1 = \pm 3u^2.$$

Because $3 \nmid (4v^6 + 1)$, we must choose the upper sign; the last equation can then be written as

$$(u + 1)^3 - (u - 1)^3 = (2v^2)^3$$

so that $|u| = 1$. Hence $x = 2$, and by (26), $dy^3 = 80 \pm 72$. The lower sign yields $d = 1$ or 8, both of which are excluded since $d > 1$ is cube-free. Thus $d = 19$, $y = 2$, and $z = -1$. The only solution in this case is

$$[(1/3)(2 + 2\sqrt[3]{19} - (\sqrt[3]{19})^2]^2 = -8 + 3\sqrt[3]{19}.$$

If x is odd, (27) implies that

$$x^3 - 2 = \pm 3u^2, \qquad x = \pm v^2.$$

Hence the lower sign must be chosen and we have $x = -v^2$. ∎

2.5.9 Lemma   *The fourth power of a positive irrational unit of R is never of the form $X + Y\alpha$.*

Proof.   Let σ be such a unit:

$$\sigma = (1/3)(x_1 + y_1\alpha + z_1\beta)$$

and assume that

$$\sigma^4 = X + Y\alpha.$$

Then, since the coefficient of β in σ^4 is 0, we obtain

$$6bx_1^2 y_1^2 + 4x_1^3 z_1 + 4ab^2 y_1^3 z_1 + 12abx_1 y_1 z_1^2 + a^2 bz_1^4 = 0. \tag{28}$$

If we put

$$\lambda = \sigma^2 = (1/3)(x + y\alpha + z\beta),$$

then

$$x = (1/3)(x_1^2 + 2aby_1z_1),$$

$$y = (1/3)(2x_1y_1 + az_1^2),$$

$$z = (1/3)(2x_1z_1 + by_1^2).$$

Because $\lambda^2 = X + Y\alpha$, we may apply Lemmas 5.7 and 5.8. The cases

$$d = 20, \qquad x = y = -z = 3,$$

$$d = 19, \qquad x = y = 2, \qquad z = -1,$$

are impossible, since in the first the above equation for z becomes $-9 = 2x_1z_1 + 2y_1^2$, while in the second the system is easily seen to be inconsistent for all choices of signs of x_1, y_1, z_1. Thus we must have $x = -v^2$, where v is odd, so that

$$3v^2 + x_1^2 = -2aby_1z_1.$$

Because v is odd, so is x_1, so that $3v^2 + x_1^2 \equiv 4 \pmod 8$. Therefore three of the numbers a, b, y_1, z_1 are odd, and the fourth is even. Owing to (28), $a^2bz_1^4$ is even, so y_1 is odd. If either a or z_1 is even, (28) ensures that $6bx_1^2y_1^2 \equiv 0 \pmod 4$, which is impossible. If b is even, (28) implies that $a^2bz_1^4 \equiv 0 \pmod 4$, which is again impossible since b is square-free. Thus the lemma is proved. ∎

2.5.10 Lemma   *The cube of a positive irrational unit of R is never of the form $X + Y\alpha$.*

Proof.   If $\lambda = (1/3)(x + y\alpha + z\beta)$, $x,y,z \in \mathbb{Z}$, is a positive unit of R, then the coefficient of β in λ^3 is

$$(1/9)(bxy^2 + x^2z + abyz^2).$$

From the equation

$$x^3 + ab^2y^3 + a^2bz^3 - 3abxyz = 27, \tag{29}$$

we deduce that $(x, b) = 1$, and hence the equation

$$bxy^2 + x^2z + abyz^2 = 0 \tag{30}$$

implies that $b \mid z$. From (29) again, $\delta = (x, y, z) = 1$ or 3. Since $\lambda \neq \pm 1$, y and z are not both zero, and we can write

$$x = \delta d_1 d_2 x_1, \qquad y = \delta d_2 d_3 y_1, \qquad z = \delta b d_1 d_3 z_1, \tag{31}$$

where

$$(x/\delta, z/\delta b) = |d_1|, \qquad (x/\delta, y/\delta) = |d_2|, \qquad (y/\delta, z/\delta b) = |d_3|,$$

and $x_1 > 0$, $y_1 > 0$, $z_1 > 0$. The numbers $d_1 x_1$, $d_2 y_1$, $d_3 z_1$ are, pairwise, relatively prime. Substituting the values from (31) into (29) and dividing by $\delta^3 b d_1 d_2 d_3$, we derive

$$d_2^2 d_3 x_1 y_1^2 + d_1^2 d_2 x_1^2 z_1 + ab^2 d_1 d_3^2 y_1 z_1^2 = 0.$$

It follows from this that $d_1 | x_1$, $d_2 | y_1$, and $d_3 | z_1$. Setting

$$x_1 = d_1 x_2, \qquad y_1 = d_2 y_2, \qquad z_1 = d_3 z_2,$$

substituting, and dividing by $d_1 d_2 d_3$, we derive

$$d_2^3 x_2 y_2^2 + d_1^3 x_2^2 z_2 + ab^2 d_3^3 y_2 z_2^2 = 0.$$

It follows that $x_2 | ab^2 d_3^3 y_2 z_2^2$, which in turn implies that $x_2 = 1$. Similarly, $y_2 = z_2 = 1$, so that

$$d_1^3 + d_2^3 + ab^2 d_3^3 = 0 \tag{32}$$

and

$$x = \delta d_1^2 d_2, \qquad y = \delta d_2^2 d_3, \qquad z = \delta b d_1 d_3^2.$$

Substituting these values into (29), we derive

$$d_1^6 d_2^3 + ab^2 d_2^6 d_3^3 + a^2 b^4 d_1^3 d_3^6 - 3ab^2 d_1^3 d_2^3 d_3^3 = 27/\delta^3.$$

Eliminating $ab^2 d_3^3$ between this equation and (32), we obtain

$$d_1^9 + 6d_1^6 d_2^3 + 3d_1^3 d_2^6 - d_2^9 = (3/\delta)^3, \tag{33}$$

and putting $d_1^3 = u$, $d_2^3 = v$, $3/\delta = w$, this becomes

$$u^3 + 6u^2 v + 3uv^2 - v^3 = w^3. \tag{34}$$

On the other hand, it is easily verified that

$$(u^3 + 6u^2 v + 3uv^2 - v^3)U^3 = V^3 + W^3,$$

where

$$U = u^2 + uv + v^2, \qquad V = u^3 + 3u^2 v - v^3, \qquad W = 3u^2 v + 3uv^2.$$

Because neither U nor V is zero for relatively prime u and v, (34) can hold with $w \neq 0$ only if $W = 0$, that is, if $u = -v$. In this case $w = v$. Because $(d_1, d_2) = 1$, it follows that $d_1 = -1$, $d_2 = 1$, $\delta = 3$. But this leads to the values $x = 3$, $y = -3$, $z = 0$ for which the coefficient of β in λ^3 is not zero, a contradiction. ∎

We are at last in a position to attain our main objective, which is to prove the following result.

2.5.11   Theorem   (Nagell 1925)   *Let $d > 1$ be a cube-free integer and R the ring of integers of $\mathbb{Q}(\sqrt[3]{d})$. Then the equation*

$$x^3 + dy^3 = 1 \qquad (x \in \mathbb{Z}, 0 \neq y \in \mathbb{Z})$$

has at most one solution. If x_0, y_0 with $y_0 \neq 0$ is a solution, then

$$x_0 + y_0\sqrt[3]{d}$$

is either the canonical fundamental unit of R or its square.

Proof.   Assume that $x_0 \in \mathbb{Z}$, $0 \neq y_0 \in \mathbb{Z}$ are such that $x_0^3 + dy_0^3 = 1$. Then $\delta = x_0 + y_0\sqrt[3]{d}$ is a unit of norm 1, by virtue of (16). Hence, by Proposition 5.3, $\delta = \varepsilon_d^n$ for some $n > 0$, where ε_d is the canonical fundamental unit of R. It therefore suffices to show that $n = 1, 2$. By Lemmas 5.9 and 5.10, the latter will follow provided that we prove the following assertion.

If $p > 3$ is a prime and

$$\lambda = (1/3)(x + y\alpha + z\beta) \qquad (x,y,z \in \mathbb{Z})$$

is a positive unit smaller than 1, then λ^p is not of the form $X + Y\alpha$.

Assume that $z = 0$. Then $3 \mid x$ and $3 \mid y$ and

$$N(\lambda) = (x/3)^3 + d(y/3)^3 = 1$$

so that $(x/3, dy/3) = 1$; by Lemma 5.5, the coefficient of β in λ^p is not zero. Hence $z \neq 0$ and by the same reasoning (applied in the field $\mathbb{Q}(\beta) = \mathbb{Q}(\alpha))y$ cannot be zero.

As we saw in the proof of Proposition 5.1, it follows from the representation

$$r = x_1 + x_2\alpha + x_3\beta$$

of an arbitrary $r \in R$ that

$$\alpha(r + \rho r' + \rho^2 r'') = 3abx_3.$$

Taking $r = \lambda^p$, we see that if the coefficient of β in λ^p is zero, then

$$[(1/3)(x + y\alpha + z\beta)]^p + \rho[(1/3)(x + y\rho\alpha + z\rho^2\beta)]^p$$
$$+ \rho^2[(1/3)(x + y\rho^2\alpha + z\rho\beta)]^p = 0. \quad (35)$$

Suppose first that $p \equiv 1 \pmod 3$. Then since $\rho^p = \rho$, we can write (35) in the form

$$[(1/3)(x\rho + y\rho^2\alpha + z\beta)]^p + [(1/3)(x\rho^2 + y\rho\alpha + z\beta)]^p$$
$$= -[(1/3)(x + y\alpha + z\beta)]^p.$$

Because p is odd, the left side is divisible by

$$(1/3)(x\rho + y\rho^2\alpha + z\beta) + (1/3)(x\rho^2 + y\rho\alpha + z\beta) = (1/3)(-x - y\alpha + 2z\beta);$$

this number is an integer, and since it divides λ^p, it is a unit. Thus

$$-x^3 - ab^2y^3 + 8a^2bz^3 - 6abxyz = \pm 27.$$

Because λ is a positive unit, we have

$$x^3 + ab^2y^3 + a^2bz^3 - 3abxyz = 27$$

and, by addition,

$$9a^2bz^3 - 9abxyz = 0 \quad \text{or} \quad 54.$$

In the first case $az^2 - xy$ must be zero. However, this number is the coefficient of α in $3\lambda^{-1}$ and, by Proposition 5.3(ii), it is not zero since $\lambda^{-1} > 1$.

In the second case we have

$$abz(az^2 - xy) = 6.$$

but then x, y, z are not divisible by 3, so that $\mathbb{Q}(\sqrt[3]{d})$ is of the second kind. This is impossible, since if $ab \mid 6$ then $a^2 - b^2 \not\equiv 0 \pmod 9$.

The case in which $p \equiv 2 \pmod 3$ proceeds similarly. Equation (35) can be written in the form

$$(1/3)(x\rho^2 + y\alpha + z\rho\beta)^p + (1/3)(x\rho + y\alpha + z\rho^2\beta)^p$$
$$= -[(1/3)(x + y\alpha + z\beta)]^p$$

which ensures that the number

$$(1/3)(x\rho^2 + y\alpha + z\rho\beta) + (1/3)(x\rho + y\alpha + z\rho^2\beta) = (1/3)(-x + 2y\alpha - z\beta)$$

is a unit. As before,

$$9ab^2y^3 - 9abxyz = 0 \quad \text{or} \quad 54.$$

Because $by^2 - xz$ is the coefficient of β in $3\lambda^{-1}$, it is not zero. On the other hand, it is impossible that $(by^2 - xz) \mid 6$ and $ab \mid 6$, since then $\mathbb{Q}(\sqrt[3]{d})$ must be both the first and second kinds. This proves the theorem. ∎

2.5.12 Corollary *Let R be the ring of integers of $\mathbb{Q}(\sqrt[3]{d})$, where $d \neq 20$ and $\mathbb{Q}(\sqrt[3]{d})$ is of the first kind. If x_0, y_0 is an integer solution of the equation $x^3 + dy^3 = 1$ and $y_0 \neq 0$, then*

$$x_0 + y_0\sqrt[3]{d}$$

is the canonical fundamental unit of R.

Proof. Since $\mathbb{Q}(\sqrt[3]{d})$ is of the first kind the canonical fundamental unit ε_d of R is of the form

$$x + y\alpha + z\beta \qquad (x, y, z \in \mathbb{Z})$$

by Proposition 5.2. Hence, if $\varepsilon_d^2 = x_0 + y_0\sqrt[3]{d}$, then by Lemma 5.7, $d = 20$. Invoking Theorem 5.11, we therefore derive $\varepsilon_d = x_0 + y_0\sqrt[3]{d}$. ∎

Remark: The above result need not be true for fields $\mathbb{Q}(\sqrt[3]{d})$ of the second kind. Indeed, let $d = 28$. Then $(x, y) = (-3, 1)$ is a non-trivial solution of

$$x^3 + dy^3 = 1$$

Hence $\delta = -3 + \sqrt[3]{28}$ is either the canonical fundamental unit or its square (Theorem 5.11). Since

$$\delta = [(1/3)(-1 - \sqrt[3]{28} + \sqrt[3]{98})]^2$$

it follows that δ is the square of the canonical fundamental unit.

2.5D   An application and examples

Let $d > 1$ be a cube-free integer, let R be the ring of integers of $\mathbb{Q}(\sqrt[3]{d})$, and let ε_d be the canonical fundamental unit of R. In this subsection, by applying Corollary 5.12, we disclose ε_d for infinitely many d. A number of examples is also provided.

2.5.13   Theorem   (Yokoi 1974)   *Assume that*

$$d = n(u^6n^2 + 3u^3n + 3) \qquad (n, u \in \mathbb{N}),$$

where $nu \equiv 0 \pmod 3$ or $n \equiv u \pmod 3$. Then

$$\varepsilon_d = (1 + u^3n) - u\sqrt[3]{d}.$$

Proof. It is easily verified that $(x, y) = (1 + u^3n, -u)$ is a non-trivial solution of the equation

$$x^3 + dy^3 = 1.$$

Therefore, by Corollary 5.12, it suffices to verify that $d \neq 20$ and that $\mathbb{Q}(\sqrt[3]{d})$ is of the first kind. If $u \neq 1$ or $n \geq 2$, then obviously $d > 20$. On the other hand, if $u = n = 1$, then $d = 7$ and thus $d \neq 20$.

Put $x = 1 + u^3n$. If $x \equiv -1 \pmod 3$, we get $u^3n \equiv 1 \pmod 3$. Hence $n \equiv u \not\equiv 0 \pmod 3$ which implies that $d \equiv \pm 2 \pmod 9$. In case $x \equiv 1 \pmod 3$, we get $u^3n \equiv 0 \pmod 3$ and therefore $n \equiv 0 \pmod 3$ or $n \not\equiv u \equiv 0 \pmod 3$. The latter implies $d \equiv 0$ or $\pm 3 \pmod 9$.

Finally, assume that $x \equiv 0 \pmod 3$ so that $u^3n \equiv -1 \pmod 3$. Then $nu \not\equiv 0 \pmod 3$ and so, by hypothesis, $n \equiv u \pmod 3$. Hence $u^4 \equiv -1 \pmod 3$, which is impossible. Thus $\mathbb{Q}(\sqrt[3]{d})$ is of the first kind and the result follows. ∎

2.5.14   Theorem   (Yokoi 1974)   *Assume that*

$$d = n(u^6n^2 - 3u^3n + 3) \qquad (n, u \in \mathbb{N}),$$

where nu $\equiv 0$(mod 3) *or* $n + u \equiv 0$(mod 3). *Then*

$$\varepsilon_d = (1 - u^3 n) + u\sqrt[3]{d}.$$

Proof. It is evident that $(x, y) = (1 - u^3 n, u)$ is a non-trivial solution of the equation

$$x^3 + dy^3 = 1.$$

Again, by Corollary 5.12, we are left to verify that $d \neq 20$ and that $\mathbb{Q}(\sqrt[3]{d})$ is of the first kind. Now if $u \geqslant 2$, then obviously $d > 20$. If $u = 1$, then we have $d = n(n^2 - n + 3) \neq 20$.

Put $x = 1 - u^3 n$. If $x \equiv 1$(mod 3), then we get $u^3 n \equiv 0$(mod 3). Therefore $n \equiv 0$(mod 3) or $n \not\equiv u \equiv 0$(mod 3), which implies that $d \equiv 0$ or ± 3(mod 9). If $x \equiv -1$(mod 3), then we get $u^3 n \equiv -1$(mod 3). Hence $nu \not\equiv 0$(mod 3) and so $n \equiv -u \equiv \pm 1$(mod 3), by hypothesis. The latter easily implies that $d \equiv \pm 2$(mod 9).

Finally, assume that $x \equiv 0$(mod 3), so that $u^3 n \equiv 1$(mod 3). Then $nu \not\equiv 0$(mod 3) and so $n \equiv -u$(mod 3), in which case $u^4 \equiv -1$(mod 3), a contradiction. Thus $\mathbb{Q}(\sqrt[3]{d})$ is of the first kind and the result follows. ∎

2.5.15 Corollary (Yokoi 1974)   *Assume that*

$$d = n(u^6 n^2 \pm 3u^3 n + 3) \qquad (n, u \in \mathbb{N}).$$

Then
$$\delta = (1 \pm u^3 n) \mp u\sqrt[3]{d}$$

is either the canonical fundamental unit of R or its square.

Proof. As has been observed in the proofs of Theorems 5.13 and 5.14,

$$(x, y) = (1 \pm u^3 n, \mp u)$$

is a non-trivial solution of the equation

$$x^3 + dy^3 = 1.$$

The desired conclusion is therefore a consequence of Theorem 5.11. ∎

2.5.16 Examples   For $d = n(u^6 n^2 + 3u^3 n + 3)$:

(a)   Assume that $u = n = 1$. Then $d = 7$ and so, by Theorem 5.13,

$$\varepsilon_7 = 2 - \sqrt[3]{7}.$$

(b)   Assume that $u = 1$, $n = 3$. Then $d = 63$ and so, by Theorem 5.13,

$$\varepsilon_6 = 4 - \sqrt[3]{63}.$$

(c)   Assume that $u = n = 2$. Then $d = 614$ and thus, by Theorem 5.13,

$$\varepsilon_{614} = 17 - 2\sqrt[3]{614}.$$

(d) Assume that $u = 3$, $n = 1$. Then $d = 813$ and therefore, by Theorem 5.13,

$$\varepsilon_{813} = 28 - 3\sqrt[3]{813}.$$

2.5.17 Examples For $d = n(u^6 n^2 - 3u^3 n + 3)$:

(a) Assume that $u = 1$, $n = 2$. Then $d = 2$ and thus, by Theorem 5.14,
$$\varepsilon_2 = -1 + \sqrt[3]{2}.$$

(b) Assume that $u = 1$, $n = 3$. Then $d = 9$ and thus, by Theorem 5.14,
$$\varepsilon_9 = -2 + \sqrt[3]{9}.$$

(c) Assume that $u = 2$, $n = 1$. Then $d = 43$ and so, by Theorem 5.14,
$$\varepsilon_{43} = -7 + 2\sqrt[3]{43}.$$

(d) Assume that $u = 3$, $n = 1$. Then $d = 651$ and hence, by Theorem 5.14,
$$\varepsilon_{651} = -26 + 3\sqrt[3]{651}.$$

We close by remarking that, with the aid of the computer, Wada (1970) calculated

$$\varepsilon_d \qquad \text{for all } 2 \leqslant d < 250.$$

3
The unit group of $\mathbb{Z}[\varepsilon_n]$

The cyclotomic fields $\mathbb{Q}(\varepsilon_n)$ $(n \geq 2)$ and their rings of integers $\mathbb{Z}[\varepsilon_n]$ play an important role in algebraic number theory. It is therefore appropriate to investigate the unit group of $\mathbb{Z}[\varepsilon_n]$ in detail. Thus we are confronted with the following problem: Given n, find a specific fundamental system of units of $\mathbb{Z}[\varepsilon_n]$. This is an exceptionally difficult task and therefore it seems more appropriate to attack the following related problem: Given n, find a specific system of independent units which generates a subgroup of finite index. A most favourable situation is the case where n is a prime power. The main result asserts that under these circumstances the units $(1 - \varepsilon_n^a)/(1 - \varepsilon_n)$, $(a, n) = 1$, $1 < a < n/2$ are independent, and hence generate subgroup of finite index. Unfortunately, if n is composite, then the above need not be true and the constructed units look more complicated. As far as the fundamental sysem of units of $\mathbb{Z}[\varepsilon_n]$ is concerned, we show that if n is a prime power and $\phi(n) \leq 66$, then the system $(1 - \varepsilon_n^a)/(1 - \varepsilon_n)$, $(a, n) = 1$, $1 < a < n/2$ is in fact fundamental. In general, the latter is not true, e.g. take $n = 163$, in which case $h_n^+ \neq 1$. The chapter ends with an important theorem due to Bass which will be applied in our future investigation of units of integral group rings.

3.1   Relations between the unit groups of $\mathbb{Z}[\varepsilon_n]$ and $\mathbb{Z}[\varepsilon_n + \varepsilon_n^{-1}]$

For any positive integer $n > 2$, denote by ε_n a primitive nth root of 1 over $\mathbb{Q}$. Then $\mathbb{Z}[\varepsilon_n]$ is the ring of integers of the cyclotomic field $\mathbb{Q}(\varepsilon_n)$.

Recall, from Proposition 1.2.20, that $\mathbb{Z}[\varepsilon_n]$ is $\mathbb{Z}$-free of rank

$$(\mathbb{Q}(\varepsilon_n) : \mathbb{Q}) = \phi(n),$$

where ϕ is the Euler function. Furthermore, by Corollary 2.2.12,

$$U(\mathbb{Z}[\varepsilon_n]) = \begin{cases} \langle \varepsilon_n \rangle \times A & \text{if } n \text{ is even,} \\ \langle -\varepsilon_n \rangle \times A & \text{if } n \text{ is odd,} \end{cases}$$

where A is a free abelian group of rank $\phi(n)/2 - 1$.

Thus effective knowledge of $U(\mathbb{Z}[\varepsilon_n])$ ultimately depends upon our ability to find a fundamental system of units of $U(\mathbb{Z}[\varepsilon_n])$, i.e. a system of $\phi(n)/2 - 1$ generators of A. The determination of such a system requires deep results and delicate computations. In fact, with the exception of some particular values of n, no specific fundamental system of units is known. Our future aim is to construct $\phi(n)/2 - 1$ independent units for the case where n is a prime power. To achieve this, it is important to tie together unit groups of $\mathbb{Z}[\varepsilon_n]$ and $\mathbb{Z}[\varepsilon_n + \varepsilon_n^{-1}]$. In this section, among other results, we prove that if n is a prime power, then any fundamental system of units of $\mathbb{Z}[\varepsilon_n + \varepsilon_n^{-1}]$ is also a fundamental system of units of $\mathbb{Z}[\varepsilon_n]$.

We begin by collecting some preliminary observations concerning $U(\mathbb{Z}[\varepsilon_n])$.

3.1.1 Lemma *Suppose r and s are integers with $(n, rs) = 1$. Then*

$$\frac{1 - \varepsilon_n^r}{1 - \varepsilon_n^s} \in U(\mathbb{Z}[\varepsilon_n]).$$

Proof. Since $(n, rs) = 1$, we have $(r, n) = (s, n) = 1$. Hence

$$r \equiv st \pmod{n} \quad \text{and} \quad s \equiv rl \pmod{n}$$

for some $t, l \in \mathbb{Z}$. Then

$$\frac{1 - \varepsilon_n^r}{1 - \varepsilon_n^s} = \frac{1 - \varepsilon_n^{st}}{1 - \varepsilon_n^s} = 1 + \varepsilon_n^s + \ldots + \varepsilon_n^{s(t-1)} \in \mathbb{Z}[\varepsilon_n]$$

and

$$\frac{1 - \varepsilon_n^s}{1 - \varepsilon_n^r} = \frac{1 - \varepsilon_n^{rl}}{1 - \varepsilon_n^r} = 1 + \varepsilon_n^r + \ldots + \varepsilon_n^{r(l-1)} \in \mathbb{Z}[\varepsilon_n]$$

as required. ∎

The units $u_a = (1 - \varepsilon_n^a)/(1 - \varepsilon_n)$ with $(n, a) = 1$ are called *cyclotomic units* and will be of great importance in our subsequent investigations.

Note that

$$u_{-a} = -\varepsilon_n^{-a} u_a$$

and hence u_a and u_{-a} differ by a root of unity. If $(a, n) = 1$, then $(1 - a)/2$ is an integer unless n is odd and a is even. In the latter case we define $\varepsilon_n^{(1-a)/2}$ by

$$\varepsilon_n^{(1-a)/2} = (\varepsilon_n^{1/2})^{1-a},$$

where $\varepsilon_n^{1/2}$ is a uniquely determined element of $\langle \varepsilon_n \rangle$ with $(\varepsilon_n^{1/2})^2 = \varepsilon_n$. Now define u_a^+ by

$$u_a^+ = \varepsilon_n^{(1-a)/2} u_a \qquad (a, n) = 1.$$

Then one immediately verifies that u_a^+ is fixed under the automorphism $\varepsilon_n \mapsto \varepsilon_n^{-1}$. Thus u_a^+ is a real unit and we call it the *real cyclotomic unit*.

3.1.2 Lemma *Suppose n is not a prime power. Then*

$$1 - \varepsilon_n \in U(\mathbb{Z}[\varepsilon_n])$$

and

$$\prod_{\substack{0<j<n \\ (j,n)=1}} (1 - \varepsilon_n^j) = 1.$$

Proof. Since $X^{n-1} + X^{n-2} + \ldots + X + 1 = \prod_{j=1}^{n-1} (X - \varepsilon_n^j)$, we may let $X = 1$ to obtain

$$n = \prod_{j=1}^{n-1} (1 - \varepsilon_n^j). \tag{1}$$

Write $n = p^a m$ with $(p, m) = 1$ and p a prime. Then ε_n^m is a primitive p^ath root of 1, and letting j run through multiples of m, we find that the product (1) contains

$$p^a = \prod_{j=1}^{p^a-1} (1 - \varepsilon_n^{mj}).$$

If we remove these factors for each prime dividing n, we obtain

$$1 = \prod_j (1 - \varepsilon_n^j), \tag{2}$$

where the product is over those j for which ε_n^j is not of prime power order. Since n is not a prime power, $1 - \varepsilon_n$ appears as a factor in (2), and hence is a unit. But $\prod_{(j,n)=1} (1 - \varepsilon_n^j)$ is the norm of $1 - \varepsilon_n$ from $\mathbb{Q}(\varepsilon_n)$ to $\mathbb{Q}$, and therefore equals ± 1. Because complex conjugation is in the Galois group, the norm of any element may be written in the form $\alpha \bar{\alpha}$, which is positive. It follows that

$$\prod_{\substack{(j,n)=1 \\ 0<j<n}} (1 - \varepsilon_n^j) = 1,$$

thus completing the proof. ∎

3.1.3 Lemma *Assume that $n = p^k$, where p is a prime. Then $(1 - \varepsilon_n)$ is a prime ideal of $\mathbb{Z}[\varepsilon_n]$ and*

$$(p) = (1 - \varepsilon_n)^{p^{k-1}(p-1)}. \tag{3}$$

Proof. Since $(\mathbb{Q}(\varepsilon_n):\mathbb{Q}) = \phi(n) = p^{k-1}(p - 1)$, it follows from Proposition 1.2.8 that (p) is a product of at most $p^{k-1}(p - 1)$ proper ideals of $\mathbb{Z}[\varepsilon_n]$. We are therefore left to verify (3).

Let $\delta_1 = \varepsilon_n, \delta_2, \ldots, \delta_s, s = p^{k-1}(p - 1)$, be all primitive nth roots of unity. Then

$$\prod_{i=1}^{s} (X - \delta_i) = (X^{p^k} - 1)/(X^{p^{k-1}} - 1) = \sum_{r=0}^{p-1} X^{rp^{k-1}}$$

so that (setting $X = 1$),

$$p = \prod_{i=1}^{s} (1 - \delta_i).$$

From Lemma 1.1 we have the equality of ideals $(1 - \delta_i) = (1 - \varepsilon_n)$, and therefore (3) is established. ∎

3.1.4 Lemma

(i) $\mathbb{Q}(\varepsilon_n + \varepsilon_n^{-1}) = \mathbb{Q}(\varepsilon_n) \cap \mathbb{R} = \mathbb{Q}(\cos 2\pi/n)$.
(ii) *The extension* $\mathbb{Q}(\varepsilon_n)/\mathbb{Q}(\varepsilon_n + \varepsilon_n^{-1})$ *is of degree* 2.
(iii) $\mathbb{Z}[\varepsilon_n + \varepsilon_n^{-1}]$ *is the ring of integers of* $\mathbb{Q}(\varepsilon_n + \varepsilon_n^{-1})$.
(iv) *The field* $\mathbb{Q}(\varepsilon_n + \varepsilon_n^{-1})$ *has* $\phi(n)/2$ *real embeddings and no complex embeddings into* $\mathbb{C}$, *while* $\mathbb{Q}(\varepsilon_n)$ *has no real embeddings and* $\phi(n)/2$ *pairs of complex embeddings.*

Proof. We first observe that (i) follows from (ii), while (ii) is a consequence of the fact that ε_n is a root of

$$X^2 - (\varepsilon_n + \varepsilon_n^{-1})X + 1.$$

Let R be the ring of integers of $\mathbb{Q}(\varepsilon_n + \varepsilon_n^{-1})$. Then, obviously, $\mathbb{Z}[\varepsilon_n + \varepsilon_n^{-1}] \subseteq R$. To prove the opposite containment, suppose

$$\alpha = a_0 + a_1(\varepsilon_n + \varepsilon_n^{-1}) + \ldots + a_s(\varepsilon_n + \varepsilon_n^{-1})^s \in R$$

$$(s \leqslant (1/2)\phi(n) - 1, \ a_i \in \mathbb{Q}).$$

If at least one of the a_i's is not in $\mathbb{Z}$, then by removing those terms with $a_i \in \mathbb{Z}$ we may assume that $a_s \notin \mathbb{Z}$. We shall show that this leads to a contradiction, which will imply that $\alpha \in \mathbb{Z}[\varepsilon_n + \varepsilon_n^{-1}]$.

Multiplying by ε_n^s and expanding the result as a polynomial in ε_n, we find that $\varepsilon_n^s \alpha$ is a $\mathbb{Z}$-linear combination of $1, \varepsilon_n, \ldots, \varepsilon_n^{2s}$, with a_s being the coefficient of ε_n^{2s}. In particular, $\varepsilon_n^s \alpha$ is an algebraic integer in $\mathbb{Q}(\varepsilon_n)$, therefore in $\mathbb{Z}[\varepsilon_n]$. Since

$$2s \leqslant \phi(n) - 2 \leqslant \phi(n) - 1,$$

$\{1, \varepsilon_n, \ldots, \varepsilon_n^{2s}\}$ forms a subset of a $\mathbb{Z}$-basis for the ring $\mathbb{Z}[\varepsilon_n]$. Therefore $a_s \in \mathbb{Z}$, which is a contradiction. This proves (iii).

Since $\mathrm{Gal}(\mathbb{Q}(\varepsilon_n)/\mathbb{Q}_n)$ is abelian, the extension $\mathbb{Q}(\varepsilon_n + \varepsilon_n^{-1})/\mathbb{Q}$ is normal. Hence any $\mathbb{Q}$-homomorphism $\mathbb{Q}(\varepsilon_n + \varepsilon_n^{-1}) \to \mathbb{C}$ is an automorphism of $\mathbb{Q}(\varepsilon_n + \varepsilon_n^{-1})$. Applying (ii), the first part of (iv) follows. The second part is a consequence of the assumption that $n > 2$ and of the fact that $1, -1$ are the only roots of unity in $\mathbb{R}$. ∎

From now on, we employ the following notation:

$$\mathbb{Q}_n = \mathbb{Q}(\varepsilon_n), \qquad \mathbb{Q}_n^+ = \mathbb{Q}_n(\varepsilon_n + \varepsilon_n^{-1}),$$
$$U_n = U(\mathbb{Z}[\varepsilon_n]), \qquad U_n^+ = U(\mathbb{Z}[\varepsilon_n + \varepsilon_n^{-1}]).$$

Thus, by Lemma 1.3(i), $\mathbb{Q}_n^+$ is the maximal real subfield of $\mathbb{Q}_n$. Furthermore, invoking Lemma 1.3(iii), we also have

$$U_n^+ = \mathbb{R} \cap U_n. \tag{4}$$

Recall also that the torsion subgroup of U_n is $\langle \pm\varepsilon_n \rangle$, which is equal to $\langle \varepsilon_n \rangle$ if n is even and $\langle -\varepsilon_n \rangle = \langle \varepsilon_{2n} \rangle$ if n is odd.

3.1.5   Lemma   *For all odd n,*

$$U_n = U_{2n} \qquad and \qquad U_n^+ = U_{2n}^+.$$

Proof.   Since n is odd, we have $\langle -\varepsilon_n \rangle = \langle \varepsilon_{2n} \rangle$, which implies that

$$\mathbb{Z}[\varepsilon_n] = \mathbb{Z}[\varepsilon_{2n}].$$

Hence $U_n = U_{2n}$ and, by (4), we also have

$$U_n^+ = \mathbb{R} \cap U_n = \mathbb{R} \cap U_{2n} = U_{2n}^+$$

as asserted.   ∎

3.1.6   Lemma

(i)   *For any $u \in U_n$, $u = \pm\varepsilon_n^k \bar{u}$ for some $k \in \mathbb{N}$.*
(ii)   *The map*

$$\begin{cases} U_n \xrightarrow{\psi} \langle \pm\varepsilon_n \rangle / \langle \pm\varepsilon_n \rangle^2, \\ u \mapsto (u/\bar{u}) \langle \pm\varepsilon_n \rangle^2, \end{cases}$$

is a homomorphism, the kernel of which is $\langle \pm\varepsilon_n \rangle U_n^+$.

Proof.

(i)   Let $\alpha = u/\bar{u}$. Then $\alpha \in U_n$ and hence α is an algebraic integer. Since complex conjugation commutes with the other elements of the Galois group, all conjugates of α have absolute value 1. Hence, by Lemma 2.2.3(ii), α is a root of unity. Thus $\alpha \in \langle \pm\varepsilon_n \rangle$ and (i) follows.

(ii)   By (i), $u/\bar{u} \in \langle \pm\varepsilon_n \rangle$ and so ψ is a homomorphism from U_n to $\langle \pm\varepsilon_n \rangle / \langle \pm\varepsilon_n \rangle^2$. Assume that $u \in \text{Ker } \psi$. Then $u/\bar{u} = \varepsilon^2$ for some $\varepsilon \in \langle \pm\varepsilon_n \rangle$, in which case $u = \varepsilon(\varepsilon^{-1}u)$ and

$$\overline{\varepsilon^{-1}u} = \varepsilon\bar{u} = \varepsilon^{-1}u \in \mathbb{R} \cap U_n = U_n^+.$$

Hence $u \in \langle \pm\varepsilon_n \rangle U_n^+$, and so $\text{Ker } \psi \subseteq \langle \pm\varepsilon_n \rangle U_n^+$. Conversely, assume that

$u = \varepsilon u_1$ for some $\varepsilon \in \langle \pm \varepsilon_n \rangle$ and $u_1 \in U_n^+$. Then

$$u/\bar{u} = \varepsilon u_1 / \varepsilon^{-1} u_1 = \varepsilon^2 \in \langle \pm \varepsilon_n \rangle^2$$

and so $u \in \operatorname{Ker} \psi$. Thus $\langle \pm \varepsilon_n \rangle U_n^+ \subseteq \operatorname{Ker} \psi$ as asserted.   ∎

Note that, by Lemma 1.4(iii), (iv) and Theorem 2.2.10, U_n^+ and U_n have the same torsion-free rank, and hence $(U_n : U_n^+)$ is finite. The next result is a sharpened version of this fact.

3.1.7 Theorem   *The following formula holds*:

$$(U_n : \langle \pm \varepsilon_n \rangle U_n^+) = \begin{cases} 1, & \textit{if } n \textit{ is a prime power or } n = 2p^\alpha, \, p \textit{ an odd prime}; \\ 2, & \textit{otherwise.} \end{cases}$$

Proof.   Let ψ be the homomorphism of Lemma 1.6(ii). By Lemmas 1.6(ii) and 1.5, it suffices to show that:

$$|\operatorname{Im} \psi| = \begin{cases} 1, & \text{if } n \text{ is a prime power}; \\ 2, & \text{if } n \text{ is composite and } n \text{ is either odd or divisible by 4.} \end{cases}$$
(5)

For the sake of clarity, we divide the rest of the proof into three steps.

Step 1:   here we treat the case where n is a power of an odd prime p. Let us note first that the required assertion in this particular case is a consequence of a general result, namely Theorem 2.3.1.

A different approach is as follows. Given $u \in U_n$, we have $u = \pm \varepsilon_n^k \bar{u}$ for some $k \geqslant 1$. It will be shown that $u = \varepsilon_n^k \bar{u}$, in which case $u/\bar{u} = \varepsilon_n^k \in \langle \varepsilon_n \rangle = \langle \varepsilon_n \rangle^2$ since n is odd. This will imply $\psi(u) = 1$ and hence (4). Assume by way of contradiction that $u = -\varepsilon_n^k \bar{u}$. Write

$$u = b_0 + b_1 \varepsilon_n + \ldots + b_s \varepsilon_n^s \qquad (s = \phi(n) - 1, \, b_i \in \mathbb{Z}).$$

Then

$$u \equiv b_0 + b_1 + \ldots + b_s (\operatorname{mod}(1 - \varepsilon_n)),$$

and therefore

$$\bar{u} = b_0 + b_1 \varepsilon_n^{-1} + \ldots + b_s \varepsilon_n^{-s} \equiv b_0 + b_1 + \ldots + b_s$$
$$\equiv u = -\varepsilon_n^k \bar{u} \equiv -\bar{u} (\operatorname{mod}(1 - \varepsilon_n)).$$

Thus $2\bar{u} \equiv 0 (\operatorname{mod}(1 - \varepsilon_n))$. But $2 \notin (1 - \varepsilon_n)$ since, by Lemma 1.3, $p \in (1 - \varepsilon_n)$. Since, by Lemma 1.3, $(1 - \varepsilon_n)$ is a prime ideal we deduce that $\bar{u} \in (1 - \varepsilon_n)$, which is impossible since $\bar{u}$ is a unit.

Step 2:   now we treat the case where $n = 2^m$ for some $m > 1$. Given $u \in U_n$, we must show that $u/\bar{u} = \varepsilon^2$ for some $\varepsilon \in \langle \pm \varepsilon_n \rangle$. Assume by way of contradiction that $u/\bar{u} \notin \langle \pm \varepsilon_n \rangle^2$. Then $u/\bar{u} = \varepsilon$, where ε is a primitive

2^mth root of unity. Let N denote the norm from $\mathbb{Q}(\varepsilon)$ to $\mathbb{Q}(i)$. Then

$$N(\varepsilon) = \varepsilon^a,$$

where

$$a = \sum_{\substack{0 < b < 2^m \\ b \equiv 1 (\bmod\, 4)}} b = \sum_{j=0}^{2^{m-2}-1} (1 + 4j) = 2^{m-2} + 2^{m-1}(2^{m-2} + 1)$$

$$\equiv 2^{m-2} (\bmod\, 2^{m-1}).$$

Therefore, ε^a is a primitive 4th root of $1 : \varepsilon^a = \pm i$. It follows that

$$N(u)/\overline{N(u)} = \pm i.$$

But $N(u)$ is a unit of $\mathbb{Z}[i]$; therefore $N(u) = \pm 1$ or $\pm i$, in which case

$$N(u)/\overline{N(u)} = \pm 1,$$

a contradiction.

Step 3:   here we consider the final case, where n is composite and is either odd or divisible by 4. Since $|\text{Im } \psi| \leqslant 2$, to prove (5), it suffices to exhibit $u \in U_n$ such that $u/\bar{u} \notin \langle \pm \varepsilon_n \rangle^2$. By Lemma 1.2, $u = 1 - \varepsilon_n \in U_n$. Furthermore, $u/\bar{u} = (1 - \varepsilon_n)/(1 - \varepsilon_n^{-1}) = -\varepsilon_n$. Assume by way of contradiction that $-\varepsilon_n = (\pm \varepsilon_n^r)^2 = \varepsilon_n^{2r}$. Then $\varepsilon_n^{2r-1} = -1$, so n is even and hence, by hypothesis, n is divisible by 4. Since $\varepsilon_n^{n/2} = -1$, we have $n/2 \equiv 2r - 1 \,(\bmod\, n)$, a contradiction. ∎

3.1.8 Corollary *Assume that n is a prime power or $n = 2p^\alpha$, p an odd prime. Then any fundamental system of units of $\mathbb{Z}[\varepsilon_n + \varepsilon_n^{-1}]$ is also a fundamental system of units of $\mathbb{Z}[\varepsilon_n]$.*

Proof.   Direct consequence of Theorem 1.7. ∎

Let R be an arbitrary Dedekind domain, let K be the quotient field of R, and let L/K be a finite separable field extension. Then the integral closure S of R in L is also a Dedekind domain, by virtue of Proposition 1.2.6. Let $I(R)$ and $I(S)$ be the groups of fractional ideals of R and S, respectively.

3.1.9 Lemma *The map $i_{S/R}: I(R) \to I(S)$, defined by*

$$i_{S/R}(A) = SA$$

is an injective homomorphism, which preserves inclusion and maps principal ideals into principal ideals.

Proof.   Every fractional ideal A of R is of the form

$$A = Rx_1 + \ldots + Rx_n \qquad (x_i \in K).$$

Hence $SA = Sx_1 + \ldots + Sx_n$ is a fractional ideal of S. It is clear that $i_{S/R}$ preserves inclusion and maps principal ideals into principal ideals. Moreover, the equality $(AB)S = (AS)(BS)$ shows that $i_{S/R}$ is a homomorphism.

We now show that

$$S \cap K = R. \tag{6}$$

Indeed, the inclusion $R \subseteq S \cap K$ is trivial. On the other hand, every element of $S \cap K$ is integral over R and lies in K. Since R is integrally closed, we conclude that $S \cap K \subseteq R$, proving (6).

Assume that $i_{S/R}(A) = S$, the identity element of $I(S)$. Then $AS = S$ and, by (6),

$$R = S \cap K = AS \cap K \supseteq A,$$

proving that A is an ideal of R. But $i_{S/R}(A^{-1}) = S^{-1} = S$ and we may repeat this argument to conclude that A^{-1} is also an ideal of R. Hence $A = R$ and the result follows. ∎

3.1.10 Corollary *Let A be a fractional ideal of R such that $SA = Sx$ for some $x \in K$. Then $A = Rx$.*

Proof. By hypothesis, $i_{S/R}(A) = Sx = i_{S/R}(Rx)$. Hence, by Lemma 1.9, $A = Rx$. ∎

By Lemma 1.9, the homomorphism $i_{S/R}$ induces a homomorphism of class groups $C(R) \rightarrow C(S)$. We shall refer to the latter homomorphism as the *natural map*.

In what follows, we denote by h_n and h_n^+ the class numbers of

$$\mathbb{Z}[\varepsilon_n] \quad \text{and} \quad \mathbb{Z}[\varepsilon_n + \varepsilon_n^{-1}]$$

respectively. Our next aim is to show that h_n^+ divides h_n. The quotient, denoted by h_n^-, is called the *relative class number* or the *first factor*.

3.1.11 Theorem *Let C and C^+ be the class groups of $\mathbb{Z}[\varepsilon_n]$ and $\mathbb{Z}[\varepsilon_n + \varepsilon_n^{-1}]$, respectively. Then the natural map $C^+ \rightarrow C$ is an injection. In particular, h_n^+ divides h_n.*

Proof. Let I be a fractional ideal of $\mathbb{Z}[\varepsilon_n + \varepsilon_n^{-1}]$ such that $\mathbb{Z}[\varepsilon_n]I = \mathbb{Z}[\varepsilon_n]\alpha$ for some $\alpha \in \mathbb{Q}_n$. Since $I \subseteq \mathbb{R}$ we must also have $\mathbb{Z}[\varepsilon_n]I = \mathbb{Z}[\varepsilon_n]\bar{\alpha}$. Thus $\bar{\alpha}/\alpha$ is a unit of $\mathbb{Z}[\varepsilon_n]$. Hence, by Lemma 1.6(i), $\bar{\alpha}/\alpha$ is a root of unity.

Assume that n is composite and n is either odd or divisible by 4. Then, by (5), ψ is surjective and so

$$\bar{\alpha}/\alpha = (\varepsilon/\bar{\varepsilon})\delta^2 = \varepsilon\delta/\overline{\varepsilon\delta}$$

for some $\varepsilon \in U_n$ and some $\delta \in \langle \pm \varepsilon_n \rangle$. Hence $\overline{\alpha \varepsilon \delta} = \bar{\alpha} \varepsilon \delta = \alpha \varepsilon \delta$ is real, and therefore

$$\mathbb{Z}[\varepsilon_n]\alpha = \mathbb{Z}[\varepsilon_n]\alpha \varepsilon \delta \qquad \text{with } \alpha \varepsilon \delta \in \mathbb{Q}_n^+.$$

Invoking Corollary 1.10, we conclude that $I = \mathbb{Z}[\varepsilon_n + \varepsilon_n^{-1}]\alpha \varepsilon \delta$. Thus I is a principal fractional ideal.

Owing to Lemma 1.5, we are left to treat the case where $n = p^k$ is a prime power. Put $\pi = 1 - \varepsilon_n$ so that, by Lemma 1.3, (π) is a prime ideal of $\mathbb{Z}[\varepsilon_n]$. We have $\pi/\bar{\pi} = -\varepsilon_n$, which generates the roots of unity in $\mathbb{Q}_n$. Therefore $\bar{\alpha}/\alpha = (\pi/\bar{\pi})^d$ for some d. By the preceding argument (with $\varepsilon = 1$ and $\delta = \bar{\pi}/\pi$), it suffices to verify that d is even.

Let v_π be the exponential valuation of $\mathbb{Q}_n$ associated with π. Then v_π takes only even values on $\mathbb{Q}_n^+$, and since $\alpha \pi^d$ and I are real, $d = v_\pi(\alpha \pi^d) - v_\pi(\alpha) = v_\pi(\alpha \pi^d) - v_\pi(I)$ is even. This completes the proof of the theorem. ∎

3.1.12 Remark   *The numbers h_n^+ seem extraordinarily difficult to compute. In fact, with very few exceptions their exact values except for 1 are not known for any n. For example, let $n = p$ be a prime. For small values of p, $h_p^+ = 1$. For some p we know $h_p^+ > 1$, e.g. $3 \mid h_{257}^+$ and $4 \mid h_{163}^+$ (see Masley 1979). But at present for each p with $h_p^+ > 1$ we do not know the exact value of h_p^+. However, the following is known [see Van der Linden (1982) for (a) and (b), and Cornell and Washington (1985) for (c)]:*

(a)   *If n is a prime power with $\phi(n) \leqslant 66$ then $h_n^+ = 1$.*
(b)   *If n is not a prime power and $n \leqslant 200$, $\phi(n) \leqslant 72$, then $h_n^+ = 1$, except for $h_{136}^+ = 2$ and the possible exceptions $n = 148$ and $n = 152$. Also we have $h_{165}^+ = 1$.*
(c)   *If l is a positive integer and n has four distinct prime factors congruent to 1 (mod 4l), then 2l divides h_n^+.*

3.1.13 Proposition   *Let $r = (1/2)\phi(n) - 1$. Then*

$$\frac{\text{reg}(\mathbb{Q}_n)}{\text{reg}(\mathbb{Q}_n^+)} = \begin{cases} 2^r, & \text{if n is a prime power or } n = 2^\alpha, \text{ p an odd prime,} \\ 2^{r-1}, & \text{otherwise.} \end{cases}$$

Proof.   Put $q = (U_n : \langle \pm \varepsilon_n \rangle U_n^+)$. Owing to Theorem 1.7, it suffices to verify that

$$\frac{\text{reg}(\mathbb{Q}_n)}{\text{reg}(\mathbb{Q}_n^+)} = \frac{2^r}{q} \tag{7}$$

To this end let $\varepsilon_1, \varepsilon_2, \ldots, \varepsilon_r$ be a fundamental system of units of U_n^+. Then, by Lemma 1.4(iv) and the definition of the regulator, we have

$$\text{reg}_{\mathbb{Q}_n}(\varepsilon_1, \varepsilon_2, \ldots, \varepsilon_r) = 2^r \text{reg}_{\mathbb{Q}_n^+}(\varepsilon_1, \varepsilon_2, \ldots, \varepsilon_r) = 2^r \text{reg}(\mathbb{Q}_n^+).$$

On the other hand, by Corollary 2.2.15,

$$\mathrm{reg}_{\mathbb{Q}_n}(\varepsilon_1, \varepsilon_2, \ldots, \varepsilon_r) = q \, \mathrm{reg}(\mathbb{Q}_n).$$

This proves (7) and hence the result.   ∎

3.2   Dirichlet *L*-series and class number formulas

In this section we review some of the basic facts about *L*-series and their relation with class numbers. All the information obtained will play an important role in our future investigation of independence of certain cyclotomic units.

Let G be a finite abelian group. Then the set

$$\hat{G} = \mathrm{Hom}(G, \mathbb{C}^*)$$

of all homomorphisms from G to $\mathbb{C}^*$ forms a group under multiplication of values:

$$(\chi_1\chi_2)(g) = \chi_1(g)\chi_2(g) \qquad (\chi_1, \chi_2 \in \hat{G}, g \in G).$$

The elements of the group $\hat{G}$ are called *characters* of G. By the *identity (or principal) character* we understand the identity element of $\hat{G}$. Thus the identity character χ_0 is defined by

$$\chi_0(g) = 1 \qquad \text{for all } g \in G.$$

If for a character χ of G we set

$$\bar{\chi}(g) = \overline{\chi(g)} \qquad (g \in G),$$

where $\overline{\chi(g)}$ is the complex conjugate of $\chi(g)$, then $\bar{\chi} \in \hat{G}$ and in fact

$$\bar{\chi} = \chi^{-1}.$$

Given a subgroup H of G, put

$$H^{\perp} = \{\chi \in \hat{G} \mid \chi(h) = 1 \quad \text{for all } h \in H\}$$

We clearly have a natural isomorphism $H^{\perp} \cong (\widehat{G/H})$. Some other elementary properties of characters are collected in

3.2.1   Lemma

(i) $G \cong \hat{G}$ *(non-canonically) and* $\hat{\hat{G}} \cong G$ *(canonically).*

(ii) $\hat{H} \cong \hat{G}/H^{\perp}$ *and* $(H^{\perp})^{\perp} = H$ *(we equate* $\hat{\hat{G}} = G$*).*

(iii) *If* $\theta \in \hat{H}$ *and* $n = (G:H)$, *then there exist exactly n characters of G extending* θ.

Proof.

(i)   To prove $G \cong \hat{G}$, it suffices to assume that $G = \mathbb{Z}_n$ since $\widehat{G_1 \times G_2} \cong \hat{G}_1 \times \hat{G}_2$. Let g be a generator for $\mathbb{Z}_n$ and let $f : \mathbb{Z}_n \to \mathbb{C}^*$ be a homomorphism. Then $f(g)^n = 1$, i.e. $f(g)$ is an nth root of unity. Let ε be a primitive nth root of unity and let $\psi : \hat{\mathbb{Z}}_n \to \langle \varepsilon \rangle$ be defined by $\psi(f) = f(g)$. It is obvious that ψ is an injective homomorphism. Since for any k, $f(g) = \varepsilon^k$ is an element of $\hat{\mathbb{Z}}_n$, ψ is also surjective.

For each $g \in G$, $\chi \in \hat{G}$, define $\psi_g(\chi) = \chi(g)$. Then $\psi_g \in \hat{\hat{G}}$, and the map $g \mapsto \psi_g$ is a homomorphism of G into $\hat{\hat{G}}$. If $\chi(g) = 1$ for all $\chi \in \hat{G}$ and $H = \langle g \rangle$, then

$$H^\perp = \hat{G} \cong (\widehat{G/H}).$$

Hence, by the foregoing, $G \cong G/H$, which implies $H = 1$. Consequently, $g = 1$ and the map $g \mapsto \psi_g$ is injective. Since $|G| = |\hat{G}| = |\hat{\hat{G}}|$, the required assertion follows.

(ii)   The map $\chi \mapsto \chi \,|\, H$ is obviously a homomorphism from $\hat{G}$ to $\hat{H}$, the kernel of which is $H^\perp$. Thus we have an injective homomorphism $\hat{G}/H^\perp \to \hat{H}$. Since $|\hat{H}| = |H|$, $|\hat{G}| = |G|$ and $|H^\perp| = |G/H|$, we also have $|\hat{G}/H^\perp| = |\hat{H}|$, proving that $\hat{G}/H^\perp \cong \hat{H}$.

Since $|H^\perp| = |G/H|$, we have $|(H^\perp)^\perp| = |H|$. If $h \in H$ then $h : \chi \mapsto \chi(h)$ maps $H^\perp \to 1$. Therefore $H \subseteq H^{\perp\perp}$ and thus $H = H^{\perp\perp}$.

(iii)   By (ii), the map $\chi \mapsto \chi \,|\, H$ is a surjective homomorphism $\hat{G} \to \hat{H}$, the kernel of which is $H^\perp$. Thus we may always choose $\chi \in \hat{G}$ with $\theta = \chi \,|\, H$. Furthermore, $\chi H^\perp$ is the set of all characters of G extending θ. Since $|H^\perp| = |G/H| = n$, the result follows.   ■

We next concentrate on characters of the group $U(\mathbb{Z}/n\mathbb{Z})$, where n is a positive integer. Note that if n divides m, then the natural ring homomorphism

$$\begin{cases} \mathbb{Z}/m\mathbb{Z} \to \mathbb{Z}/n\mathbb{Z}, \\ a + m\mathbb{Z} \mapsto a + n\mathbb{Z}, \end{cases}$$

restricts to a surjective group homomorphism

$$\psi_{m,n} : U(\mathbb{Z}/m\mathbb{Z}) \to U(\mathbb{Z}/n\mathbb{Z})$$

with

$$\mathrm{Ker}\ \psi_{m,n} = \{a + m\mathbb{Z} \mid a \equiv 1 (\mathrm{mod}\ n),\ (a, m) = 1\}.$$

Hence, for $n \mid m$, any character χ of $U(\mathbb{Z}/n\mathbb{Z})$ lifts in an obvious way to a unique character of $U(\mathbb{Z}/m\mathbb{Z})$. Conversely, given $k \mid n$, χ is lifted from a character of $\mathbb{Z}/k\mathbb{Z}$ if and only if

$$\mathrm{Ker}\ \psi_{n,k} \subseteq \mathrm{Ker}\ \chi,$$

i.e. if and only if

$$\chi(a + n\mathbb{Z}) = 1 \qquad \text{for all } a \in \mathbb{N} \text{ with } a \equiv 1 (\text{mod } k) \text{ and } (a, n) = 1.$$

The smallest positive divisor d of n for which χ is lifted from a character of $U(\mathbb{Z}/d\mathbb{Z})$ is called the *conductor* of χ. Thus χ is the identity character if and only if its conductor is equal to 1.

The character χ of $U(\mathbb{Z}/n\mathbb{Z})$ is called *primitive* if the conductor of χ is n. Expressed otherwise, χ is primitive if χ cannot be lifted from a character of $U(\mathbb{Z}/d\mathbb{Z})$, where $d \mid n$ and $d \neq n$. The above shows that if d is the conductor of χ, then χ is lifted from a unique primitive character χ' of $U(\mathbb{Z}/d\mathbb{Z})$. We shall refer to χ' as the *primitive character corresponding to* χ.

It is convenient to classify characters of $U(\mathbb{Z}/n\mathbb{Z})$ into two types: if $\chi(-1 + n\mathbb{Z}) = 1$, then χ is called *even*; if $\chi(-1 + n\mathbb{Z}) = -1$, then χ is called *odd*. In the future, we shall often identity $U(\mathbb{Z}/n\mathbb{Z})$ with $\mathrm{Gal}(\mathbb{Q}(\varepsilon_n)/\mathbb{Q})$ by means of the isomorphism which sends $a + n\mathbb{Z}$ to $\varepsilon_n \mapsto \varepsilon_n^a$. With this identification, χ is even if and only if $\chi \mid H = 1$, where H is the subgroup of $\mathrm{Gal}(\mathbb{Q}(\varepsilon_n)/\mathbb{Q})$ generated by complex conjugation. Of course, H can also be characterized as the Galois correspondent of the maximal real subfield $\mathbb{Q}(\varepsilon_n + \varepsilon_n^{-1})$ of $\mathbb{Q}(\varepsilon_n)$.

To each character χ of $U(\mathbb{Z}/n\mathbb{Z})$, we associate a map

$$\chi^* : \mathbb{Z} \to \mathbb{C}^*$$

by setting

$$\chi^*(a) = \begin{cases} \chi(a + n\mathbb{Z}) & \text{if } a \text{ and } n \text{ are coprime,} \\ 0 & \text{otherwise.} \end{cases}$$

We refer to χ^* as a *Dirichlet character modulo* n. It is clear that χ^* satisfies the following properties:

(a) $\chi^* = (n + m) = \chi^*(m)$ for all $m \in \mathbb{Z}$;
(b) $\chi^*(km) = \chi^*(k)\chi^*(m)$ for all $k, m \in \mathbb{Z}$;
(c) $\chi^*(m) \neq 0$ if and only if $(m, n) = 1$.

Conversely, given any χ^* satisfying (a), (b), and (c), the map defined by

$$\chi(a + n\mathbb{Z}) = \chi^*(a) \qquad ((a, n) = 1)$$

defines a character of $U(\mathbb{Z}/n\mathbb{Z})$. Thus the characters χ and χ^* determine each other and, for this reason, we shall use the same symbol to denote both of them. All the introduced terminology pertaining to characters of $U(\mathbb{Z}/n\mathbb{Z})$ will be applied to the corresponding Dirichlet characters.

3.2.2 Lemma *Let G be a finite abelian group. If χ, $\psi \in \hat{G}$ and $a, b \in G$,* *then*

$$\sum_{a \in G} \chi(a)\overline{\psi(a)} = |G|\,\delta(\chi, \psi),$$

where $\delta(\chi, \chi) = 1$ *and* $\delta(\chi, \psi) = 0$ *if* $\chi \neq \psi$.

Proof. Since $\sum_{a \in G} \chi(a)\overline{\psi(a)} = \sum_{a \in G} (\chi\psi^{-1})(a)$, it suffices to show that $\sum_{a \in G} \chi(a) = |G|$ if $\chi = \chi_0$, and $\sum_{a \in G} \chi(a) = 0$ if $\chi \neq \chi_0$. The first assertion is clear by definition. Assume $\chi \neq \chi_0$. Then there is a $b \in G$ such that $\chi(b) \neq 1$. We have

$$\sum_a \chi(a) = \sum_a \chi(ba) = \chi(b)\sum_a \chi(a)$$

and thus $\sum_a \chi(a) = 0$, as required. ∎

3.2.3 Corollary *Let χ and ψ be Dirichlet characters modulo m and $a, b \in \mathbb{Z}$. Then*

$$\sum_{a=1}^m \chi(a)\overline{\psi(a)} = \phi(m)\delta(\chi, \psi).$$

Proof. Apply Lemma 2.2 for $G = U(\mathbb{Z}/m\mathbb{Z})$. ∎

Let χ be a primitive Dirichlet character modulo m. Then the *Gauss sum* $\tau(\chi)$ attached to χ is defined by

$$\tau(\chi) = \sum_{a=1}^m \chi(a)e^{2\pi i a/m}.$$

3.2.4 Lemma

(i) *For every integer b, $\sum_{a=1}^m \bar{\chi}(a)e^{2\pi i ab/m} = \chi(b)\tau(\bar{\chi})$. In particular, by taking $b = -1$, $\tau(\chi) = \chi(-1)\tau(\bar{\chi})$.*

(ii) $|\tau(\chi)| = \sqrt{m}$.

Proof.

(i) First assume that $(b, m) = 1$. If b' is the inverse of b modulo m, then the substitution $c = ab$ yields

$$\sum_{a=1}^m \bar{\chi}(a)e^{2\pi i ab/m} = \sum_{c=1}^m \bar{\chi}(cb')e^{2\pi i c/m} = \sum_{c=1}^m \bar{\chi}(c)\chi(b)e^{2\pi i c/m} = \chi(b)\tau(\bar{\chi}).$$

Now assume that $(b, m) = d > 1$. Then, by definition, $\chi(b) = 0$ and hence $\chi(b)\tau(\bar{\chi}) = 0$. Since m is the conductor of χ, there exists $y \equiv 1(\mathrm{mod}(m/d))$, $(y, m) = 1$ such that $\chi(y) \neq 1$. Because $dy \equiv d(\mathrm{mod}\, m)$

we have $by \equiv b \pmod{m}$ and so

$$\sum_{a=1}^{m} \bar{\chi}(a)e^{2\pi iab/m} = \sum_{a=1}^{m} \bar{\chi}(a)e^{2\pi iaby/m}$$

$$= \chi(y) \sum_{a=1}^{m} \bar{\chi}(a)e^{2\pi iab/m}.$$

Since $\chi(y) \neq 1$, the sum is 0 as required.

(ii)   We have

$$\phi(m)\,|\tau(\chi)|^2 = \sum_{b=1}^{m} |\chi(b)\tau(\chi)|^2 \qquad \text{(by Corollary 2.3)}$$

$$= \sum_{b=1}^{m} \sum_{a=1}^{m} \chi(a)e^{2\pi iab/m} \sum_{c=1}^{m} \bar{\chi}(c)e^{-2\pi ibc/m} \qquad \text{(by Lemma 2.4)}$$

$$= \sum_{a} \sum_{c} \chi(a)\bar{\chi}(c) \sum_{b} e^{2\pi ib(a-c)/m}$$

$$= \sum_{a} \chi(a)\bar{\chi}(a)m \qquad \text{(the sum over b is 0 unless $a = c$)}$$

$$= m\phi(m)$$

since $\chi(a)\bar{\chi}(a) = 1$ if $(a, m) = 1$, and 0 otherwise. This completes the proof. ∎

Let χ be a primitive Dirichlet character modulo m. We define the *Dirichlet L-function associated to χ* by the formula

$$L(s, \chi) = \sum_{n=1}^{\infty} \chi(n)n^{-s}, \qquad \text{Re}(s) > 1.$$

For $\chi = 1$, this is the usual Riemann zeta function $\zeta(s)$. It is well known that $L(s, \chi)$ may be continued analytically to the whole complex plane, except for a simple pole at $s = 1$ when $\chi = 1$. We now evaluate $L(1, \chi)$. In what follows we ignore questions of convergence, most of which may be treated by partial summation techniques.

3.2.5   Theorem   *Let χ be a primitive Dirichlet character modulo m and let $\varepsilon_m = e^{2\pi i/m}$. Then*

$$L(1, \chi) = -\frac{\chi(-1)\tau(\chi)}{m} \sum_{a=1}^{m} \bar{\chi}(a) \log(1 - \varepsilon_m^a).$$

In particular, if χ is even and $\chi \neq 1$, then

$$L(1, \chi) = -\frac{\tau(\chi)}{m} \sum_{a=1}^{m} \bar{\chi}(a) \log |1 - \varepsilon_m^a|.$$

Proof.   We have

$$L(1, \chi) = \sum_{n=1}^{\infty} \frac{\chi(n)}{n} = \sum_{n=1}^{\infty} \frac{1}{n} \frac{1}{\tau(\bar{\chi})} \sum_{a=1}^{m} \bar{\chi}(a) e^{2\pi i a n/m} \qquad \text{(by Lemma 2.4(i))}$$

$$= \frac{1}{\tau(\bar{\chi})} \sum_{a=1}^{m} \bar{\chi}(a) \sum_{n=1}^{\infty} \frac{1}{n} e^{2\pi i a n/m}$$

$$= -\frac{1}{\tau(\bar{\chi})} \sum_{a=1}^{m} \bar{\chi}(a) \log(1 - \varepsilon_m^a).$$

On the other hand, by Lemma 2.4, we also have

$$\tau(\bar{\chi}) = \chi(-1)\overline{\tau(\chi)} = \chi(-1)m/\tau(\chi).$$

Thus

$$L(1, \chi) = -\frac{\chi(-1)\tau(\chi)}{m} \sum_{a=1}^{m} \bar{\chi}(a) \log(1 - \varepsilon_m^a).$$

Now

$$\log(1 - \varepsilon_m^a) + \log(1 - \varepsilon_m^{-a}) = 2 \log |1 - \varepsilon_m^a|.$$

Hence, if χ is even, then since $\chi(a) = \chi(-a)$, we derive

$$L(1, \chi) = -\frac{\tau(\chi)}{m} \sum_{a=1}^{m} \bar{\chi}(a) \log |1 - \varepsilon_m^a|,$$

thus completing the proof.   ∎

The *Dedekind zeta-function* of an algebraic number field K is defined by

$$\zeta_K(s) = \sum_I N(I)^{-s} \qquad (1)$$

in the open half-plane $\text{Re}(s) > 1$, the sum being taken over all ideals of the ring R of integers of K. It is well known that in the half-plane $\text{Re}(s) > 1$, the series (1) is absolutely convergent. The function $\zeta_K(s)$ can be continued analytically to a meromorphic function with a simple pole at $s = 1$, where it has residue

$$2^{r_1}(2\pi)^{r_2} \text{reg}(K) |d(K)|^{-1/2} \omega(K)^{-1} h(K) \qquad (2)$$

where

$\omega(K)$ is the number of roots of unity in K

$\text{reg}(K)$ is the regulator of K

$d(K)$ is the discriminant of K

$h(K)$ is the class number of K

$[r_1, r_2]$ is the signature of K

The proof of this standard fact can be found in Lang (1970, Chapter VII). Assume that K is a subfield of $\mathbb{Q}(\varepsilon_n)$ and let H be the Galois correspondent of K, i.e.

$$H = \{\tau \in \mathrm{Gal}(\mathbb{Q}(\varepsilon_n)/\mathbb{Q}) \mid \tau(x) = x \quad \text{for all } x \in K\}.$$

Then the restriction homomorphism $\mathrm{Gal}(\mathbb{Q}(\varepsilon_n)/\mathbb{Q}) \to \mathrm{Gal}(K/\mathbb{Q})$ is surjective and its kernel is H. Thus the group of characters of $\mathrm{Gal}(K/\mathbb{Q})$ can be identified with the group of those characters of $\mathrm{Gal}(\mathbb{Q}(\varepsilon_n)/\mathbb{Q})$ for which $\chi \mid H = 1$. For example, if $K = \mathbb{Q}(\varepsilon_n + \varepsilon_n^{-1})$ then H is generated by complex conjugation. Hence, given a character χ of $\mathrm{Gal}(\mathbb{Q}(\varepsilon_n)/\mathbb{Q})$, $\chi \mid H = 1$ if and only if χ is even.

We assume known the following standard fact (see Lang 1978, p. 75):

$$\zeta_K(s) = \prod_{\chi} L(s, \chi'), \tag{3}$$

where χ runs over all characters of $\mathrm{Gal}(\mathbb{Q}_n/\mathbb{Q})$ which are trivial on the Galois correspondent of K, and χ' denotes the primitive character corresponding to χ.

3.2.6   Theorem   (Class number formula)   *Let K be a subfield of $\mathbb{Q}(\varepsilon_n)$, let H be the Galois correspondent of K and let X be the set of all non-identity characters of $\mathrm{Gal}(\mathbb{Q}(\varepsilon_n)/\mathbb{Q})$ which are trivial on H. Then*

$$h(K) = \frac{\omega(K) \, |d(K)|^{1/2}}{2^{r_1 + r_2} \pi^{r_2} \, \mathrm{reg}(K)} \prod_{\chi \in X} L(1, \chi').$$

Proof.   The proof is a direct consequence of (2), (3) and the fact that $\zeta(s)$ has a simple pole at $s = 1$ with residue 1. For the latter fact refer to Apostol (1976, p. 255). ∎

Let $\mathbb{Q}_n = \mathbb{Q}(\varepsilon_n)$, let $\mathbb{Q}_n^+ = \mathbb{Q}(\varepsilon_n + \varepsilon_n^{-1})$ and let h_n and h_n^+ be the class numbers of $\mathbb{Q}_n$ and $\mathbb{Q}_n^+$, respectively. For future use, we now record the following consequences of Theorem 2.6.

3.2.7   Corollary   *For all $n > 2$, we have*

$$h_n = \frac{\omega(\mathbb{Q}_n) \, |d(\mathbb{Q}_n)|^{1/2}}{(2\pi)^{\phi(n)/2} \, \mathrm{reg}(\mathbb{Q}_n)} \prod_{\chi} L(1, \chi'),$$

where χ runs over all non-identity characters of $\mathrm{Gal}(\mathbb{Q}_n/\mathbb{Q})$.

Proof.   If $K = \mathbb{Q}_n$, then $r_1 = 0$, $r_2 = \phi(n)/2$ and $H = 1$. Now apply Theorem 2.6. ∎

3.2.8 Corollary *For all $n > 2$, we have*

$$h_n^+ = \frac{2^{1-\phi(n)/2} |d(\mathbb{Q}_n^+)|^{1/2}}{\mathrm{reg}(\mathbb{Q}_n^+)} \prod_\chi L(1, \chi'),$$

where χ runs over all non-identity even characters of $\mathrm{Gal}(\mathbb{Q}_n/\mathbb{Q})$.

Proof. If $K = \mathbb{Q}_n^+$, then $r_2 = 0$, $r_1 = \phi(n)/2$, $\omega(\mathbb{Q}_n^+) = 2$ and H is generated by complex conjugation. The required assertion is therefore a consequence of Theorem 2.6. ■

We close this section by quoting the following standard fact.

3.2.9 Theorem *Let K be a subfield of $\mathbb{Q}_n$, let H be the Galois correspondent of K and let G be the group of characters of $\mathrm{Gal}(\mathbb{Q}_n/\mathbb{Q})$ which are trivial on H. Then:*

(i) $\displaystyle \prod_{\chi \in G} \tau(\chi) = \begin{cases} \sqrt{|d(K)|} & \text{if K is totally real,} \\ i^{r_2}\sqrt{|d(K)|} & \text{if K is totally complex;} \end{cases}$

(ii) $d(K) = (-1)^{r_2} \displaystyle\prod_{\chi \in G} c(\chi);$

where $c(\chi)$ is the conductor of χ.

Proof. See Lang (1978, Theorem 3.1) and Washington (1982, Theorem 3.11). ■

3.3 Cyclotomic units

Throughout this section, we shall employ the following notation:

ε_n is a primitive nth root of 1

$U_n = U(\mathbb{Z}[\varepsilon_n])$ and $U_n^+ = U(\mathbb{Z}[\varepsilon_n + \varepsilon_n^{-1}])$

$u_a = (1 - \varepsilon_n^a)/(1 - \varepsilon_n), \quad (a, n) = 1$

$u_a^+ = \varepsilon_n^{(1-a)/2} u_a$ (as we have seen in Section 1, $u_a^+ \in U_n^+$)

$C_n = \langle \pm\varepsilon_n, u_a \mid 1 < a < n/2, (a, n) = 1 \rangle$

$C_n^+ = \langle -1, u_a^+ \mid 1 < a < n/2, (a, n) = 1 \rangle$

Our aim is to show that if n is a prime power, then the cyclotomic units u_a, $1 < a < n/2$, $(a, n) = 1$, are independent and hence generate a subgroup of finite index in U_n. It will also be shown that in this case

$$(U_n : C_n) = (U_n^+ : C_n^+) = h_n^+$$

where h_n^+ is the class number of $\mathbb{Q}_n^+$.

The problem whether the set of units

$$u_a = (1 - \varepsilon_n^a)/(1 - \varepsilon_n), \qquad 1 < a < n/2, \ (a, n) = 1$$

is independent in the general case was posed by Milnor (see Rama-chandra 1966). It turns out that for composite n the units u_a need not be independent, as was proved by Ramachandra (1966) (the simplest example is the case where $n = 55$). Note also that Ramachandra (1966) discovered a new set of $(1/2)\phi(n) - 1$ units of U_n (see Theorem 3.5) which are independent and hence generate a subgroup of finite index. The treatment of the general case is similar in many ways to that of the prime power case (Theorem 3.3 below), but is much more technical. For this reason the general case is quoted without proof.

3.3.1 Lemma   *Let n be a prime power and let A be a subgroup of* U_n^+ *containing* -1. *Then*

$$U_n^+/A \cong U_n/\langle \pm \varepsilon_n \rangle A$$

and, in particular,

$$U_n/C_n \cong U_n^+/C_n^+.$$

Proof. By Theorem 1.7, $U_n = \langle \pm \varepsilon_n \rangle U_n^+$ and so the natural homomorphism

$$U_n^+ \to U_n/\langle \pm \varepsilon_n \rangle A$$

induced by inclusion $U_n^+ \subseteq U_n$ is surjective. Since $U_n^+ \cap \langle \pm \varepsilon_n \rangle A = A$, the result follows. ∎

3.3.2 Lemma   *Let G be a finite abelian group, let* $\hat{G} = \mathrm{Hom}(G, \mathbb{C}^*)$ *and let f be any complex valued function on G. Then:*

(i) $\det(f(xy^{-1}))_{x,y \in G} = \displaystyle\prod_{\chi \in G} \sum_{x \in G} \chi(x) f(x);$

(ii) $\det(f(xy^{-1}) - f(x))_{x, y \neq 1} = \displaystyle\prod_{\chi \neq 1} \sum_{x \in G} \chi(x) f(x).$

Proof.

(i) Let V be the space of all functions $G \to \mathbb{C}$. It is a finite-dimensional vector space whose dimension is the order of G. The space V has two natural bases. First the elements of $\hat{G}$, and second the functions $\{\delta_g\}$, $g \in G$, where

$$\delta_g(x) = 1 \qquad \text{if } x = g,$$
$$\delta_g(x) = 0 \qquad \text{if } x \neq g.$$

For each $x \in G$, let $T_x f : G \to \mathbb{C}$ be defined by $(T_x f)(y) = f(xy)$. Then

$$(T_x \chi)(y) = \chi(xy) = \chi(x)\chi(y) = [\chi(x)\chi](y),$$

so that
$$T_x \chi = \chi(x)\chi.$$

This shows that χ is an eigenvector of the linear transformation T_x of V. Let the linear transformation T of V be defined by

$$T = \sum_{x \in G} f(x) T_x.$$

Then, for each $\chi \in \hat{G}$, we have

$$T(\chi) = \left[\sum_{x \in G} \chi(x) f(x) \right] \chi.$$

Therefore χ is an eigenvector of T and thus

$$\det T = \sum_{\chi \in \hat{G}} \sum_{x \in G} \chi(x) f(x).$$

On the other hand, we have

$$T_x \delta_y(g) = \delta_y(xg) = \delta_{x^{-1}y}(g),$$

so that $T_x \delta_y = \delta_{x^{-1}y}$. Hence

$$T\delta_y = \sum_{x \in G} f(x)\delta_{x^{-1}y} = \sum_{g \in G} f(yg^{-1})\delta_g,$$

which shows that $(f(xy^{-1}))_{x,y \in G}$ is the matrix of T with respect to the basis $\{\delta_g\}$. Thus $\det T = \det(f(xy^{-1}))_{x,y \in G}$, proving (i).

(ii)   Let W be the subspace of V consisting of all functions $\psi : G \to \mathbb{C}$ with $\sum_g \psi(g) = 0$. Then the functions ψ_g, $g \neq 1$, defined by

$$\psi_g(x) = \delta_g(x) - |G|^{-1}$$

form a basis for W. Using the fact that

$$\psi_1(x) = -\sum_{g \neq 1} \psi_g(x)$$

we easily find that

$$(f(xy^{-1}) - f(x))_{x,y \neq 1}$$

is the marix of $T \mid W$ with respect to the basis ψ_g, $g \neq 1$. As before, the non-trivial characters diagonalize T/W, so (ii) follows.   ∎

3.3.3 Theorem   (Kummer 1851) *Let $n = p^m$ be a prime power. Then:*

(i) *The cyclotomic units u_a, $1 < a < n/2$, $(a, n) = 1$, are independent and generate a subgroup of finite index in U_n.*

(ii) $(U_n : C_n) = (U_n^+ : C_n^+) = h_n^+.$

Proof. Let us show first that (i) is a consequence of (ii). Indeed, if (ii) is true, then the units u_a^+, $1 < a < n/2$, $(a, n) = 1$ generate a subgroup of U_n^+ of finite index. Since their number is precisely the torsion-free rank of U_n^+, the units u_a^+ and hence u_a are independent. The latter obviously implies that the units u_a, $1 < a < n/2$, $(a, n) = 1$ generate a subgroup of finite index in U_n, proving (i).

To prove (ii), it suffices, by Lemma 3.1, to show that $(U_n^+ : C_n^+) = h_n^+$. Owing to Corollary 2.15, the latter will follow provided we prove that

$$\text{reg}(\{u_a^+\}) = h_n^+ \, \text{reg}(\mathbb{Q}_n^+) \tag{1}$$

(since (1) implies $\text{reg}(\{u_a^+\}) \neq 0$, the units u_a^+ are linearly independent).

Put $\varepsilon = \varepsilon_n$, $\lambda_a = u_a^+$ and let $\mu_a : \varepsilon \to \varepsilon^a$ be in $\text{Gal}(\mathbb{Q}_n/\mathbb{Q})$. The elements

$$\mu_a, \; 1 \leq a < n/2, \qquad (a, n) = 1$$

yield $G = \text{Gal}(\mathbb{Q}_n^+/\mathbb{Q})$. We may write

$$\lambda_a = \frac{(\varepsilon^{-1/2}(1 - \varepsilon))^{\mu_a}}{\varepsilon^{-1/2}(1 - \varepsilon)}$$

(if $p = 2$, extend μ_a to $\mathbb{Q}(\varepsilon_{2^{m+1}})$. Everything below works, since

$$|(\varepsilon^{-1/2}(1 - \varepsilon))^{\mu_a}|$$

is all that matters, and it is independent of the choice of the extension). Consider the map

$$f : G \to \mathbb{C}$$

defined by

$$f(\sigma) = \log |(\varepsilon^{-1/2}(1 - \varepsilon))^\sigma| = \log |(1 - \varepsilon)^\sigma| .$$

Then we have

$$
\begin{aligned}
\text{reg}(\{\lambda_a\}) &= \pm \det[\log |\lambda_a^\tau|]_{a, \tau \neq 1} \\
&= \pm \det[f(\sigma\tau) - f(\tau)]_{\sigma, \tau \neq 1} \\
&= \pm \det[f(\tau\sigma^{-1}) - f(\tau)]_{\sigma, \tau \neq 1} \\
&= \pm \prod_{\substack{\chi \neq 1 \\ \chi \in \hat{G}}} \sum_{\sigma \in G} \chi(\sigma) \log |(1 - \varepsilon)^\sigma| \qquad \text{(by Lemma 3.2(ii))} \\
&= \pm \prod \sum_{1 \leq a < n/2} \chi(a) \log |1 - \varepsilon^a| \\
&= \pm \prod \tfrac{1}{2} \sum_{a=1}^{n} \chi(a) \log |1 - \varepsilon^a| . \tag{2}
\end{aligned}
$$

If the conductor $c(\chi)$ of χ is p^k with $1 \leq k \leq m$, then using the relation

$$\prod_{\substack{1 < a < n \\ a \equiv b \,(\text{mod}\, p^k)}} (1 - \varepsilon_n^a) = 1 - \varepsilon_{p^k}^b$$

we obtain

$$\sum_{a=1}^{n} \chi(a)\log|1 - \varepsilon^a| = \sum_{b=1}^{p^k} \chi(b)\log|1 - \varepsilon_{p^k}^b|$$

$$= -\frac{c(\chi)}{\tau(\chi)} L(1, \bar{\chi}) \qquad \text{(by Theorem 2.5)}$$

$$= -\tau(\chi)L(1, \bar{\chi}) \qquad \text{(by Lemma 2.4)}.$$

Invoking (2), we derive

$$\text{reg}(\{\lambda_a\}) = \pm \prod_{\chi \neq 1} -\tfrac{1}{2}\tau(\chi)L(1, \bar{\chi})$$

$$= h_n^+ \text{reg}(\mathbb{Q}_n^+) \qquad \text{(by Corollary 2.8 and Theorem 2.9(i))}$$

proving (1) and hence the result. ∎

3.3.4 Corollary *If n is a prime power, then the following conditions are equivalent*:

(i) $\{u_a \mid 1 < a < n/2, (a, n) = 1\}$ *is a fundamental system of units of* U_n;
(ii) $\{u_a^+ \mid 1 < a < n/2, (a, n) = 1\}$ *is a fundamental system of units of* U_n^+;
(iii) $h_n^+ = 1$.

Proof. The proof is a direct consequence of Theorem 3.3(ii). ∎

For the practical determination of the unit group of $\mathbb{Z}[\varepsilon_n]$ the vital question is, of course, how to find a specific fundamental system of units. To this effect we have the following result.

3.3.5 Corollary *Let n be a prime power with* $\phi(n) \leqslant 66$. *Then*

$$(1 - \varepsilon_n^a)/(1 - \varepsilon_n) \qquad (1 < a < n/2, (a, n) = 1)$$

is a fundamental system of units of $U(\mathbb{Z}[\varepsilon_n])$.

Proof. Apply Corollary 3.4 and Remark 1.12. ∎

Turning to the general case, we quote the following result in which restriction $n \not\equiv 2 \pmod 4$ is inessential in view of Lemma 1.5.

3.3.6 Theorem *Let* $n \not\equiv 2 \pmod 4$ *and let* $n = \prod_{i=1}^{s} p_i^{e_i}$ *be its prime factorization. Let I run through all subsets of* $\{1, \ldots, s\}$, *except* $\{1, \ldots, s\}$ *and let*

$$n_I = \prod_{i \in I} p_i^{e_i}.$$

Put $d_a = (1/2)(1-a)\sum_l n_l$ *and consider the set*

$$\lambda_a = \varepsilon_n^{d_a} \prod_l \frac{1-\varepsilon_n}{1-\varepsilon_n^{n_l}}, \qquad 1 < a < n/2,\ (a,n)=1.$$

Finally, let C_n' *denote the group generated by* -1 *and the* λ_a's. *Then*:

(i)   *The units* λ_a, $1 < a < n/2$, $(a,n)=1$ *are independent and generate a subgroup of finite index in* U_n.

(ii) $(U_n^+ : C_n') = h_n^+ \prod\limits_{\chi \neq 1} \prod\limits_{p_i \, \nmid \, c(\chi)} (\phi(p_i^{e_i}) + 1 - \chi(p_i)) \neq 0,$

   where $c(\chi)$ *is the conductor of* χ *and* χ *runs through the non-trivial even characters of* $U(\mathbb{Z}/n\mathbb{Z})$.

Proof.   See Theorem 8.3 in Washington (1982).   ∎

3.4   Bass independence theorem

The main result of this section (Theorem 4.3) will play an important role in our future investigation of unit groups of integral group rings. Throughout, we fix the following notation:

$$\mathbb{Q}_n = \mathbb{Q}(\varepsilon_n) \quad \text{is the } n\text{th cyclotomic field}$$

$$\Psi_n = \langle \varepsilon_n \rangle \quad \text{is the group of } n\text{th roots of 1 over } \mathbb{Q}$$

$$\Phi_n \quad \text{is the set of all primitive } n\text{th roots of 1 over } \mathbb{Q}$$

By a character of Ψ_n we shall mean a homomorphism $\chi : \Psi_n \to \Psi_n$. Recall that, by Lemmas 1.2 and 1.3, if $\varepsilon \in \Phi_n$ and $N = N_{\mathbb{Q}_n/\mathbb{Q}}$, then

$$N(1-\varepsilon) = \begin{cases} p & \text{if } n = p^m, \ p \text{ a prime,} \\ 1 & \text{if } n \text{ is not a prime power.} \end{cases} \qquad (1)$$

We begin by calculating some norms.

3.4.1   Lemma   *Let* $n = pn_1$, *where* p *is a prime, and let* $N = N_{\mathbb{Q}_n/\mathbb{Q}_{n_1}}$:

(i)   *Assume* $p \nmid n_1$, $1 \neq \varepsilon \in \Psi_{n_1}$, *and* $\delta \in \Phi_p$. *Then*

$$N(1-\varepsilon) = (1-\varepsilon)^{p-1},$$

$$N(1-\varepsilon\delta) = (1-\varepsilon^p)(1-\varepsilon)^{-1}.$$

(ii)   *Assume* $p \mid n_1$, *and let* $1 \neq \varepsilon \in \Psi_n$. *Then*

$$N(1-\varepsilon) = \begin{cases} (1-\varepsilon)^p & \text{if } \varepsilon \in \Psi_{n_1}, \\ 1-\varepsilon^p & \text{if } \varepsilon \notin \Psi_{n_1}. \end{cases}$$

Proof.   In both cases the first formula is simply a question of degree:

$$(\mathbb{Q}_n : \mathbb{Q}_{n_1}) = \frac{\phi(n)}{\phi(n_1)} = \begin{cases} p - 1 & \text{in case (i),} \\ p & \text{in case (ii).} \end{cases}$$

For the proof of the second formula we need the following identity:

$$1 - \theta^n = \prod_{\eta \in \Psi_n} (1 - \theta\eta), \qquad \theta \text{ a root of 1.} \tag{2}$$

To establish (2), we set $X = \theta^{-1}$ in

$$\prod_{\eta \in \Psi_n} (X - \eta) = X^n - 1$$

to infer that

$$\theta^{-n} - 1 = \prod_{\eta \in \Psi_n} (\theta^{-1} - \eta) = \prod_{\eta \in \Psi_n} \theta^{-1}(1 - \theta\eta) = \theta^{-n} \sum_{\eta \in \Psi_n} (1 - \theta\eta).$$

Now multiply by θ^n and we obtain (2).

Returning to the second formula of (i), we note that $(\varepsilon\delta)^p = \varepsilon^p \in \mathbb{Q}_{n_1}$. It follows that the conjugates of $\varepsilon\delta$ are all $\varepsilon\tau$, $\tau \in \Phi_p$. Therefore

$$N(1 - \varepsilon\delta) = \prod_{\tau \in \Phi_p} (1 - \varepsilon\tau)$$

$$= \left[\prod_{\tau \in \Psi_p} (1 - \varepsilon\tau)\right](1 - \varepsilon)^{-1}$$

$$= (1 - \varepsilon^p)(1 - \varepsilon)^{-1}$$

using (2). In case (ii), $\varepsilon^p \in \mathbb{Q}_{n_1}$ and $\Psi_p \subseteq \mathbb{Q}_{n_1}$. Hence, if $\varepsilon \notin \mathbb{Q}_{n_1}$, the conjugates of ε are all $\varepsilon\tau$, $\tau \in \Psi_p$. Applying (2), we therefore conclude that

$$N(1 - \varepsilon) = \prod_{\tau \in \Psi_p} (1 - \varepsilon\tau) = 1 - \varepsilon^p$$

as required.   ∎

We need a deep result due to Franz (1935). We shall omit the proof since it relies on analytic methods. If ε is a root of 1, we shall write

$$e(\varepsilon) = \begin{cases} 1 - \varepsilon & \text{if } \varepsilon \neq 1, \\ 1 & \text{if } \varepsilon = 1. \end{cases}$$

3.4.2 Lemma (Franz Independence Lemma)   *Suppose $m > 1$ and that a_ε, $\varepsilon \in \Phi_m$ are integers such that (i) $a_\varepsilon = a_{\varepsilon^{-1}}$, (ii) $\sum a_\varepsilon = 0$, and (iii) for all characters χ of Ψ_m,*

$$\prod_{\varepsilon \in \Phi_m} e(\chi(\varepsilon))^{a_\varepsilon} = 1.$$

Then all $a_\varepsilon = 0$.

We are now ready to prove the following fundamental result.

3.4.3   Theorem   (Bass Independence Theorem, Bass 1966)   *Let m be an integer* $\geqslant 1$. *Suppose* $a_\varepsilon(\varepsilon \in \Psi_m)$ *are integers such that*

$$a_\varepsilon = a_{\varepsilon^{-1}} \tag{3}$$

and such that, for all characters χ *of* Ψ_m,

$$\prod_{\varepsilon \in \Psi_m} e(\chi(\varepsilon))^{a_\varepsilon} = 1. \tag{4}$$

Then $a_\varepsilon = 0$ *for all* $\varepsilon \neq 1$.

Proof.   Our plan is to demonstrate that conditions (3) and (4) imply that $a_\varepsilon = 0$ if $\varepsilon \notin \Phi_m$, and that $\sum_{\varepsilon \in \Phi_m} a_\varepsilon = 0$. Once this is accomplished, the result will follow by virtue of Lemma 4.2. A remark that will be used implicitly is the following. If χ is the identity character, then (4) becomes

$$\prod_{\varepsilon \in \Psi_m} e(\varepsilon)^{a_\varepsilon} = 1. \tag{5}$$

Suppose that χ has trivial kernel. Then (5) implies (4) since χ extends to an automorphism of the field $\mathbb{Q}_m$.

The theorem being vacuous for $m = 1$, we argue by induction on m. For the sake of clarity, we shall carry out the proof in three steps.

Step 1:   here we prove

$$\sum_{\substack{\varepsilon \in \Psi_m \\ \varepsilon \neq 1}} a_\varepsilon = 0 \tag{6}$$

by showing that

$$\text{if}\ \ k \,|\, m,\, k > 1, \ \ \text{and if}\ \ \varepsilon \notin \Psi_k, \ \ \text{then}\ \ \sum_{\delta \in \Psi_k} a_{\varepsilon\delta} = 0 \tag{7}$$

and that

$$\text{if}\ \ p \,|\, m,\, p \text{ prime, then}\ \sum_{\delta \in \Phi_p} a_\delta = 0. \tag{8}$$

To prove (7) write $m = ks$ and choose a character χ such that $\text{Ker}\,\chi = \Psi_k$ and such that $\text{Im}\,\chi = \Psi_s$. Setting

$$b_\theta = \sum_{\chi(\varepsilon) = \theta} a_\varepsilon \ \ \ (\theta \in \Psi_s)$$

it follows that (7) is equivalent to $b_\theta = 0$, for all $\theta \neq 1$. But $b_\theta (\theta \in \Psi_s)$ satisfy the hypotheses of the theorem; hence by induction $b_\theta = 0$ for all $\theta \neq 1$.

To prove (8), write $m = p^n q$, with $p \nmid q$, and set $N = N_{\mathbb{Q}_m/\mathbb{Q}}$. It is a consequence of (1) that if $r_i = (\mathbb{Q}_m : \mathbb{Q}_{p^i})(1 \leqslant i \leqslant n)$, then

$$N(e(\varepsilon)) = \begin{cases} 1 & \text{if } \varepsilon \in \Phi_d, \ d \text{ composite,} \\ p^{r_i} & \text{if } \varepsilon \in \Phi_{p^i}. \end{cases}$$

Applying N to (5), we therefore can neglect all but prime power order ε's. Segregating the latter according to the relevant prime, we deduce that

$$\prod_{\varepsilon \in \Psi_{p^n}} N(e(\varepsilon))^{a_\varepsilon} = 1.$$

Hence

$$\sum_{i=1}^{n} r_i \sum_{\varepsilon \in \Phi_{p^i}} a_\varepsilon = 0.$$

If $i > 1$, Φ_{p^i} is a union of non-trivial cosets of Ψ_p, so it follows from (7) (with $k = p$) that

$$\sum_{\varepsilon \in \Phi_{p^i}} a_\varepsilon = 0.$$

Assertion (8) is therefore established.

Step 2: we now prove that

$$\text{if} \quad m = pt, \ p \nmid t, \ p \text{ prime, then for} \quad \varepsilon \neq 1 \quad \text{in} \quad \Psi_t, \ a_\varepsilon = 0. \tag{9}$$

Let $N = N_{\mathbb{Q}_m/\mathbb{Q}_t}$. Applying N to (5) and using Lemma 4.1(i) and (1), we obtain

$$1 = \prod_{\varepsilon \in \Psi_m} N(e(\varepsilon))^{a_\varepsilon}$$

$$= \left\{ \prod_{\substack{\varepsilon \in \Psi_t \\ \varepsilon \neq 1}} e(\varepsilon)^{(p-1)a_\varepsilon} \prod_{\delta \in \Phi_p} [e(\varepsilon^p)e(\varepsilon)^{-1}]^{a_{\varepsilon\delta}} \right\} \prod_{\delta \in \Phi_p} p^{a_\delta}$$

$$= \left[\prod_{\substack{\varepsilon \in \Psi_t \\ \varepsilon \neq 1}} e(\varepsilon)^{c_\varepsilon} \right] p^d,$$

where

$$c_{\varepsilon^p} = (p-1)a_{\varepsilon^p} - \sum_{\delta \in \Phi_p} a_{\varepsilon^p\delta} + \sum_{\delta \in \Phi_p} a_{\varepsilon\delta}$$

and

$$d = \sum_{\delta \in \Phi_p} a_\delta.$$

It follows from (8) that $d = 0$. By (7) (with $k = p$) we must also have

$$c_{\varepsilon^p} = (p-1)a_{\varepsilon^p} + a_{\varepsilon^p} - a_\varepsilon = pa_{\varepsilon^p} - a_\varepsilon.$$

Hence

$$1 = \prod_{\substack{\varepsilon \in \Psi_t \\ \varepsilon \neq 1}} e(\varepsilon). \tag{10}$$

We now claim that to prove (9), it suffices to verify that the hypotheses of
the theorem for t and c_ε. For then we can apply induction and deduce
$c_\varepsilon = 0$ for $\varepsilon \neq 1$, i.e. $a_\varepsilon = pa_{\varepsilon^p}$ for $\varepsilon \neq 1$ in Ψ_t. Choosing r such that
$p^r \equiv 1 (\mathrm{mod}\, t)$ and setting $q = p^r$ it follows that $\varepsilon^q = \varepsilon$, whence $a_\varepsilon = qa_{\varepsilon^q} = qa_\varepsilon$. This shows that $a_\varepsilon = 0$ and substantiates our claim. That t and c_ε
satisfy (3) is obvious. For (4), assume χ is a character of Ψ_t. If Ker $\chi = 1$
we can apply the remark of the first paragraph. Therefore assume
Ker $\chi = \Psi_k$, $k > 1$. Extend χ to a character of $\Psi_m = \Psi_t \times \Psi_p$ by the
identity map on Ψ_p. Invoking (4) we have

$$1 = \prod_{\varepsilon \in \Psi_m} e(\chi(\varepsilon))^{a_\varepsilon}$$

$$= \left\{ \prod_{\substack{\varepsilon \in \Psi_t \\ \varepsilon \notin \Psi_k}} \left[e(\chi(\varepsilon))^{a_\varepsilon} \prod_{\delta \in \Phi_p} e(\chi(\varepsilon)\delta)^{a_{\varepsilon\delta}} \right] \right\} \left[\prod_{\varepsilon \in \Psi_k} \prod_{\delta \in \Phi_p} e(\delta)^{a_{\varepsilon\delta}} \right].$$

Since $k > 1$ we have $\sum_{\varepsilon \in \Psi_k} a_{\varepsilon\delta} = 0$ $(\delta \in \Phi_p)$ by (7); therefore the second
factor is 1. We now apply N to the first factor and calculate as above:

$$1 = \left(N \prod_{\substack{\varepsilon \in \Psi_t \\ \varepsilon \notin \Psi_k}} e(\chi(\varepsilon))^{a_\varepsilon} \prod_{\delta \in \Phi_p} e(\chi(\varepsilon)\delta)^{a_{\varepsilon\delta}} \right)$$

$$= \prod_{\substack{\varepsilon \in \Psi_t \\ \varepsilon \notin \Psi_k}} e(\chi(\varepsilon))^{c_\varepsilon} \prod_{\substack{\varepsilon \in \Psi_t \\ \varepsilon \neq 1}} e(\chi(\varepsilon))^{c_\varepsilon}$$

(since $e(1) = 1$). This verifies (4) for the $c_\varepsilon (\varepsilon \in \Psi_t)$, so (9) is established.
Step 3:   here we complete the proof by demonstrating

$$\text{if} \quad m = pt, p \mid t, p \text{ prime, then for } \varepsilon \neq 1 \quad \text{in} \quad \Psi_t, a_\varepsilon = 0. \quad (11)$$

Again, we apply $N = N_{\mathbb{Q}_m/\mathbb{Q}_t}$ to (5), this time applying Lemma 4.1(ii):

$$1 = \left[\prod_{\varepsilon \in \Psi_t} N(e(\varepsilon))^{a_\varepsilon} \right] \left[\prod_{\substack{\varepsilon \in \Psi_m \\ \varepsilon \notin \Psi_t}} N(e(\varepsilon))^{a_\varepsilon} \right]$$

$$= \left[\prod_{\varepsilon \in \Psi_t} e(\varepsilon)^{pa_\varepsilon} \right] \left[\prod_{\substack{\varepsilon \in \Psi_m \\ \varepsilon \notin \Psi_t}} e(\varepsilon^p)^{a_\varepsilon} \right].$$

Collecting exponents in the second factor leads to sums over non-trivial
cosets of Ψ_p, and these sums are 0 by (7). Therefore

$$1 = \prod_{\varepsilon \in \Psi_t} e(\varepsilon)^{pa_\varepsilon}$$

and these data clearly satisfy (3). For (4), suppose χ is a character of Ψ_t
and write $m = p^n s, p \nmid s$. Extending χ to a character of Ψ_m, we deduce

from (4) that

$$1 = \left[\prod_{\varepsilon \in \Psi_t} e(\chi(\varepsilon))^{a_\varepsilon} \right] \left[\prod_{\substack{\varepsilon \in \Psi_m \\ \varepsilon \notin \Psi_t}} e(\chi(\varepsilon))^{a_\varepsilon} \right].$$

If $\mathrm{Ker}\, \chi \cap \Psi_{p^n} \neq 1$, then we can collect exponents modulo Ψ_p to eliminate the second factor. Otherwise, we apply N to the last equation and calculate as above, using Lemma 4.1(ii). It therefore follows, by induction, that $pa_\varepsilon = 0$ for $\varepsilon \neq 1$ in Ψ_t, and so (11) is established. The following observation now completes the proof. Given $\varepsilon \in \Psi_m - \Phi_m$, we can choose a prime p, $m = pt$, so that $p \in \Psi_t$. Then $a_\varepsilon = 0$, owing to (9) and (11); hence the theorem reduces to Lemma 4.2, by applying (6) and (4). ∎

4
Multiplicative groups of fields

In this chapter we confine our attention to the study of multiplicative groups of fields. After presenting some preparatory results, we investigate the isomorphism class of F^*, where F is a distinguished field such as local or global. We then concentrate on field extensions E/F and study the structure of the factor group E^*/F^*. In particular, it is shown that if F is infinite, then E^*/F^* is not finitely generated. The rest of the chapter, apart from Kneser's theorem, is based on works of May who made fundamental contributions to the subject. The chapter provides an exhaustive account of the papers of May. Special attention is drawn to the investigation of circumstances under which F^* is free modulo torsion. Among other important results, we prove that if F is a field such that for every finite field extension E, E^* is free modulo torsion, and if K is any field generated over F by algebraic elements whose degrees over F are bounded, then K^* is free modulo torsion. Numerous examples are given to illustrate that the results of such type are the best possible.

4.1 Multiplicative structure of some classical fields

We intend to characterize the multiplicative groups of some important classes of fields and their finite extensions. In what follows, all direct products are assumed to be *restricted direct products*. For a cardinal λ and a group A, A^λ will denote the direct product of λ copies of A. The cardinality of $\mathbb{N}$ is denoted by $\aleph_0$ (aleph-null). A set is *countable* if it is either finite or of cardinality $\aleph_0$. For convenience, we divide the remainder of this section into subsections.

4.1A Preparatory results

In order not to interrupt future discussions at an awkward stage, we provide here a number of useful observations. A weaker version of some of the results presented would be sufficient for our immediate purposes. However, the general version will be needed later.

We start by recalling the following piece of information. Let R be an arbitrary commutative ring. As in $\mathbb{Z}$ we define $a \mid b$ (for any $a, b \in R$) to mean

$$b = ar \qquad \text{for some} \qquad r \in R.$$

This is equivalent to the requirement $(b) \subseteq (a)$. Observe that an element $u \in R$ is a unit if and only if $u \mid 1$. The units are trivial divisors since they are divisors of every element of R. If $a \mid b$ and $b \mid a$ (or equivalently, if $(a) = (b)$), then we shall say that a and b are *associated*. It is easy to see that a and b are associated if and only if $a = bu$ for some unit u of R.

A non-zero element p of R is said to be a *prime* if (p) is a prime ideal. Expressed otherwise, a non-zero $p \in R$ is a prime if p is a non-unit such that

$$p \mid ab \qquad \text{implies} \qquad p \mid a \qquad \text{or} \qquad p \mid b.$$

A *unique factorization domain* (*UFD*) is an integral domain in which every element not zero or a unit can be written as a product of primes and, given two such factorizations of the same element,

$$x = y_1 y_2 \ldots y_r = z_1 z_2 \ldots z_t \qquad (y_i, z_j \text{ are primes})$$

then $r = t$ and, after suitable renumbering of the z_i's, y_i is associated to z_i, $1 \leq i \leq r$.

4.1.1 Lemma *Let F be the quotient field of a unique factorization domain R, and let A be a complete set of non-associated prime elements of R. Then F^* is a direct product of $U(R)$ and the free abelian group of rank $|A|$.*

Proof. By hypothesis, each non-zero element of R is either a unit or of the form

$$up_1 p_2 \ldots p_n, \qquad u \in U(R), \, p_i \in A, \, n \geq 1.$$

This shows that $F^* = U(R) \cdot \langle A \rangle$. Assume that $\lambda_1, \lambda_2, \ldots, \lambda_n$ are non-negative integers such that

$$p_1^{\lambda_1} p_2^{\lambda_2} \ldots p_k^{\lambda_k} p_{k+1}^{-\lambda_{k+1}} \ldots p_n^{-\lambda_n} = u \in U(R).$$

Then $p_1^{\lambda_1} p_2^{\lambda_2} \ldots p_k^{\lambda_k} = u p_{k+1}^{\lambda_{k+1}} \ldots p_n^{\lambda_n}$, which is possible only in the case $\lambda_i = 0$, $1 \leq i \leq n$, since R is a *UFD* and the primes $p_1, \ldots, p_n$ are non-associated. Thus $F^* = U(R) \times \langle A \rangle$ and, by taking $u = 1$, $\langle A \rangle$ is free of rank $|A|$. So the lemma is true. ∎

4.1.2 Lemma *Let F be the quotient field of the Dedekind domain R, and let $P(R)$ be the group of all principal fractional ideals of R. Then $P(R)$ is free and*

$$F^* \cong U(R) \times P(R).$$

In particular, if the class group of R is a torsion group, then F^* is a direct product of $U(R)$ and a free abelian group whose rank is equal to the cardinality of the set of non-zero prime ideals of R.

Proof. Let $I(R)$ be the group of all fractional ideals of R. Since R is a Dedekind domain, $I(R)$ is a free group freely generated by non-zero prime ideals of R (Proposition 1.2.5). Hence $P(R)$ is a free group. Furthermore, if $C(R)$ is torsion then the ranks of $I(R)$ and $P(R)$ coincide. We are therefore left to verify that $F^* \cong U(R) \times P(R)$.

The map

$$\begin{cases} F^* \to P(R), \\ a \mapsto Ra \end{cases}$$

is obviously a surjective homomorphism. Since R is the identity element of $P(R)$, the kernel of this map is $U(R)$. Thus

$$1 \to U(R) \to F^* \to P(R) \to 1$$

is an exact sequence, which splits since $P(R)$ is free. This shows that $F^* \cong U(R) \times P(R)$ and the results follows. ∎

Let $(a_i)_{i \in I}$ be a family of cardinals, say $a_i = |A_i|$ $(i \in I)$, and assume that $A_i \cap A_j = \varnothing$ for $i \neq j$. Then we define

$$\sum_{i \in I} a_i = \left| \bigcup_{i \in I} A_i \right|.$$

Similarly, we put

$$\prod_{i \in I} a_i = \left| \prod_{i \in I} A_i \right|.$$

Then one immediately verifies that:

1. If a and b are non-zero cardinals and at least one of a, b is infinite, then $a + b = ab = \max\{a, b\}$. In particular, if a is infinite, then $a^n = a$ for any natural number n.
2. If $(a_i)_{i \in I}$ is a family of cardinals, where $a_i = a$ for every $i \in I$ and at least one of a, $|I|$ is infinite, then $\sum_{i \in I} a_i = \max\{a, |I|\}$.
3. If $(a_i)_{i \in I}$ and $(b_i)_{i \in I}$ are two families of cardinals indexed by the same set I and $a_i \leq b_i$ for every $i \in I$, then $\sum_{i \in I} a_i \leq \sum_{i \in I} b_i$.
4. If A is the set of all finite subsets of an infinite set S, then $|S| = |A|$.

Applying (1)–(4), we next establish the following useful observation.

4.1.3 Lemma   *For any group A and any cardinal λ,*

$$|A^\lambda| = \begin{cases} |A|^\lambda & \text{if both } |A| \text{ and } \lambda \text{ are finite,} \\ \max\{|A|, \lambda\} & \text{otherwise.} \end{cases}$$

In particular, for any field F:

(i) $|F[X]| = \{\max |F|, \aleph_0\}$ *(since $F[X] \cong F^{\aleph_0}$ as additive groups).*

(ii)   *If $F^* = t(F^*) \times A^\lambda$, where A is countable and λ is infinite, then $\lambda = |F|$ (since $t(F^*)$ is countable and $|F| = |F^*|$).*

Proof.   We distinguish two cases. Suppose first that λ is finite. If $|A|$ is finite, then the result is trivial; otherwise, (1) yields

$$|A^\lambda| = |A|^\lambda = |A|,$$

as required. Suppose that λ is infinite. Let S be a set with $|S| = \lambda$ and let $\{S_i \mid i \in I\}$ be the set of all finite subsets of S. For each $s \in S$, put $A_s = A$, so that $A^\lambda = \coprod_{s \in S} A_s$. Setting $A_i = \prod_{s \in S_i} A_s$, we then have $A^\lambda = \cup_{i \in I} A_i$. Since $|S| = \lambda$ is infinite, property (4) tells us that $|I| = \lambda$. Obviously $|A^\lambda| \leqslant \sum_{i \in I} |A_i|$: but the finite case proved above dictates that $|A_i| = |A|$ if $|A|$ is infinite, and $|A_i| < \aleph_0$ if $|A|$ is finite. Consequently, properties (2) and (3) may be employed to infer that

$$|A^\lambda| \leqslant \max\{|A|, \lambda\}.$$

The opposite inequality being trivial, the lemma is proved.   ∎

4.1.4   Lemma   *Let $F(X)$ be the field of rational functions in an indeterminate X over a field F. Then $F(X)^*$ is the direct product of F^* and a free abelian group of rank $\max(|F|, \aleph_0)$.*

Proof.   Put $R = F[X]$, so that $F(X)$ is the quotient field of R. If A denotes the set of irreducible monic polynomials in R, then since R is a *PID* (and hence *UFD*) it follows from Lemma 1.1 that $F(X)^* \cong U(R) \times L$, where L is free of rank $|A|$. Since $U(R) \cong F^*$ we are left to verify that $|A| = \max(|F|, \aleph_0)$ or, by Lemma 1.3, that $|A| = |F(X)|$. If F is infinite, then $|A| \geqslant |F| = |F(X)|$ and hence $|A| = |F(X)|$. If F is finite, then $|F(X)| = \aleph_0 = |A|$.   ∎

We close by providing some results pertaining to infinite abelian groups.

Let p denote a prime. The p^nth complex roots of unity, with n running over all integers $\geqslant 0$, form an infinite multiplicative group. This group is called *quasi-cyclic* or a *group of type p^∞*, and is denoted by $Z(p^\infty)$.

If ε_{p^n} is a primitive p^nth root of 1, $n = 1, 2, \ldots$, then $Z(p^\infty)$ is a union of the chain

$$1 \subset \langle \varepsilon_p \rangle \subset \langle \varepsilon_{p^2} \rangle \subset \ldots \subset \langle \varepsilon_{p^n} \rangle \subset \ldots$$

and every proper subgroup of $Z(p^\infty)$ is of the form $\langle \varepsilon_{p^n} \rangle$ for some $n \geqslant 0$.

Under addition the rational numbers form a group called the *full rational group*, denoted by $\mathbb{Q}$. The groups $Z(p^\infty)$ and $\mathbb{Q}$ are typical examples of divisible groups. In fact, it will be shown that there are no divisible groups other than direct products of $Z(p^\infty)$ and $\mathbb{Q}$.

Let A be an arbitrary abelian group. Then the *socle* $S(A)$ of A consists of all $a \in t(A)$, the torsion subgroup of A, for which the order of a is a square-free integer. Thus $S(A) = 1$ if and only if A is torsion-free, and $S(A) = A$ if and only if A is an *elementary* group in the sense that every element has a square-free order. For a p-group A, we obviously have $S(A) = A[p]$, where $A[p] = \{a \in A \mid a^p = 1\}$.

4.1.5 Lemma   *If A is a divisible subgroup of an abelian group G, then G splits over A.*

Proof.   Owing to Zorn's lemma, we may choose a maximal subgroup B of G having trivial intersection with A. It therefore suffices to show that $G = A \cdot B$. Assume by way of contradiction that there is an element $g \in G$ such that $g \notin A \cdot B$. If $\langle g \rangle \cap A \cdot B = 1$, then $(B \cdot \langle g \rangle) \cap A = 1$, which is impossible since $B \cdot \langle g \rangle \supset B$. Therefore, there is a positive integer n such that $g \notin A \cdot B$ but $g^n \in A \cdot B$.

Choose m to be the least positive integer such that $g^m \in A \cdot B$. Setting $g_1 = g^{m/p}$, where p is a prime divisor of m, it follows that $g_1^p \in A \cdot B$ but $g_1 \notin A \cdot B$. Since A is divisible, there exists $a \in A$, $b \in B$ such that $g_1^p = a^p b$; hence $g_2^p = b$ where $g_2 = g_1 a^{-1} \notin A \cdot B$. By our choice of B, we may choose a non-identity element $a' \in A \cap \langle g_2, B \rangle$. Since $A \cap B = 1$, and since $g_2^p \in B$, a' can be written in the form

$$a' = g_2^k b', \qquad 0 < k < p, b' \in B.$$

Because $(k, p) = 1$, there exist $l, s \in \mathbb{Z}$ such that $lk + sp = 1$. But $g_2^{sp} \in B$ and $g_2^k = a'(b')^{-1} \in A \cdot B$; hence $g_2 = g_2^{lk} \cdot g_2^{sp} \in A \cdot B$. This provides the contradiction we have been seeking and so the proof is complete. ∎

4.1.6 Theorem   *Let G be a divisible abelian group. Then*:

(i)   *G is a direct product of quasi-cyclic and full rational groups.*
(ii)   *The cardinal numbers of the sets of components $Z(p^\infty)$ (for every p) and $\mathbb{Q}$ form a complete and independent system of invariants for G.*

Proof.   Since $t(G)$ is obviously divisible, it follows from Lemma 1.5 that $G = t(G) \times A$, where A is torsion-free and, clearly, again divisible. Put $T = t(G)$ and let T_p be the p-component of T. Then $G = (\amalg_p T_p) \times A$, and hence it suffices to show that T_p is a direct product of groups $Z(p^\infty)$, and A is a direct product of groups $\mathbb{Q}$.

Let $\{g_i\}_{i \in I}$ be a maximal independent system in the socle of T_p. Owing to divisibility, for each $i \in I$, there is an infinite sequence $g_{i1}, \ldots, g_{in}, \ldots$ in T_p, such that

$$g_{i1} = g_i, \quad g_{i,n+1}^p = g_{in} \qquad (n = 1, 2, \ldots).$$

It follows that every g_i can be embedded in a subgroup $B_i \cong Z(p^\infty)$ of T_p, namely, $B_i = \langle g_{i1}, \ldots, g_{in}, \ldots \rangle$. Because $\langle g_i \rangle$ is the socle of B_i and the g_i $(i \in I)$ are independent, the B_i generate their direct product $B = \coprod_{i \in I} B_i$ in T_p. Now B is divisible, and hence a direct factor of T_p. Since B contains the socle of T_p, we have $B = T_p$ and T_p is a direct product of groups $Z(p^\infty)$.

Now choose a maximal independent system $\{a_j\}$, $j \in J$ in A. Since A is torsion-free and divisible, for every positive $n \in \mathbb{Z}$ there is exactly one $x \in A$ satisfying $x^n = a_j$, which shows that every a_j can be embedded in a subgroup $A_j \cong \mathbb{Q}$ of A. Because $\{a_j\}$ is a independent system, the A_j generate their direct product $X = \coprod_{j \in J} A_j$ in A. This is a direct factor of A containing a maximal independent system of A, hence $X = A$, and A is a direct product of groups $\mathbb{Q}$.

Let $r(G)$ and $r_p(G)$ be the torsion-free rank and the p-rank of G, respectively. Then the numbers of direct factors isomorphic to $Z(p^\infty)$ or to $\mathbb{Q}$ in a decomposition of G into groups $Z(p^\infty)$, $\mathbb{Q}$ are obviously equal to $r_p(G)$ and $r(G)$. Now the ranks $r(G)$ and $r_p(G)$ are obviously invariants of G. Moreover, they form a complete system of invariants, since if given $r_p(G)$ and $r(G)$, we can uniquely reconstruct G as the direct product of $r_p(G)$ copies of $Z(p^\infty)$ for every p and $r(G)$ copies of $\mathbb{Q}$. Their independence being obvious, the result follows.   ∎

Let G be an abelian p-group. The *height* of an element g in G is the largest integer r such that $g \in G^{p^r}$, if a largest r exists; otherwise the height of g is infinity.

4.1.7 Theorem (Kulikov 1945) *Let G be an abelian p-group. Then G is a direct product of cyclic groups if and only if G is the union of an ascending chain*

$$G_1 \subseteq G_2 \subseteq \ldots \subseteq G_n \subseteq \ldots$$

of subgroups such that the heights of elements $\neq 1$ in G_n are less than a finite bound k_n.

Proof. If G is a direct product of cyclic groups, then as G_n we may take the product of all direct factors of order $\leq p^n$. To prove the converse, denote by x_1 an element of order p in G_1 which has maximal height in G. Suppose that for all ordinals $\alpha < \beta$ we have chosen in G the elements x_α satisfying the following conditions:

1. All elements x_α are of order p.
2. If $x_\alpha \in G_n - G_{n-1}$, and if $A_\alpha = \langle x_{\alpha'} \mid \alpha' < \alpha \rangle$, then:
   (a) $A_\alpha \supseteq G_{n-1}[p]$;
   (b) $x_\alpha \notin A_\alpha$, and among all elements in $G_n - A_\alpha$, x_α is of maximal height in G.

If $A_\beta \neq G[p]$, then x_β can be chosen as follows. Owing to 2(a), there exists n such that all $x_\alpha \in G_n$ but not all $x_\alpha \in G_{n-1}$. Assume that $A_\beta \neq G_n[p]$. Then among the elements of order p in $G_n - A_\beta$ pick x_β with maximal height in G. If $A_\beta = G_n[p]$, then pick a similar element in the smallest subgroup G_k which satisfies $G_k[p] \supseteq G_n[p]$. Hence there exists an ordinal γ such that $G[p] = \langle x_\alpha \mid \alpha < \gamma \rangle$.

Applying 1 and 2(b), we infer that

3. $\quad G[p] = \coprod_{\alpha < \gamma} \langle x_\alpha \rangle.$

Denote by h_α the height of x_α in G and let $y_\alpha \in G$ be such that $y_\alpha^{p^{h_\alpha}} = x_\alpha$. In view of 3, the cyclic groups $\langle y_\alpha \rangle$ generate a subgroup F such that

4. $\quad F = \coprod_{\alpha < \gamma} \langle y_\alpha \rangle \quad$ and $\quad G[p] \subseteq F.$

We now claim that any element $z \in G[p] \subseteq F$ has height in F equal to that in G. Indeed, by 3, $z = x'_{\alpha_1} x'_{\alpha_2} \ldots x'_{\alpha_n}$, where $x'_{\alpha_i} \neq 1$ $(1 \leqslant i \leqslant n)$ and x'_{α_i} is a power of x_{α_i}. Consequently, x'_{α_i} has height h_{α_i} both in G and in F. Now, the height h of z in F is equal to $\min\{h_{\alpha_i} \mid 1 \leqslant i \leqslant n\}$ by 4. The height of z in G cannot be less than h, and we shall show that it cannot be greater. To this end, let k be such that $h = h_{\alpha_k}$, but $h_{\alpha_i} > h$ for $i > k$. Then, in the product

$$z = (x'_{\alpha_1} \ldots x'_{\alpha_k})(x'_{\alpha_{k+1}} \ldots x'_{\alpha_n})$$

the second factor is either 1 (when $k = n$) or its height in G is strictly greater than h; the first factor is not contained in A_{α_k} and its height in G cannot be greater than $h_{\alpha_k} = h$, by 2(b). Thus the height of z in G cannot be greater than h, so z has the same height both in G and in F.

To complete the proof, we are left to verify that $G = F$. Assume by way of contradiction that $G \neq F$. Then we may choose an element g of minimal order, say p^s, in $G - F$ (plainly, $s \geqslant 2$). The element $g^{p^{s-1}}$ is of order p, and hence belongs to F. Because $g^{p^{s-1}}$ has the same height in F and in G, there exists $f \in F$ such that $f^{p^{s-1}} = g^{p^{s-1}}$. It follows that the order of gf^{-1} is at most p^{s-1}, so $gf^{-1} \in F$. The latter, however, is contrary to the assumption that $g \notin F$. Hence $G = F$ and the result follows. ∎

4.1.8 Corollary (Baer 1934, Prüfer 1923). *Any bounded abelian group G is a direct product of cyclic groups.*

Proof. Since G is bounded, G is torsion and the p-components of G are again bounded. Thus we may assume that $G^{p^n} = 1$ for some $n \geqslant 1$. It follows that the heights of all non-identity elements in G are $\leqslant n$. Hence the sequence $\{G_n\}_{n=1}^{\infty}$, where $G_n = G$ for all n, satisfies the hypothesis of Theorem 1.7. ∎

A subgroup H of an abelian group G is said to be *pure* in G if $G^n \cap H = H^n$ for all $n \geq 1$. The following properties are direct consequence of the definition:

(a) Any direct factor of G is pure in G.
(b) If G/H is torsion-free, then the H is pure in G.
(c) If G is torsion-free, then a subgroup H of G is pure if and only if G/H is torsion-free.
(d) Purity is transitive, i.e. if K is pure in H and H is pure in G, then K is pure in G.

4.1.9 Lemma

(i) *If A is a pure subgroup of G such that G/A is a direct product of cyclic groups, then G splits over A.*
(ii) *If A is a bounded pure subgroup of G, then G splits over A.*

Proof.
  (i) We first show that every coset of G modulo A contains an element of the same order as this coset. Fix $\bar{g} \in G/A$. If $\bar{g}$ is of infinite order, then so is every representative of the coset $\bar{g}$. If $\bar{g}$ is of finite order n, then for any representative $t \in \bar{g}$, we have $t^n \in A$. Purity implies $a^n = t^n$ for some $a \in A$. Then $h = ta^{-1} \in \bar{g}$ is of order $\leq n$, and hence of order n. Write $G/A = \coprod \langle g_\alpha \rangle$, where $g_\alpha \in \bar{g}_\alpha$ is of the same order as $\bar{g}_\alpha$. Then $G = A \cdot B$, where B is the subgroup of G generated by the g_α. If

$$g_{\alpha_1}^{k_1} \cdots g_{\alpha_n}^{k_n} \in A$$

then

$$(\bar{g}_{\alpha_1})^{k_1} \cdots (\bar{g}_{\alpha_n})^{k_n} = 1$$

and so $(\bar{g}_{\alpha_i})^{k_i} = 1$ ($1 \leq i \leq n$). It follows that $g_{\alpha_i}^{k_i} = 1$, whence $A \cap B = 1$, as required.
  (ii) By hypothesis, there exists $n \geq 1$ such that $A^n = 1$. Let $H = A \cdot G^n$, $\bar{H} = H/G^n$, and $\bar{G} = G/G^n$. It is routine to verify that $\bar{H}$ is a pure subgroup of $\bar{G}$. Now $\bar{G}/\bar{H}$ is a bounded group, and hence a direct product of cyclic groups (Corollary 1.8). We may therefore apply (i) to deduce that $\bar{G} = \bar{H} \cdot \bar{F}$, $\bar{H} \cap \bar{F} = 1$, for a suitable subgroup $\bar{F}$ of $\bar{G}$. Let F be the inverse image of $\bar{F}$ in G. Then $G = H \cdot F = A \cdot F$ and $H \cap F = G^n$. Taking into account that A is a pure subgroup, we have $A \cap G^n = A^n = 1$. From this it follows that $A \cap F \subseteq H \cap F \subseteq G^n \cap A = 1$, whence the result. ■

As a final preparatory result, we now prove

4.1.10 Theorem (Baer 1936, Fomin 1937) *Let $t(G)$ be the torsion*

subgroup of an abelian group G. If t(G) is a direct product of a divisible and bounded group, then G splits over t(G).

Proof.   By hypothesis, there exist subgroups A and B of G, where A is divisible and B bounded, such that $t(G) = A \times B$. Thanks to Lemma 1.5, there is a subgroup C of G such that $G = A \times C$. This implies at once that $t(G) = A \times t(C)$, so $t(C) \cong t(G)/A \cong B$. Since $t(C)$ is a pure subgroup of C, Lemma 1.9 (ii) may be employed to infer that $C = t(C) \times D$ for some subgroup D of C. Hence

$$G = A \times t(C) \times D = t(G) \times D$$

as required.   ∎

4.1B   *The isomorphism class of t(F*)*

Throughout, $t(F^*)$ denotes the torsion subgroup of F^*.

4.1.11   Proposition   *Let F be an algebraically closed field:*

(i)   *If char $F = 0$, then $t(F^*) \cong \coprod_p Z(p^\infty) \cong \mathbb{Q}/\mathbb{Z}$.*
(ii)   *If char $F = q > 0$, then $t(F^*) \cong \coprod_{p \neq q} Z(p^\infty)$.*

Proof.
(i)   Assume that char $F = 0$. Then we obviously have

$$t(F^*) \cong t(\mathbb{C}^*) = \coprod_p Z(p^\infty).$$

Moreover, the map $\mathbb{Q} \to t(\mathbb{C}^*)$, $r \mapsto e^{2ir\pi}$ is clearly a surjective homomorphism. Since the kernel of this map is $\mathbb{Z}$, the assertion follows.
(ii)   Assume that char $F = q > 0$ and let p be an arbitrary prime. If $p = q$, then

$$X^{p^n} - 1 = (X - 1)^{p^n}$$

for all $n \geq 1$ and hence the p-component of $t(F^*)$ is 1. On the other hand, if $p \neq q$ then $t(F^*)$ and $t(\mathbb{C}^*)$ have isomorphic p-components. Now apply (i).   ∎

4.1.12   Lemma   *Let p be a prime and let F be the field of p^n elements. Then F^* is cyclic of order $p^n - 1$.*

Proof.   Let $\mathbb{F}_p$ be the field of p elements. Then F is isomorphic to the splitting field of the polynomial $X^{p^n} - X \in \mathbb{F}_p[X]$. Hence F^* consists precisely of the roots of the polynomial $X^{p^n - 1} - 1$ and therefore is cyclic.   ∎

4.1.13 Proposition *A torsion group is isomorphic to $t(F^*)$ for some field F of characteristic 0 if and only if it is isomorphic to a subgroup of $\mathbb{Q}/\mathbb{Z}$ with non-trivial 2-component.*

Proof. Let F be a field of characteristic 0. Since $-1 \in t(F^*)$, $t(F^*)$ contains an element of order 2. If E is the algebraic closure of F then, by Proposition 1.11(i),

$$t(F^*) \subseteq t(E^*) \cong \mathbb{Q}/\mathbb{Z}.$$

Thus $t(F^*)$ is isomorphic to a subgroup of $\mathbb{Q}/\mathbb{Z}$ with non-trivial 2-component.

Conversely, let G be a subgroup of $\mathbb{Q}/\mathbb{Z}$ with non-trivial 2-component. We realize G as a multiplicative group of complex roots of unity, and select a finite or infinite ascending chain of cyclic groups $\langle \varepsilon_k \rangle$ of even order n_k such that

$$G = \bigcup_k \langle \varepsilon_k \rangle.$$

Define the field F as the union of the tower

$$\mathbb{Q}(\varepsilon_1) \subseteq \ldots \subseteq \mathbb{Q}(\varepsilon_k) \subseteq \ldots.$$

Then $t(F^*)$ is the union of $t(\mathbb{Q}(\varepsilon_k)^*)$, $k = 1, 2, \ldots$. Since n_k is even, we have $t(\mathbb{Q}(\varepsilon_k)^*) = \langle \varepsilon_k \rangle$ and thus $G = t(F^*)$, as required. ∎

4.1.14 Proposition *A finite group is isomorphic to $t(F^*)$ for some field F if and only if it is cyclic and its order is either even or of the form $2^n - 1$ for some integer $n \geq 1$.*

Proof. If G is a cyclic group of even order n and ε_n a primitive complex nth root of 1, then G is isomorphic to the torsion subgroup of $\mathbb{Q}(\varepsilon_n)^*$. On the other hand, if G is cyclic of order $2^n - 1$, then by Lemma 1.12, $G \cong F^* = t(F^*)$, where F is the field of 2^n elements.

Conversely, let F be a field such that $t(F^*)$ is finite. Since $t(F^*)$ is a subgroup of $\mathbb{Q}/\mathbb{Z}$ (Proposition 1.11), $t(F^*)$ must be cyclic. If char $F = 0$, then $-1 \in t(F^*)$ and so $t(F^*)$ is of even order. Finally, assume that char $F = p > 0$, and let ε be a generator of $t(F^*)$. Denote by E the smallest subfield of F containing ε. Then we have $t(E^*) = t(F^*)$, which proves the result by appealing to Lemma 1.12. ∎

4.1.15 Proposition

(i) *If a subgroup of $\mathbb{Q}/\mathbb{Z}$ with trivial 2-component is isomorphic to $t(F^*)$ for some field F, then it is isomorphic to K^* for some subfield K of F.*

(ii)   *A subgroup G of $\mathbb{Q}/\mathbb{Z}$ with trivial 2-component is isomorphic to the multiplicative group of some field if and only if G is the union of an ascending chain of cyclic groups whose orders are of the form $2^n - 1$.*

Proof.

(i)   Let G be a subgroup of $\mathbb{Q}/\mathbb{Z}$ with trivial 2-component, and with $G \cong t(F^*)$ for some field F. Since G has trivial 2-component, char $F = 2$. Let K be the smallest subfield of F containing $t(F^*)$. Then, clearly, $K^* = t(F^*) \cong G$.

(ii)   Let G be a subgroup of $\mathbb{Q}/\mathbb{Z}$ with trivial 2-component. Assume that $G \cong F^*$ for some field F. Then char $F = 2$ and F^* is torsion. Hence F is a union of an ascending chain of its finite subfields. This shows that G has the required property, by appealing to Lemma 1.12. Conversely, assume that G is the union of an ascending chain of cyclic groups G_k of order $2^k - 1$, $k = 1, 2, \ldots$. Then the field F which is the union of an ascending chain of Galois fields F_k of order 2^k, $k = 1, 2, \ldots$, is such that $G \cong F^*$.   ∎

4.1C   Global fields

For any field F, let $F(X)$ be the field of rational functions in an indeterminate X over F. A *function field* is a finite extension of $\mathbb{F}_q(X)$, where $\mathbb{F}_q$ is the Galois field of q elements. A *global field* is either an algebraic number field or else a function field.

4.1.16   Lemma $\mathbb{Q}^* \cong \mathbb{Z}_2 \times \mathbb{Z}^{\aleph_0}$.

Proof.   The field $\mathbb{Q}$ is the quotient field of a unique factorization domain $\mathbb{Z}$. As a complete set of non-associated primes of $\mathbb{Z}$ we can choose all prime natural numbers. Since their cardinality is $\aleph_0$ and since $U(\mathbb{Z}) = \{\pm 1\} \cong \mathbb{Z}_2$, the result follows by virtue of Lemma 1.1.   ∎

4.1.17   Lemma   (Skolem 1947)   *Let F be an algebraic number field. Then*

$$F^* \cong \mathbb{Z}_m \times \mathbb{Z}^{\aleph_0} \qquad (\text{for some even } m)$$

and, conversely, for any even m, $\mathbb{Z}_m \times \mathbb{Z}^{\aleph_0}$ is isomorphic to F^* for some algebraic number field F.

Proof.   Let R be the ring of integers of F. Then F is the quotient field of R, and R is a Dedekind domain. Hence $F^* \cong U(R) \times P(R)$ by Lemma 1.2. Since $P(R)$ is free, it follows from the Dirichlet's Unit Theorem (Theorem 2.2.10) that $F^* \cong \mathbb{Z}_m \times \mathbb{Z}^\lambda$ for some even m and some cardinal λ. Since $\mathbb{Q}^* \subseteq F^*$, $\lambda \geq \aleph_0$ by Lemma 1.16. On the other hand, $|F| = \aleph_0$ and so $\lambda \leq \aleph_0$. Thus $\lambda = \aleph_0$ and $F^* \cong \mathbb{Z}_m \times \mathbb{Z}^{\aleph_0}$ (the fact that $\lambda = \aleph_0$ can also be deduced from Theorem 2.1.4 and Lemma 1.2).

Conversely, let $G = \mathbb{Z}_m \times \mathbb{Z}^{\aleph_0}$ for some even m. Then $G \cong F^*$, where F is the mth cyclotomic field $\mathbb{Q}_m$. ∎

Turning to function fields, we first provide two auxiliary results.

4.1.18 Lemma *Let E/F be an arbitrary field extension and let v be a valuation of E which is trivial on F. Then v is trivial on the algebraic closure of F in E.*

Proof. Let $a \neq 0$ be an element of E such that

$$a^n = \lambda_0 + \lambda_1 a + \ldots + \lambda_{n-1} a^{n-1} \qquad (\lambda_i \in F,\ n \geqslant 1).$$

Since v is trivial on F, v is non-archimedean by Proposition 1.2.23, and so

$$v(a)^n \leqslant \max\{v(\lambda_0),\ v(\lambda_1)v(a),\ \ldots,\ v(\lambda_{n-1})v(a)^{n-1}\}$$
$$= \max\{1,\ v(a),\ \ldots,\ v(a)^{n-1}\}.$$

Hence $v(a) \leqslant 1$, for if $v(a) > 1$ then $v(a)^n > v(a)^i > 1$, $1 \leqslant i \leqslant n - 1$, contrary to the above inequality. Applying the same argument to a^{-1}, we have $v(a^{-1}) = 1/v(a) \leqslant 1$, which shows that $v(a) \geqslant 1$. Thus $v(a) = 1$ and the result follows. ∎

4.1.19 Lemma *Let F be an arbitrary field. Given $c > 1$ in $\mathbb{R}$ and $0 \neq a = f(X)/g(X) \in F(X)$, define*

$$\deg a = \deg f(X) - \deg g(X)$$

and put

$$v_c(a) = c^{\deg a}, \qquad v_c(0) = 0.$$

(i) *The map $v_c : F(X) \to \mathbb{R}$ is a valuation which is trivial on F, and whose equivalence class is independent of the choice of $c > 1$ in $\mathbb{R}$.*

(ii) *If v is a non-trivial valuation of $F(X)$ which is trivial on F, then either $v = v_c$ for some $c > 1$ in $\mathbb{R}$ or $v = v_P$ for some P-adic valuation v_P, where P is a prime ideal of $F[X]$.*

Proof.

(i) Straightforward.

(ii) Suppose first that $v(X) > 1$ and put $c = v(X)$. Given

$$f(X) = c_0 + c_1 X + \ldots + c_n X^n \in F[X] \qquad (c_n \neq 0)$$

we have

$$v(f(X)) = v(X)^n = c^{\deg f(X)}$$

since v is non-archimedean and $v(c_n X^n) > v(c_i X^i)$, $0 \leqslant i \leqslant n - 1$. Thus $v = v_c$.

Now assume that $v(X) \leqslant 1$. Then, for any $f(X) \in F[X]$, $v(f(X)) \leqslant 1$. Since v is non-trivial, we must have $v(f(X)) < 1$ for some $f(X) \in F[X]$. The set of all such polynomials is an ideal P generated by an irreducible polynomial $p(X)$. If $v(p(X)) = c$ and $f(X) = p(X)^v g(X)$, $(p(X), g(X)) = 1$, then $v(f(X)) = c^v$. Hence $v = v_P$ as required. ∎

4.1.20 Lemma *Let K be a finite field extension of $F(X)$. Then there exists a full set $\{v_i \mid i \in I\}$ of non-equivalent non-trivial valuations of K which are trivial on F and such that*:

(i) *each v_i is discrete;*
(ii) *for every $a \neq 0$ in K, $v_i(a) \neq 1$ for only finitely many $i \in I$;*
(iii) *for every $a \neq 0$ in K, $\prod_{i \in I} v_i(a) = 1$;*
(iv) *the set $K_0 = \{0\} \cup \{x \in K \mid v_i(x) = 1$ for all $i \in I\}$ is a subfield of K.*

Proof. Let $\{v_i \mid i \in I\}$ be any full set of non-equivalent non-trivial valuations of K which are trivial on F. Owing to Lemma 1.19, the restriction w_i of v_i to $F(X)$ is discrete. Since the field extension $K/F(X)$ is finite, it follows from Proposition 1.2.28 that each v_i is discrete. From Lemma 1.19 it also follows that, given $a \in F(X)$, $w_i(a) = 1$ for all but a finite number of $i \in I$. Let $a \neq 0$ be an element of K. Then

$$a^n = \lambda_0 + \lambda_1 a + \ldots + \lambda_{n-1} a^{n-1} \qquad (\lambda_i \in F(X), \ n \geqslant 1)$$

and therefore

$$v_i(a^n) \leqslant \max\{w_i(\lambda_0), w_i(\lambda_1)v_i(a), \ldots, w_i(\lambda_{n-1})v_i(a)^{n-1}\} \qquad (i \in I)$$
$$= \max\{1, v_i(a), \ldots, v_i(a)^{n-1}\}$$

where the latter equality holds for all but a finite number of $i \in I$. Hence $v_i(a) \leqslant 1$ for all but a finite number of $i \in I$. Since the same argument is applicable to a^{-1}, we conclude $v_i(a) \neq 1$ for only finitely many $i \in I$.

We shall prove (iii) in the special case where $K = F(X)$ [for the general case, we refer to a classic book of Artin (1967, pp. 235–6)]. Select a real number d, $0 < d < 1$. Let v_c be as in Lemma 1.19 and let $\{v_i \mid i \in I\}$ consist of v_c for $c = d^{-1}$ and all the P-adic valuations v_P, where for a fixed prime ideal $P = (p(X))$ of $F[X]$, v_P is determined by

$$v_P(p(X)) = d^{\deg p(X)}.$$

To prove the required formula, we may assume that $a = p(X)$ is an irreducible polynomial. If v_i is distinct from v_c and v_P, then $v_i(a) = 1$. On the other hand,

$$v_c(a)v_p(a) = c^{\deg p(X)} d^{\deg p(X)} = 1,$$

proving (iii).

Consider the set K_0 of all $a \in K$ for which $v_i(a) \leqslant 1$ for all $i \in I$. Let x

and y be two elements of K_0. It follows at once that $-x$, xy, and $x + y$ are also in the set K_0. If $x \in K_0$ and $x \neq 0$, then (iii) implies that $v_i(x) = 1$ for all $i \in I$. It now follows that x^{-1} is in K_0. Thus K_0 forms a subfield of K, and it consists of 0 and those elements x of K which satisfy $v_i(x) = 1$ for all $i \in I$. This completes the proof of the lemma. ∎

We have now accumulated all the information necessary to prove the following result.

4.1.21 Theorem *Let F be an arbitrary field, let K be a finite field extension of $F(X)$ and let L be the algebraic closure of F in K. Then the extension L/F is finite and*

$$K^* \cong L^* \times A,$$

where A is a free abelian group.

Proof. Assume that L/F is an infinite extension. Then there exists an infinite chain

$$F(\lambda_1) \subset F(\lambda_1, \lambda_2) \subset \ldots \subset F(\lambda_1, \ldots, \lambda_n) \subset \ldots \subset L.$$

Since $F(X)(\lambda_1, \ldots, \lambda_n) = F(\lambda_1, \ldots, \lambda_n)(X)$, we see that $F(\lambda_1, \ldots, \lambda_n)$ is the algebraic closure of F in $F(X)(\lambda_1, \ldots, \lambda_n)$. Thus

$$F(X)(\lambda_1) \subset F(X)(\lambda_1, \lambda_2) \subset \ldots \subset F(X)(\lambda_1, \ldots, \lambda_n) \subset \ldots \subset K$$

is a strictly increasing chain of subfields of K containing $F(X)$, contrary to the assumption that the extension $K/F(X)$ is finite.

Let $\{v_i \mid i \in I\}$ be the set of valuations of K exhibited in Lemma 1.20. Consider the homomorphism

$$K^* \xrightarrow{\psi} \prod_{i \in I} v_i(K),$$

$$k \mapsto (v_i(k)).$$

Since each v_i is discrete and $v_i(k) \neq 1$ for only finitely many $i \in I$, the image of ψ is a free abelian group. We are therefore left to verify that $K_0 = L$.

By Lemma 1.18, we have $L \subseteq K_0$. To prove the opposite containment, assume by way of contradiction that $k \in K_0$ is transcendental over F. Then, from the equation satisfied by k with respect to $F(X)$, it follows that X must be algebraic over $F(k)$. Since $F(k) \subseteq K_0$, we conclude, from Lemma 1.18, that $X \in K_0$. But, by Lemma 1.19, $v_c(X) > 1$ and hence $v_i(X) > 1$ for some $i \in I$, a contradiction. Thus $K_0 = L$ and the result follows. ∎

4.1.22 Corollary *Let F be an algebraic number field and let K be a* *finite field extension of $F(X)$. Then*

$$K^* \cong \mathbb{Z}_m \times \mathbb{Z}^{\aleph_0} \quad \text{from some even } m.$$

Proof.   Owing to Theorems 1.17 and 1.21,

$$K^* \cong \mathbb{Z}_m \times \mathbb{Z}^\lambda$$

for some even m and some infinite cardinal λ. Since $|F| = \aleph_0$ and $|F(X)| = |K|$, it follows from Lemma 1.3 that $\lambda = |K| = \aleph_0$, as required. ∎

4.1.23   Corollary   *Let K be a finite extension of $\mathbb{F}_q(X)$, $q = p^n$, p prime. Then*

$$K^* \cong \mathbb{Z}_{p^m-1} \times \mathbb{Z}^{\aleph_0}$$

for some $m \geq n$.

Proof.   By Theorem 1.21, there exists $m \geq n$ such that for $l = p^m$

$$K^* \cong \mathbb{F}_l^* \times \mathbb{Z}^\lambda$$

for some cardinal λ. By Lemma 1.12, $F_l^* \cong \mathbb{Z}_{p^m-1}$ and, by Lemma 1.4, λ is infinite. Hence, by Lemma 1.3, $\lambda = |K| = \aleph_0$ and the result follows. ∎

4.1D   Algebraically closed and real closed fields

A field F is called *formally real* if the only relations of the form $\sum_{i=1}^r x_i^2 = 0$ in F are those for which every $x_i = 0$. It is immediate that F is formally real if and only if -1 is not a sum of squares of elements of F. If char $F = p > 0$, then $0 = 1^2 + \ldots + 1^2$ (p terms); hence it is clear that formally real fields are necessarily of characteristic 0.

A field F is called *real closed* if F is formally real and no proper algebraic extension of F is formally real. For example, the real number field $\mathbb{R}$ is real closed. The following properties of real closed fields are standard (see Jacobson 1964):

(a)   If F is real closed, then any element of F is either a square or the negative of the square.

(b)   If F is real closed, then every polynomial of odd degree with coefficients in F has a root belonging to F.

(c)   If F is real closed, then $\sqrt{-1} \notin F$ and $F(\sqrt{-1})$ is algebraically closed. Conversely, every algebraically closed field E of characteristic 0 contains a real closed subfield F such that $E = F(\sqrt{-1})$.

4.1.24   Theorem   *Let F be an algebraically closed field.*

(i)   *If* char $F = 0$, *then*

$$F^* \cong (\mathbb{Q}/\mathbb{Z}) \times \mathbb{Q}^{|F|}.$$

(ii)   *If char $F = q > 0$ and F is algebraic over its prime subfield, then*

$$F^* \cong \coprod_{p \neq q} Z(p^\infty).$$

(iii) *If char $F = q > 0$ and F is not algebraic over its prime subfield, then*

$$F^* \cong \left(\coprod_{p \neq q} Z(p^\infty) \right) \times \mathbb{Q}^{|F|}.$$

Conversely, for any infinite cardinal λ and any prime q, the groups

$$(\mathbb{Q}/\mathbb{Z}) \times \mathbb{Q}^\lambda \qquad and \qquad \left(\coprod_{p \neq q} Z(p^\infty) \right) \times \mathbb{Q}^\lambda$$

are isomorphic to multiplicative groups of suitable algebraically closed fields.

Proof.   Since F is algebraically closed, for every $a \in F$ and every prime p, the equation $x^p - a = 0$ is solvable. Hence F^* is divisible and, by the structure theorem of divisible groups (Theorem 1.6), we have

$$F^* \cong t(F^*) \times \mathbb{Q}^\lambda$$

for some cardinal λ. Since $\mathbb{Q}$ is of torsion-free rank 1, λ is the torsion-free rank of F^*. If char $F = 0$, then $\mathbb{Q}^* \subseteq F^*$ and so λ is infinite by Lemma 1.16. Hence, by Lemma 1.3(ii), $\lambda = |F|$. This proves (i), by applying Proposition 1.11(i).

Assume that char $F = q > 0$. If the field F is algebraic over its prime subfield, then F^* is a torsion group, and hence $F^* = t(F^*)$. Thus (ii) is also verified, by virtue of Proposition 1.11(ii). If F is not algebraic over its prime subfield, then λ is infinite by Lemma 1.4. Hence $\lambda = |F|$, by Lemma 1.3(ii). This proves (iii), by appealing to Proposition 1.11(ii).

Conversely, let λ be any infinite cardinal and q a prime. Choose any set S with $|S| = \lambda$, and define K to be $\mathbb{Q}$ or $\mathbb{F}_q$, the field of q elements. Denote by F the algebraic closure of the purely transcendental extension $K(S)$ of K generated by S. Since the algebraic closure of any field of infinite cardinality is of the same cardinality, we have

$$|F| = |K(S)| = |S| = \lambda.$$

The desired conclusion is therefore a consequence of (i) and (iii).   ∎

Turning to real closed fields, we prove the following result which is a slightly sharpened version of a theorem of Fuchs (1958).

4.1.25   Theorem   *If F is a real closed field, then*

$$F^* \cong \mathbb{Z}_2 \times \mathbb{Q}^{|F|}.$$

Conversely, for any infinite cardinal λ, the group $\mathbb{Z}_2 \times \mathbb{Q}^\lambda$ is isomorphic to the multiplicative group of a suitable real closed field.

Proof. Let F be a real closed field and let E be the algebraic closure of F. If ε is a primitive nth root of unity in E, then for $n \neq 2$

$$1^2 + \varepsilon^2 + \ldots + \varepsilon^{2(n-1)} = (\varepsilon^{2n} - 1)/(\varepsilon^2 - 1) = 0$$

and hence $\varepsilon \notin F$. Consequently, ± 1 are the only roots of unity in F, i.e. $t(F^*) \cong \mathbb{Z}_2$. Since $t(F^*)$ is finite, $F^* \cong \mathbb{Z}_2 \times A$ for some torsion-free group A (Theorem 1.10). It is a consequence of properties (a) and (b) listed at the beginning of the subsection that the group A is divisible. Hence, by the argument of Theorem 1.24, $A \cong \mathbb{Q}^{|F|}$ and therefore $F^* \cong \mathbb{Z}_2 \times \mathbb{Q}^{|F|}$.

Conversely, given an infinite cardinal λ, we may find an algebraically closed field E with $|E| = \lambda$. By the property (c), $E = F(\sqrt{-1})$ for some real closed subfield F of E. Since $|E| = |F| = \lambda$, we have $F^* \cong \mathbb{Z}_2 \times \mathbb{Q}^\lambda$ and the result follows. ∎

4.1E The rational p-adic field

Throughout, p denotes a fixed rational prime $\mathbb{Z}_p$ the ring of p-adic integers, and $\mathbb{Q}_p$ the rational p-adic field. To avoid confusion, we temporarily abandon our convention that $\mathbb{Z}_n$ denotes the cyclic group of order n, and write $\mathbb{Z}/n\mathbb{Z}$ instead of $\mathbb{Z}_n$.

Recall that, previously, $\mathbb{Q}_p$ was defined as the completion of $\mathbb{Q}$ with respect to the p-adic valuation $v_p : \mathbb{Q} \to \mathbb{R}$ given by

$$v_p(p^\nu b) = p^{-\nu} \qquad (b \neq 0),$$

where the numerator and denominator of b are prime to p. Recall also that $\mathbb{Z}_p$ was defined as the valuation ring of $\tilde{v}_p$, where $\tilde{v}_p$ is the corresponding valuation of $\mathbb{Q}_p$, and that $\mathbb{Q}_p$ is the quotient field of $\mathbb{Z}_p$.

One of the concrete realizations of the above construction is by using the notion of a projective limit defined below.

Let I be an ideal of an arbitrary ring S and let $\bigcap_{n=1}^{\infty} I^n = 0$. From the descending chain

$$I \supseteq I^2 \supseteq I^3 \supseteq \ldots$$

of ideals of S we obtain a chain of natural ring epimorphisms:

$$\longrightarrow S/I^n \xrightarrow{\alpha_n} S/I^{n+1} \longrightarrow \ldots \longrightarrow S/I^2 \xrightarrow{\alpha_2} S/I \longrightarrow 0.$$

The corresponding *projective limit ring* $\varprojlim S/I^n$ is defined as the set

$$\{(\ldots, a_n, a_{n-1}, \ldots, a_2, a_1) \mid a_n \in S/I^n, \; \alpha_n(a_n) = a_{n-1}$$

$$\text{for} \qquad n = 2, 3, \ldots \}$$

with component-wise addition and multiplication. There is a natural embedding of S into $\varprojlim S/I^n$, namely

$$s \mapsto (\ldots, s + I^n, \ldots, s + I^2, s + I)$$

and we identify S with its image in $\varprojlim S/I^n$. The ring $\mathbb{Z}_p$ of p-adic integers is defined by

$$\mathbb{Z}_p = \varprojlim \mathbb{Z}/p^n\mathbb{Z}.$$

Thus $\mathbb{Z}_p$ is a subring of $\prod_{n\geqslant 1}(\mathbb{Z}/p^n\mathbb{Z})$, and we denote by π_n the restriction to $\mathbb{Z}_p$ of the natural projection

$$\prod_{n\geqslant 1}(\mathbb{Z}/p^n\mathbb{Z}) \to \mathbb{Z}/p^n\mathbb{Z}.$$

4.1.26   Lemma   *The sequence*

$$0 \longrightarrow p^n\mathbb{Z}_p \longrightarrow \mathbb{Z}_p \xrightarrow{\ \pi_n\ } \mathbb{Z}/p^n\mathbb{Z} \longrightarrow 0$$

is exact (in particular, $\mathbb{Z}_p/p^n\mathbb{Z}_p \cong \mathbb{Z}/p^n\mathbb{Z}$.

Proof.   Let $\phi_n : \mathbb{Z}/p^n\mathbb{Z} \to \mathbb{Z}/p^{n-1}\mathbb{Z}$ be the natural epimorphism. Assume that $x = (x_n) \in \mathbb{Z}_p$ is such that $px = 0$. Then $px_{n+1} = 0$ for all n, whence $x_{n+1} = p^n y_{n+1}$ for some $y_{n+1} \in \mathbb{Z}/p^{n+1}\mathbb{Z}$. Since $x_n = \phi_{n+1}(x_{n+1})$, x_n is multiple of p^n and hence is 0.

It is clear that $p^n\mathbb{Z}_p \subseteq \operatorname{Ker} \pi_n$. Conversely, if $x = (x_m) \in \operatorname{Ker} \pi_n$, then $x_m \equiv 0 \pmod{p^n}$ for all $m \leqslant n$, and hence $x_m = p^n y_{m-n}$ for some $y_{m-n} \in \mathbb{Z}/p^{m-n}\mathbb{Z}$. The elements y_i determine an element y of $\mathbb{Z}_p = \varprojlim \mathbb{Z}/p^i\mathbb{Z}$, which satisfies $p^n y = x$. So the lemma is true.   ∎

4.1.27   Lemma
(i) $U(\mathbb{Z}_p) = \mathbb{Z}_p - p\mathbb{Z}_p$.
(ii)   *Each non-zero element of $\mathbb{Z}_p$ can be uniquely written in the form*

$$p^n u \qquad (u \in U(\mathbb{Z}_p),\ n \geqslant 0).$$

Proof.

(i)   It suffices to show that $U(A_n) = A_n - pA_n$, where $A_n = \mathbb{Z}/p^n\mathbb{Z}$. If $x \in A_n$ and $x \notin pA_n$, then its image in $A_1 = \mathbb{F}_p$ is distinct from 0 and hence a unit. Then there exist $y, z \in A_n$ such that $xy = 1 - pz$, whence

$$xy(1 + pz + \ldots + p^{n-1}z^{n-1}) = 1$$

proving that $x \in U(A_n)$.
(ii)   Let x be a non-zero element of $\mathbb{Z}_p$. Then there exists a greatest integer n such that $x_n = \pi_n(x)$ is 0. Then by Lemma 1.26, we have $x = p^n u$, where u is not divisible by p. But then, by (i), $u \in U(\mathbb{Z}_p)$ and the uniqueness of the given decomposition being obvious, the result follows.   ∎

It is an immediate consequence of Lemma 1.27(ii) that $\mathbb{Z}_p$ is an integral domain. Its quotient field $\mathbb{Q}_p$ is the field of the p-adic numbers. It is clear that

$$\mathbb{Q}_p = \mathbb{Z}_p[p^{-1}]$$

and that each element of $\mathbb{Q}_p^*$ can be uniquely written in the form

$$p^n u, \qquad n \in \mathbb{Z}, u \in U(\mathbb{Z}_p).$$

This shows that

$$\mathbb{Q}_p^* = \langle p \rangle \times U(\mathbb{Z}_p) \cong \mathbb{Z} \times U(\mathbb{Z}_p)$$

and hence the multiplicative structure of $\mathbb{Q}_p$ is entirely reflected by that of $U(\mathbb{Z}_p)$.

4.1.28 Theorem *The following holds*:

$$\mathbb{Q}_p^* \cong \begin{cases} \mathbb{Z} \times \mathbb{Z}_p \times \mathbb{Z}/(p-1)\mathbb{Z}, & \text{if } p \neq 2; \\ \mathbb{Z} \times \mathbb{Z}_2 \times \mathbb{Z}/2\mathbb{Z}, & \text{if } p = 2. \end{cases}$$

Proof. It suffices to show that

$$U(\mathbb{Z}_p) \cong \begin{cases} \mathbb{Z}_p \times \mathbb{Z}/(p-1)\mathbb{Z}, & \text{if } p \neq 2; \\ \mathbb{Z}_2 \times \mathbb{Z}/2\mathbb{Z}, & \text{if } p = 2. \end{cases} \tag{1}$$

Put $U = U(\mathbb{Z}_p)$ and $U_n = 1 + p^n \mathbb{Z}_p$, $n \geq 1$. Then U_n is the kernel of the homomorphism $U \to U(\mathbb{Z}/p^n\mathbb{Z})$ induced by π_n. In particular, U/U_1 is isomorphic to $\mathbb{F}_p^*$ and hence cyclic of order $p - 1$. The groups U_n form a decreasing chain of open subgroups of U and we have $U = \lim_{\leftarrow} U/U_n$. If $n \geq 1$, then the map $(1 + p^n x) \mapsto x \pmod{p}$ determines an isomorphism

$$U_n/U_{n+1} \cong \mathbb{Z}/p\mathbb{Z}$$

since

$$(1 + p^n x)(1 + p^n y) \equiv 1 + p^n(x + y) \pmod{p^{n+1}}.$$

By induction, it follows that $|U_1/U_n| = p^{n-1}$. Since the order of U/U_1 is coprime to p, we obtain exact splitting sequence

$$1 \to U_1/U_n \to U/U_n \to \mathbb{F}_p^* \to 1.$$

Hence U/U_n contains a unique subgroup $V_n \cong \mathbb{F}_p^*$, and the projection

$$U/U_n \to U/U_{n-1}$$

isomorphically maps V_n onto V_{n-1}. Because $U = \lim_{\leftarrow} U/U_n$, passing to limit, we obtain a subgroup V of U such that $U = V \times U_1$ and $V \cong \mathbb{F}_p^*$. Thus

$$U \cong \mathbb{Z}/(p-1)\mathbb{Z} \times U_1. \tag{2}$$

Let $x \in U_n - U_{n+1}$, where $n \geqslant 1$ if $p \neq 2$ and $n \geqslant 2$ if $p = 2$. We claim that

$$x^p \in U_{n+1} - U_{n+2}. \tag{3}$$

Indeed, by hypothesis, $x = 1 + kp^n$, where $k \not\equiv 0 \pmod{p}$. Applying binomial theorem, we have

$$x^p = 1 + kp^{n+1} + \ldots + k^p p^{np}$$

and the powers of p which are not indicated are $\geqslant 2n + 1$ and hence $\geqslant n + 2$. Moreover, since $n \geqslant 2$ if $p = 2$, we have $np \geqslant n + 2$. Thus

$$x^p \equiv 1 + kp^{n+1} \pmod{p^{n+2}},$$

proving (3).

It will next be shown that

$$U_1 \cong \begin{cases} \mathbb{Z}_p, & \text{if } p \neq 2, \\ (\mathbb{Z}/2\mathbb{Z}) \times \mathbb{Z}_2, & \text{if } p = 2, \end{cases} \tag{4}$$

which will complete the proof, by virtue of (1) and (2). Suppose first that $p \neq 2$. Choose any element $x \in U_1 - U_2$, say $x = 1 + p$. Then, by (3), $x^{p^i} \in U_{i+1} - U_{i+2}$. Let x_n be the image of x in U_1/U_n. Then $x_n^{p^{n-2}} \neq 1$ and $x_n^{p^{n-1}} = 1$. But U_1/U_n is of order p^{n-1}, and hence U_1/U_n is a cyclic group generated by x_n. Let $\theta_{n,x} : \mathbb{Z}/p^{n-1}\mathbb{Z} \to U_1/U_n$, $z \mapsto x_n^z$ be an isomorphism. The diagram

$$
\begin{array}{ccc}
\mathbb{Z}/p^n\mathbb{Z} & \xrightarrow{\theta_{n+1,x}} & U_1/U_{n+1} \\
\downarrow & & \downarrow \\
\mathbb{Z}/p^{n-1}\mathbb{Z} & \xrightarrow{\theta_{n,x}} & U_1/U_n
\end{array}
$$

is commutative. Hence the isomorphisms $\theta_{n,x}$ determine an isomorphism of $\mathbb{Z}_p = \varprojlim \mathbb{Z}/p^{n-1}\mathbb{Z}$ onto

$$U_1 = \varprojlim U_1/U_n,$$

which proves (4) for $p \neq 2$.

Finally, assume that $p = 2$. Choose $x \in U_2 - U_3$, or equivalently, put $x \equiv 5 \pmod 8$. We may determine, in the above manner, isomorphisms

$$\theta_{n,x} : \mathbb{Z}/2^{n-2}\mathbb{Z} \to U_2/U_n$$

and hence an isomorphism $\theta_x : \mathbb{Z}_2 \to U_2$. The homomorphism

$$U_1 \to U_1/U_2 \cong \mathbb{Z}/2\mathbb{Z}$$

induces an isomorphism of $\{\pm 1\}$ and $\mathbb{Z}/2\mathbb{Z}$. Accordingly,

$$U_1 = \{\pm 1\} \times U_2 \cong \mathbb{Z}/2\mathbb{Z} \times \mathbb{Z}_2,$$

proving (4) and hence the result. ∎

4.2 Multiplicative groups of local fields

A field F is *local* if there is a complete discrete valuation v of F such that the residue class field of v is finite. As we shall see below, F is local if and only if it is one of the following types:

(i) a finite extension of the rational p-adic field $\mathbb{Q}_p$;
(ii) a formal power series field in one variable over a finite field.

The cases (i) and (ii) occur according as to whether F and the residue class field of v have unequal or equal characteristic. For this reason, we shall refer to (i) and (ii) as 'unequal characteristic case' and 'equal characteristic case', respectively.

Since we shall deal exclusively with non-archimedean valuations, it will be more convenient to write these valuations additively, i.e. to replace them by exponential valuations. Thus, in accordance with our previous conventions, we use the term 'principal valuation' instead of 'discrete valuation'.

4.2A Complete discrete valued fields

Throughout, F denotes a field and $v: F \to \mathbb{R}$ a fixed principal valuation of F. In case v can be chosen such that F is complete with respect to v, then we say that F is a *complete discrete valued field* (or simply that F is complete). By definition, the value group of v is $a\mathbb{Z}$ for some $a > 0$ in $\mathbb{R}$. If we replace v by the equivalent valuation v' define by $v'(x) = (1/a)v(x)$, then the value group of v' is $\mathbb{Z}$ and we say that v' is *normalized*. In what follows, R, P, and K denote the valuation ring, the unique maximal ideal and the residue class field of v, respectively. Thus, by definition,

$$R = \{x \in F \mid v(x) \geq 0\}, \qquad P = \{x \in F \mid v(x) > 0\}, \qquad \text{and} \qquad K = R/P.$$

Recall that the unit group $U(R)$ of R coincides with $R - P$, i.e.

$$U(R) = \{x \in F \mid v(x) = 0\}.$$

The limit of any sequence of elements of F with respect to v will also be called a *P-adic limit*.

By a *residue system* of K in R we mean any set S of elements of R such that the $s + P$, $s \in S$ are all distinct elements of K. A residue system S is said to be *multiplicative* if S is closed under multiplication. Of course, in general, a multiplicative system need not exist. But if it does exist, then $0 \in S$ and the set S^* of non-zero elements of S forms a multiplicative group isomorphic to K^*.

In what follows, we write K^+ for the additive group of K. By a *prime element* of R (or of F) we understand an element $\pi \in R$ of least positive value. The element π is determined uniquely, up to multiplication by a unit of R. In what follows π always denotes a prime element of R.

4.2.1 Lemma

(i) $P = (\pi)$ *and all ideals of* R *are of the form* (π^n), $n \geqslant 0$.
(ii) $F^* = \langle \pi \rangle \times U(R) \cong \mathbb{Z} \times U(R)$.

Proof. By taking v normalized, we may assume that $v(\pi) = 1$. Hence, for any $x \in F^*$, if $v(x) = n$, then $v(x\pi^{-n}) = v(x) - nv(\pi) = 0$, so $x\pi^{-n} = u \in U(R)$. Thus we have

$$x = \pi^n u \qquad (n = v(x),\ u \in U(R))$$

and it is clear that such representation is unique once π have been chosen. The above obviously implies (i) and (ii). ∎

The natural ring epimorphism $R \to R/P = K$ induces a group epimorphism

$$U(R) \to K^*,$$

the kernel of which is

$$1 + P = \{1 + x \mid x \in P\}.$$

Thus we have an exact sequence

$$1 \to 1 + P \to U(R) \to K^* \to 1.$$

The elements of $1 + P$ are called 1-units and, for convenience, in future we write $U_1(R)$ instead of $1 + P$. More generally, for any $n \geqslant 1$, we put

$$U_n(R) = 1 + P^n.$$

4.2.2 Lemma

(i)   *For any* $n \geqslant 1$, $U_n(R)/U_{n+1}(R) \cong K^+$.
(ii) *For each* $u_n \in U_n(R)$ *there exists a unique* $u_{n+1} \in U_{n+1}(R)$ *and* $a_n \in R$ *determined uniquely* mod P, *such that*

$$u_n = (1 + a_n\pi^n)u_{n+1}.$$

(iii) *If* F *is complete,* $u = u_1 \in U_1(R)$ *and, for each* $n \geqslant 1$, S_n *is a residue system of* K *in* R, *then* u *has a unique representation*

$$u = \prod_{n=1}^{\infty} (1 + a_n\pi^n) \qquad (a_n \in S_n).$$

Proof.
(i) By Lemma 2.1(i), $P^n = (\pi^n)$ for all $n \geqslant 1$, so a typical element of

$U_n(R)$ is $1 + \pi^n x$, $x \in R$. Consider the map

$$\begin{cases} U_n(R) \xrightarrow{\psi} K^+, \\ 1 + \pi^n x \mapsto xP. \end{cases}$$

Since

$$(1 + \pi^n x)(1 + \pi^n y) = 1 + \pi^n(x + y + \pi^n xy),$$

the map ψ is a homomorphism. By definition of ψ, ψ is surjective and Ker ψ is obviously equal to $U_{n+1}(R)$.

(ii) By (i), any residue system S of K in R yields a transversal $1 + \pi^n s$, $s \in S$, for $U_{n+1}(R)$ in $U_n(R)$. Hence, given $u_n \in U_n(R)$, we may write uniquely $u_n = (1 + \pi^n s)u_{n+1}$, with $s \in S$, and $u_{n+1} \in U_{n+1}(R)$, proving (ii).

(iii) Put $\lambda_n = \prod_{i=1}^{n}(1 + a_i \pi^i)$. Then, by (ii), $u = \lambda_n u_{n+1}$ and, since $\lim_{n \to \infty} u_n = 1$, we have $u = \lim_{n \to \infty} \lambda_n$. Since each $a_n \in S_n$ is uniquely determined by u, the assertion follows. ∎

4.2.3 Lemma *Let F be complete and let S be a residue system of K in R. Then*:

(i) *Every element $r \in R$ can be written uniquely as a convergent series*

$$r = \sum_{n=0}^{\infty} r_n \pi^n \qquad (r_n \in S).$$

and, conversely, every such series converges to an element of R

(ii) *Every element $x \in F$ can be written uniquely as a convergent series*

$$x = \sum_{n \gg -\infty} x_n \pi^n, \qquad x_n \in S,$$

where the summation is formally over all integers n, but for $n < 0$ only finitely many terms are non-zero.

Proof. Property (ii) follows from (i) by multiplying by a suitable negative power of π. If $r \in R$, then by definition of S, there is an $r_0 \in S$ such that $r \equiv r_0 \pmod P$; if one writes $r = r_0 + \pi x_1$ and applies the same procedure to x_1, one obtains an $r_1 \in S$ such that $r = r_0 + r_1 \pi + x_2 \pi^2$ and so on. Thus r defines a unique infinite series $\sum_{n=0}^{\infty} r_n \pi^n$, with $r_n \in S$ for all n. The series converges to r and one sees easily that it is unique. Conversely, every series of the form $\sum_{n=0}^{\infty} r_n \pi^n$ is convergent, since its general term converges to zero and F is complete. ∎

Turning our attention to the case where char $K = p > 0$, we first prove the following useful observation.

4.2.4 Lemma   *Assume that* char $K = p > 0$. *Then, for all* $n, m \in \mathbb{N}$,

$$a \equiv b \,(\mathrm{mod}\ P^n) \qquad \text{implies} \qquad a^{p^m} \equiv b^{p^m} (\mathrm{mod}\ P^{n+m}).$$

Proof.   We may clearly assume that $m = 1$. If $a = b + \lambda \pi^n$ with $\lambda \in R$, then

$$a^p = b^p + \binom{p}{1} b^{p-1} \lambda \pi^n + \ldots + \binom{p}{p-1} b \lambda^{p-1} \pi^{(p-1)n} + \lambda^p \pi^{pn}$$

$$\equiv b^p + p b^{p-1} \lambda \pi^n (\mathrm{mod}\ P^{n+1}).$$

Since char $K = p$, we have $p \equiv 0 (\mathrm{mod}\ P)$ and so $p\pi^n \equiv 0 (\mathrm{mod}\ P^{n+1})$. Thus $a^p \equiv b^p (\mathrm{mod}\ P^{n+1})$, as required.   ∎

Let $\mathbb{Z}_p$ be the ring of p-adic integers. Then any $g \in \mathbb{Z}_p$ can be represented as the p-adic limit of a convergent sequence of rational integers g_n:

$$g = \lim_{n \to \infty} g_n.$$

For example, by Lemma 2.3(i), we can take for the g_n the partial sums in the p-adic series representation

$$g = \sum_{n=0}^{\infty} a_n p^n,$$

with coefficients a_n belonging to $\{0, 1, \ldots, p-1\}$. The p-adic convergence of the sequence g_n is described by a system of congruences

$$g_{n+1} \equiv g_n (\mathrm{mod}\ p^{\lambda_n}) \qquad \text{for all} \qquad n \geq 0,$$

where $\lambda_n \to \infty$ as $n \to \infty$. Moreover, if

$$g = \lim_{n \to \infty} g_n'$$

then

$$g_n' \equiv g_n (\mathrm{mod}\ p^{\mu_n}) \qquad \text{for all} \qquad n \geq 0,$$

where $\mu_n \to \infty$ and $n \to \infty$. Applying Lemma 2.4, we now provide the following simple observation which will play a crucial role in our subsequent investigations.

4.2.5 Lemma   *Assume that F is complete and* char $K = p > 0$. *Given $g \in \mathbb{Z}_p$, choose any sequence g_n of rational integers such that $g = \lim_{n \to \infty} g_n$.*

(i)   *For any $\lambda \in U_1(R)$, $\lim_{n \to \infty} \lambda^{g_n}$ exists and is independent of the choice of the sequence g_n.*

(ii)   *For all $n \geq 1$, $U_n(R)$ is a $\mathbb{Z}_p$-module, where for $\lambda \in U_n(R)$ and*

$g \in \mathbb{Z}_p,$

$$\lambda^g = \lim_{n \to \infty} \lambda^{g_n}.$$

(iii) *For all rational integers $m \not\equiv 0 (\bmod p)$ and all $n \geq 1$, the map $U_n(R) \to U_n(R)$, $x \mapsto x^m$ is an automorphism.*

Proof.

(i) Given $\lambda \in U_1(R)$, it follows from Lemma 2.4 that

$$\lambda^{p^n} \equiv 1 (\bmod P^{n+1}) \qquad \text{for all } n \geq 0.$$

Thus, if $h \in \mathbb{Z}$ is such that $h \equiv 0 (\bmod p^n)$, then

$$\lambda^h \equiv 1 (\bmod P^{n+1}) \qquad \text{for all } n \geq 0.$$

Therefore, using the notation preceding Lemma 2.5, we derive

$$\lambda^{g_{n+1}} \equiv \lambda^{g_n} (\bmod P^{\lambda_n + 1}),$$

$$\lambda^{g'_n} \equiv \lambda^{g_n} (\bmod P^{\mu_n + 1}),$$

for all $n \geq 0$. This shows that the sequence λ^{g_n} converges P-adically, and hence has a limit in F depending only on g, and not on the choice of the particular sequence g_n converging to g.

(ii) If $\lambda \equiv 1 (\bmod P^n)$, then all the $\lambda^{g_n} \equiv 1 (\bmod P^n)$ as well and hence

$$\lambda^g \equiv 1 (\bmod P^n),$$

proving that $\lambda^g \in U_n(R)$. The rest follows from the laws for operating with limits, and from the power rules for rational integral exponents.

(iii) Let $m \in \mathbb{Z}$ be such that $m \not\equiv 0 (\bmod p)$. Then m is a unit of $\mathbb{Z}_p$, and hence the action of m on $U_n(R)$ induces an automorphism of $U_n(R)$. But, by definition of the action of $\mathbb{Z}_p$ on $U_n(R)$, for all $\lambda \in U_n(R)$, λ^m is the ordinary mth power of λ. So the lemma is true. ■

Recall that a field F is called *perfect* if either char $F = 0$, or char $F = p > 0$ and $F^p = F$. In the latter case, the map $F \to F$, $\lambda \mapsto \lambda^p$ is an automorphism. Thus, for any $n \geq 1$ and any $\lambda \in F$, there is a unique element $\lambda^{p^{-n}} \in F$ defined by $(\lambda^{p^{-n}})^{p^n} = \lambda$.

When does $U(R)$ split over $U_1(R)$? It turns out that this is always the case if F is complete and K is perfect.

4.2.6 Theorem *Assume that F is complete and K is perfect.*

(i) *If char $K = $ char F, then there is a residue system of K in R which is a field.*

(ii) *If char $K = p > 0$ and char $F = 0$, then there is a unique multiplicative residue system of K in R (this system consists precisely of* *those elements r of R for which r is a p^nth power for all $n \geq 0$). Thus, in any case, there is a subgroup L of U(R) such that*

$$U(R) = L \times U_1(R) \cong K^* \times U_1(R).$$

In particular, by Lemma 2.1(ii),

$$F^* = \langle \pi \rangle \times L \times U_1(R) \cong \mathbb{Z} \times K^* \times U_1(R).$$

Proof. We divide the proof into two steps.

Step 1: here we consider the case where char $K = 0$ (and hence char $F = 0$). As $\mathbb{Z} \to R \to K$ is injective, the homomorphism $\mathbb{Z} \to R$ extends to $\mathbb{Q}$, and we see that R contains $\mathbb{Q}$. By Zorn's lemma, there exist a maximal subfield L of R. If S denotes the image of L in K, we shall show that $S = K$. This will prove the required assertion. We first claim that K is algebraic over S. Indeed, otherwise there would exist an $r \in R$, the image $\bar{r}$ of which in K is transcendental over S. Hence the subring $L[r]$ of R maps into $S[\bar{r}]$, so is isomorphic to $L[X]$, and $L[r] \cap P = 0$. But then R contains the field $L(r)$ of rational functions in r, contradicting the maximality of L. By the above, any $\lambda \in K$ has a minimal polynomial $\bar{f}(X)$ over S. Since char $K = 0$, λ is a simple root of $\bar{f}$. Let $f \in L[X]$ be the polynomial corresponding to $\bar{f}$ under the isomorphism $S \to L$. By Proposition 1.2.25, there is an $x \in R$ such that $\bar{x} = \lambda$ and $f(x) = 0$, and one can lift $S[\lambda]$ into R by sending λ to x. Applying the maximality of L, we deduce that $\lambda \in S$. Thus $S = K$, as we wished to show.

Step 2: here we treat the case where char $K = p > 0$ and K is perfect. Fix $\lambda \in K$ and, for all $n \geq 0$, let λ_n be a fixed inverse image of $\lambda^{p^{-n}}$ in R. Then the powers $\lambda_n^{p^n}$ all belong to the original residue class λ. Let us show that the sequence $\lambda_n^{p^n}$, $n \geq 0$, converges. Since λ_{n+1}^p and λ_n both belong to $\lambda^{p^{-n}}$, we have

$$\lambda_{n+1}^p \equiv \lambda_n (\mathrm{mod}\, P)$$

and hence, by Lemma 2.4,

$$\lambda_{n+1}^{p^{n+1}} \equiv \lambda_n^{p^n} (\mathrm{mod}\, P^{n+1}).$$

This proves the convergence. Because F is complete, the limit

$$f(\lambda) = \lim_{n \to \infty} \lambda_n^{p^n} = \lambda_0 + (\lambda_1^p - \lambda_0) + (\lambda_2^{p^2} - \lambda_1^p) + \dots$$

exists in R and we have

$$f(\lambda) \equiv \lambda_0 (\mathrm{mod}\, P).$$

Thus $f(\lambda)$ belongs to λ, proving that $S = f(K)$ is a residue system of K in R. We next show that $f(\lambda)$ is independent of the choice of the sequence λ_n. Indeed, let λ_n' denote another such sequence. Then

$$\lambda_n' \equiv \lambda_n (\mathrm{mod}\, P)$$

and so, by Lemma 2.4,

$$(\lambda'_n)^{p^n} \equiv \lambda_n^{p^n} (\text{mod } P^{n+1}),$$

proving that

$$\lim_{n \to \infty} (\lambda'_n)^{p^n} = \lim_{n \to \infty} \lambda_n^{p^n}.$$

Fix $x, y \in K$ and let $f(x) = \lim x_n$, $f(y) = \lim y_n$. Then as $(xy)_n$ we can choose $x_n y_n$ and, since $(x_n y_n)^{p^n} = x_n^{p^n} y_n^{p^n}$, we obtain

$$f(xy) = \lim(x_n^{p^n} y_n^{p^n}) = (\lim x_n^{p^n})(\lim y_n^{p^n}) = f(x)f(y).$$

This proves that S is a multiplicative residue system of K in R. In particular, we have $f(\lambda^p) = f(\lambda)^p$ for all $\lambda \in K$. Since K is perfect, the latter implies that $S^p = S$. Let S_1 be any residue system of K in R with $S_1^p = S_1$. For all $n \geq 0$, let α_n denote a representative of $\lambda^{p^{-n}}$ in S_1. Since $S_1^p = S_1$, we then have $\alpha_n^{p^n} = \alpha_0$. Therefore, we also have $\lim_{n \to \infty} \alpha_n^{p^n} = \alpha_0$. On the other hand, α_n can be chosen as a sequence for the calculation of the representative of λ in S. Thus we have $\alpha_0 = f(\lambda) \in S$ and therefore $S = S_1$, proving that S is a unique multiplicative residue system of K in R. If $r \in R$ is a p^nth power for all $n \geq 0$, then r is the limit of an appropriate sequence $r_n^{p^n}$ and hence $r \in S$. Conversely, since $S^p = S$, each $r \in S$ is a p^nth power for all $n \geq 0$, proving that S consists precisely of those elements r of R for which r is a p^nth power for all $n \geq 0$. Finally, if R is itself of characteristic p, then $(a + b)^{p^n} = a^{p^n} + b^{p^n}$, which implies that $f(x + y) = f(x) + f(y)$ for all $x, y \in K$. Thus S is a field and the result follows. ■

Let A be a commutative ring and let $A[[X]]$, be the set of all formal sums

$$\sum_{n=0}^{\infty} a_n X^n \qquad (a_n \in A).$$

Given two elements of $A[[X]]$, say

$$\sum_{n=0}^{\infty} a_n X^n \qquad \text{and} \qquad \sum_{n=0}^{\infty} b_n X^n,$$

define their sum and product to be, respectively,

$$\sum_{n=0}^{\infty} (a_n + b_n) X^n \qquad \text{and} \qquad \sum_{n=0}^{\infty} c_n X^n,$$

where

$$c_n = \sum_{i+j=n} a_i b_j.$$

Then we see that $A[[X]]$ becomes a ring, to which we refer as the ring of *formal power series* in one variable over A. Let

$$f = \sum_{n=0}^{\infty} a_n X^n$$

be a non-zero element of $A[[X]]$. Then the smallest n for which $a_n \neq 0$ is called the *order* of f, and will be denoted by $o(f)$. We agree to attach the order $+\infty$ to the zero element of $A[[X]]$. The following properties are easy consequences of the definitions:

$$o(f+g) \geq \min\{o(f), o(g)\}, \qquad o(fg) \geq o(f) + o(g); \tag{1}$$

$$o(fg) = o(f) + o(g) \qquad \text{if } A \text{ is an integral domain;} \tag{2}$$

$$f \text{ is a unit of } A[[X]] \qquad \text{if and only if } a_0 \text{ is a unit of } A. \tag{3}$$

Now assume that A is a field. Then, by (3), each non-zero element f of $A[[X]]$ of order n is of the form $f = X^n g$, where g is a unit. Furthermore, by (2) $A[[X]]$ is an integral domain. If follows that $P = (X)$ is a prime ideal of $A[[X]]$, and all other ideals are powers of P. Thus $A[[X]]$ is a discrete valuation ring with residue class field equal to A. We denote by $A((X))$ the quotient field of $A[[X]]$, and refer to $A((X))$ as the field of *formal power series* in one variable over A. Every non-zero element of $A((X))$ can be uniquely written in the form

$$f = \sum_{n \geq n_0} a_n X^n \qquad (n_0 \in \mathbb{Z}, \ a_{n_0} \neq 0),$$

and one defines the *order* $o(f)$ of f to be the integer n_0. Then one obtains a principal valuation of $A((X))$, the valuation ring of which is $A[[X]]$, and the residue class field is A. An easy verification shows that the given valuation of $A((X))$ is complete. Thus $A((X))$ is a complete discrete valued field with residue class field A. Of course, since A is a subfield of $A((X))$, A and $A((X))$ have the same characteristic. It turns out that the latter property guarantees the converse, namely

4.2.7 Theorem   *Let F be a complete discrete-valued field with valuation ring R and residue class field K. If char F = char K, then*

$$R \cong K[[X]]$$

and

$$F \cong K((X)).$$

Proof.   Since F is the quotient field of R and $K((X))$ is the quotient field of $K[[X]]$, it suffices to prove that $R \cong K[[X]]$. If the field K is perfect, then the required assertion is a consequence of Lemma 2.3(i) and Theorem 2.6(i). For the case where K is not necessarily perfect we refer to Cohen (1946) and Roquette (1959). ∎

4.2B   Residue degree, ramification index and related results

In what follows, E/F denotes a finite field extension of degree n, and v an exponential valuation on F, with an extension w to E. We shall write R_v, P_v, Γ_v for the valuation ring, maximal ideal and the value group of v, and R_w, P_w, Γ_w for the corresponding entities for w. In case both v and w are principal, we write p and π for the prime elements of R_v and R_w, respectively. We say that E/F is an extension of complete discrete-valued fields if v and w are principal and complete.

Since w restricts to v on F, we have $\Gamma_v \subseteq \Gamma_w$. The index

$$e = e_{E/F} = (\Gamma_w : \Gamma_v)$$

is called the *ramification index* of w (in the extension E/F). Since $R_v \cap P_w = P_v$, the natural homomorphism $R_v/P_v \to R_w/P_w$ is an embedding so, denoting the residue class fields by K_v, K_w, respectively, we see that K_w is an extension of K_v. The degree

$$f = f_{E/F} = (K_w : K_v)$$

is called the *residue degree* of w (in the extension E/F). In the special case where $F = \mathbb{Q}_p$ we refer to e and f as the *absolute ramification index* and *absolute residue degree*, respectively. Both the residue degree and ramification index are finite and, in fact, we have

4.2.8   Theorem

(i) $ef \leqslant n$.
(ii) If v is principal and complete, then:
  (a) w is principal and complete, $e = w(p)/w(\pi)$, $n = ef$ and $R_w p = P_w^e$;
  (b) R_w is a free R_v-module of rank n with basis $\{u_i \pi^j \mid 1 \leqslant i \leqslant f, 0 \leqslant j \leqslant e - 1\}$, where $u_1, \ldots, u_f$ are any elements of R_w such that their images in K_w form a basis of K_w over K_v. Furthermore, the $u_i \pi^j$ form a basis of E over F.

Proof.   For the sake of clarity, we divide the proof into three steps.
  Step 1:   let $u_1, \ldots, u_r \in R_w$ be such that their images $\bar{u}_1, \ldots, \bar{u}_r$ in K_w are linearly independent over K_v. We claim that for any $\lambda_1, \ldots, \lambda_r \in F$,

$$w(\lambda_1 u_1 + \ldots + \lambda_r u_r) = \min\{v(\lambda_1), \ldots, v(\lambda_r)\}. \tag{1}$$

To prove (1), we may assume, after appropriate renumbering, that

$$v(\lambda_1) = \ldots = v(\lambda_k) < v(\lambda_i) \qquad (i > k).$$

Since $\bar{u}_i \neq 0$, we have $w(u_i) = 0$, and so the left-hand side of (1) has value $\geqslant v(\lambda_1)$. If the value is $> v(\lambda_1)$, then after dividing by λ_1 and reducing mod P_w we would get

$$\bar{u}_1 + \overline{(\lambda_2/\lambda_1)}\bar{u}_2 + \ldots + \overline{(\lambda_r/\lambda_1)}\bar{u}_r = 0$$

contrary to the linear independence of the $\bar{u}_i$ over K_v. This establishes (1).

Step 2: let $x_1, \ldots, x_s \in E$ be such that the $w(x_j)$ are incongruent mod Γ_v. We assert that the rs elements $u_i x_j$, $1 \le i \le r$, $1 \le j \le s$, are F-linearly independent. This will prove that $rs \le n$; in particular, taking $s = e$, $r = f$ we obtain (i). Assume that there is a relation

$$\sum \lambda_{ij} u_i x_j = 0 \qquad (\lambda_{ij} \in F). \qquad (2)$$

Setting $a_j = \sum u_i \lambda_{ij}$, we then have $\sum a_j x_j = 0$. If follows that $w(a_h x_h) = w(a_k x_k)$ for some $h, k, 1 \le h < k \le s$. This means that

$$w(x_h/x_k) = w(x_h) - w(x_k) = w(a_k) - w(a_h). \qquad (3)$$

But the right-hand side of (3) is in Γ_v, by (1), and the left-hand side is not, by our choice of the x_i. Hence all the a_j vanish and so, again by (1) the λ_{ij} must also vanish, which shows that the $u_i x_j$ are F-linearly independent.

Step 3: completion of the proof. Assume that v is principal and complete. Then w is principal and complete, by Proposition 1.2.28 and Proposition 1.2.26. Since both v and w are principal,

$$\Gamma_v = v(p)\mathbb{Z} \qquad \text{and} \qquad \Gamma_w = w(\pi)\mathbb{Z}.$$

Let $k \ge 1$ be such that $R_w p = P_w^k$. Since $P_w^k = (\pi^k)$, we have $w(p) = kw(p)$. Hence

$$e = (\Gamma_w : \Gamma_v) = w(p)/w(\pi) = k.$$

We now claim that every element of R_w is an R_v-linear combination of the elements $u_i \pi^j$, $1 \le i \le f$, $0 \le j \le e - 1$. If sustained, it will follow (by multiplying by a negative power of π) that every element of E is an F-linear combination of the $u_i \pi^j$. Hence $n \le ef$ and, by (i) $n = ef$. Furthermore, it will follow that the $u_i \pi^j$ form an F-basis of E and hence an R_v-basis of R_w, which will complete the proof. Choose $x \in R_w$. Then $w(x) = w(\pi^k)$ for some $k \ge 0$. We can write $k = m_1 e + j_1$ with $m_1 \ge 0$, $0 \le j_1 \le e - 1$. Then $w(x) = w(p^{m_1}\pi^{j_1})$ (since $e = w(p)/w(\pi)$), so $y = (p^{m_1}\pi^{j_1})^{-1}x$ satisfies $w(y) = 0$. Then the definition of the u_i shows that there exist elements $\alpha_{1i} \in R_v$ such that $y - \sum_{i=1}^f \alpha_{1i} u_i \in P_w$. Then

$$w\left(\sum_{i=1}^f \alpha_{1i} u_i\right) = w(y) = 0$$

and, if $x = p^{m_1}\pi^{j_1}(y - \sum_{i=1}^{f} \alpha_{1i}u_i)$, then $w(x_1) > w(x)$. We have

$$x = p^{m_1}\pi^{j_1}y = p^{m_1}\pi^{j_1}\left(\sum_{i=1}^{f} \alpha_{1i}u_i\right) + x_1. \tag{4}$$

We may repeat this argument with x_1 and obtain a sequence $x_1, x_2, \ldots$, such that

$$x_{k-1} = p^{m_k}\pi^{j_k}\left(\sum_{i=1}^{f} \alpha_{ki}u_i\right) + x_k \qquad (\alpha_{ki} \in R_v, \ m_k \geq 0, \ 0 \leq j_k \leq e - 1), \tag{5}$$

where

$$w\left(\sum_{i=1}^{f} \alpha_{ki}u_i\right) = 0 \qquad \text{and} \qquad w(x_k) > w(x_{k-1}).$$

Then (5) implies that $w(x_{k-1}) = w(p^{m_k}\pi^{j_k})$. If follows that $x_k \to 0$, $p^{m_k} \to 0$, and $(\sum_{i=1}^{f} \alpha_{ki}u_i)p^{m_k} \to 0$. The last implies that every infinite series whose terms form a subsequence of the sequence $(\sum_{i=1}^{f} \alpha_{ki}u_i)p^{m_k}$, $k = 1, 2, \ldots$, converges. By (4) and (5), we have

$$x = p^{m_1}\pi^{j_1}\left(\sum \alpha_{1i}u_i\right) + \ldots + p^{m_k}\pi^{j_k}\left(\sum \alpha_{ki}u_i\right) + x_k. \tag{6}$$

Since $x_k \to 0$ and the coefficients of the various powers of π^j, $0 \leq j \leq e - 1$, in (6) converge, we obtain from (6) that

$$x = \sum \beta_{ij}u_i\pi^j \qquad (0 \leq j \leq e - 1)$$

for some $\beta_{ij} \in R_v$. This completes the proof of the theorem. ∎

Let E/F be an extension of complete discrete valued fields. We say that E/F is *unramified* if $e = 1$ (equivalently, if $f = n$). In case $e = n$ (equivalently, $f = 1$), E/F is said to be *totally ramified*.

4.2.9 Proposition   Let E/F be an extension of complete discrete-valued fields.

(i) If E/F is totally ramified of degree e, then $E = F(\pi)$ and π satisfies an equation:

$$x^e + a_1 px^{e-1} + \ldots + a_{e-1}px + a_e p = 0$$

$$(a_i \in R_v, \ 1 \leq i \leq e - 1, \ a_e \in U(R_v))$$

(such an equation is called an *Eisenstein equation of degree* e).

(ii) If $E = F(\pi_0)$, where π_0 satisfies an Eisenstein equation of degree e, then E/F is totally ramified of degree e and π_0 is a prime element of E.

Proof.

(i) Assume that E/F is totally ramified of degree e. Then, by Theorem 2.8(ii), R_w is a free R_v-module with basis $\{\pi^j \mid 0 \leqslant j \leqslant e-1\}$ (which is also an F-basis for E), and $\pi^e = pu$ for some unit u of R_w. Hence $E = F(\pi)$ and

$$\pi^e = p(a_1\pi^{e-1} + \ldots + a_{e-1}\pi + a_e)$$

for some $a_i \in R_v$, $1 \leqslant i \leqslant e$. Furthermore, since u is a unit of R_w, a_e is a unit of R_v, which proves the required assertion.

(ii) Let $n = (E:F)$ and let $n = e^*f^*$ be the momentarily unknown decomposition of n into ramification index e^* and residue degree f^*. We must show that $n = e^* = e$ and that $w(\pi_0) = w(\pi)$. Since π_0 satisfies an equation of degree e, we have $n \leqslant e$ and thus $e^* \leqslant n \leqslant e$. On the other hand, note that $w(\pi_0) > 0$. This is so since otherwise, $1/\pi_0 \in R_w$ and the Eisenstein equation, rewritten in $1/\pi_0$, would yield a contradiction:

$$0 = 1 + a_1 p(1/\pi_0) + \ldots + a_{e-1} p(1/\pi_0)^{e-1} + a_e p(1/\pi_0)^e \equiv 1 \pmod{P_v}.$$

Hence, for any $x \in P_v$, $w(\pi_0 x) > w(x) \geqslant w(p)$. Since $w(-a_e p) = w(p)$, we have $w(-a_e p + \pi_0 x) = w(p)$. But, by hypothesis, $\pi_0^e \equiv -a_e p$ $(\bmod\ \pi_0 P_v)$ and hence $ew(\pi_0) = w(p)$. Since, by Theorem 2.8(ii), $e^* = w(p)/w(\pi)$ and, by definition of π, $w(\pi_0) \geqslant w(\pi)$ we conclude that $e^* \geqslant e$. Hence $n = e^* = e$ and $w(\pi_0) = w(\pi)$, as required. ∎

Now assume that E/F is an extension of degree n of complete discrete valued fields, that char $F = 0$ and that K_v is a perfect field of characteristic p (hence so is K_w). For simplicity, we further assume that p is a prime element of R_v. Owing to Theorem 2.6, there is a unique multiplicative residue system S_v (respectively S_w) of K_v in R_v (respectively, of K_w in R_w). Furthermore, it is a standard fact (see Hasse 1980) that for any finite extension K_w/K_v, there exists an unramified extension E/F such that K_w and K_v are the residue class fields attached to E and F, respectively. By a p'-*root of unity* we understand a root of unity, the order of which is not divisible by p. It is now easy to see that if E/F is unramified and the field K_w consists of p^s elements then

$$E = F(\varepsilon) \qquad \text{and every } p'\text{-root of unity of } E \text{ is contained in } \langle \varepsilon \rangle \quad (6)$$

where ε is a primitive $(p^s - 1)$-th root of unity. Furthermore, for any $t \geqslant 1$,

$$F(\delta)/F \qquad \text{is unramified} \tag{7}$$

where δ is a primitive $(p^t - 1)$-th root of unity. Indeed, to prove (6) we need only apply Theorem 2.8(ii) with $u_1 = \varepsilon$, $u_2, \ldots, u_f$ being chosen from the set S_w, and the characterization of S_w given by Theorem 2.6(ii).

Property (7) is a consequence of (6) by choosing s large enough to ensure that $F(\delta) \subseteq F(\varepsilon)$.

Keeping the above notation and assumptions, we now introduce the notion of the inertia field. By the *inertia field* of E/F we understand an intermediate field E_0 with the following property:

(i) E_0/F is unramified of degree $f = f_{E/F}$.

Note that (i) implies that $f = f_{E_0/F}$, and hence that $f_{E/E_0} = 1$. Thus the field E_0 also satisfies:

(ii) E/E_0 is totally ramified of degree $e = e_{E/F}$.

Let us show that if E_0 exists, then it is uniquely determined within E. Indeed, assume that E_0 exists. Then, since $f = f_{E_0/F}$, E_0 has the same residue class field as E. Hence, if $\lambda = w \mid E_0$, then S_λ is a multiplicative residue system of K_w in R_w, and thus $S_\lambda = S_w$. Furthermore, since E_0/F is unramified, p is a prime element of E_0. Therefore, E_0 consists of the elements

$$\sum_{i \gg -\infty} a_i p^i \qquad (a_i \in S_w),$$

which shows that E_0 is uniquely determined within E. Furthermore, by Proposition 2.9(i), we also have

$$E = E_0(\pi) \qquad \text{with} \qquad \{1, \pi, \ldots, \pi^{e-1}\} \qquad \text{as an } E_0\text{-basis.}$$

Choosing $u_1, u_2, \ldots, u_f$ to be the elements of S_w such that their images in K_w form a basis of K_w over K_v, it follows, from Theorem 2.8(ii), that $u_1, \ldots, u_f$ are F-linearly independent elements of E_0. Since $(E_0 : F) = f$, we conclude that

$$\{u_1, \ldots, u_f\} \qquad \text{is an } F\text{-basis of } E_0.$$

In particular, if K_v is a finite field of p^m elements, then by (6)

$$E_0 = F(\varepsilon),$$

where ε is a primitive $(p^{mf} - 1)$-th root of unity.

That such a field E_0 actually exists can be verified by showing that the elements

$$\sum_{i \gg -\infty} a_i p^i \qquad (a_i \in S_w)$$

constitute a subfield of E (see Hasse 1980).

In the series representations of the elements, the insertion of the inertia field looks as follows:

$$F: \quad \sum_{i \gg -\infty} a_i p^i \quad (a_i \in S_v).$$

$$E_0: \quad \sum_{i \gg -\infty} a_i p^i \quad (a_i \in S_w).$$

$$E: \quad \sum_{i \gg -\infty} a_i \pi^i \quad (a_i \in S_w).$$

Turning our attention to local fields, we now prove

4.2.10   Theorem   *A field F is local if and only if it is one of the following types*:

(i)   *A finite extension of the rational p-adic field $\mathbb{Q}_p$.*
(ii)   *A formal power series field in one variable over a finite field.*

Proof.   We have already observed, in our discussion prior to Theorem 2.7, that if F satisfies (ii), then F is local. By Theorem 2.8, F is local if it satisfies (i). Conversely, assume that F is local and let R, K be the valuation ring and residue class field attached to F. If char $F = $ char K, then F satisfies (ii) by virtue of Theorem 2.7. So assume that char $F = 0$ and char $K = p > 0$. Then the injection $\mathbb{Z} \to R$ extends by continuity to an injection of the ring $\mathbb{Z}_p$ of p-adic integers into R and therefore $\mathbb{Q}_p \subseteq F$. Now the residue field K is finite with, say $q = p^f$, elements. Hence the proof of Theorem 2.8(ii) shows that $(F : \mathbb{Q}_p) = ef$, where $e = w(p)$ and w is the normalized principal valuation attached to F. Thus F satisfies (i) and the result follows.   ∎

We next record the following consequence of the foregoing discussion.

4.2.11   Proposition   *Let $F/\mathbb{Q}_p$ be a finite field extension, let R and P be the valuation ring and its maximal ideal determined by the valuation v of F and let e, f be the absolute ramification index and absolute residue degree of v, respectively. Denote by F_0 the inertia field of $F/\mathbb{Q}_p$.*

(i) $(F : \mathbb{Q}_p) = ef.$
(ii) $Rp = P^e.$
(iii)   *R is a free $\mathbb{Z}_p$-module of rank ef.*
(iv) $F_0 = \mathbb{Q}_p(\varepsilon)$, *where ε is a primitive $(p^f - 1)$-th root of unity.*

Proof.   Apply Theorem 2.8(ii) and (6).   ∎

Let E/F be a field extension and let v_1, v_2 be two valuations of E.

Then v_1 *and* v_2 are said to be *F-conjugate* if $v_1 = v_2^\sigma$ for some $\sigma \in \mathrm{Gal}(E/F)$, where

$$v_2^\sigma(x) = v_2(\sigma(x)) \qquad \text{for all } x \in E.$$

It is clear that conjugate valuations restrict to the same valuation of F. The following result shows that the converse is also true, provided that E/F is a finite Galois extension.

4.2.12   Theorem   *Let E/F be a finite field extension of degree n, let v be an exponential valuation of F and let $v_1, \ldots, v_r$ be all distinct valuations of E which extend v. Let e_i, f_i be the ramification index and residue degree of v_i in the extension E/F, $1 \leqslant i \leqslant r$.*

(i)   *$\sum_{i=1}^{r} e_i f_i \leqslant n$.*
(ii)   *$\sum_{i=1}^{r} e_i f_i = n$, if E/F is separable and v is principal.*
(iii)   *If E/F is Galois, then all the v_i are F-conjugate and hence $e_i = e$, $f_i = f$ independently of i. In particular, $efr \leqslant n$ and, if E/F is Galois and v is principal, then $efr = n$.*

Proof.
   (i), (ii)   Let E_i be the completion of E at v_i and let $\bar{F}$ be the completion of F at v. Owing to Proposition 1.2.27, we have

$$n \geqslant \sum_{i=1}^{r} (E_i : \bar{F}),$$

with equality if E/F is separable. By Proposition 1.2.24, E_i and E have the same value group relative to v_i and $\bar{F}$ and F have the same value group relative to v. Hence the ramification index e_i of E over F relative to v_i is the same as that of E_i over $\bar{F}$. Similarly, Proposition 1.2.27 and the definitions show that the residue degree f_i of E/F relative to v_i is the same as that of E_i over $\bar{F}$. By Theorem 2.8, we have

$$e_i f_i \leqslant (E_i : \bar{F}) \qquad \text{and} \qquad e_i f_i = (E_i : \bar{F}) \qquad \text{if } v \text{ is principal.}$$

Hence

$$\sum_{i=1}^{r} e_i f_i \leqslant \sum_{i=1}^{r} (E_i : \bar{F}) \leqslant n$$

in every case with equalities if E/F is separable and v is principal.
   (iii)   Fix $i, j \in \{1, 2, \ldots, r\}$ and consider the valuations of E defined by

$$x \mapsto v_i(\sigma(x)), \quad x \mapsto v_j(\sigma(x)) \qquad \text{for all} \qquad \sigma \in G = \mathrm{Gal}(E/F).$$

If these two sets of valuations are not disjoint, we have $v_i^\sigma = v_j^\tau$ for some $\sigma, \tau \in G$ and so v_i and v_j are F-conjugate. If they are disjoint, then by

Proposition 1.2.22, there exists $\lambda \in E$ such that $v_i(\sigma(\lambda)) > 0$, $v_j(\sigma(\lambda) - 1) > 0$ for all $\sigma \in G$. Hence

$$v_j(\sigma(\lambda)) = 0 \qquad \text{and} \qquad v(N(\lambda)) = \sum_\sigma v_j(\sigma(\lambda)) = 0,$$

but also $v(N(\lambda)) = \sum_\sigma v_i(\sigma(\lambda)) > 0$, which is a contradiction.   ∎

4.2C   Cyclotomic extension of p-adic fields

Let $\mathbb{Q}_p$ be the rational p-adic field. A finite field extension F of $\mathbb{Q}_p$ is said to be a *P-adic field* if P is the maximal ideal of the valuation ring R attached to F. In case $F/\mathbb{Q}_p$ is unramified (equivalently, P is generated by p) we say that F is a *p-adic field*. Thus a finite extension F of $\mathbb{Q}_p$ is a p-adic field if and only if p is a prime element of F. Any p-adic field F is completely determined by its degree f over $\mathbb{Q}_p$. Indeed, by Proposition 2.11(iv), we have

$$F = \mathbb{Q}_p(\varepsilon),$$

where ε is a primitive $(p^f - 1)$th root of unity.

For any natural number n and any given field F, with char $F \nmid n$, we denote by ε_n a primitive nth root of 1 over F. Our main objective is to investigate the structure of $F(\varepsilon_n)$ in case F is a p-adic field. The following general observation will clear our path.

4.2.13 Proposition *Let $F/\mathbb{Q}_p$ be a finite field extension of absolute residue degree f, let $n \not\equiv 0 (\mathrm{mod}\, p)$ and let m be the order of p^f modulo n, i.e. the smallest exponent such that*

$$p^{fm} \equiv 1 (\mathrm{mod}\, n).$$

Then the extension $F(\varepsilon_n)/F$ is unramified of degree m.

Proof. Since n divides $p^{fm} - 1$, we have $F(\varepsilon_n) \subseteq F(\varepsilon_{p^{fm}-1})$. Hence, by properties (6) and (7) of Section B, $F(\varepsilon_n)/F$ is unramified and

$$F(\varepsilon_n) = F(\varepsilon_{p^{fk}-1}), \quad \langle \varepsilon_n \rangle \subseteq \langle \varepsilon_{p^{fk}-1} \rangle,$$

where $k = (F(\varepsilon_n):F)$. The containment $F(\varepsilon_{p^{fk}-1}) \subseteq F(\varepsilon_{p^{fm}-1})$ shows that $k \mid m$, while the containment $\langle \varepsilon_n \rangle \subseteq \langle \varepsilon_{p^{fk}-1} \rangle$ implies that n divides $p^{fk} - 1$, i.e. $p^{fk} \equiv 1 (\mathrm{mod}\, n)$. Hence $m = k$ and the result follows.   ∎

By the *p'-component* of a torsion abelian group A we understand the (restricted) direct product of all q-components of A for which the prime q is distinct from p. Thus the p'-component of the torsion subgroup $t(F^*)$ of F^* is just the subgroup of $t(F^*)$ generated by all p'-roots of unity in F.

4.2.14 Corollary   *Let $F/\mathbb{Q}_p$ be a finite field extension of absolute residue degree f. Then $\langle \varepsilon_{p^f-1} \rangle$ is the p'-component of $t(F^*)$.*

Proof.   Let $n \not\equiv 0 (\bmod p)$ and let $\varepsilon_n \in F$. Since $F(\varepsilon_n) = F$, it follows from Proposition 2.13 that $p^f \equiv 1 (\bmod n)$. Thus n divides $p^f - 1$ and hence $\varepsilon_n \in \langle \varepsilon_{p^f-1} \rangle$, as required.   ∎

4.2.15 Corollary   *Let F be a p-adic field of degree f over $\mathbb{Q}_p$. Then:*

(i) *For any $n \not\equiv 0 (\bmod p)$, $F(\varepsilon_n)/F$ is unramified of degree m, where m is the order of p^f modulo n.*

(ii) *$\langle \varepsilon_{p^f-1} \rangle$ is the p'-component of $t(F^*)$.*

Proof.   Since F is a p-adic field, f coincides with the absolute residue degree of F. Now apply Proposition 2.13 and Corollary 2.14.   ∎

So far we have considered only adjoining of p'-roots of unity to F. The next result examines the opposite case.

4.2.16 Proposition   *Let F be a p-adic field. Then, for any $n \geqslant 1$, $F(\varepsilon_{p^n})/F$ is a totally ramified extension of degree $\phi(p^n) = p^{n-1}(p-1)$ and*

$$\pi_n = 1 - \varepsilon_{p^n}$$

is a prime element of $F(\varepsilon_{p^n})$.

Proof.   Put $\varepsilon_{p^{n-1}} = \varepsilon_{p^n}^p$, $\varepsilon_{p^{n-2}} = \varepsilon_{p^{n-1}}^p$, ..., $\varepsilon_p = \varepsilon_{p^2}^p$. Using induction on n, it suffices to show that:

(a) $F(\varepsilon_p)/F$ is totally ramified of degree $p - 1$ with $\pi_1 = 1 - \varepsilon_p$ as a prime element.

(b) $F(\varepsilon_{p^n})/F(\varepsilon_{p^{n-1}})$, $n > 1$, is totally ramified of degree p with $\pi_n = 1 - \varepsilon_{p^n}$ as a prime element, provided $\pi_{n-1} = 1 - \varepsilon_{p^{n-1}}$ is a prime element of $F(\varepsilon_{p^{n-1}})$.

To prove (a), note that $F(\varepsilon_p) = F(\pi_1)$ and

$$\frac{1-\varepsilon_p^p}{1-\varepsilon_p} = 1 + \varepsilon_p + \ldots + \varepsilon_p^{p-1} = 0.$$

The latter implies that

$$\frac{1-(1-\pi_1)^p}{\pi_1} = p - \binom{p}{2}\pi_1 + \ldots + (-1)^{p-1}\pi_1^{p-1} = 0, \qquad (1)$$

and thus π_1 satisfies an Eisenstein equation over F of degree $p - 1$. This proves (a), by applying Proposition 2.9(ii).

To prove (b), we first observe that $F(\varepsilon_{p^n}) = F(\varepsilon_{p^{n-1}})(\varepsilon_{p^n})$ and $\varepsilon_{p^n}^p = \varepsilon_{p^{n-1}}$. Hence $F(\varepsilon_{p^n}) = F(\varepsilon_{p^{n-1}})(\pi_n)$ and

$$(1 - \pi_n)^p - (1 - \pi_{n-1}) = \pi_{n-1} - \binom{p}{1}\pi_n + \binom{p}{2}\pi_n^2 + \ldots + (-1)^p\pi_n^p = 0.$$

Therefore, π_n satisfies an Eisenstein equation over $F(\varepsilon_{p^{n-1}})$ of degree p which yields the assertion, by applying Proposition 2.9(ii). ■

By combining Corollary 2.15 and Proposition 2.16, we derive

4.2.17 Corollary   *Let $n = kp^s$ with $k \not\equiv 0 \pmod p$ and let F be a p-adic field of degree f over $\mathbb{Q}_p$. Then $F(\varepsilon_k)$ is the inertia field of the extension $F(\varepsilon_n)/F$ with*

$$e_{F(\varepsilon_n)/F} = \phi(p^s) \qquad and \qquad f_{F(\varepsilon_n)/F} = m,$$

where m is the order of p^f modulo k.

Proof. By Corollary 2.15, $F(\varepsilon_k)/F$ is unramified of degree m, while by Proposition 2.16, the field $F(\varepsilon_n) = F(\varepsilon_k)F(\varepsilon_{p^s}) = F(\varepsilon_k)(\varepsilon_{p^s})$ is totally ramified of degree $\phi(p^s)$ over $F(\varepsilon_k)$. ■

As a supplement of Proposition 2.16, we now establish two properties of the pth roots of unity which will be needed later.

4.2.18 Proposition   *Let F be a p-adic field. Then*

$$F(\varepsilon_p) = F(\lambda),$$

where λ is a prime element of $F(\varepsilon_p)$ such that:

(i) $\lambda^{p-1} = -p$;
(ii) $\lambda = (1 - \varepsilon_p)u$ for some 1-unit u of $F(\varepsilon_p)$.

Proof. Put $\pi_1 = 1 - \varepsilon_p$ and $P_1 = (\pi_1)$. Then, by Proposition 2.16, π_1 is a prime element of $F(\varepsilon_p)$ and hence $F(\varepsilon_p)$ is a P_1-adic field. If $p = 2$ then the result is obvious by taking $\lambda = -2$. Hence we may assume that p is odd, in which case $(-1)^{p-1} = 1$. Applying eqn (1), we deduce that

$$-p \equiv (-1)^{p-1}\pi_1^{p-1} = \pi_1^{p-1} \pmod{P_1^p}$$

which in turn implies

$$-p = \pi_1^{p-1} u_0, \qquad \text{for some 1-unit } u_0 \text{ in } F(\varepsilon_p).$$

By Lemma 2.5(iii), there exists a 1-unit u in $F(\varepsilon_p)$ such that $u^{p-1} = u_0$. Setting $\lambda = \pi_1 u$, we therefore derive $\lambda^{p-1} = -p$. This shows that λ is a prime element of $F(\varepsilon_p)$ which satisfies (i) and (ii). Since $F(\varepsilon_p)/F$ is totally ramified (Proposition 2.16), we also have $F(\varepsilon_p) = F(\lambda)$, by appealing to Proposition 2.9(i). ■

4.2.19 Proposition   *Let F be a P-adic field with prime element π and write the prime decomposition of p in F in the form*

$$-p = \pi^e u \qquad (u \text{ is a unit of } R).$$

Then $\varepsilon_p \in F$ if and only if the following two conditions hold:

(i)   *$p - 1$ divides e;*
(ii) *$x^p - ux \in P$ for some unit x of R.*

Proof.   Let F_0 denote the inertia field of the extension $F/\mathbb{Q}_p$. Assume that conditions (i) and (ii) hold. By (ii), $x^{p-1} \equiv u \pmod{p}$, so that

$$-p \equiv \pi^e x^{p-1} \pmod{p}$$

and hence

$$-p = \pi^e x^{p-1} u_0 \qquad \text{for some 1-unit } u_0 \text{ in } F.$$

Since $p - 1$ divides e, $\pi^e = (\pi^k)^{p-1}$ for some $k \geqslant 1$ and, by Lemma 2.5(iii), $y^{p-1} = u_0$ for some 1-unit y in F. Setting $\lambda = \pi^k x y$ we therefore derive $-p = \lambda^{p-1}$. Hence, by Proposition 2.18, $F_0(\varepsilon_p) = F_0(\lambda) \subseteq F$ and hence $\varepsilon_p \in F$.

Conversely, assume that $\varepsilon_p \in F$ and hence that $F_0(\varepsilon_p) \subseteq F$. Since F_0 is unramified and, by Proposition 2.16, $F_0(\varepsilon_p)$ is totally ramified of degree $p - 1$, it follows that $p - 1$ divides e. Furthermore P^k, $k = e/(p - 1)$, is the prime ideal attached to $F_0(\varepsilon_p)$. Owing to Proposition 2.18, we now have

$$F_0(\varepsilon_p) = F(\lambda) \qquad \text{with} \qquad \lambda^{p-1} = -p,$$

where $\lambda \in P^k$. From this it follows that $\lambda = \pi^k u_0$ for some $u_0 \in U(R)$, and hence $-p = \pi^e u_0^{p-1}$. Thus $u = u_0^{p-1}$, and so $x = u_0$ satisfies (ii), as required.   ∎

As a final result of this section, we prove

4.2.20 Proposition   *Let F be a P-adic field with the valuation ring R and residue class field $K = R/P$, let S be a unique multiplicative residue system of K in R and let $U_k(R) = 1 + P^k$, $k \geqslant 1$. Denote by e the absolute ramification index of F.*

(i)   *If $\varepsilon_{p^n} \in F$, then $\phi(p^n)$ divides e and $\varepsilon_{p^n} \in U_k(R) - U_{k+1}(R)$, $k = e/\phi(p^n)$.*
(ii) *$S^* = S - \{0\}$ consists precisely of all p'-roots of unity of F.*

Proof.   Property (ii) is a consequence of the fact that $S^* \cong K^*$, hence S^* is of p'-order, and of the characterization of S given by Theorem 2.6(ii).

Let $\varepsilon_{p^n} \in F$ and let F_0 be the inertia group of F. Then, by the argument of Proposition 2.19, $\phi(p^n)$ divides e and P^k, $k = e/\phi(p^n)$, is the prime ideal attached to $F_0(\varepsilon_{p^n})$. But, by Proposition 2.16, $1 - \varepsilon_{p^n}$ is a prime element of $F_0(\varepsilon_{p^n})$. Hence

$$1 - \varepsilon_{p^n} \in P^k - P^{k+1}$$

and the result follows. ■

4.2D   The equal characteristic case

Throughout this subsection, F denotes a local field and R, P, and K are the valuation ring, the unique maximal ideal and the residue field attached to it. Thus, by definition, K is a finite field, say of p^f elements. Our aim is to examine the structure of F^* in the equal characteristic case, i.e. when char $F = p$. By Theorem 2.6(i), we may choose a residue system S of K in R which is a field. Hence $S \cong K$ and, in view of Theorem 2.6, we have

$$F^* \cong \mathbb{Z} \times \mathbb{Z}/(p^f - 1)\mathbb{Z} \times U_1(R). \tag{1}$$

Here we again abandon the convention that $\mathbb{Z}_n$ denotes a cyclic group of order n in order to avoid a possible confusion with the ring $\mathbb{Z}_p$ of p-adic integers. In view of (1), we are left to investigate $U_1(R)$. By Lemma 2.5, $U_1(R)$ is a $\mathbb{Z}_p$-module, and the precise structure of this module is given by the following result.

4.2.21   Theorem   (Hensel 1916)   *Let $\lambda_1, \ldots, \lambda_f$ be an $\mathbb{F}_p$-basis of S and let*

$$e_{ik} = 1 + \lambda_i \pi^k \qquad (1 \leqslant i \leqslant f, \, k \not\equiv 0 \,(\mathrm{mod}\, p)),$$

where π is a prime element of F. Then each $u \in U_1(R)$ has a unique representation in the form

$$u = \prod_{k \not\equiv 0 (\mathrm{mod}\, p)} \prod_{i=1}^{f} e_{ik}^{\alpha_{ik}} \qquad (\alpha_{ik} \in \mathbb{Z}_p) \tag{2}$$

where the infinite power products appearing in (2) are to be understood as limits of the sequences of their finite partial products with respect to the valuation of F. Thus $U_1(R)$ is isomorphic to the (unrestricted) direct product of $\aleph_0$ copies of the additive group of $\mathbb{Z}_p$.

Proof. We first observe that if T_m is a transversal for $U_{m+1}(R)$ in $U_m(R)$, then $T_m^p = \{x^p \mid x \in T_m\}$ is a transversal for $U_{pm+1}(R)$ in $U_{pm}(R)$. Indeed, if $t = 1 + r\pi^m$, $r \in R$, then since char $F = p$, we have

$$t^p = 1 + r^p \pi^{mp} \in U_{mp}(R).$$

Since the map $rP \mapsto r^p P$ is an automorphism of K, our observation follows by virtue of Lemma 2.2(i). Starting with the transversal T_k for all

$k \not\equiv 0 \pmod p$, we thus obtain specific transversals T_m:

$$T_k, \ T_{pk} = T_k^p, \ \ldots, \ T_{p^n k} = T_k^{p^n}, \ \ldots \tag{3}$$

for all possible m, each exactly once.

By hypothesis, each element s of S can be uniquely written in the form

$$s = \sum_{i=1}^{f} \alpha_i \lambda_i \qquad (\alpha_i \in \mathbb{Z}, \ 0 \leqslant \alpha_i < p).$$

Let us fix $k \not\equiv 0 \pmod p$. Then, in view of the isomorphism $sP \mapsto (1 + \pi^k s) U_{k+1}(R)$ of K^+ onto $U_k(R)/U_{k+1}(R)$ established in Lemma 2.2(i), the power products

$$\prod_{i=1}^{f} e_{ik}^{\alpha_i} \equiv 1 + s\pi^k \pmod{U_{k+1}(R)} \qquad (\alpha_i \in \mathbb{Z}, \ 0 \leqslant \alpha_i < p)$$

constitute a transversal, say T_k, for $U_k(R)$ in $U_{k+1}(R)$. With the aid of (3), we now obtain specific transversals T_m for $U_{m+1}(R)$ in $U_m(R)$ for all possible m, each exactly once.

We now apply Lemma 2.2(iii), to express each $u \in U_1(R)$ uniquely in the form

$$u = \prod_{v=0}^{\infty} \prod_{k \not\equiv 0 \pmod p} \prod_{i=1}^{f} e_{ik}^{\alpha_{ik}^{(v)} p^v} \qquad (\alpha_{ik}^{(v)} \in \mathbb{Z}, \ 0 \leqslant \alpha_{ik}^{(v)} < p).$$

By rearrangement, we obtain, from the nature of action of $\mathbb{Z}_p$ on $U_1(R)$, the unique representation

$$u = \prod_{k \not\equiv 0 \pmod p} \prod_{i=1}^{f} e_{ik}^{\alpha_{ik}} \qquad \left(\alpha_{ik} = \sum_{v=0}^{\infty} \alpha_{ik}^{(v)} p^v \ \text{in} \ \mathbb{Z}_p \right),$$

thus completing the proof. ∎

4.2E   *The unequal characteristic case*

Let F be a local field and R, P, and K the valuation ring, the unique maximal ideal and the residue class field attached to it. We wish to examine the structure of F^* in the unequal characteristic case, i.e. when char $F = 0$ and char $K = p > 0$. By Theorem 2.10, F is a finite extension of the rational p-adic field $\mathbb{Q}_p$. Throughout, we fix the following notation:

$$n = (F : \mathbb{Q}_p)$$

$$e = e_{F/\mathbb{Q}_p}, \quad f = f_{F/\mathbb{Q}_p}$$

π is a prime element of F

$\mathbb{Z}_p$ is the ring of p-adic integers

Then, by definition of f, K is a finite field of p^f elements and, by

Proposition 2.11, we have

$$n = ef \quad \text{and} \quad Rp = P^e.$$

Owing to Theorem 2.6(ii), there is a unique multiplicative residue system S of K in R. We denote by S^* the set of non-zero elements of S. Thus

$$S^* \cong K^* \cong \mathbb{Z}/(p^f - 1)\mathbb{Z}$$

and, by Theorem 2.6(ii), S^* contains all p'-roots of unity in F. Again, by Theorem 2.6,

$$F^* \cong \mathbb{Z} \times \mathbb{Z}/(p^f - 1)\mathbb{Z} \times U_1(R) \tag{1}$$

and so we need only to investigate the structure of the $\mathbb{Z}_p$-module $U_1(R)$.

4.2.22 Lemma Let $x, r \in R$ be such that $x \equiv 1 + r\pi^k \pmod{P^{k+1}}$ and write $-p = \pi^e u$ for some $u \in U(R)$. Then:

(i) If $k < e/(p-1)$, then

$$x^p \equiv 1 + r^p \pi^{pk} \pmod{P^{pk+1}}.$$

(ii) If $k = e/(p-1)$, then

$$x^p \equiv 1 + r p \pi^k + r p \pi^{pk} = 1 + (r^p - ur)\pi^{pk} \pmod{P^{pk+1}}.$$

(iii) If $k > e/(p-1)$, then

$$x^p \equiv 1 + r p \pi^k = 1 - ur\pi^{k+e} \pmod{P^{k+e+1}}.$$

Proof. We may harmlessly assume that $x = 1 + r\pi^k$. Indeed, since $x \equiv 1 + r\pi^k \pmod{P^{k+1}}$, we have $x = 1 + r'\pi^k$ with $r' \equiv r \pmod{P}$ and the expressions appearing in the assertions (i), (ii), and (iii) on the right depend only on the residue class r mod P.

Applying the binomial theorem, we have

$$x^p = 1 + p r\pi^k + \binom{p}{2} r^2 \pi^{2k} + \ldots + \binom{p}{p-1} r^{p-1} \pi^{(p-1)k} + r^p \pi^{pk}.$$

Now $p = -\pi^e u$ and the middle binomial coefficients are divisible by p. Hence the terms following the 1 are contained, respectively, in

$$P^{k+e}, P^{2k+e}, \ldots, P^{(p-1)k+e}, P^{pk}.$$

Therefore, for $pk < k + e$, that is for $k < e/(p-1)$, all these terms, except for the final $r^p \pi^{pk}$, are contained in P^{pk+1}, proving (i). In case $pk = k + e$, that is, for $k = e/(p-1)$, all the terms, except for the initial $p r\pi^k$ and the final $r^p \pi^{pk}$, are contained in P^{pk+1}, proving (ii). Finally, for $pk > k + e$, that is, for $k > e/(p-1)$, all terms, except for the initial $p r\pi^k$, are contained in P^{k+e+1}, proving (iii). ∎

We have now come to the demonstration of the main result.

4.2.23   Theorem   (Hensel 1916)

(i)   *For any $t > e/(p-1)$, $(U_1(R):U_t(R)) = p^{(t-1)f}$ and $U_t(R)$ is a free $\mathbb{Z}_p$-module of rank n.*

(ii) *$U_1(R)$ is a direct product of a cyclic group of order p^μ for some $\mu \geq 0$ and a free $\mathbb{Z}_p$-module of rank n.*

Proof.   We first show that (ii) is a consequence of (i). Indeed, $\mathbb{Z}_p$ is a principal ideal domain and so by (i), $U_1(R)$ is a finitely generated $\mathbb{Z}_p$-module, the free factor of which is of rank n. Let x be an element of the torsion submodule of $U_1(R)$ and let $a = p^n u$, $n \geq 0$, $u \in U(\mathbb{Z}_p)$, be an arbitrary element of $\mathbb{Z}_p$. Then $x^a = 1$ implies $(x^{p^n})^u = 1$, and hence $x^{p^n} = 1$, since u acts as an automorphism of $U_1(R)$. Thus torsion submodule of $U_1(R)$ consists of roots of unity in F whose order is a power of p. By Proposition 2.16, there are only finitely many of them and hence they constitute a cyclic group of order p^μ for some $\mu \geq 0$. This shows that (ii) is a consequence of (i), by applying the fact that $U_1(R)$ is a direct product of the torsion submodule and a free module.

To prove (i), we first note that the equality $(U_1(R):U_t(R)) = p^{(t-1)f}$ is a consequence of the isomorphism $U_i(R)/U_{i+1}(R) \cong K^+$ (Lemma 2.2(i)) and the fact that K is a finite field of p^f elements.

Choose an $\mathbb{F}_p$-basis $\omega_1, \omega_2, \ldots, \omega_f$ of K, and let $\lambda_1, \lambda_2, \ldots, \lambda_f$ be elements of S such that ω_i is the image of λ_i, $1 \leq i \leq f$. For any $u \in U(R)$, the map $r + P \mapsto -ur + P$ is an automorphism of the group K^+. Therefore, by Lemma 2.22(iii), for any $m \geq t > e/(p-1)$, if T_m is a transversal for $U_{m+1}(R)$ in $U_m(R)$, then T_m^p is a transversal for $U_{m+e+1}(R)$ in $U_{m+e}(R)$. Thus, if T_k is a transversal for $U_{k+1}(R)$ in $U_k(R)$ for $t \leq k < t + e$, then

$$T_k, \; T_k^p, \; T_k^{p^2}, \ldots, \qquad (t \leq k < t+e)$$

exhaust all transversals for $U_{m+1}(R)$ in $U_m(R)$ with $m \geq t$. Applying the argument employed in the proof of Theorem 2.21, we conclude that each element of $U_t(R)$ can be uniquely written in the form

$$\prod_{t \leq k < t+e} \prod_{i=1}^{f} e_{ik}^{\alpha_{ik}} \qquad (\alpha_{ik} \in \mathbb{Z}_p),$$

with

$$e_{ik} = 1 + \lambda_i \pi^k \qquad (1 \leq i \leq f, \; t \leq k < t+e).$$

Hence the e_{ik} constitute a free basis for $U_t(R)$ and since their number is $n = ef$, the result follows.   ∎

4.2.24 Corollary Let F be a finite extension of $\mathbb{Q}_p$. Then, with the above notation, there exists $\mu \geq 0$ such that:

(i) $F^* \cong [\mathbb{Z}/p^{\mu}(p^f - 1)\mathbb{Z}] \times \mathbb{Z} \times (\mathbb{Z}_p \times \ldots \times \mathbb{Z}_p);$
$$\underset{\leftarrow n \text{ factors} \rightarrow}{}$$

(ii) $t(F^*) \cong \mathbb{Z}/p^{\mu}(p^f - 1)\mathbb{Z};$

(iii) $\mu = 0$ if and only if $(p - 1) \nmid e$ or $x^{p-1} + u \in U(R)$ for all $x \in U(R)$, where $p = \pi^e u$ with $u \in U(R);$

(vi) $\phi(p^{\mu})$ divides e and the p-component of $t(F^*)$ is contained in $U_k(R)$, where $k = e/\phi(p^{\mu})$, but not contained in $U_{k+1}(R)$.

Proof. Property (i) is a consequence of (1) and Theorem 2.23(ii), while (ii) is a consequence of (i). Since $\mu = 0$ if and only if F does not contain a primitive pth root of unity, the assertion (iii) follows from Proposition 2.19. Finally, (iv) is a consequence of Proposition 2.20(i). ∎

4.2.25 Corollary With the preceding notation, for any $m = p^k m_0$, $(p, m_0) = 1$,

$$F^*/(F^*)^m \cong [\mathbb{Z}/(m, p^{\mu}(p^f - 1))\mathbb{Z}] \times (\mathbb{Z}/m\mathbb{Z}) \times (\mathbb{Z}/p^k\mathbb{Z} \times \ldots \times \mathbb{Z}/p^k\mathbb{Z}) .$$
$$\underset{\leftarrow n \text{ factors} \rightarrow}{}$$

In particular

$$(F^* : (F^*)^m) = (m, p^{\mu}(p^f - 1))mp^{nk}.$$

Proof. Let $G = \langle g \rangle$ be a cyclic group of order s. Then $G^m = \langle g^m \rangle$ is of order $s/(m, s)$ and hence $(G : G^m) = (m, s)$. If G is an infinite cyclic group, then clearly $(G : G^m) = m$. On the other hand, we have $m\mathbb{Z}_p = p^k\mathbb{Z}_p$ since m_0 is a unit of $\mathbb{Z}_p$. Hence $\mathbb{Z}_p/m\mathbb{Z}_p \cong \mathbb{Z}/p^k\mathbb{Z}$ is cyclic of order p^k. The desired conclusion is therefore a consequence of Corollary 2.24(i). ∎

4.2.26 Corollary Let ε be a primitive mth root of unity over $\mathbb{Q}_p$ with $m = p^s k$, $(k, p) = 1$, and let t be the order of p modulo k. Then

$$\mathbb{Q}_p(\varepsilon)^* \cong [\mathbb{Z}/p^s(p^t - 1)\mathbb{Z}] \times \mathbb{Z} \times (\mathbb{Z}_p \times \ldots \times \mathbb{Z}_p) \qquad \text{if } p \text{ is odd or } s \geq 1;$$
$$\underset{\leftarrow \phi(p^s)t \text{ factors} \rightarrow}{}$$

$$\mathbb{Q}_2(\varepsilon)^* \cong [\mathbb{Z}/2(2^t - 1)\mathbb{Z}] \times \mathbb{Z} \times (\mathbb{Z}_2 \times \ldots \times \mathbb{Z}_2) \qquad \text{if } p = 2 \text{ and } s = 0.$$
$$\underset{\leftarrow t \text{ factors} \rightarrow}{}$$

Proof. Let ε_k and ε_{p^s} be primitive kth and p^sth roots of unity, respectively. By Corollary 2.17, $\mathbb{Q}_p(\varepsilon_k)$ is the inertia field of the extension $\mathbb{Q}_p(\varepsilon)/\mathbb{Q}_p$ and

$$e_{\mathbb{Q}_p(\varepsilon)/\mathbb{Q}_p} = \phi(p^s) \qquad \text{and} \qquad f_{\mathbb{Q}_p(\varepsilon)/\mathbb{Q}_p} = t. \qquad (2)$$

Now $\mathbb{Q}_p(\varepsilon) = \mathbb{Q}_p(\varepsilon_k)(\varepsilon_{p^s})$ and $\mathbb{Q}_p(\varepsilon_k)$ is a p-adic field. Hence

$$(\mathbb{Q}_p(\varepsilon):\mathbb{Q}_p(\varepsilon_k)) = \phi(p^s) \tag{3}$$

by Proposition 2.16. If $\mathbb{Q}_p(\varepsilon)$ contains $\varepsilon_{p^{s+1}}$, then $\mathbb{Q}_p(\varepsilon) = \mathbb{Q}_p(\varepsilon_k)(\varepsilon_{p^{s+1}})$ and, by (3), $\phi(p^s) = \phi(p^{s+1})$, which is possible only in the case $p = 2$ and $s = 0$. Thus, if p is odd or $s \geqslant 1$ then, in the notation of Corollary 2.24, $\mu = s$. This proves the case where p is odd or $s \geqslant 1$, by applying (2). In case $p = 2$ and $s = 0$, $\mathbb{Q}_2(\varepsilon)$ contains -1, but does not contain a primitive 4th root of unity by virtue of (3). Thus, in the notation of Corollary 2.24, $\mu = 1$ and the result follows again by applying (2).   ∎

4.3   Intermediate fields

Let E/F be a field extension with $E \neq F$. Our aim is to investigate the structure of the factor group E^*/F^*. We start by considering the case where E is an algebraic number field. In this case our result, essentially due to May, asserts that $E^*/F^* \cong A \times B$, where A is free abelian of rank $\aleph_0$ and B is finite. Turning to the general case, we then prove a theorem due to Brandis which states that if F is infinite, then E^*/F^* is not finitely generated.

In what follows, unless explicitly stated otherwise, $F \subset E$ are algebraic number fields and $R \subset S$ are their rings of integers. We say that a prime ideal P of R *splits completely* in S if PS is a product of distinct prime ideals of residue degree 1. Thus, if P splits completely in S, then P does not ramify in S and PS is not a prime ideal of S. We known that almost all prime ideals of R do not ramify in S (see Proposition 1.2.10). Our first aim is to show that there are infinitely many prime ideals of R which split completely in S.

4.3.1 Lemma   *There exists $s \in S$ such that $E = F(s)$.*

Proof.   Since E/F is a separable extension, $E = F(\lambda)$ for some $\lambda \in E$. Let

$$r_n X^n + r_{n-1} X^{n-1} + \ldots + r_0 = 0 \qquad (r_i \in R, \, r_n \neq 0)$$

be an equation satisfied by λ. Then the element $s = r_n \lambda$ satisfies the equation

$$X^n + r_{n-1} X^{n-1} + r_{n-2} r_n X^{n-2} + \ldots + r_0 r_n^{n-1} = 0,$$

which shows that $s \in S$ and $E = F(s)$.   ∎

Let M be any non-zero R-module contained in R. The set

$$M^* = \{x \in E \mid Tr_{E/F}(xM) \subseteq R\}$$

is called the *codifferent* of M over F. It is obvious that M^* is an R-module which may be equal to 0 (e.g. take $M = E$).

4.3.2 Lemma *Let s be an element of S with $E = F(s)$ and let $f(X)$ be the minimal polynomial over R for s. Then $R[s]^*$ is generated as an R-module by the set*

$$s^j/f'(s), \qquad 0 \leq j \leq n-1, \quad n = (E:F).$$

Proof. Let A be the R-module generated by the above elements. Applying the Lagrange interpolation formula successively to the polynomials $X, X^2, \ldots, X^{n-1}$ and $X^n - f(X)$, we obtain the identities

$$\sum_{j=1}^{n} \frac{s_d^k}{f'(s_j)} \frac{f(X)}{X - s_j} = \begin{cases} X^{1+k}, & k = 0, 1, \ldots, n-2, \\ X^n - f(X), & k = n-1, \end{cases}$$

in which s_j ($j = 1, 2, \ldots, n$) denote the conjugates of s, and $s = s_1$. Putting $X = 0$, we immediately derive $Tr_{E/F}(s^j/f'(s)) \in R$. Hence $s^j/f'(s) \in R[s]^*$, which shows that $A \subseteq R[s]^*$.

To prove the opposite containment, *fix b in $R[s]^*$*. If we put

$$g(X) = \sum_{i=1}^{n} b_i f(X)/(X - s_i),$$

where the b_i's are all conjugates of b, then we have

$$g(X) = \sum_{j=1}^{n} c_j \sum_{k=0}^{j-1} X^k Tr_{E/F}(bs^{j-k-1}),$$

where the c_i are defined by $c_n = 1$ and $f(X) = X^n + c_{n-1}X^{n-1} + \ldots + c_0$. Since, by hypothesis, $Tr_{E/F}(bR[s]) \subseteq R$ we see that the coefficients of $g(X)$ lie in R. But then $bf'(s) = g(s) \in R[s]$, i.e. $b \in A$ and therefore $R[s]^* \subseteq A$. ∎

4.3.3 Corollary *With the assumptions and notations of Lemma 3.2,*

$$f'(s)S \subseteq R[s].$$

Proof. Since $S \subseteq R[s]^*$, we have $f'(s)S \subseteq f'(s)R[s]^* \subseteq R[s]$. ∎

Let $I(R)$ and $I(S)$ be the groups of fractional ideals of R and S, respectively. Recall that the *norm homomorphism* $N_{E/F}: I(S) \to I(R)$ is defined as follows. Let Q be a prime ideal of S with residue degree f over R. Then we put

$$N_{E/F}(Q) = P^f,$$

where $P = Q \cap R$. This defines a unique homomorphism from $I(S)$ to $I(R)$ since $I(S)$ is freely generated by all such Q. Basic properties of the norm homomorphism are given by Propositions 1.2.11–1.2.14. Let A be a subring of S containing R. Then the greatest common divisor of ideals of S contained in A is called the *conductor* of A.

4.3.4 Lemma *Let P be a prime ideal in R, s an element of S with $E = F(s)$, $A = R[s]$ and C the conductor of A. Then the following conditions are equivalent:*

(i) $PS \cap A = P[s]$.
(ii) *The embedding of A in S induces an isomorphism $S/PS \cong A/(A \cap PS)$.*
(iii) *The embedding of A in S induces isomorphism*

$$S/P^m S \cong A/(A \cap P^m S) \qquad (m = 1, 2, \dots).$$

(iv) *For all $m \geqslant 1$, $P^m S \cap A = P^m[s]$.*
(v) *The prime ideal P does not divide the norm $N_{E/F}(C)$.*

Proof.

(i) $\Rightarrow$ (ii)   The induced homomorphism $A/(A \cap PS) \to S/PS$ is obviously injective. Since $|S/PS| = N(PS) = N(P)^n$, where $n = (E:F)$, and, by hypothesis,

$$|A/(A \cap PS)| = |A/P[s]| = N(P)^n$$

the assertion follows.

(ii) $\Rightarrow$ (iii)   The case $m = 1$ being true by hypothesis, we argue by induction on m. So assume that (iii) holds for m and choose $x \in P - P^2$. We may write $xR = PI$, where $P \nmid I$. Let $y \in I - P$, let y' be defined by $yy' \equiv 1 \pmod{P^m}$ and let $b \in S$. By our assumption, there is a polynomial $f(X)$ over R such that $b \equiv f(s) \pmod{P^m S}$. Therefore the element $b_1 = (b - f(s))y^m x^{-m}$ lies in S. If we now choose a polynomial $f_1(X) \in R[X]$ with $b_1 \equiv f_1(s) \pmod{PS}$, then

$$b = f(s) + b_1 x^m y^{-m} \equiv f(s) + (xy')^m f_1(s) \pmod{P^{m+1} S},$$

proving (iii).

(iii) $\Rightarrow$ (iv)   By hypothesis, the index of $A \cap P^m S$ in A is $N(P)^{mn}$. On the other hand, the index of $P^m[s]$ in A is again $N(P)^{mn}$. The desired conclusion is therefore a consequence of $P^m[s] \subseteq A \cap P^m S$.

(iv) $\Rightarrow$ (i)   Trivial.

(iii) $\Rightarrow$ (v)   Let $f(X)$ be the minimal polynomial for s over R; write

$$N_{E/F}(f'(s))R = P^m I \qquad (P \nmid I)$$

and fix $b \in I - P$. By assumption, for every $x \in S$ there is an element $y \in A$ such that $b(x - y)/N_{E/F}(f'(s))$ lies in S. Therefore $b(x - y)/f'(s)$ is in S; hence

$$b(x - y) \in f'(s)S \subseteq A$$

by Corollary 3.3. Thus $by \in A$ and we obtain $bS \subseteq A$, proving that $b \in C$. Because $b \notin P$, we also have $N_{E/F}(b) = b^n \notin P$, but $N_{E/F}(b) \subseteq N_{E/F}(C)$ and so P does not divide $N_{E/F}(C)$.

(v) $\Rightarrow$ (ii)   If P does not divide $N_{E/F}(C)$ and $PS = \prod_{i=1}^{r} P_i^{e_i}$ then no P_i can divide C. Owing to the Chinese remainder theorem, there exists an element b satisfying

$$b \equiv 0 (\mathrm{mod}\, C), \quad b \equiv 1 (\mathrm{mod}\, P_i) \qquad (1 \leqslant i \leqslant r).$$

Having such a $b = f(s)$, $f(X) \in R[X]$, we may write every $x \in S$ in the form $x = g(s)/b = g(s)/f(s)$ for a suitable $g(X) \in R[X]$. Because the ideal bS is relatively prime to PS, there exists $r \in \mathbb{N}$ with $f(s)^r \equiv 1 (\mathrm{mod}\, PS)$. Thus

$$x \equiv g(s)/f(s)^{r-1} (\mathrm{mod}\, PS)$$

as required.   ∎

We have now accumulated all the information necessary to prove the following result.

4.3.5   Theorem   *Let $F \subset E$ be algebraic number fields, $R \subset S$ their rings of integers, let $s \in S$ be such that $E = F(s)$, and let $f(X) \in R[X]$ be the minimal polynomial of s. Let C be the conductor of $R[s]$, let P be a prime ideal of R not dividing $N_{E/F}(C)$, and let $\pi : R[X] \to R[X]/P[X] = K[X]$ be the homomorphism induced by the natural map $R \to R/P = K$. Write*

$$\pi(f) = f_1^{a_1} f_2^{a_2} \ldots f_m^{a_m}$$

where the f_i's are distinct irreducible mononic polynomials in $K[X]$. Then

$$PS = P_1^{a_1} P_2^{a_2} \ldots P_m^{a_m},$$

where the P_i's are distinct prime ideals of S, the residue degree $f_{E/F}(P_i)$ of P_i is equal to the degree n_i of f_i, and

$$P_i = PS + F_i(s)S \qquad (1 \leqslant i \leqslant m),$$

where $F_i(X) \in R[X]$ is such that $\pi(F_i) = f_i$.

Proof.   Put $K_i = K[X]/f_i K[X]$ and consider the natural homomorphism

$$\pi_i : K[X] \to K_i \qquad (1 \leqslant i \leqslant m).$$

Then the kernel of $\psi_i = \pi_i \cdot \pi$ is the ideal $I_i = P[X] + F_i(X)R[X]$ of $R[X]$. The principal ideal generated by f in $R[X]$ is divisible by I_i and so ψ_i induces a surjective homomorphism $R[X]/fR[X] \to K_i$. Because $R[X]/fR[X] \cong R[s]$, we obtain a surjective homomorphism

$$\lambda_i : R[s] \to K_i \qquad (1 \leqslant i \leqslant m)$$

with $\mathrm{Ker}\, \lambda_i = P[s] + F_i(s)R[s]$.

Let $b \in S$ and choose (using Lemma 3.4(ii)) a polynomial $v \in R[X]$ such that $b \equiv v(s) \pmod{PS}$. Define $\mu_i : S \to K_i$ by

$$\mu_i(b) = \lambda_i(v(s)) \qquad (1 \leq i \leq m).$$

This is well defined since from $v(s) \equiv 0 \pmod{PS}$ we obtain, by applying Lemma 3.4(i), $v(s) \in PS \cap R[s] = P[s]$, and so $\lambda_i(v(s)) = 0$.

The homomorphism μ_i is surjective, since so is λ_i. Setting $P_i = \mathrm{Ker}\, \mu_i$ we see that P_i is a non-zero prime ideal in S. Furthermore, we have

$$P_i = PS + F_i(s)S \qquad (1 \leq i \leq m).$$

Indeed,

$$\begin{aligned}
P_i &= \{x \in S \mid x \equiv v(s) (\mathrm{mod}\, PS),\ v \in R[X],\ v(s) \in P[s] + F_i(s)R[s]\} \\
&= PS + P[s] + F_i(s)R[s] \\
&= PS + F_i(s)R[s] \\
&= PS + F_i(s)S,
\end{aligned}$$

since Lemma 3.4(ii) implies $S = R[s] + PS$, which in turn gives

$$F_i(s)S \subseteq F_i(s)R[s] + PF_i(s)S \subseteq F_i(s)R[s] + PS.$$

Assume by way of contradiction that $P_i = P_j$ for $i \neq j$. Since $(f_i, f_j) = 1$, there exist polynomials $g_1, g_2 \in R[X]$ and $g_3 \in P[X]$ such that

$$g_1(X)f_i(X) + g_2(X)f_j(X) = 1 + g_3(X). \tag{1}$$

But $F_i(s) \in P_i$, $F_j(s) \in P_j$, and $g_3(s) \in P_i$, so putting $X = s$ in (1) we obtain $1 \in P_i$, a contradiction. Thus all the P_i's, $1 \leq i \leq m$, are distinct.

Because the element

$$F_1(s)^{a_1} \ldots F_m(s)^{a_m} \tag{2}$$

belongs to PS, the prime ideals $P_1, \ldots, P_m$ are the only possible prime ideals dividing PS. Thus there exist positive integers $e_1, \ldots, e_m$ such that

$$PS = P_1^{e_1} \ldots P_m^{e_m}.$$

The residue degree of P_i equals $(K_i : K)$ and this is equal to n_i. We are therefore left to verify that $a_i = e_i$ for all $i \in \{1, \ldots, m\}$.

The ideal $P_i^{e_i}$ divides the product (2) but, as we have seen, P_i can divide only the factor $F_i(s)^{a_i}$ of that product. Hence $P_i^{e_i}$ divides $F_i(s)^{a_i}S$. It follows that the ideal $PS + F_i(s)^{a_i}S$ is divisible by $P_i^{e_i}$, and since it divides $P_i^{a_i} = P^{a_i}S + F_i(s)^{a_i}S$ we have $a_i \geq e_i$, $1 \leq i \leq m$. On the other hand if $n = (E : F)$, then $n = e_1 n_1 + \ldots + e_m n_m$ (see Proposition 1.2.9) and obviously $a_1 n_1 + \ldots + a_m n_m = n$. Thus we must have $a_i = e_i$ and the result follows. ∎

As an application of the preceding result, we prove

4.3.6 Theorem   *Let $F \subset E$ be algebraic number fields and let $R \subset S$ be their rings of integers. Then there exist infinitely many prime ideals of R which split completely in S.*

Proof.   Let $M/\mathbb{Q}$ be a finite normal extension with $E \subseteq M$, and let L be the ring of integers of M. If $p\mathbb{Z}$ is a prime ideal in $\mathbb{Z}$ which splits completely in L, then every prime ideal of R lying above $p\mathbb{Z}$ splits completely in S, by the multiplicative property of the residue degree. Thus we may harmlessly assume that $F = \mathbb{Q}$ (hence $R = \mathbb{Z}$) and that the extension E/F is normal. Since almost all prime ideals P of $\mathbb{Z}$ do not ramify in S, almost all prime ideals P of $\mathbb{Z}$ do not ramify in S, and satisfy the hypothesis of Theorem 3.5. Furthermore, since E/F is normal, then in the notation of Theorem 3.5 the polynomials f_i, $1 \leq i \leq m$, have the same degree (see Proposition 1.2.9). Applying Theorem 3.5, we are therefore left to verify the following assertion: If $P(X)$ is a non-constant polynomial over $\mathbb{Z}$, then its values are divisible by infinitely many rational primes.

Assume by way of contradiction that the values of some $P(X) \in \mathbb{Z}[X]$ are divisible only by a finite number of primes, say $p_1, p_2, \ldots, p_r$. Select x_0 so that $P(x_0)$ is different from 0, -1 and 1. Then we have

$$P(x_0) = \pm p_1^{a_1} \ldots p_r^{a_r}$$

for suitable $a_i \geq 0$, not all of which are zero. If now t is an arbitrary rational integer and $x = x_0 + t p_1 \ldots p_r P(x_0)$, then for large t we have

$$P(x) = \pm p_1^{b_1} \ldots p_r^{b_r} \equiv P(x_0) (\mathrm{mod}\, p_1 \ldots p_r P(x_0)),$$

which implies $a_i = b_i$ for $i = 1, 2, \ldots, r$. Hence $P(x) = \pm P(x_0)$ for all large t's, which is a clear contradiction.   ∎

4.3.7 Corollary   *Let $F \subset E$ be algebraic number fields and let $R \subset S$ be their rings of integers. Then there exists infinitely many prime ideals P of R for which SP is a product of some $n \geq 2$ distinct prime ideals of S.*

Proof.   Since every prime ideal P of R which splits completely in S satisfies the above property, the result follows by virtue of Theorem 3.6.   ∎

We are now ready to prove the following result, essentially due to May (1979a).

4.3.8 Theorem   *Let $F \subset E$ be algebraic number fields. Then*

$$E^*/F^* \cong A \times B,$$

where A is free abelian of rank $\aleph_0$ and B is finite.

Proof. Let R and S be rings of integers of F and E, respectively, and let $I(R)$ and $I(S)$ be their groups of fractional ideals. Denote by $P(R)$ and $P(S)$ the subgroups of principal fractional ideals of $I(R)$ and $I(S)$, respectively. Owing to Lemma 3.1.9, the map $i_{S/R}:I(R)\to I(S)$ defined by $i_{S/R}(X)=SX$ is an injective homomorphism which preserves inclusion and maps $P(R)$ into $P(S)$. We may therefore regard $I(R)$ as embedded in $I(S)$ and $P(R)$ in $P(S)$. Let $C\subseteq I(R)$ be the free subgroup generated by the prime ideals of R that do not ramify in S. For each such prime ideal of R choose a prime ideal of S lying above it. The remaining prime ideals of S generate a complement D of C in $I(S)$. Furthermore, by Corollary 3.7, D is free of infinite rank. Since only finitely many prime ideals of R ramify in S, $I(R)/C$ if finitely generated. Therefore $I(S)/I(R)$ is the product of a finite group and a free group of infinite rank. Now $P(S)\cap I(R)=P(R)$ (Corollary 3.1.10), so

$$P(S)/P(R)\cong P(S)\cdot I(R)/I(R)\subseteq I(S)/I(R).$$

Furthermore, $I(S)/P(S)\cdot I(R)$ is finite, since so is $I(S)/P(S)$. Thus $P(S)/P(R)$ is the product of a finite group and a free group of infinite rank.

By Lemma 1.2, there are natural isomorphisms

$$F^*/U(R)\cong P(R)\qquad\text{and}\qquad E^*/U(S)\cong P(S).$$

Moreover, $F^*\cap U(S)=U(R)$, and the induced inclusion $F^*/U(R)\subseteq E^*/U(S)$ agrees with $P(R)\subseteq P(S)$. Since $P(S)/P(R)$ is the product of a finite group and a free group of infinite rank, the same applies to the isomorphic group $E^*/U(S)F^*$. But

$$U(S)F^*/F^*\cong U(S)/U(R)$$

and $U(S)$ is finitely generated, by the Dirichlet unit theorem. It follows that E^*/F^* is the product of a finite group and a free group of infinite rank. Since E^*/F^* is countable, this rank must be equal to $\aleph_0$ and the result follows. ∎

For future use, we next establish the following two results concerning simple transcendental extensions.

4.3.9   Lemma   *Let $F(\lambda)/F$ be a simple transcendental extension where F is an arbitrary field, and let μ be an element of $F(\lambda)$ not in F. Write $\mu=f(\lambda)g(\lambda)^{-1}$, where $f(\lambda)$ and $g(\lambda)$ are polynomials in λ with no common factor of positive degree in λ. If $n=\max(\deg f,\deg g)$, then λ is algebraic over $F(\mu)$ and $(F(\lambda):F(\mu))=n$.*

Proof. We may write

$$f(\lambda)=\alpha_0+\alpha_1\lambda+\ldots+\alpha_n\lambda^n,\qquad g(\lambda)=\beta_0+\beta_1\lambda+\ldots+\beta_n\lambda^n,$$

where either $\alpha_n \neq 0$ or $\beta_n \neq 0$. The relation $\mu = f(\lambda)g(\lambda)^{-1}$ gives $f(\lambda) - \mu g(\lambda) = 0$ and

$$0 = (\alpha_n - \mu\beta_n)\lambda^n + (\alpha_{n-1} - \mu\beta_{n-1})\lambda^{n-1} + \ldots + (\alpha_0 - \mu\beta_0).$$

Moreover, $\alpha_n - \mu\beta_n \neq 0$ since α_n or $\beta_n \neq 0$ and $\mu \notin F$. Hence λ is a root of the polynomial of degree $n : \sum_{i=0}^{n} (\alpha_i - \mu\beta_i)X^i$ with coefficients in $F(\mu)$. We are therefore left to verify that the given polynomial is irreducible.

First, it is clear that μ is transcendental over F, since λ is algebraic over $F(\mu)$; hence μ algebraic over F implies λ algebraic over F, contrary to assumption. The ring $F[\mu, X] = F[\mu][X]$ is the polynomial ring in two indeterminates μ, X and hence the theorem on unique factorization into irreducible elements holds in $F[\mu, X]$. Note also that a polynomial in $F[\mu, X]$ of positive degree in X is irreducible in $F(\mu)[X]$ if it is irreducible in $F[\mu, X]$. Now

$$f(\mu, X) = \sum (\alpha_i - \mu\beta_i)X^i = f(X) - \mu g(X)$$

is of degree 1 in μ. Therefore, if $f(\mu, X)$ is reducible in $F(\mu)[X]$, then it has a factor $h(X)$ of positive degree in X. But this implies that $f(X)$ and $g(X)$ are divisible by $h(X)$, contrary to assumption. ∎

4.3.10 Theorem (Lüroth) *Let F be an arbitrary field and let $F(\lambda)/F$ be a simple transcendental extension. Then any field K with $F \subset K \subseteq F(\lambda)$ is also a simple transcendental extension: $K = F(\gamma)$, γ transcendental.*

Proof. By hypothesis, K contains an element β not in F, so by Lemma 3.9 $P = F(\lambda)$ is algebraic over $F(\beta)$ and hence is algebraic over $K \supseteq F(\beta)$. Let the minimum polynomial of λ over K be

$$f(x) = X^n + \gamma_1 X^{n-1} + \ldots + \gamma_n.$$

The γ_i have the form $\mu_i(\lambda)v_i(\lambda)^{-1}$, where μ_i, v_i are polynomials in the transcendental element λ. Multiplication of $f(X)$ by a suitable polynomial in λ will give a polynomial

$$f(\lambda, X) = c_0(\lambda)X^n + c_1(\lambda)X^{n-1} + \ldots + c_n(\lambda) \tag{3}$$

in $F[\lambda, X]$, which is a primitive polynomial in X in the sense that the highest common factor of the $c_i(\lambda)$ is 1. Note also that $\gamma_i = c_i(\lambda)c_0(\lambda)^{-1} \in K$, and not all of these are in F since λ is transcendental over F. Hence one of the γ's has the form $\gamma = g(\lambda)h(\lambda)^{-1}$, where $g(\lambda), h(\lambda)$ have no common factor of positive degree in λ and $\max(\deg g, \deg h) = m > 0$. By Lemma 3.9, $g(X) - \gamma h(X)$ is irreducible in $F(\gamma)[X]$ and $[P:F(\gamma)] = m$. Because $K \supseteq F(\gamma)$ and $(P:K) = n$, clearly $m \geqslant n$. We shall show that $m = n$ and this will prove that $K = F(\gamma)$.

Because λ is a root of $g(X) - \gamma h(X)$ and the coefficients of this polynomial are contained in K, we have $g(X) - \gamma h(X) = f(X)q(X)$ in $K(X)$. We have $\gamma = g(\lambda)h(\lambda)^{-1}$ and we can replace the coefficients of f and q by their rational expressions in λ and then multiply by a suitable polynomial in λ, to obtain a relation in $F[\lambda, X]$ of the form

$$k(\lambda)[g(X)h(\lambda) - g(\lambda)h(X)] = f(\lambda, X)q(\lambda, X), \tag{4}$$

where $f(\lambda, x)$ is a primitive polynomial given in (3). We may therefore deduce that $k(\lambda)$ is a factor of $q(\lambda, X)$ and so, cancelling this, we may assume the relation is

$$g(X)h(\lambda) - g(\lambda)h(X) = f(\lambda, X)q(\lambda, X). \tag{5}$$

Now the degree in λ of the left-hand side is at most m. Since $\gamma = g(\lambda)h(\lambda)^{-1}$ with $(g(\lambda), h(\lambda)) = 1$ and $\max(\deg g, \deg h) = m$, the λ-degree of $f(\lambda, X)$ is at least m. It follows that it is exactly m, and $q(\lambda, X) = q(x) \in F[X]$. Then the right-hand side of (5) is primitive as a polynomial in X. This holds also for the left-hand side. By symmetry, the left-hand side is primitive as a polynomial in λ also, and this ensures that $q(X) = q$ is a non-zero element of F. Then (5) implies that the X-degree and λ-degree of $f(\lambda, X)$ are the same. Thus $m = n$ and the result follows. ∎

We now apply the preceding result to prove

4.3.11 Theorem (Brandis 1965) *Let E/F be a field extension, where F is an arbitrary infinite field and $E \neq F$. Then E^*/F^* is not finitely generated.*

Proof. Since $E \neq F$, we may choose $\lambda \in E$ such that $1, \lambda$ are F-linearly independent. Hence the elements $1 + \alpha\lambda$, $\alpha \in F$, of E^* are mutually incongruent modulo F^*. Since F is infinite, we conclude that E^*/F^* is also infinite. Assume, by way of contradiction, that E^*/F^* is finitely generated, say $E^*/F^* = \langle u_1 F^*, \ldots, u_k F^* \rangle$. Since E^*/F^* is infinite we may assume that $u_1 F^*$ is of infinite order. Let F_0 be the prime subfield of F, and put $E_1 = F_0(u_1)$, $F_1 = E_1 \cap F$. Then $F_1 \subset E_1$ and

$$E_1^*/F_1^* \cong E_1^* F^*/F^* \subseteq E^*/F^*,$$

which shows that E_1^*/F_1^* is also finitely generated. Since no power of u_1 is in F_0, E_1 is an infinite field. Thus, if $\operatorname{char} F_0 = p > 0$, then u_1 is transcendental over F_0. We are therefore left to verify the following assertion: E^*/F^* is not finitely generated, provided that $E = F_0(u)$, $F \subset E$, F_0 is a prime subfield of E, and u is transcendental over F_0 if $\operatorname{char} F_0 = p > 0$.

If $F_0 = \mathbb{Q}$ and u is algebraic over $\mathbb{Q}$, then the result follows by virtue of Theorem 3.8. If u is transcendental over F_0 and $F = F_0$, then the result follows by applying Lemma 1.4. Thus, by Theorem 3.10, we may assume that $E = F_0(u)$ and $F = F_0(v)$, where both u and v are transcendental over the prime field F_0.

Write $v = v(u) = p(u)/q(u)$, where $p(u), q(u) \in F_0[u]$, $(p(u), q(u)) = 1$, and where the highest common factor of coefficients of $p(u)$ and $q(u)$ is 1. The embedding of F into E is given by the map

$$f(v) \mapsto f^*(u) = f(v(u)).$$

Let $g(v) = a_0 + a_1 v + \ldots + a_n v^n \in F_0[v]$ with $a_n \neq 0$. Then

$$g^*(u) = \{a_0 q(u)^n + a_1 q(u)^{n-1} p(u) + \ldots + a_n p(u)^n\}/q(u)^n.$$

Setting

$$z_g(u) = a_0 q(u)^n + a_1 q(u)^{n-1} p(u) + \ldots + a_n p(u)^n \tag{6}$$

is suffices, in view of Lemma 1.1, to show that there exist infinitely many irreducible polynomials $g(v) \in F_0[v]$ for which $z_g(u)$ is not irreducible.

To this end, put $r = \deg p$, $s = \deg q$, and $h = \max(r, s)$. We claim that, if $g(v)$ is irreducible, then

$$h \geq 2; \tag{7}$$

$$\deg z_g(u) = hn \quad \text{if} \quad g(v) \neq v, v - 1. \tag{8}$$

Indeed, by Lemma 3.9, $(E:F) = h$ and hence $h \geq 2$, since $E \neq F$. To prove (8), it suffices to show that $\deg z_g(u) > hn$ since the inequality $\deg z_g(u) \leq hn$ is obvious.

Suppose firstly that $r \neq s$. Then the term u^{hn} in (6) comes exactly from one of the summands $a_0 q(u)^n$ or $a_n p(u)^n$. By hypothesis, $a_n \neq 0$ and since $g(v) \neq v$ and $g(v)$ is irreducible, we also have $a_0 \neq 0$. Thus $\deg z_g(u) \geq hn$, proving the case $r \neq s$. In case $r = s$, the coefficient of u^{hn} is equal to $a_0 + a_1 + \ldots + a_n = g(1) \neq 0$, since $g(v)$ is irreducible and $g(v) \neq v - 1$. This proves (7) and (8).

Let $w(u)$ be an irreducible polynomial in $F_0[u]$ with $w(u) \nmid q(u)$. We claim that there is an irreducible polynomial $g(v) \in F_0[v]$ with $w(u) \mid z_g(u)$. Indeed, let λ be a root of $w(u)$ in the algebraic closure $\bar{F}_0$ of F_0. Because $w(u) \nmid q(u)$, $v(\lambda) = \mu \in \bar{F}_0$. We choose $g(v)$ to be irreducible and such that $g(\mu) = 0$. Then we have $g(\mu) = g(v(\lambda)) = g^*(\lambda) = 0$, and therefore $z_g(\lambda) = 0$. It follows that $w(u) \mid z_g(u)$ as claimed.

For any natural number m, there exists an irreducible polynomial of degree m over F_0 (e.g. if $F_0 = \mathbb{Q}$, then take $X^m - 2$ and if $f_0 = \mathbb{F}_p$ then take the minimal polynomial corresponding to the extension $\mathbb{F}_{p^m}/\mathbb{F}_p$). We choose a sequence of infinitely many distinct irreducible polynomials $w_1(u), w_2(u), \ldots$, the degrees of which are not divisible by h; for example, of degrees $1, h + 1, 2h + 1, \ldots$ By the foregoing, for each polynomial $w_i(u)$ we obtain a polynomial $g_i(v)$ such that $w_i(u) \mid z_{g_i}(u)$.

When $g_i(v) \neq v$, $v - 1$ it follows from (8) that $\deg w_i(u) \neq \deg z_{g_i}(u)$, and thus $z_{g_i}(u)$ is reducible. Each $z_{g_i}(u)$ contains only a finite number of prime factors, and therefore there are infinitely many distinct $z_{g_i}(u)$. Since all the $z_{g_i}(u)$ are reducible, the result follows. ∎

4.4 Kneser's theorem and related results

Let E/F be a field extension and let $t(E^*/F^*)$ be the torsion subgroup of E^*/F^*. Our aim in this section is to provide some information on the structure of $t(E^*/F^*)$. We begin by proving some general results which are of independent interest. Let M be a subgroup of E^* such that F^*M/F^* is finite. When is it the case that $|f^*M/F^*| = (F(M):F)$? To answer this question, we need the following preliminary observation.

4.4.1 Lemma *Let p be a prime and λ an element in F with no pth root in F. Then $X^p - \lambda$ is irreducible over F.*

Proof. Suppose the contrary, and let f be an irreducible factor of $X^p - \lambda$ of degree k $(1 \leq k < p)$. Let c be the constant term of f. The roots of $X^p - \lambda$ (in some splitting field) all have the form εu, where u is one fixed root and $\varepsilon^p = 1$. Because $\pm c$ is a product of k of these roots, we have $\pm c = \delta u^k$, $\delta^p = 1$. Since $(k, p) = 1$, there exist integers r and s such that $rk + sp = 1$. We have

$$u = u^{rk}u^{sp} = (\pm c/\delta)^r \lambda^s.$$

Hence $u\delta^r$ lies in F. Since its pth power is λ, we have the desired contradiction. ∎

We are now ready to prove the following important result.

4.4.2 Theorem (Kneser 1975) *Let E/F be a separable field extension and let M be a subgroup of E^* such that F^*M/F^* is finite. Assume that for all odd primes p, each pth root of 1 which lies in F^*M also lies in F and that $i = \sqrt{-1} \in F$ if $1 \pm i \in F^*M$. Then*

$$|F^*M/F^*| = (F(M):F).$$

Proof. We first note that $F(M)$ consists of all F-linear combinations of F^*M. Hence we may choose finitely many $x_1, \ldots, x_n$ in F^*M, which is an F-basis for $F(M)$. Then $x_1F^*, \ldots, x_nF^*$ are distinct elements of F^*M/F^*, and thus

$$(F(M):F) \leq |F^*M/F^*|. \tag{1}$$

Assume that the result is true for all Sylow subgroups of F^*M/F^*. If

N/F^* is a Sylow p-subgroup of F^*M/F^* of order p^t, then

$$p^t = (F(N):F) \,|\, ((F(M):F),$$

and hence $|F^*M/F^*| \leqslant (F(M):F)$. In view of (1), we may therefore assume that F^*M/F^* is a p-group.

Choose a chain of subgroups

$$F^* = N_0 \subset N_1 \subset \ldots \subset N_t = F^*M$$

with $[N_s : N_{s-1}] = p$, $1 \leqslant s \leqslant t$. We prove, by induction on s, that

$$(F(N_s):F(N_{s-1})) = p, \tag{2}$$

and that an element of $F(N_s)$ (respectively, $F(N_s) \cap F^*M$ if $p = 2$ and $i \in F(N_s)$), the pth power of which is in N_s, is itself in N_s. Assume that the statement is true for $s-1$. If $a \in N_s$ is such that aN_{s-1} generates N_s/N_{s-1}, then a is a root of the polynomial $f(X) = X^p - a^p$ with coefficients in $F(N_{s-1})$. Hence, if f is irreducible, then (2) holds. If f is reducible, then $f(X)$ has a root, say b, in $F(N_{s-1})$, by Lemma 4.1. Hence $b = a\varepsilon$ with $\varepsilon^p = 1$ and $b^p = a^p \in N_{s-1}$. By induction, $b \in N_{s-1}$ and so $\varepsilon \in N_s \subseteq F^*M$, which implies, by hypothesis, that $\varepsilon \in F$. It follows that $a \in N_{s-1}$, contrary to the assumption that aN_{s-1} is of order $p > 1$. This proves (2).

Now assume that $c \in F(N_s)$ satisfies $c^p \in N_s$ (and $c \in F^*M$, when $p = 2$ and $i \in F(N_s)$). Then $c^p = a^q d$ with $0 \leqslant q < p$ and $d \in N_{s-1}$. We wish to show that $c \in N_s$. This will be achieved by proving the case $q = 0$ and by showing that the case $q > 0$ cannot occur.

Assume that $q = 0$ so that $c^p \in N_{s-1}$. Since E/F is separable, so is $F(N_s)/F(N_{s-1})$. Invoking (2) proposition 1.2.15(ii), we deduce that there exists an $F(N_{s-1})$-homomorphism f of $F(N_s)$ into the normal closure of $F(N_s)/F(N_{s-1})$ such that $f(a) \neq a$. Since $f(a^p) = a^p = f(a)^p$, we have $f(a) = a\varepsilon$, with ε being a primitive pth root of 1. Similarly, $f(c^p) = c^p = f(c)^p$ and so $f(c) = c\varepsilon^r$ for some $0 \leqslant r \leqslant p-1$. The latter implies that $f(a^{-r}c) = a^{-r}c$ and hence that $a^{-r}c = b \in F(N_{s-1})$. Furthermore, $b^p = (a^p)^{-r}c^p \in N_{s-1}$ (and $b \in F^*M$ in case $c \in F^*M$), which shows, by induction, that $b \in N_{s-1}$ and hence that $c \in N_s$.

Assume by way of contradiction that $q > 0$, and denote by N the norm of $F(N_s)$ over $F(N_{s-1})$. Since $N(a) = (-1)^{p-1}a^p$, we have

$$((-1)^{p-1}a^p)^q = N(c)^p d^{-p}.$$

For odd p, a^p is then a pth power of an element in $F(N_{s-1})$, which contradicts (2). In case $p = 2$, we have $-a^2 = \lambda^2$ with $\lambda \in F(N_{s-1})$, which shows that $i \in F(N_s)$, $i \notin F(N_{s-1})$ and $c^2 = ad = \pm i\lambda d$. Let us write

$$c = g + ih \qquad \text{with} \qquad g, h \in F(N_{s-1}).$$

Then $c^2 = (g^2 - h^2) + 2ghi = \pm i\lambda d$, which implies $g^2 = h^2$ and hence

$c = (1 \pm i)g$. We conclude that

$$g^4 = -c^4/4 \in N_{s-1},$$

which ensures, by applying the induction hypothesis twice, that $g \in N_{s-1}$. But then $1 \pm i \in F^*M$ and hence, by hypothesis, $i \in F \subseteq F(N_{s-1})$. This contradiction completes the proof of the theorem. ∎

Let $T = T(E/F)$ denote the subgroup of E^* containing F^* such that T/F^* is the torsion subgroup $t(E^*/F^*)$ of E^*/F^*. Therefore

$$T = \{x \in E^* \mid x^m \in F^* \text{ for some } m \geq 1\}.$$

Let W denote the subgroup of E^* generated by F^* and all roots of unity that lie in E^*. Thus $E^* \supseteq T \supseteq W \supseteq F^*$. It is possible for W/F^* to be infinite even when E is a finite extension of F, and hence we shall restrict our considerations to T/W. For each prime number p, let T_p (respectively, W_p) denote the subgroup of E^* containing F^* such that T_p/F^* (respectively, W_p/F^*) is the p-component of T/F^* (respectively, W/F^*).

It is natural to provide a link between the order of the group T/W and the degree of the extension E/F. This will be achieved with the aid of the following preliminary result.

4.4.3 Lemma   *Let p be a prime distinct from the characteristic of F, and let ε be a primitive pth root of unity if p is odd, or a primitive 4th root of unity if $p = 2$. Put $E = F(\varepsilon)$. Then $T_p = W_p$ except when $p = 2$, $\varepsilon \notin F$, $\sqrt{-2} \notin F$, and E contains only finitely many roots of unity whose orders are powers of 2. In the exceptional case, we have $|T_2/W_2| = 2$.*

Proof.   Let p be a prime and let $n = (E:F)$. Given $\alpha \in T_p$, we have $\alpha^{p^r} = \gamma$ for some $r \geq 1$ and $\gamma \in F$. If N denotes the norm mapping from E^* to F^*, then $N(\alpha)^{p^r} = \gamma^n$. Thus $(\alpha^n N(\alpha)^{-1})^{p^r} = 1$ and therefore $\alpha^n \in W_p$, proving that T_p/W_p has exponent n. In particular, if p is an odd prime, then n divides $p - 1$; hence we deduce that $T_p = W_p$ in this case.

We may now assume that $p = 2$ and that $\varepsilon \notin F$. Let V be the 2-component of the group of roots of unity that lie in E. Because $(E:F) = 2$, it follows from the above that if $\alpha \in T_2$, then $\alpha^2 = v\gamma$ for some $v \in V$ and $\gamma \in F$. We claim that if $v = v_1^2$ for some $v_1 \in V$, then $\alpha \in W_2$. Indeed, $(\alpha v_1^{-1})^2 \in F$, hence $\alpha v_1^{-1} = \varepsilon^i \gamma_1$ for some integer i and $\gamma_1 \in F$. Therefore $\alpha \in W_2$, as claimed. In particular, if V is infinite, then $T_2 = W_2$.

We may now further assume that V is finite, say $V = \langle \delta \rangle$. Let σ be the non-trivial automorphism of E over F. First assume that $\sqrt{-2} \in F$. Then it is easy to show that $\sigma(\delta) = -\delta^{-1}$ since $\varepsilon \notin F$ and $\sqrt{-2} \in F$. If $\alpha \in T_2$, then $\alpha^2 = \delta^i \gamma$ for some integer i and $\gamma \in F$. Hence $\sigma(\alpha)^2 = -\delta^i \gamma$ and therefore $N(\alpha)^2 = (-1)^i \gamma^2$. Thus $(-1)^i$ is a square in F. Because $\varepsilon \notin F$, we deduce that i is even. Accordingly, $\alpha \in W_2$ by the claim in the previous paragraph. Hence $T_2 = W_2$ in this case.

Finally, we now assume that $\sqrt{-2} \notin F$. In this situation, we obtain $\sigma(\delta) = \delta^{-1}$. If we put $\alpha_0 = 1 + \delta$ then $\alpha_0^2 = \delta\gamma_0$, where $\gamma_0 = \delta + \delta^{-1} + 2 \in F$. Consequently $\alpha_0 \in T_2$. We claim hat $\alpha_0 \notin W_2$. Assume that $\alpha_0 = \delta^i\gamma$ for some integer i and $\gamma \in F$. Then $1 + \delta = \delta^i\gamma$ and, applying σ, we obtain $1 + \delta^{-1} = \delta^{-1}\gamma$. Thus

$$\delta = (1 + \delta)(1 + \delta^{-1})^{-1} = \delta^{2i},$$

contrary to the fact that δ has even order. Thus $\alpha_0 \notin W_2$ as claimed. To prove that α_0 generates T_2 modulo W_2, let $\alpha \in T_2$. Then $\alpha^2 = \delta^i\gamma$ for some integer i and $\gamma \in F$. Hence $(\alpha_0^i\alpha^{-1})^2 = \gamma_0^i\gamma^{-1}$, and again we must have $\alpha_0^i\alpha^{-1} \in W_2$. Thus $|T_2/W_2| = 2$ in this case and the result follows. ∎

We now combine Theorem 4.2 and Lemma 4.3 to obtain

4.4.4 Theorem (May 1980) *Let E be a finite separable extension field of F. Then $|T/W|$ divides $(E:F)$.*

Proof. Let p be a prime. It suffices to show that $|T_p/W_p|$ divides $(E:F)$. Choose ε as in the statement of Lemma 4.3, and let $F_0 = F$ if $\varepsilon \notin E$, or $F_0 = F(\varepsilon)$ if $\varepsilon \in E$. Then F_0 satisfies the hypothesis of Theorem 4.2 applied to $T_pF_0^*$. Therefore $|T_pF_0^*/F_0^*|$ divides $(E:F_0)$. Put $K = T_p \cap F_0^*$. Then $|T_p/K|$ divides $(E:F_0)$. Observe further that $K = T_p(F_0/F)$ and $K \cap W_p = W_p(F_0/F)$. Invoking Lemma 4.3, we conclude that $|K/(K \cap W_p)|$ divides $(F_0:F)$. Accordingly, $|T_p/W_p|$ divides $(E:F)$ and the result follows. ∎

As a preparation for the proof of the final result, we provide a number of auxiliary assertions.

4.4.5 Lemma *Let p be a prime, $X^p - a$ be irreducible over a field F, and λ be a root of $X^p - a$. Then:*

(i) *If p is odd, or if $p = 2$ and char $F = 2$, then λ is not a pth power in $F(\lambda)$.*

(ii) *If $p = 2$ and char $F \neq 2$, then λ is a square in $F(\lambda)$ if and only if $a \in -4F^4$.*

Proof.

(i) Assume by way of contradiction that $\lambda = w^p$ for some $w \in F(\lambda)$. The case where char $F = p$ is straightforward: since w is a polynomial in λ and pth powers are taken termwise we have $w^p \in F$, which is impossible. Now assume that char $F \neq p$ and adjoin to $F(\lambda)$ a primitive pth root of unity, say ε. The resulting field E is a splitting field of $X^p - a$ over F and therefore E/F is a normal extension. Any automorphism of E/F sends λ into some $\varepsilon^i \lambda$, $0 \leqslant i \leqslant p - 1$, and for every $i \in \{0, 1, \ldots, p - 1\}$ there is an automorphism f_i sending λ into $\varepsilon^i \lambda$. Put $w_i = f_i(w)$. Then $\varepsilon^i \lambda = w_i^p$. The element w is in $F(\lambda)$ but not in F, hence its irreducible polynomial (say f) over F has degree p and has p distinct roots in E. If w' is any of these roots, then there is an automorphism ψ of E/F sending w into w'. If $\psi(\lambda) = \varepsilon^j \lambda$ we must have $\psi(w) = w_j$. Hence the elements $w_0, w_1, \ldots, w_{p-1}$ are all the roots of f, which implies that $z = w_0 w_1 \ldots w_{p-1} \in F$. We now multiply together the equations $\varepsilon^i \lambda = w_i^p$, to obtain $\mu \lambda^p = \mu a = z^p$, where $\mu = 1 \cdot \varepsilon \cdot \varepsilon^2 \ldots \varepsilon^{p-1}$. If p is odd, $\mu = 1$, and we have a contradiction $a = z^p$ proving (i).

(ii)   Assume that $\lambda = w^2$ with $w = \alpha + \beta \lambda$ ($\alpha, \beta \in F$). From $\lambda = (\alpha + \beta \lambda)^2$ we obtain the equations $\alpha^2 + \beta^2 a = 0$, $2\alpha\beta = 1$. Eliminating β, we find $a = -4\alpha^4$, so that $a \in -4F^4$. Conversely, if $a = -4\alpha^4$, we take $\beta = (1/2)\alpha$ and verify $\lambda = (\alpha + \beta \lambda)^2$. ∎

4.4.6 Lemma   *Let F be an arbitrary field, p a prime and a an element in F with no pth root in F.*

(i)   *If p is odd, then $X^{p^n} - a$, is irreducible over F for any $n \geqslant 0$.*

(ii)   *If $p = 2$ and char $F = 2$, then $X^{2^n} - a$ is irreducible over F for any $n \geqslant 0$.*

(iii)   *If $p = 2, n \geqslant 2$, and char $F \neq 2$, then $X^{2^n} - a$ is irreducible over F if and only if $a \notin - 4F^4$.*

Proof.

(i), (ii)   Let λ be a root of $X^{p^n} - a$ and write $\mu = \lambda^{p^{n-1}}$. Then $\mu^p = a$ and hence, by Lemma 4.1, $(F(\mu):F) = p$. We claim that λ has degree p^{n-1} over $F(\mu)$; if sustained, it will follow that λ has degree p^n over F and $X^{p^n} - a$ is irreducible over F. That λ has degree p^{n-1} over $F(\mu)$ will be true by induction on n provided that μ is not a pth power in $F(\mu)$. The desired conclusion is therefore a consequence of Lemma 4.5(i).

(iii)   Suppose first that $a \in -4F^4$ and write $a = -4\alpha^4$, $Y = X^{2^{n-2}}$. Then

$$X^{2^n} - a = Y^4 + 4\alpha^4 = (Y^2 + 2\alpha Y + 2\alpha^2)(Y^2 - 2\alpha Y + 2\alpha^2).$$

Conversely, assume that $a \notin - 4F^4$. Then $-4a$ is not a fourth power in F. Again, let λ be a root of $X^{2^n} - a$ and $\mu = \lambda^{2^{n-1}}$. We have $(F(\mu):F) = 2$ since $\mu^2 = a$, and we must show that $(F(\lambda):F(\mu)) = 2^{n-1}$. For $n = 2$ this will be true if μ is not a square in $F(\mu)$, and for $n > 2$ this will be true by induction on n, provided that μ is not a square in $F(\mu)$ and -4μ is not a fourth power in $F(\mu)$. In the latter case, $-\mu$ is a square in $F(\mu)$. So it will suffice to rule out the possibility that either μ or $-\mu$ is a square in $F(\mu)$. Now these two statements are equivalent since $\mu \mapsto -\mu$ induces an automorphism of $F(\mu)$ over F. Hence, by applying Lemma 4.5(ii), the result follows. ∎

Let F be a field and m a positive integer. A Galois extension E/F is said to be of *exponent* m if $\sigma^m = 1$ for all $\sigma \in \mathrm{Gal}(E/F)$. In what follows we assume that F contains a primitive mth root of unity (denoted by ε_m), and hence that m is prime to char F. All algebraic extensions below are contained in a fixed algebraic closure $\bar{F}$ of F. We denote by $(F^*)^m$ the subgroup of F^* consisting of all mth powers of elements of F^*. Let H be a subgroup of F^* containing $(F^*)^m$. We denote by F_H the composite of all fields $F(\sqrt[m]{a})$ with $a \in H$. It is uniquely determined by H as a subfield of $\bar{F}$. Keeping the above notation and hypotheses, we now prove

4.4.7 Proposition

(i)   If H is a subgroup of F^* containing $(F^*)^m$, then:
  - (a) F_H/F is Galois and abelian of component m.
  - (b)   the extension F_H/F is finite if and only if the group $H/(F^*)^m$ is finite and, if that is the case, then $\mathrm{Gal}(F_H/F) \cong H/(F^*)^m$ and, in particular, $|H/(F^*)^m| = (F_H : F)$.

(ii)   The map $H \mapsto F_H$ gives a bijection of the set of subgroups of F^* containing $(F^*)^m$ and the abelian extensions of F of exponent m.

Proof.

(i)   Let $a \in H$ and let α be an mth root of a. Then the polynomial $X^m - a$ splits into linear factors in F_H and thus F_H is Galois over F, because this holds for all $a \in H$. Let G be the Galois group of F_H over F. Then, given $\sigma \in G$, we have $\sigma(\alpha) = \omega_\sigma \alpha$ for some $\omega_\sigma \in \langle \varepsilon_m \rangle \subseteq F$. Hence, if λ is another element of G, then

$$(\sigma\lambda)(\alpha) = \sigma(\omega_\lambda \alpha) = \omega_\lambda \omega_\sigma \alpha = \omega_\sigma \omega_\lambda \alpha = (\lambda\sigma)(\alpha).$$

Because F_H is generated over F by all such α, we conclude that G is abelian. Furthermore, since $\sigma^i(\alpha) = \omega_\sigma^i \alpha$, the extension F_H/F is of exponent m. The map $\sigma \mapsto \omega_\sigma$ is obviously a homomorphism of G into $\langle \varepsilon_m \rangle$. Furthermore, it is independent of the choice of mth root of a. Indeed, $\omega_\sigma = \sigma(\alpha)/\alpha$ and if α' is another mth root, then $\alpha' = \delta\alpha$ for some $\delta \in \langle \varepsilon_m \rangle$, whence

$$\sigma(\alpha')/\alpha' = \delta\sigma(\alpha)/\delta\alpha = \delta(\alpha)/\alpha.$$

Let $f_a \in \mathrm{Hom}(G, \langle \varepsilon_m \rangle)$ be defined by $f_a(\sigma) = \omega_\sigma$. Consider the map

$$\begin{cases} H \xrightarrow{\psi} \mathrm{Hom}(G, \langle \varepsilon_n \rangle), \\ a \mapsto f_a. \end{cases}$$

If $a, b \in H$ and $\alpha^m = a$, $\beta^m = b$ then $(\alpha\beta)^m = ab$, and hence

$$\sigma(\alpha\beta)/\alpha\beta = (\sigma(\alpha)/\alpha)(\sigma(\beta)/\beta),$$

proving that ψ is a homomorphism. If $a = \lambda^m$ with $\lambda \in F^*$, then $f_a(\sigma) = \sigma(\lambda)/\lambda = 1$ and thus $(F^*)^m \subseteq \mathrm{Ker}\ \psi$. Conversely, assume that $a \in \mathrm{Ker}\ \psi$ so that $\sigma(\alpha) = \alpha$ for all $\sigma \in G$, where $\alpha^m = a$. Consider the subfield $F(\sqrt[m]{a})$ of F_H. If α is not in F, there exists an automorphism of $F(\sqrt[m]{a})$ over F which is not identity. Extend this automorphism to F_H and call this extension σ_0. Then clearly $\sigma_0(\alpha) \neq \alpha$, a contradiction. Hence $\mathrm{Ker}\ \psi = (F^*)^m$ and so there is an exact sequence

$$1 \to H/(F^*)^m \to \mathrm{Hom}(G, \langle \varepsilon_m \rangle).$$

If $f_a(\sigma) = 1$ for all $a \in H$, then for every generator α of F_H with $\alpha^m = a \in H$ we have $\sigma(\alpha) = \alpha$, and hence $\sigma = 1$. Since the map $G \to \mathrm{Hom}(H/(F^*)^n \langle \varepsilon_m \rangle)$, $\sigma \mapsto \lambda_\sigma$, where $\lambda_\sigma(a(F^*)^n) = f_a(\sigma)$ is a homomorphism, we have an exact sequence

$$1 \to G \to \mathrm{Hom}(H/(F^*)^m, \langle \varepsilon_m \rangle),$$

which proves (i).

(ii) Let H_1, H_2 be subgroups of F^* containing $(F^*)^m$. If $H_1 \subseteq H_2$ then $F_{H_1} \subseteq F_{H_2}$. Conversely, assume that $F_{H_1} \subseteq F_{H_2}$ and fix $b \in H_1$. Then $F(\sqrt[m]{b}) \subseteq F_{H_2}$ and $F(\sqrt[m]{b})$ is contained in a finitely generated subextension of F_{H_2}. Thus to prove that $H_1 \subseteq H_2$ we may assume that $H_2/(F^*)^m$ is finitely generated, and hence finite. Let H_3 be the subgroup of F^* generated by H_2 and b. Then $F_{H_2} = F_{H_3}$ and from what we saw above, the degree of this field over F is precisely

$$|H_2/(F^*)^m| \quad \text{or} \quad |H_3/(F^*)^m|.$$

Thus the above numbers are equal, and $H_2 = H_3$ proving that $H_1 \subseteq H_2$. Finally, assume that E is an abelian extension of F of exponent m. Any finite subextension is a composite of cyclic extensions of exponent m because any finite abelian group is a direct product of cyclic groups, and we can apply Proposition 1.2.18. By Proposition 1.2.17, every cyclic extension can be obtained by adjoining an mth root. Therefore E can be obtained by adjoining a family of mth roots, say mth roots of elements $\{b_j\}_{j \in J}$ with $b_j \in F^*$. Let H be the subgroup of F^* generated by all b_j and $(F^*)^m$. If $b' = ba^m$ with $a, b \in F$, then obviously

$$F(\sqrt[m]{b'}) = F(\sqrt[m]{b}).$$

Thus $F_H = E$ and the result follows. ∎

4.4.8 Corollary *Let p be a prime, let F be a field with* char $F \neq p$ *and let F contain a primitive pth root of unity. Suppose $\alpha^p \in F$ and that $\beta \in F(\alpha)$ with $\beta^p \in F$. Then $\beta = \alpha^k a$ for some k and some $a \in F$.*

Proof. The assertion is trivial in case $\alpha \in F$. Assume that $\alpha \notin F$. Thus $\alpha^p \notin F^p$ since F contains all pth roots of unity. Thus $(F(\alpha):F) = p$ and $F(\alpha) = F_H$, where $H = \langle \alpha^p \rangle (F^*)^p$. We may assume $\beta \notin F$, thus $F(\alpha) = F(\beta)$ and $F(\beta) = F_{H_1}$, where $H_1 = \langle \beta^p \rangle (F^*)^p$. Thus, by Proposition 4.7, $H = H_1$ and hence $\beta^p = \alpha^{pk} c^p$ for some k and some $c \in F^*$. Taking pth roots, we obtain $\beta = \alpha^k a$ for some $a \in F$. ∎

We are now ready to prove the following result.

4.4.9 Theorem (May 1979a) *Let F be a field, E an extension field, and p a prime different from the characteristic of F. Assume that $E = F(\alpha)$, where $\alpha^p \in F - F^p$. If $p = 2$, further assume that $E \neq F(i)(i^2 = -1)$. Then*

$$t(E^*/F^*) = (\langle \alpha \rangle t(E^*)F^*)/F^*.$$

Proof. By Lemma 4.1, $X^p - a^p$ is irreducible over F and hence $(E:F) = p$. Let $\beta \in E^*$ be an element of prime-power order modulo F^*. First assume that the order is q^r for some prime $q \neq p$, and let $\beta^{q^r} = \gamma \in F^*$. If N is the norm map from E to F, then $N(\beta)^{q^r} = \gamma^p$, hence $\gamma = \gamma_1^{q^r}$ for some $\gamma_1 \in F^*$. It follows that $\beta = \gamma_1 \varepsilon$ for some root of unity $\varepsilon \in F^*$, which shows that $\beta F^* \in t(E^*)F^*$.

Now assume that β has order p^r modulo F^*, and let $\beta^{p^r} = \gamma \in F^*$. We claim that $\gamma \notin F^p$. Assume the contrary and write $\gamma = \gamma_1^p$ for some $\gamma_1 \in F^*$. Let ε be a primitive pth root of unity. Then

$$\beta^{p^{r-1}} = \gamma_1 \varepsilon^m, \qquad \text{for some} \qquad m.$$

We must have $p \nmid m$, for otherwise the order of β modulo F^* would be less than p^r. For the same reason, we must have $\varepsilon \notin F^*$. It follows that

$$E \supseteq F(\varepsilon) \supseteq F$$

with $(F(\varepsilon):F) > 1$. But $(E:F) = p$ and $(F(\varepsilon):F)$ divides $p - 1$. This contradiction substantiates our claim.

Except possibly when $p = 2$, $r > 1$ and $i \notin F$, we have $(F(\beta):F) = p^r$, by virtue of Lemma 4.6, and thus $r = 1$. Consider the exceptional case, i.e. assume that $p = 2, r > 1$, $i \notin F$ and $(F(\beta):F) < 2^r$. Under these circumstances we know, from Lemma 4.6, that $\gamma = -4\delta^4$ for some $\delta \in F$. From $\beta^{2^r} = -4\delta^4$ we see that $i \in E$ and hence $E = F(i)$. Since this case is excluded in the hypothesis, we may therefore return to the situation where

$$\beta^p = \gamma \qquad \text{and} \qquad (F(\beta):F) = p.$$

Then we have

$$F(\beta) = E = F(\alpha).$$

Let L be generated over F by a primitive pth root of unity. Since $(L:F)$

divides $p - 1$, it follows that $E \cap L = F$, hence $(EL:L) = p$. Therefore, EL is a cyclic extension of L of degree p and we may apply Corollary 4.8. Since $EL = L(\alpha) = L(\beta)$, it follows from Corollary 4.8 that $\beta = \alpha^m \lambda$ for some m and some $\lambda \in L$. But $\lambda = \beta \alpha^{-m} \in E$ also, and hence $\lambda \in E \cap L = F$. Thus $\beta F^* \in \langle \alpha \rangle F^*$ and the result follows. ∎

4.5   Fields with free multiplicative groups modulo torsion

Let G be an abelian group and let $t(G)$ be the torsion subgroup of G. For convenience, the identity group will be regarded as a free group (of rank 0). We say that G is *free modulo torsion* if $G/t(G)$ is free. Thus G is free modulo torsion if and only if $G = t(G) \times A$, where A is a free subgroup of G.

Let E/F be a field extension. Then E/F is said to be *abelian* (*solvable*) if E/F is a Galois extension, the Galois group of which is abelian (solvable).

The problem that motivates this section is to discover conditions under which the multiplicative group F^* of a field F is free modulo torsion. A typical example of such a field F is an algebraic number field (see Theorem 1.17). More generally, if F is finitely generated (i.e. finitely generated over its prime subfield), then F^* is free modulo torsion, by virtue of Theorem 5.1 below. What can be said about infinitely generated fields? The main results of this section assert the following:

(i)   Assume that F is a field such that for every finite field extension E, E^* is free modulo torsion. If K is any field generated over F by algebraic elements, the degrees of which over F are bounded, then K^* is free modulo torsion.

(ii) Assume that F is a field such that for every finite field extension E, E^* is free modulo torsion and E contains only finitely many roots of unity. If K is any abelian field extension of F, then K^* is free modulo torsion.

We present below (see Theorems 5.12 and 5.21) an improved version of the original proof of (i) and (ii). This version was furnished by May in a private communication to the author.

4.5.1 Theorem (May 1972) *Let E/F be a finitely generated field extension and let F_1 be the algebraic closure of F in E. Then $(F_1:F) < \infty$ and there exists a free abelian group A such that $E^* \cong F_1^* \times A$. In particular, suppose that E is a finitely generated field. Then $E^* \cong W \times A$, where W is a finite cyclic group and A is a free abelian group.*

Proof.   Since E/F is finitely generated, so is F_1/F. Hence F_1/F must be finite. If E/F is algebraic, then $E = F_1$ and there is nothing to prove. We may therefore assume that the transcendence degree n of E over F is ≥ 1. If $n = 1$, then E is finite extension of $F(X)$ and the result follows by virtue of Theorem 1.21. We may choose a subfield L of E with $L \supseteq F$, and with L being a purely transcendental extension of degree 1 over F. Then E has transcendence degree $n - 1$ over L and hence, by induction, $E^* \cong L_1^* \times A$, where L is the algebraic closure of L in E and A is a free abelian group.

Now consider the chain $F \subseteq L \subseteq L_1$. Since $L \cong F(X)$, L_1 is a finite extension of $F(X)$; hence by Theorem 1.21, $L_1^* \cong L_2^* \times A_1$, where L_2 is the algebraic closure of F in L_1 and A_1 is a free abelian group. But L_2 obviously equals F_1; hence the first part of the theorem.

Suppose that E is a finitely generated field. Then we may apply the foregoing to the special case where $F = \mathbb{Q}$ or $F = \mathbb{F}_p$, the field of p elements. If $F = \mathbb{Q}$, then F_1 is an algebraic number field, and the required assertion follows by virtue of Theorem 1.17. If $F = \mathbb{F}_p$, then F_1 is a finite field and hence F_1^* is cyclic. This completes the proof. ∎

The rest of this section will be devoted to providing circumstances under which an infinitely generated field F has the property that F^* is free modulo torsion. The following observation is very useful.

4.5.2 Lemma *Let S be an arbitrary set and let $F(S)$ be a purely transcendental extension of F generated by S. Then*

$$F(S)^* \cong F^* \times A$$

for some free abelian group A.

Proof. Let $R = F[S]$ be the polynomial ring generated by S. Then R is a UFD (see Bourbaki 1965, Ch. 7). Since $F(S)$ is the quotient field of R, it follows from Lemma 1.1 that

$$F(S)^* \cong U(R) \times A,$$

where A is a free abelian group. Since $U(R) \cong F^*$, the result follows. ∎

4.5.3 Corollary *Assume that F is a field such that F^* is free modulo torsion. If E/F is an arbitrary purely transcendental extension of F, then E^* is free modulo torsion.*

Proof. The proof is a direct consequence of Lemma 5.2. ∎

4.5.4 Lemma *Let E/F be a finite field extension and let T/F^* be the torsion subgroup of E^*/F^*. If E^* is free modulo torsion, then*

$$E^* = A \times T$$

for some free abelian subgroup A of E^*.

Proof. Let N denote the norm mapping from E^* to F^*, let H denote the kernel of N, and let $n = (E:F)$. If $\lambda \in E^*$, then $N(\lambda^n N(\lambda)^{-1}) = 1$ and thus $(E^*)^n \subseteq HF^*$. If $\lambda \in H \cap T$, then $\lambda^m \in F^*$ for some positive integer m, hence
$$1 = N(\lambda^m) = \lambda^{mn}.$$
Thus we conclude that $H \cap T = t(H)$.

Let $f: E^* \to E^*/T$ be the natural homomorphism. Because E^* is free modulo torsion, it follows that $H/t(H)$ is free. But

$$f(H) = HT/T \cong H/H \cap T = H/t(H)$$

and what we have shown above implies that $(E^*/T)^n \subseteq f(H)$. Since E^*/T is torsion-free, we deduce that E^*/T is free and the result follows. ∎

The following example shows that the above result fails for infinite algebraic extensions.

4.5.5 Example (May 1980) *Let $\{x_p \mid p$ prime or $p = 0\}$ be algebraically independent transcendentals over $\mathbb{Q}$, let F be the field generated over $\mathbb{Q}$ by $\{x_p\}$, and let E be the field generated over $\mathbb{Q}$ by $\{y_p\}$, where*

$$y_0^2 = x_0 \quad and \quad y_p^p = x_p(y_0 + 1), \qquad for \ p \ prime.$$

Then $y_0 + 1 \notin T$, but
$$y_p^p \equiv y_0 + 1 (\mathrm{mod}\ T)$$

for every prime p. Hence E^*/T is not free. But, by Corollary 5.3, E^* is free modulo torsion since $\{y_p\}$ are algebraically independent transcendentals over $\mathbb{Q}$. Thus T cannot have a complement in E^* since such a complement would necessarily be free. ∎

If F is any field, let $\bar{F}$ denote $F^*/t(F^*)$ and let $\bar{\lambda}$ be the image of $\lambda \in F^*$ in $\bar{F}$. Note that if E/F is a field extension, then by identifying $F^* t(E^*)/t(E^*)$ with $\bar{F}$ we may regard $\bar{F}$ as a subgroup of $\bar{E}$.

4.5.6 Lemma *Let E/F be a field extension. Suppose that $\lambda^m = \mu$ with $\lambda \in E^*$, $\mu \in F^*$ and $m \geq 1$. Assume that m is the order of $\bar{\lambda}$ in $\bar{E}$ modulo $\bar{F}$. Then the polynomial $X^m - \mu$ is irreducible over F.*

Proof. Assume that $f(X) \in F[X]$ is a divisor of $X^m - \mu$ of degree $j \in \{1, 2, \ldots, m - 1\}$. Then all the roots of $f(X)$ (in a splitting field of f) are of the form $\lambda \delta^i$ with $i \in \{0, 1, \ldots, m - 1\}$ and with $\delta^m = 1$. Hence the product of these roots is of the form $\lambda^j \varepsilon$, where ε is a root of unity. But then $\lambda^j \varepsilon \in F$, and this would imply that $\bar{\lambda}^j \in \bar{F}$, contrary to the way in which m was chosen. ∎

The discussion has now reached a point where, in order to make further progress, we need to bring in certain group-theoretic facts.

Let B, H be subgroups of an abelian group A. Then H is said to be a *B-high subgroup* if

$$H \cap B = 1, \quad \text{and if} \quad H \subset H_1 \subseteq A \quad \text{implies} \quad H_1 \cap B \neq 1.$$

That is, H is maximal with respect to the property of having trivial intersection with B. Then, in particular, $HB = H \times B$. The existence of B-high subgroups, for every B, is guaranteed by Zorn's lemma. Furthermore, H may be chosen so as to contain a prescribed subgroup G of A with $G \cap B = 1$. Indeed, the set of all subgroups of A that contain G and have trivial intersection with B is non-empty and is inductive; thus it contains a maximal member H.

4.5.7 Lemma *Let B, C be subgroups of an abelian group A, where C is a B-high subgroup of A. Then*:

(i) $a \in A$, $a^p \in C$ (p prime) *implies* $a \in B \times C \subseteq A$;
(ii) $A = B \times C$ *if and only if* $a^p = bc$ ($a \in A$, $b \in B$, $c \in C$) *implies* $(b')^p = b$ *for some* $b' \in B$.

Proof.
(i) If $a \in C$, there is nothing to prove. If $a \notin C$, then $\langle C, a \rangle$ contains, by the choice of C, an element $1 \neq b \in B$, i.e. $b = ca^k$ for some $c \in C$ and integer k. Therefore $(k, p) = 1$ because of $a^p \in C$ and $B \cap C = 1$. It follows that $rk + sp = 1$ for some $r, s \in \mathbb{Z}$, and so

$$a = (a^k)^r (a^p)^s = (bc^{-1})^r (a^p)^s \in B \times C$$

as required.

(ii) Assume that $A = B \times C$. Then, given $a \in A$, we have $a = b'c'$ ($b' \in B$, $c' \in C$), and hence $a^p = (b')^p (c')^p = bc$ implies $(b')^p = b$. Conversely, if $a^p = bc$ implies $(b')^p = b$ for some $b' \in B$, then $a(b')^{-1}$ satisfies the hypothesis of (i), and so $a(b')^{-1} \in B \times C$, $a \in B \times C$. This shows that the guotient group $A/(B \times C)$ contains no elements of prime order, and hence is torsion-free. But if $x \in A$ is arbitrary, not in $B \times C$, then $\langle C, x \rangle$ intersects B in a non-identity element b'', $c''x^l = b''$ for some $c'' \in C$ and integer l. Moreover, $l \neq 0$ because of $B \cap C = 1$, thus $x^l = b''(c'')^{-1} \in B \times C$, and $A/(B \times C)$ is a torsion group. The conclusion is that $A = B \times C$, as asserted. ∎

4.5.8 Lemma *Let A be an abelian group and let n be a positive integer. If A^n splits over its torsion subgroup, then A splits over its torsion subgroup.*

Proof. We may harmlessly assume that $n = p$ is a prime. By hypothesis,

$A^p = t(A^p) \times B$, where B is torsion-free. Obviously we have $t(A^p) = t(A)^p$. Let G be a $t(A)$-high subgroup of A containing B. If $a \in A$ and $a^p = bc$ $(b \in t(A),\ c \in G)$, then $b \in t(A^p) = t(A)^p$. Hence, by Lemma 5.7(ii), $A = t(A) \times G$ and the result follows. ■

We now apply Lemma 5.8 to prove

4.5.9 Corollary *Let B be a subgroup of an abelian group A. If A/B is bounded and B is free modulo torsion, then A is free modulo torsion.*

Proof. By hypothesis, $A^n \subseteq B$ for some positive integer n. Since B is free modulo torsion, so is A^n. Hence A^n splits over $t(A^n)$ and, by Lemma 5.8, A splits over $t(A)$, say $A = t(A) \times F$, where F is torsion-free. Since $A^n = t(A)^n \times F^n$, we see that F^n is free. But then the map $F \to F^n$, $x \mapsto x^n$ is an isomorphism, since F is torsion-free. Hence $F \cong F^n$ is free, as asserted. ■

Let K be an extension field of F, and let τ be an ordinal. Following May (1980), we shall say that a family of subfields of K

$$\{E_\alpha \mid \alpha < \tau\}$$

forms a *continuous chain of finite extension from F to K* if the following conditions hold:

$$E_\beta \subseteq E_\alpha \qquad \text{for } \beta \leqslant \alpha < \tau,$$
$$(E_{\alpha+1} : E_\alpha) < \aleph_0 \qquad \text{for } \alpha + 1 < \tau,$$
$$E_\alpha = \bigcup_{\beta < \alpha} E_\beta \qquad \text{for } \alpha \text{ a limit ordinal,}$$

and

$$K = \bigcup_{\alpha < \tau} E_\alpha.$$

4.5.10 Lemma *Let $\{E_\alpha \mid \alpha < \tau\}$ be a continuous chain of finite extensions from F to K. For each $\alpha < \tau$, put $W_\alpha = E_\alpha^* t(K^*)$ and let T_α / E_α^* be the torsion subgroup of K^*/E_α^*. Assume that E_α^* is free modulo torsion and that T_α / W_α is bounded for every α. Then K^* is free modulo torsion.*

Proof. Using the obvious analogue of the definition above, we note that $\{T_\alpha \mid \alpha < \tau\}$ forms a continuous chain of subgroups from T_0 to K^*. (We do not require $T_{\alpha+1}/T_\alpha$ to be finite.) Since E_0^* is free modulo torsion, so is W_0. Because T_0/W_0 is bounded, it follows from Corollary 5.9 that T_0 is free modulo torsion.

Let B_0 be a free complement in T_0 for the torsion subgroup of T_0 (which in fact equals the torsion subgroup of K^*). It suffices to show that

$T_{\alpha+1}/T_\alpha$ is free for every α such that $\alpha + 1 < \tau$, for then we can choose a free abelian group $B_{\alpha+1}$ such that $T_{\alpha+1} = B_{\alpha+1} \times T_\alpha$. The group $\prod_{\alpha<\tau} B_\alpha$ is then easily seen to be a free complement in K^* for the torsion subgroup of K^*.

To show that $T_{\alpha+1}/T_\alpha$ is free, let T'_α/E^*_α be the torsion subgroup of $E^*_{\alpha+1}/E^*_\alpha$. Then

$$T_\alpha \cap E^*_{\alpha+1} = T'_\alpha$$

and Lemma 5.4 implies that $E^*_{\alpha+1}/T'_\alpha$ is free, hence $T_\alpha E^*_{\alpha+1}/T_\alpha$ is free. But $T_{\alpha+1}/T_\alpha$ is torsion-free, and $T_{\alpha+1}/T_\alpha E^*_{\alpha+1}$ is bounded since $T_{\alpha+1}/W_{\alpha+1}$ is bounded by hypothesis. Hence $T_{\alpha+1}/T_\alpha$ is free, by Corollary 5.9. The proof is thus complete. ∎

4.5.11 Lemma *Let $F \subseteq E \subseteq K$ be a chain of fields, let K be generated over F by a family of elements that are algebraic over F of degrees bounded by some positive integer n, and assume that $i \in F(i^2 = -1)$ if char $F \neq 2$. If T/E^* is the torsion subgroup of K^*/E^*, then $T/E^*t(K^*)$ is bounded.*

Proof. Enlarge K by adjoining all conjugates of the given generators. Then K is still generated by elements of degree $\leq n$, and it suffices to prove the lemma for this new K, which is then normal over F.

If σ is an automorphism of some field extension between K and F, then it extends to K by normality. Since the extension permutes conjugates of the generators, the order of σ must divide $n!$ To complete the proof, it suffices to show that every element of $T/E^*t(K^*)$ has order dividing $(n!)^2$.

Let $v \in T$ be of order p^r modulo $E^*t(K^*)$, where p is a prime. Then $v^{p^r} = u\varepsilon$ for some $u \in E^*$ and some root of unity ε. We may assume that ε has p-power order $< p^r$ by absorbing an appropriate root of unity factor into v.

First consider the case where $p = $ char K. Since K is generated over E by elements of degree $\leq n$, there is a uniform power $p^s \leq n$ such that the p^s-power of each generator is separable over E. Then every element of K^{p^s} is separable over E. If $s < r$, then v^{p^s} is separable over E and is a root of $X^{p^{r-s}} - u$. Thus $v^{p^s} \in E$, contrary to the order of v modulo $E^*t(K^*)$. Therefore $r \leq s$ and hence $p^r \mid n!$

Now we consider the case where $p \neq $ char K. Let ε_{p^s} be a primitive p^sth root of unity, and let s be the largest integer $\leq r$ such that $\varepsilon_{p^s} \in K$, and put $\eta = \varepsilon_{p^s}$. Put $d = (E(\eta, v):E)$ and let N be the norm mapping for this extension. Then $N(v)^{p^r} = u^d N(\eta)$, where $N(\eta)$ is a root of unity. We claim that $p^r \mid d$. Assume the contrary. Then

$$p^{r-1} = xp^r + yd$$

for some integers x and y. Therefore

$$u^{p^{r-1}} = (u^x)^{p^r}(N(v)^y)^{p^r}N(\eta)^{-y};$$

hence $u = w^p\eta_1$ for some $w \in E$ and some root of unity η_1. But then $v^{p^{r-1}} = w\eta_2$ for some root of unity η_2, contrary to the order of v modulo $E^*t(K^*)$. Thus $p^r \mid d$. Let t be the smallest integer $\geq (1/2)r$. Since

$$d = (E(\eta, v):E(\eta))(E(\eta):E),$$

it follows that p^t divides at least one of the two degrees. In the second case, the extension $E(\eta)/E$ is cyclic (recall that $\varepsilon_4 \in F$ if $p = 2$), and hence there is an automorphism of order p^t. The first case will require more work.

Let p^k be the order of v modulo $E(\eta)^*$. Then $v^{p^k} \notin E(\eta)^p$ since, otherwise, $\varepsilon_p \in K - E(\eta)$, contrary to the choice of η. Thus $X^{p^k} - v^{p^k}$ is irreducible over $E(\eta)$. But normality of K over $E(\eta)$ implies that $\varepsilon_{p^k} \in K$, which in turn implies that $\varepsilon_{p^k} \in E(\eta)$ by choice of η. Hence the Galois group of $E(v, \eta)/E(\eta)$ is cyclic of order p^k. But p^t divides the degree in this case, and hence there is an automorphism of order p^t in this case also. But then $p^t \mid n!$, hence $p^r \mid (n!)^2$ and the result follows. ∎

We have done most of the work to demonstrate the following remarkable result.

4.5.12   Theorem   (May 1980)   *Assume that F is a field such that for every finite field extension E, E^* is free modulo torsion. If K is any field generated over F by algebraic elements whose degrees over F are bounded, then K^* is free modulo torsion.*

Proof.   Let K and F satisfy the hypotheses of the theorem. We may add $i(i^2 = -1)$ to K if the characteristic of K is $\neq 2$, since the original K will be free modulo torsion if the enlarged one is. Now enlarge F by adding i to it. The hypothesis on F will carry over to a finite extension. We shall use induction on $(K:F)$. Clearly, we may assume that $(K:F) \geq \aleph_0$. Let τ be the first ordinal with $|\tau| = (K:F)$. If K is generated over F by elements whose degrees are bounded by n, then we may choose a family $\{x_\alpha \mid \alpha < \tau\}$ of such generators. For each $\alpha < \tau$, let E_α be generated over F by $\{x_\beta \mid \beta < \alpha\}$. Then it is clear that $\{E_\alpha \mid \alpha < \tau\}$ is a continuous chain of finite extensions from F to K. Because $(E_\alpha:F) < |\tau|$, it follows by induction that E_α^* is free modulo torsion. Lemma 5.11 now applies to show that T_α/W_α is bounded, hence Lemma 5.10 finishes the proof. ∎

4.5.13   Corollary   (May 1972)   *Let F be a finitely generated field, and assume that K is a field extension generated over F by algebraic elements of bounded degrees. Then K^* is free modulo torsion.*

Proof. The proof is a direct consequence of Theorems 5.1 and 5.12. ∎

4.5.14 Corollary *Assume that F is a field such that for every finite field extension E, E^* is the (restricted) direct product of cyclic groups. If K is any field generated over F by algebraic elements whose degrees over F are bounded, then K^* is the direct product of cyclic groups.*

Proof. By Theorem 5.12, K^* is free modulo torsion. Hence it suffices to show that there is no prime number $p \neq$ char K such that $\varepsilon_{p^n} \in K$ for every n, where ε_{p^n} is a primitive p^n-th root of unity. This is true for every finite extension of F by hypothesis, hence its failure for K would imply the existence of automorphisms of K/F or arbitrary large order. The latter being impossible, by virtue of the first paragraph in proof of Lemma 5.11, the result follows. ∎

4.5.15 Corollary (Schenkman 1964) *Let K be any field generated over $\mathbb{Q}$ by algebraic elements of bounded degree. Then K^* is the direct product of cyclic groups.*

Proof. Put $F = \mathbb{Q}$ and observe that, by Theorem 1.17, F satisfies the hypothesis of Corollary 5.14. ∎

4.5.16 Corollary (Schenkman 1964) *A multiplicative group generated by algebraic numbers of bounded degree is the direct product of cyclic groups.*

Proof. The proof is a direct consequence of Corollary 5.15 and the fact that a subgroup of the direct product of cyclic groups is the direct product of cyclic groups. ∎

4.5.17 Corollary (Čarin 1954) *Suppose that K is an algebraic extension of $\mathbb{Q}$ and that the degrees of elements of K over $\mathbb{Q}$ are bounded. Then K^* is the direct product of cyclic groups.*

Proof. This is a special case of Corollary 5.16. ∎

Returning to Theorem 5.12, we now show that it is not sufficient to assume that just F^* is free modulo torsion.

4.5.18 Theorem (May 1972) *There exists a field F and a quadratic field extension K such that the following holds:*

(i) *F^* is free modulo torsion, but K^* is not free modulo torsion;*

(ii)   *F is algebraic over $\mathbb{Q}$ and every finite field extension of F contains only finitely many roots of unity.*

Proof.   Let i denote the usual complex number. Construct a sequence $\{\alpha_n\}$ of complex numbers by letting

$$\alpha_0 = (2 + i)(2 - i)^{-1},$$

and by choosing α_n, $n \geqslant 1$, such that

$$\alpha_n^4 = \alpha_{n-1}.$$

Put $K_0 = \mathbb{Q}(i)$ and define K_n, $n \geqslant 1$, by $K_n = K_0(\alpha_n)$. Then we have an ascending chain of fields

$$K_0 \subseteq K_1 \subseteq \ldots \subseteq K_n \subseteq \ldots$$

and we define K by

$$K = \bigcup_{n=0}^{\infty} K_n.$$

Clearly, K^* is not free modulo torsion. Denote complex conjugation by σ. Let us show that σ induces an automorphism of each K_n, and hence an automorphism of K. Taking into account that $\sigma(\alpha_0) = \alpha_0^{-1}$ and $\alpha_n^{4n} = \alpha_0$, we have $(\alpha_n \sigma(\alpha_n))^{4n} = 1$. Since $\alpha_n \sigma(\alpha_n)$ is a real number, $\sigma(\alpha_n) = \pm\alpha_n^{-1}$ and therefore $\sigma(\alpha_n) \in K_n$ for every n. Thus σ induces an automorphism of K_n or order 2.

Let F (respectively, F_n) be the fixed field of σ on K (respectively, K_n). Then

$$(K : F) = (K_n : F_n) = 2 \qquad (n \geqslant 0)$$

and $F \cap K_n = F_n$. Furthermore, it is obvious that $F_{n-1} \subseteq F_n (n \geqslant 1)$ and that

$$F = \bigcup_{n=0}^{\infty} F_n.$$

Because α_0 is not a square in K_0, it follows that $X^{4^n} - \alpha_0$ is irreducible over K_0 (see Lemma 4.6(iii)). Thus $(K_n : K_0) = 4^n$, and therefore $(K_n : K_{n-1}) = 4 \ (n \geqslant 1)$. Moreover, by looking at the extension K_n/F_{n-1}, it is clear that $(F_n : F_{n-1}) = 4$.

Let the field C be obtained from $\mathbb{Q}$ by adjoining all roots of unity. It will be shown that

$$K \cap C = K_0 \qquad \text{or} \qquad K \cap C = K_0(\sqrt{5}). \tag{1}$$

Since $(K : \mathbb{Q}) = 2^\infty$ (in the sense that every finite subextension of $K/\mathbb{Q}$ has degree of power of 2 and these degrees are unbounded) the Galois theory of $C/\mathbb{Q}$ tells us that $K \cap C$ is generated over K_0 by 2^rth roots of unity for appropriate integers r, and by elements of form $\sqrt{p}$ for appropriate odd primes p.

We next examine how primes of K_0 ramify in K. To this end, we first observe that $5\alpha_n$ is an integral generator for K_n over K_0, the minimal polynomial f of which is $X^{4^n} - 5^{4^n-1}(2+i)$. Bearing in mind that

$$f'(5\alpha_n) = 4^n(5\alpha_n)^{4^n-1},$$

we conclude that the only primes of K_0 that can ramify in K are

$$1+i, 2+i, \quad \text{and} \quad 2-i.$$

However, the presence of $\sqrt{p}$ in K would imply the ramification of the Gaussian prime(s) lying above p, thus only $\sqrt{5}$ can lie in K. To examine the 2^rth roots of unity, it suffices to show that $\varepsilon_8 \notin K$, where ε_8 is a primitive 8th root of unity. Assume that contrary. Then there is an n such that $K_0 \subset \mathbb{Q}(\varepsilon_8) \subseteq K_n$. But it is clear that the primes $2+i$ and $2-i$ ramify totally from K_0 to K_n. Therefore they would also ramify from K_0 to $\mathbb{Q}(\varepsilon_8)$. This is contrary to the standard fact that every odd prime is unramified from $\mathbb{Q}$ to $\mathbb{Q}(\varepsilon_8)$. Thus (1) holds, and it follows directly from (1) that every finite extension of K contains only finitely many roots of unity.

We now show that F^* is free modulo torsion. Owing to Lemma 5.10 and Theorem 5.1, we need only verify that the torsion subgroup of $\bar{F}/\bar{F}_n$ is bounded. Assume that $v \in F^*$ and $\bar{v}$ has order m modulo $\bar{F}_n$. Then $v^m = u\varepsilon$ for some root of unity ε and some $u \in F_n^*$. Since F is contained in $\mathbb{R}$, we have $\varepsilon = \pm 1$. Therefore we may assume that $v^m = u$, in which case Lemma 5.6 guarantees that $X^m - u$ is irreducible over F_n. Because $(F:F_n) = 2^\infty$, we deduce that m is a power of 2. It will be shown that it is impossible for $m = 4$, which will imply that 2 is the exponent of the torsion subgroup of $\bar{F}/\bar{F}_n$.

Assume by way of contradiction that $m = 4$. Then $(F_n(v):F_n) = 4$ and so $(K_n(v):K_n) = 4$ since $F \cap K_n = F_n$. We argue that

$$K_n(v) = K_{n+1}. \tag{2}$$

Choose t such that K_t contains both $K_n(v)$ and K_{n+1}. To prove (2), it suffices to verify that the intermediate fields of the extension K_t/K_n are nested (hence there is exactly one of degree 4 over K_n). Let ε be a primitive 4^tth root of unity. It follows from (1) that $K \cap \mathbb{Q}(\varepsilon) = K_0$. Hence, by translation, it will suffice to show that the intermediate fields of the extension $K_t(\varepsilon)/K_n(\varepsilon)$ are nested. But $K_t(\varepsilon) = K_n(\varepsilon, \alpha_t)$, where $\alpha_t^{4^t} \in K_n(\varepsilon)$. Therefore the Galois group of this extension is cyclic of order a power of two. Hence the subgroups are nested and thus the same applies to the intermediate fields, proving (2).

We may apply Proposition 4.7 to the extension K_{n+1}/K_n since it is cyclic of degree 4. Then $u = \alpha_n^k \beta^4$, where $\beta \in K_n$, and where k is relatively prime to 4. Let N denote the norm mapping from K_n to K_0. Because u is a real number, one easily sees that $N(u)$ must also be real, hence $N(u) \in \mathbb{Q}$. We may write $N(u) = 5^s b$, where s is some integer and b is a non-zero rational number whose prime factorization does not involve 5. Now the minimal polynomial of α_n over K_0 is $X^{4^n} - \alpha_0$, and hence $N(\alpha_n) = -\alpha_0$. It follows that

$$5^s b = N(u) = N(\alpha_n)^k N(\beta)^4 = -\alpha_0^k N(\beta)^4$$

and thus

$$N(\beta)^4 = -(2+i)^{s-k}(2-i)^{s+k}b.$$

Applying the unique factorization in $\mathbb{Q}(i)$, we infer that 4 divides $s - k$ and also $s + k$. But then 2 divides k, contrary to the fact that $(k, 4) = 1$. This contradiction establishes the theorem. ∎

We now head towards our second major result which is Theorem 5.21 below.

4.5.19 Lemma Let $F \subseteq E \subseteq K$ be a chain of fields such that K/F is abelian, $i \in F (i^2 = -1)$ if char $F \neq 2$ and such that E contains only finitely many roots of unity. If T/E^* is the torsion subgroup of K^*/E^*, then $T/E^* t(K^*)$ is bounded.

Proof. Enlarge K by adding all roots of unity. This new K is an abelian extension of F, and it suffices to prove the lemma for this new K.

Let p be a prime and let $v \in T$ be of order p^r modulo $E^* t(K^*)$. Denote by ε_{p^r} a primitive p^rth root of unity. It suffices to show that $\varepsilon_{p^r} \in E$ because of the assumption on E. We have $v^{p^r} = u$, $u \in E$, where any root of unity factor on the right could be absorbed in v since K contains all roots of unity. If $u = u_1^p$ for some $u_1 \in E$, then $v^{p^{r-1}} = u_1 \varepsilon_p^j$, contrary to the order of v modulo $E^* t(K^*)$. Thus $u \notin E^p$ and hence, by Lemma 4.6, $X^{p^r} - u$ is irreducible over E.

Since K/E is separable, we have $p \neq$ char K. Since K/E is abelian, $E(v)/E$ is normal, hence $X^{p^r} - u$ splits in $E(v)$. Consequently, $\varepsilon_{p^r} \in E(v)$. Note that $\varepsilon_p \in E$ since the degree of ε_p over E is relatively prime to p. To complete the proof, we assume that $\varepsilon_{p^r} \notin E$ and derive a contradiction. The polynomial $X^{p^r} - u$ must be reducible then over $E(\varepsilon_{p^r})$, hence $u = u_1^p$ with $u_1 \in E(\varepsilon_{p^r})$. Therefore

$$v^{p^{r-1}} = u_1 \varepsilon_p^j,$$

which implies that $E(v^{p^{r-1}}) \subseteq E(\varepsilon_{p^r})$. Because $(E(v^{p^{r-1}}):E) = p$, the Galois theory of $E(\varepsilon_{p^r})/E$ implies that $E(v^{p^{r-1}}) = E(\varepsilon_{p^s})$ for some s such that $\varepsilon_{p^{s-1}} \in E$. (The Galois group of $E(\varepsilon_{p^r})/E$ is cyclic of p-power order.) By Corollary 4.8, $v^{p^{r-1}} = \varepsilon_{p^s}^k u_2$ for some k and some $u_2 \in E$. This contradicts the order of v modulo $E^* t(K^*)$, and the proof is therefore complete. ∎

4.5.20 Lemma Let $F_1 \subseteq F_2 \subseteq \ldots$ be subfields of a field K, and put $F_\infty = \bigcup_{n \geq 1} F_n$. Then there exist subfields $K_1 \subseteq K_2 \subseteq \ldots$ such that

$$K_n \cap F_\infty = F_n \quad \text{for} \quad n \geq 1 \quad \text{and} \quad K = \bigcup_{n \geq 1} K_n.$$

Proof. Applying Zorn's lemma, we may choose a maximal subfield K_1 of K such that

$$K_1 \cap F_\infty = F_1 \quad \text{and} \quad K_1 F_2 \cap F_\infty = F_2.$$

Again, invoking Zorn's lemma, we may choose a maximal subfield K_2 of K such that

$$K_2 \supseteq K_1, \quad K_2 \cap F_\infty = F_2 \quad \text{and} \quad K_2 F_3 \cap F_\infty = F_3.$$

We may therefore continue by induction to construct $K_1 \subseteq K_2 \subseteq \ldots$ such that

$$K_n \cap F_\infty = F_n, \quad K_n F_{n+1} \cap F_\infty = F_{n+1}$$

and K_n is maximal with respect to these two properties. Given $\alpha \in K$, we must show that $\alpha \in \bigcup_{n \geq 1} K_n$. Note that α cannot be transcendental over $\bigcup_{n \geq 1} K_n$, since otherwise $K_1(\alpha)$ is easily seen to contradict the maximality of K_1. Thus, there is some $m \geq 1$ such that α is algebraic over K_m. Let $n \geq m$. We may assume that $K_n(\alpha) \neq K_n$ since otherwise there is nothing to prove. Owing to the maximality of K_n, we must have either $K_n(\alpha) \cap F_\infty \supset F_n$ or $K_n(\alpha) F_{n+1} \cap F_\infty \supset F_{n+1}$. First assume that

$$K_n(\alpha) \cap F_\infty \supset F_n.$$

Choose $\beta \in K_n(\alpha) \cap F_\infty$, $\beta \notin F_n$. Then $\beta \in K_j$ for some $j > n$. Observe that $\beta \notin K_n$, hence

$$(K_n(\alpha) : K_n) > (K_n(\alpha) : K_n(\beta))$$

and therefore

$$(K_n(\alpha) : K_n) > (K_j(\alpha) : K_j).$$

Now assume that $K_n(\alpha) F_{n+1} \cap F_\infty \supset F_{n+1}$. Then

$$K_{n+1}(\alpha) \cap F_\infty \supset F_{n+1}$$

since $K_{n+1}(\alpha) \supset K_n(\alpha) F_{n+1}$. Applying the previous case, we conclude that

$$(K_n(\alpha) : K_n) \geq (K_{n+1}(\alpha) : K_{n+1}) > (K_j(\alpha) : K_j)$$

for some $j > n$. Therefore, in either case, this reduction of the degree of α implies that $\alpha \in \bigcup_{n \geq 1} K_n$. So the lemma is true. ∎

4.5.21 Theorem (May 1980) *Assume that F is a field such that for every finite field extension E, E^* is free modulo torsion and E contains*

only finitely many roots of unity. If K is any abelian field extension of F,
then K* is free modulo torsion.

Proof. Let K and F satisfy the hypotheses of the theorem. We may
enlarge both K and F by adding i $(i^2 = -1)$ (if char $K \neq 2$) since the
hypothesis on F carries over to a finite extension, and it suffices to prove
that the enlarged K^* is free modulo torsion.

We shall use induction on $(K:F)$, which we may assume is infinite.
First we shall suppose that K contains only finitely many roots of unity.
Let τ be the first ordinal with $|\tau| = (K:F)$, and choose a family of
generators $\{x_\alpha \mid \alpha < \tau\}$ of K over F. For each $\alpha < \tau$, let E_α be generated
over F by $\{x_\beta \mid \beta < \alpha\}$. Then $\{E_\alpha \mid \alpha < \tau\}$ is a continuous chain of finite
extensions from F to K. Because E_α is an abelian extension of F, and
because $(E_\alpha : F) < |\tau|$, it follows by induction that E_α^* is free modulo
torsion. Hence Lemmas 5.19 and 5.10 finish the proof in this case.

We now suppose that K may contain infinitely many roots of unity.
Denote by F_∞ the field generated over F by the roots of unity that lie in
K. The hypotheses on F imply that we may choose subfields

$$F \subseteq F_1 \subseteq F_2 \subseteq \dots$$

such that

$$\bigcup_{n \geqslant 1} F_n = F_\infty$$

and F_n contains only finitely many roots of unity for each n. Invoking
Lemma 5.20, we obtain subfields $K_1 \subseteq K_2 \subseteq \dots$ such that

$$\bigcup_{n \geqslant 1} K_n = K \quad \text{and} \quad K_n \cap F_\infty = F_n \qquad \text{for all } n \geqslant 1.$$

Put $K_0 = F$. Then it is clear that we may choose a continuous chain of
finite extensions from K_n to K_{n+1} for every $n \geqslant 0$. Hence we obtain a
continuous chain $\{E_\alpha \mid \alpha < \tau\}$ (for some ordinal τ) of finite extensions
from F to K such that, given $\alpha < \tau$, there exists $n \geqslant 1$ such that $E_\alpha \subseteq K_n$.
Therefore each E_α contains only finitely many roots of unity. Because E_α
is an abelian extension of F with $(E_\alpha : F) \leqslant (K:F)$, what we have proved
in the previous paragraph guarantees that E_α^* is free modulo torsion.
Applying Lemmas 5.19 and 5.10, the result follows. ■

4.5.22 Corollary (May 1972) *Let F be a finitely generated field and
let K be any abelian field extension of F. Then K* is free modulo torsion.*

Proof. Any finite field extension of F is again finitely generated. The
desired assertion is therefore a consequence of Theorems 5.1 and
5.21. ■

We now give an example which illustrates that in the statement of Theorem 5.21 neither the hypothesis on the commutativity of the Galois group nor the hypothesis concerning roots of unity can be dropped.

4.5.23 Example (May 1972)   *Let the field K be generated over $\mathbb{Q}$ by*

$$\{\varepsilon_p, \alpha_p \mid p \text{ a prime}\},$$

where ε_p is a primitive pth root of unity and $\alpha_p^p = 2$. Let M be the subfield of K generated over $\mathbb{Q}$ by ε_p, where p ranges over all prime numbers.

Then $K/\mathbb{Q}$ is a Galois extension, the Galois group of which is solvable, since K/M is abelian and $M/\mathbb{Q}$ is abelian. Furthermore, any finite extension of $\mathbb{Q}$ contains only finitely many roots of unity. However, K^* is clearly not free modulo torsion. Thus we cannot replace 'abelian' by 'solvable' in the statement of Theorem 5.21.

We have K/M abelian, and further, if M_1 is any finite extension of M, then M_1^* is free modulo torsion. This follows from Corollary 5.22 since M_1 is an abelian extension of a finite extension of $\mathbb{Q}$. But K^* is not free modulo torsion. This shows that we cannot drop the assumption on roots of unity. ∎

Let $F/\mathbb{Q}$ be an algebraic extension. The above example shows that F^* need not be free modulo torsion. It is therefore appropriate to ask whether a weaker property holds, namely whether F^* splits over its torsion subgroup. The following result was furnished by May in a private communication to the author.

4.5.24 Theorem   *There exists an algebraic field extension $F/\mathbb{Q}$ such that F^* does not split over its torsion subgroup.*

Proof.   Let F be generated over $\mathbb{Q}$ by $\{\varepsilon_p, \alpha_p \mid p \text{ odd prime}\}$, where ε_p is a primitive pth root of unity and $\alpha_p^p = 2\varepsilon_p$ for every odd prime p. Then F is obviously an algebraic extension of $\mathbb{Q}$. Denote by M the subfield generated over $\mathbb{Q}$ by all ε_p, p odd prime.

Assume that F^* splits over $t(F^*)$. The residue class of 2 modulo torsion has a pth root for all odd primes p. Thus some coset representative in F^* must have this property. Therefore there exists a root of unity ε such that 2ε has a pth root in F^* for all odd primes p. Now fix p not dividing the order of ε. Then 2 has a pth root in F^*. Since $2\varepsilon_p$ has a pth root in F^* (namely, α_p), we see that ε_p has a pth root, that is $\varepsilon_{p^2} \in F$.

We claim that $\alpha_p \in M(\varepsilon_{p^2})$. We may suppose $\alpha_p \notin M$. Since $\varepsilon_p \in M$, it follows that $\alpha_p^p \notin M^p$, hence $(M(\alpha_p):M) = p$, by virtue of Lemma 4.1. But by the same argument for all odd $q \neq p$, $M(\alpha_q) = M$ or else $(M(\alpha_q):M) = q$. Since in the latter case, $M(\alpha_a)/M$ is an abelian extension of degree q, the Galois group of F/M is $\coprod_{\alpha_q \notin M} \mathbb{Z}_q$. Now the only roots of unity in $\mathbb{Q}(\varepsilon_n)$ are those of orders dividing n (if n is even) or $2n$ (if n is odd). Hence

$$\varepsilon_{p^2}^p = \varepsilon_p \notin M^p$$

and so

$$(M(\varepsilon_{p^2}):M) = p.$$

Therefore the Galois theory of F/M implies $M(\varepsilon_{p^2}) = M(\alpha_p)$. Thus $\alpha_p \in M(\varepsilon_{p^2})$. But $M(\varepsilon_{p^2})/\mathbb{Q}$ is an abelian extension (it is generated by roots of unity). Thus any subextension is normal. In particular, $\mathbb{Q}(\alpha_p \varepsilon_{p^2}^{-1})/\mathbb{Q}$ is normal. However,

$$(\alpha_p \varepsilon_{p^2}^{-1})^p = \alpha_p^p \varepsilon_p^{-1} = 2\varepsilon_p \varepsilon_p^{-1} = 2$$

and thus

$$\alpha_p \varepsilon_{p^2}^{-1} = \sqrt[p]{2}.$$

On the other hand, $\mathbb{Q}(\sqrt[p]{2})/\mathbb{Q}$ is not a normal extension. Indeed, it has depree p, but normality would imply $\varepsilon_p \in \mathbb{Q}(\sqrt[p]{2})$, since $\sqrt[p]{2}\,\varepsilon_p$ is a conjugate of $\sqrt[p]{2}$. This cannot occur since ε_p has degree $p-1$ over $\mathbb{Q}$, which provides the desired contradiction. ∎

We close by remarking that P. M. Cohn (1962) was the first person to discover a field, the multiplicative group of which does not split over its torsion subgroup. Cohn's example utilizes his embedding theorem, and thus introduces transcendentals.

4.6   Embedding groups

Throughout this section, all groups are assumed to be *abelian* and, for any given group G, $t(G)$ denotes the torsion subgroup of G. We say that G is *locally cyclic* if every finitely generated subgroup of G is cyclic. By Proposition 1.11, if G is contained in the multiplicative group of a field, then $t(G)$ is locally cyclic. Owing to P. M. Cohn (1962), the converse is also true. In fact, Cohn proved that if $t(G)$ is locally cyclic, then G can be embedded in a special manner as a subgroup of the multiplicative group of a field. Subsequently, May (1972) provided a refinement of Cohn's theorem which we present below.

We begin by establishing some group-theoretic results.

4.6.1 Lemma   *Every group can be embedded as a subgroup of a divisible group.*

Proof.   The infinite cyclic group $\mathbb{Z}$ can clearly be embedded in a divisible group, namely $\mathbb{Q}$. Hence every free group is embeddable in a divisible group. Given an arbitrary group G, we may write $G \cong F/N$ for a suitable free group F and a subgroup N of F. If we embed F in a divisible group D, then G will be isomorphic to the subgroup F/N of the divisible group D/N. ∎

4.6.2   Lemma   *Let G_1 be a group. Then there exists a group G such that $G_1 \subseteq G$, $t(G_1) = t(G)$, $G/t(G)$ is divisible and $G_1/t(G)$ contains a basis for $G/t(G)$.*

Proof.   Suppose A is torsion-free with $A \subseteq D$, where D is divisible. We claim that there exists a torsion-free divisible group D_1 with $A \subseteq D_1 \subseteq D$ and D_1/A is torsion. Indeed, $t(D)$ is divisible and $A \cap t(D) = 1$. The projection $A \times t(D) \to t(D)$ extends to a homomorphism $D \to t(D)$ since $t(D)$ is divisible. Let X be the kernel of $D \to t(D)$. Then $A \subseteq X$. Let D_1 be the group with $A \subseteq D_1 \subseteq X$ such that D_1/A is the torsion subgroup of X/A. Then D_1 is divisible and D_1 is torsion-free since $D_1 \cap t(D) = 1$, as claimed.

By Lemma 6.1, we have $G_1 \subseteq D$, where D is divisible. Then $G_1/t(G_1) \subseteq D/t(G_1)$, where $D/t(G_1)$ is divisible. Since $G_1/t(G_1)$ is torsion-free, the preceding paragraph implies the existence of D_1 such that $G_1 \subseteq D_1 \subseteq D$ and such that $D_1/t(G_1)$ is torsion-free divisible and D_1/G_1 is torsion. Put $G = D_1$. Then $t(G) = t(G_1)$ and $G/t(G)$ is divisible. If $\{x_i\}$ is a basis for $G/t(G)$, then since G/G_1 is torsion, for any $i \in I$ there exists $m_i > 0$ such that $\{x_i^{m_i}\}$ is a basis for $G_1/t(G)$. ∎

We are now ready to prove

4.6.3   Theorem   (May 1972)   *Let G be an abelian group such that $t(G)$ is locally cyclic. Then there is a field L and a group H such that $L^* \cong G \times H$, where H is a free abelian group if $t(G)$ has a non-trivial 2-component, and H is the direct product of a free abelian group with a cyclic group of order 2 if $t(G)$ has a trivial 2-component.*

Proof.   If $t(G)$ has a trivial 2-component, we may replace G by the direct product of G with a cyclic group of order 2. Therefore we may harmlessly assume that $t(G)$ has a non-trivial 2-component. For the sake of clarity, we divide the rest of the proof into three steps. The first two steps treat the special case where $G/t(G)$ is divisible. The third step utilizes the construction given in the previous steps to complete the proof in the general case.

Step 1:   assume that $G/t(G)$ is divisible. Then, by Theorem 1.6(i), there exists a well ordered set I and an inner direct product decomposition

$$G/t(G) = \coprod_{\alpha \in I} Q_\alpha,$$

where each Q_α is isomorphic to the additive group of $\mathbb{Q}$. Given $\alpha \in I$, let G_α denote the inverse image modulo $t(G)$ of $\coprod_{\beta \leqslant \alpha} Q_\beta$. Then

$$G = \bigcup_{\alpha \in I} G_\alpha.$$

Now fix a sequence of positive integers $n_1, n_2, \ldots$ such that $n_1 = 1$, $n_i \mid n_{i+1}$ for every i and such that any given positive integer divides some n_i. For every $\alpha \in I$, choose an identification of Q_α with the additive group of $\mathbb{Q}$. Then Q_α is generated by $\{n_i^{-1} \mid i \geqslant 1\}$. Let $g_{\alpha,i} \in G_\alpha$ be such that $g_{\alpha,i}$ reduces modulo $t(G)$ to n_i^{-1} in Q_α. Our choice of $g_{\alpha,i}$ then ensures that G is generated by $t(G)$ and $\{g_{\alpha,i} \mid \alpha \in I, i \geqslant 1\}$, subject to relations of the form

$$g_{\alpha,i+1}^{m_{i+1}} = g_{\alpha,i} u_{\alpha,i},$$

where $u_{\alpha,i} \in t(G)$ and $m_{i+1} = n_{i+1}/n_i$. Owing to Proposition 1.13, we may choose a cyclotomic extension of $\mathbb{Q}$, say K, such that $t(K^*) \cong t(G)$. Fix an isomorphism of $t(K^*)$ with $t(G)$, and let $\varepsilon_{\alpha,i}$ be the root of unity in K that corresponds to $u_{\alpha,i}$. Next choose elements $t_{\alpha,1}$ $(\alpha \in I)$ that are algebraically independent over K. For each $\alpha \in I$, inductively choose elements $t_{\alpha,i}$ $(i \geqslant 2)$ such that

$$t_{\alpha,i}^{m_i} = t_{\alpha,i-1} \varepsilon_{\alpha,i-1}.$$

If $j \geqslant 1$, then $t_{\alpha,j}^{n_j} = t_{\alpha,1} \eta_{\alpha,j}$ for some root of unity $\eta_{\alpha,j}$. Therefore, if we choose a positive integer i_α for each α, then $\{t_{\alpha,i_\alpha} \mid \alpha \in I\}$ is obviously algebraically independent over K. Put

$$L = K(\{t_{\alpha,i} \mid \alpha \in I, i \geqslant 1\})$$

and observe that the isomorphism of $t(G)$ with $t(K^*)$ may be extended to a homomorphism $\phi: G \to L^*$. Since ϕ is injective on $t(G)$ and on a basis $\{g_{\alpha,1} \mid \alpha \in I\}$ of G modulo $t(G)$, we see that ϕ is injective. We may therefore regard G as a subgroup of L^*, where G is generated by $\{\varepsilon_{\alpha,i} t_{\alpha,i} \mid \alpha \in I, i \geqslant 1\}$. Note that

$$L = K(G) = \bigcup_{\alpha \in I} K(G_\alpha).$$

Step 2:   keeping the assumption that $G/t(G)$ is divisible, we now prove that L^*/G is free. To this end, note that $G_\alpha \subseteq K(G_\alpha)^* \cap G$. To prove the opposite containment, fix $x \in K(G_\alpha)^* \cap G$. Then $x = \omega\varepsilon$, where ε is a root of unity, and where ω is a word in $\{t_{\beta,i_\beta} \mid \beta \in I\}$ for a suitable choice i_β for every β. Put

$$S = \{t_{\beta,i_\beta} \mid \beta \leqslant \alpha\} \qquad \text{and} \qquad S' = \{t_{\beta,i_\beta} \mid \beta > \alpha\}.$$

Then S' is algebraically independent over $K(S)$. Because each $t_{\beta,j}$ ($\beta \le \alpha$, $j \ge 1$) is algebraic over $K(S)$, $K(G_\alpha)$ is algebraic over $K(S)$ and thus S' is algebraically independent over $K(G_\alpha)$. The conclusion is that ω is a word in S since $\omega \in K(G_\alpha)$ and so $x \in G_\alpha$, proving that $K(G_\alpha)^* \cap G = G_\alpha$. We may therefore regard $\{K(G_\alpha)^*/G_\alpha \mid \alpha \in I\}$ as a linearly ordered family of subgroups of L^*/G. Let us construct a free basis for L^*/G by induction over I. At the initial step, we have $K^*/t(G)$. Since K is an abelian field extension of $\mathbb{Q}$, Theorem 5.21 guarantees that $K^*/t(G)$ is free. Choose a free basis for $K^*/t(G)$, fix $\alpha \in I$ and put $H_\alpha = \cup_{\beta<\alpha} G_\beta$. Then

$$(K(H_\alpha)^*/H_\alpha) = \bigcup_{\beta<\alpha}(K(G_\beta)^*/G_\beta).$$

At the inductive step, we may assume that we have constructed bases for

$$K(G_\beta)^*/G_\beta \qquad (\beta < \alpha)$$

such that their union gives a basis for $K(H_\alpha)^*/H_\alpha$. We are left to verify that this basis can be extended to a basis for $K(G_\alpha)^*/G_\alpha$. To this end, put $H_{\alpha,i} = \langle H_\alpha, t_{\alpha,i} \rangle$, $i \ge 1$. Then

$$H_{\alpha,1} \subseteq H_{\alpha,2} \subseteq \ldots, \quad G_\alpha = \bigcup_{i=1}^{\infty} H_{\alpha,i}$$

and

$$K(G_\alpha) = \bigcup_{i=1}^{\infty} K(H_{\alpha,i}).$$

Setting

$$R_i = K(H_\alpha)[t_{\alpha,i}]$$

we see that R_i is a principal ideal domain with quotient field $K(H_{\alpha,i})$, since $t_{\alpha,i}$ is transcendental over $K(H_\alpha)$. Because $t_{\alpha,i+1}$ is integral over R_i and R_{i+1} is integrally closed, it follows that R_{i+1} is the integral closure of R_i in $K(H_{\alpha,i+1})$. Let D_i be the group of fractional ideals of R_i. Then the inclusion $R_i \subseteq R_{i+1}$ induces an injection $D_i \to D_{i+1}$ for every i. We denote by D the direct limit of $\{D_i \mid i \ge 1\}$ under these mappings. Let $\phi : K(G_\alpha)^* \to D$ be a homomorphism induced by the natural mappings $K(H_{\alpha,i})^* \to D_i$, $i \ge 1$. Then ϕ induces a homomorphism

$$\psi : K(G_\alpha)^*/G_\alpha \to D/\phi(G_\alpha).$$

The kernel of ϕ is precisely $K(H_\alpha)^*$, hence $\operatorname{Ker} \psi = (K(H_\alpha)^* G_\alpha)/G_\alpha$. But

$$K(H_\alpha)^* \cap G_\alpha = H_\alpha;$$

hence $\operatorname{Ker} \psi$ may be identified with $K(H_\alpha)^*/H_\alpha$. We claim that $D/\phi(G_\alpha)$ is free; if sustained, it will follow that the free basis of $K(H_\alpha)^*/H_\alpha$ could be extended to a free basis of $K(G_\alpha)^*/G_\alpha$. To substantiate our claim, note that the image of $H_{\alpha,i}$ in D_i is the subgroup T_i generated by the prime ideal $\langle t_{\alpha,i} \rangle$. In D_i we have $\langle t_{\alpha,i} \rangle = \langle t_{\alpha,i+1} \rangle^{m_{i+1}}$, so we obtain an injective mapping

$$\psi_i : D_i/T_i \to D_{i+1}/T_{i+1} \qquad \text{for every } i.$$

Because $G_\alpha = \bigcup_{i=1}^{\infty} H_{\alpha,i}$, we have $D/\phi(G_\alpha) = \bigcup_{i=1}^{\infty} (D_i/T_i)$. Each D_i/T_i is free on the prime ideals distinct from $\langle t_{\alpha,i} \rangle$; thus to prove that $D/\phi(G_\alpha)$ is free, we need only show that the mapping ψ_i splits. Choose a prime ideal P of R distinct from $\langle t_{\alpha,i} \rangle$, and let $P_1, P_2, \ldots$ be the prime ideals of R_{i+1} extending P. Note that $R_{i+1} = R_i[t_{\alpha,i+1}]$ and that $f(t_{\alpha,i+1}) = 0$, where $f = X^{m_{i+1}} - t_{\alpha,i}\varepsilon_{\alpha,i}$. Because $f'(t_{\alpha,i+1}) \in \langle t_{\alpha,i+1} \rangle$, it follows that $P_1, P_2, \ldots$ are all unramified over P. Consequently, $P = P_1 P_2 \ldots$ is in D_{i+1}. Define $\theta_i : D_{i+1}/T_{i+1} \to D_i/T_i$ by $\theta_i(P_1) = P$, $\theta_i(P_j) = 1$ ($j \geqslant 2$). Then $\theta_i \circ \psi_i$ is the identity map on D_i/T_i, and hence ψ_i splits as desired.

Step 3: we now consider the case of a group G_1 such that $t(G_1)$ is locally cyclic. Owing to Lemma 6.2, there exists an abelian group G such that

$$G_1 \subseteq G, \qquad t(G_1) = t(G), \qquad G/t(G) \qquad \text{is divisible}$$

and such that

$$G_1/t(G) \qquad \text{contains a basis for} \qquad G/t(G).$$

Note that the last condition ensures that G/G_1 is a torsion group. Let L be constructed from G as before, and let $L_1 = K(G_1)$. To complete the proof, it suffices to show that L_1^*/G_1 is free. The latter will follow provided that we prove that $L_1 \cap G = G_1$ since L^*/G is free. The inclusion $G_1 \subseteq L_1 \cap G$ is trivial; hence assume that $x \in L_1 \cap G$. There exist distinct indices $\alpha_1, \ldots, \alpha_k$, and positive integers $l_1, \ldots l_k$ such that, if H is a subgroup generated by $t(G)$ and by the elements $t_{\alpha_i, l_i}, \ldots, t_{\alpha_k, l_k}$, then $x \in H$, and also $x \in K(H \cap G_1)$. Observe that the transcendence degree of $K(H)$ over K is k. Because G/G_1 is a torsion group, there exists a positive integer r such that $H^r \subseteq H \cap G_1$. However, H is free modulo $t(G)$ of rank k, and the same holds for H^r, thus $H \cap G_1$ is free modulo $t(G)$ of rank k. Hence

$$H \cap G_1 = \langle t(G), h_1, \ldots, h_k \rangle \qquad (h_i \in H)$$

and therefore $K(H \cap G_1) = K(h_1, \ldots, h_k)$. But $K(H)$ is an algebraic field extension of $K(H \cap G_1)$ since $H^r \subseteq H \cap G_1$. Thus, k is also the transcendence degree of $K(H \cap G_1)$ over K, from which it follows that $h_1, \ldots, h_k$ are algebraically independent over K. Because $x^r \in H \cap G_1$, we have $x^r = \omega\varepsilon$, where ε is a root of unity and ω is a word in $h_1, \ldots, h_k$. Bearing in mind that $x \in K(h_1, \ldots, h_k)$, it follows by unique factorization that x is a product of a root of unity and word in $h_1, \ldots, h_k$. This shows that $x \in G_1$ and completes the proof. ∎

We close by remarking that the following problem is still wide open:

Problem. Let G be a group whose torsion subgroup is locally cyclic. What are necessary and sufficient conditions for G to be isomorphic to the multiplicative group of a field?

4.7 Multiplicative groups under field extensions

Let E/F be a field extension. Our aim is to examine the change in multiplicative groups in going from F^* to E^*. The problem of relating the group-theoretic structure of E^* to that of F^* is quite complicated, even when F^* has particularly simple structure and the extension is quadratic. The simple structure that we shall consider for F^* is the direct product of a finite cyclic group and a free abelian group. What can be said about the structure of E^*? First let us note a trivial fact. If $(E:F) = n < \infty$ and A is a free direct factor of F^*, then E^* has a free direct factor isomorphic to A. Indeed, let ψ be the composite

$$E^* \to F^* \to A$$

of the norm followed by the projection map. Then E^* has a free direct factor isomorphic to $\psi(E^*)$. But the image of the norm map contains $(F^*)^n$, and hence $\psi(E^*) \cong A$. Thus, to relate the structure of F^* to that of E^*, we must compare the sizes of complementary factors. It will be shown that even in the case where E is a quadratic extension of F, the complementary factor for E^* may be essentially as arbitrary as possible for a subgroup of the multiplicative group of a field.

4.7.1 Theorem (May 1979a) *Let G be any abelian group, the torsion subgroup of which is locally cyclic with non-trivial 2-component. Then there exist a field K and a quadraic field extension L such that $K^* \cong \mathbb{Z}_2 \times A$ for some free abelian group A, while $L^* \cong G \times B$ for some free abelian group B.*

Proof. Let L be the field constructed in the proof of Theorem 6.3. Then G is a subgroup of L^*, $L = \mathbb{Q}(G)$, L^*/G is a free abelian group, and there exists a torsion-free basis for G consisting of elements that are algebraically independent over $\mathbb{Q}(t(G)$ (called K in the proof of Theorem 6.3).

Let H be any finitely generated subgroup of G and write

$$H = \langle \varepsilon \rangle \times \langle h_1 \rangle \times \ldots \times \langle h_n \rangle,$$

where ε is a root of unity and $h_1, \ldots, h_n$ are torsion-free elements of G. We assert that $h_1, \ldots, h_n$ are algebraically independent over $\mathbb{Q}(\varepsilon)$. Indeed, by the remark on the torsion-free basis for G, there exist elements $g_1, \ldots, g_m \in G$ that are algebraically independent over $\mathbb{Q}(\varepsilon)$, and a positive integer k such that $\langle h_1^k, \ldots, h_n^k \rangle \subseteq \langle g_1, \ldots, g_m \rangle$. But any basis for $\langle g_1, \ldots, g_m \rangle$ is clearly algebraically independent. Applying the stacked basis theorem (see Lang 1965; Theorem 5, p. 393), it therefore follows that $\langle h_1^k, \ldots, h_n^k \rangle$ has a basis that is algebraically independent. The conclusion is that $h_1, \ldots, h_n$ are algebraically independent over $\mathbb{Q}(\varepsilon)$.

It follows from the above that the automorphism of $\mathbb{Q}(\varepsilon)$ sending ε to ε^{-1} can be extended to a unique automorphism σ_H of $\mathbb{Q}(H)$ such that $\sigma_H(h) = h^{-1}$ for any $h \in H$. If $H_1 \subseteq H_2$ are two finitely generated subgroups of G, then σ_{H_1} is the restriction of σ_{H_2} to $\mathbb{Q}(H_1)$. Let $\sigma = \bigcup \sigma_H$, where the union is taken over all finitely generated subgroups H of G. Then σ is an automorphism of L such that $\sigma(g) = g^{-1}$ for all $g \in G$. Denote by K the fixed field of σ. Since σ has order 2, we have $(L:K) = 2$. Furthermore, $K^* \cap G = \langle -1 \rangle$ and so $K^*/\langle -1 \rangle$ is isomorphic to a subgroup of the free abelian group A and $L^* \cong G \times B$ for a suitable free abelian group B. ∎

Let p be a prime and n a non-zero integer. The symbol (n/p), defined by

$$\left(\frac{n}{p}\right) = \begin{cases} 1 & \text{if } n \equiv x^2 \ (\mathrm{mod}\, p), \\ -1 & \text{if } n \not\equiv x^2 \ (\mathrm{mod}\, p), \end{cases}$$

is known as the quadratic symbol, and depends only on the residue class of n modulo p.

4.7.2 Lemma   *Let p be a prime, let ε be a primitive pth root of unity, and let*

$$a = \sum_n \left(\frac{n}{p}\right) \varepsilon^n,$$

the sum being taken over non-zero residue classes modulo p. Then

$$a^2 = \left(\frac{-1}{p}\right) p$$

and, in particular,

$$\mathbb{Q}(\sqrt{p}) \subseteq \mathbb{Q}(\varepsilon, i).$$

Proof.   We have

$$a^2 \sum_{n,m} \left(\frac{n}{p}\right)\left(\frac{m}{p}\right) \varepsilon^{n+m} = \sum_{n,m} \left(\frac{nm}{p}\right) \varepsilon^{n+m}.$$

If n ranges over non-zero residue classes, so does nm for any fixed m, and thus replacing n by nm gives

$$a^2 = \sum_{n,m} \left(\frac{nm^2}{p}\right) \varepsilon^{m(n+1)} = \sum_{n,m} \left(\frac{n}{p}\right) \varepsilon^{m(n+1)}$$

$$= \sum_m \left(\frac{-1}{p}\right) \varepsilon^0 + \sum_{n \neq -1} \left(\frac{n}{p}\right) \sum_m \varepsilon^{m(n+1)}.$$

However, $1 + \varepsilon + \ldots + \varepsilon^{p-1} = 0$, and the sum on the right over m therefore yields -1. Accordingly,

$$a^2 = \left(\frac{-1}{p}\right)(p - 1) + (-1) \sum_{n \neq -1} \left(\frac{n}{p}\right)$$

$$= p\left(\frac{-1}{p}\right) - \sum_n \left(\frac{n}{p}\right)$$

$$= p\left(\frac{-1}{p}\right)$$

as asserted. ∎

Let G be an abelian group and let p be a prime. Given $g \in G$, the greatest non-negative integer r for which $x^{p^r} = g$ is solvable for some $x \in G$, is called the *p-height* (or simply the *height*) of g. If $x^{p^r} = g$ is solvable for all r, we say that g is of *infinite p-height*. By the *circle group* we mean the unit circle in the complex numbers

$$\{\alpha \in \mathbb{C} \mid |\alpha| = 1\}.$$

4.7.3 Theorem (May 1979a) *Let G be any countable torsion-free abelian group. Then there exist a field K, algebraic over $\mathbb{Q}$, and a quadratic field extension L such that $K^* \cong \mathbb{Z}_2 \times A$, for some free abelian group A, while $L^* \cong \mathbb{Z}_4 \times G \times B$ for some free abelian group B.*

Proof. We shall construct L as a subfield of $\mathbb{C}$. By Dirichlet's theorem on primes in arithmetic progressions, there are infinitely many primes p with $p \equiv 1 \pmod{12}$. Because such primes also satisfy $p \equiv 1 \pmod 4$, there is a factorization $p = (a + bi)(a - bi)$, where $a \pm bi$ are non-associated primes in the Gaussian integers. Observe further that

$$(a + bi)(a - bi)^{-1} \tag{1}$$

lies on the unit circle, and that these elements for various p are independent free generators. Let G_0 be the subgroup of G generated by some basis for G. Because G_0 is of countable rank, we may assume that G_0 is generated by a subset of the elements given above. Bearing in mind that the circle group is divisible, we may in fact assume that G is realized as a subgroup of the unit circle. Let us choose a chain of subgroups

$$G_0 \subseteq G_1 \subseteq G_2 \subseteq \ldots$$

such that

$$(G_{n+1} : G_n) = p_n$$

for some prime p_n $(n \geqslant 0)$ and such that

$$\bigcup_n G_n = G.$$

Given $n \geqslant 0$, choose $\alpha_n \in G_{n+1}$ with $G_{n+1} = \langle G_n, \alpha_n \rangle$. Then $\alpha_n \notin G_n$ and $\alpha_n^{p_n} \in G_n$. Now define

$$L_0 = \mathbb{Q}(i), \qquad L_n = L_0(G_n), \quad n \geqslant 1, \qquad \text{and} \qquad L = L_0(G).$$

Then we must have

$$L_{n+1} = L_n(\alpha_n) \qquad \text{and} \qquad \bigcup_n L_n = L.$$

Let the field C be obtained from $\mathbb{Q}$ by adjoining all roots of unity. It will be shown, by induction on n, that there exist free groups $B_0, B_1, \ldots$ such that

$$L_n^* = \langle i \rangle \times G_n \times \prod_{j \leqslant n} B_j \tag{2}$$

and such that

$$((L_n \cap C) : L_0) \text{ is a power of 2.} \tag{3}$$

By the unique factorization in the Gaussian integers, the elements in (1) used earlier to give a basis for G_0 are part of a free basis for $L_0^* / \langle i \rangle$. In fact, it is immediate that there exists a free group B_0 such that $L_0^* = \langle i \rangle \times G_0 \times B_0$, and such that all odd primes $p \not\equiv 1 \pmod{12}$ form part of a set of free generators for B_0. Since $L_0 \cap C = L_0$, the initial step of the induction is completed.

   Now assume that (2) and (3) hold for n. We shall write $\alpha = \alpha_n$, $p = p_n$, and $\alpha^p = \beta$. Because G is torsion-free, β has p-height 0 in G_n, hence $\beta \notin L_n^p$, and therefore $(L_{n+1} : L_n) = p$. We claim that $((L_{n+1} \cap C) : L_0)$ is a power of 2, and that $t(L_{n+1}^*) = \langle i \rangle$. Assume that $((L_{n+1} \cap C) : (L_n \cap C)) = m > 1$. This is a normal extension (being contained in C), so $L_n(L_{n+1} \cap C) \supseteq L_n$ is a normal extension of degree m. Because the extension is contained in L_{n+1}, we deduce that $m = p$, and that L_{n+1} is a normal extension of L_n. Now the polynomial $X^p - \beta$ splits in L_{n+1}, so L_{n+1} contains a primitive pth root of unity. This root of unity must therefore lie in L_n since its degree over L_n divides $p - 1$. Invoking the decomposition of L_n^*, we infer that $p = 2$, which shows that $((L_{n+1} \cap C) : L_0)$ is a power of 2. To prove that $t(L_{n+1}^*) = \langle i \rangle$, first assume that $\varepsilon \in t(L_{n+1}^*)$ is a primitive qth root of unity for q an odd prime. Then

$\mathbb{Q}(\varepsilon) \subseteq L_{n+1} \cap C$ and hence, by what we have shown above, $q - 1$ is a power of 2. Therefore $q \not\equiv 1 \pmod{12}$. Furthermore, by Lemma 7.2, $\sqrt{q} \in \mathbb{Q}(i, \varepsilon)$ and thus $\sqrt{q} \in L_{n+1}$. By the way B_0 was chosen, q has height 0 in B_0, and hence height 0 in L_n^*. This shows that $(L_n(\sqrt{q}) : L_n) = 2$, so $p = 2$ and

$$L_n(\sqrt{q}) = L_{n+1} = L_n(\alpha).$$

Hence, by Proposition 4.7, $\sqrt{q} = \alpha\gamma$ for some $\gamma \in L_n^*$, which in turn implies that $q = \beta\gamma^2$, where $\beta \in G_n$. However, this is impossible since q has height 0 in L_n^*/G_n. The conclusion is that $t(L_{n+1}^*)$ has trivial q-component for odd primes q. Now assume that a primitive 8th root of unity ε_8 lies in $t(L_{n+1}^*)$. Because $\varepsilon_8 \notin L_n$, we must have $p = 2$ and $L_{n+1} = L_n(\varepsilon_8)$. Again, by Proposition 4.7, we have $\varepsilon_8 = \alpha\gamma$, and hence $1 = \beta^4\gamma^8$. Taking into account that G is torsion-free and the 2-height of β in G_n is 0, we conclude that the 2-height of β^4 in G_n is 2. Hence the 2-height of β^4 in L_n^* is 2. However, the 2-height of γ^8 in L_n^* is at least 3. This contradiction shows that $\varepsilon_8 \notin t(L_{n+1}^*)$. Therefore $t(L_{n+1}^*) = \langle i \rangle$, thus substantiating our claim.

It is now an easy matter to prove the remaining part of the induction step. Indeed, by Theorem 4.9 and by what we have just shown, we have

$$t(L_{n+1}^*/L_n^*) = G_{n+1}L_n^*/L_n^*.$$

Invoking Theorem 3.8, we therefore deduce that $L_{n+1}^*/G_{n+1}L_n^*$ is free. Hence

$$L_{n+1}^* = (G_{n+1}L_n^*) \times B_{n+1}$$

for some free group B_{n+1}. Consider the group G_{n+1}/L_n^*. Because G_{n+1} is torsion-free and G_{n+1}/G_n is torsion, we have

$$G_{n+1} \cap \left(\langle i \rangle \times \prod_{j \leq n} B_j \right) = 1,$$

and thus

$$G_{n+1}L_n^* = \langle i \rangle \times G_{n+1} \times \prod_{j \leq n} B_j.$$

It follows that

$$L_{n+1}^* = \langle i \rangle \times G_{n+1} \times \prod_{j \leq n+1} B_j,$$

and the induction is finished.

Finally, define $B = \bigoplus_{j < \omega} B_j$. Then B is free and it is clear that

$$L^* = \langle i \rangle \times G \times B.$$

As in the proof of Theorem 7.1, L is closed under complex conjugation because G is contained in the unit circle and $L = L_0(G)$. Put

$$K = L \cap \mathbb{R}$$

so that $(L:K) = 2$. Bearing in mind that $K^* \cap G = 1$, it follows that the projection

$$\langle i \rangle \times G \times B \to \langle i \rangle \times B$$

is injective on K^*. Hence

$$K^* \cong \mathbb{Z}_2 \times A$$

for some free group A, thus completing the proof. ∎

5
Multiplicative groups of division rings

In this chapter we present a number of results pertaining to the multiplicative structure of division rings. After proving Wedderburn's theorem and its generalization due to Kaplansky, we concentrate on the study of subnormal subgroups. Among other results, it is shown that every subnormal solvable subgroup of the multiplicative group of a division ring is necessarily central. In particular, for any division ring D, all non-abelian subnormal subgroups of D^* are not solvable. Turning to multiplicative commutators, we then prove Herstein's theorem which asserts that if all $[x, y]$, $x, y \in D^*$, are of finite order, then D is commutative. Our next result shows that all periodic subnormal subgroups of D^* are contained in the centre of D. Finally, the following conjecture is investigated: a non-central subnormal subgroup of the multiplicative group of a division ring contains a non-cyclic free subgroup. Among other results, we prove Lichtman's theorem, which asserts that if G is a subnormal (originally stated normal) subgroup of D^* and G has a non-abelian nilpotent-by-finite subgroup, then G contains a non-cyclic free subgroup.

5.1 Commutativity conditions

Let D be a division ring, i.e. a ring in which every non-zero element is invertible. In what follows we denote by $Z(D)$ the centre of D and by D^* the multiplicative group of D. Our aim in this section is to provide circumstances under which D is commutative (equivalently, under which D^* is an abelian group). Our point of departure is the following celebrated theorem of Wedderburn (1905).

5.1.1 Theorem   *Every finite division ring is commutative.*

Proof. (Witt 1931). Let D be a finite division ring. Then the centre $Z(D)$ of D is a finite field of, say, q elements. If n denotes the dimension of the vector space D over the field $Z(D)$, then we must have

$$|D| = q^n \qquad \text{and} \qquad |D^*| = q^n - 1.$$

Note that if m is a positive integer and $q^m - 1$ divides $q^n - 1$, then m divides n. Indeed, we can write $n = ma + b$ with $0 \leq b < m$, in which case

$$q^n = q^{ma}q^b \equiv q^b (\mathrm{mod}(q^m - 1)).$$

Hence $q^m - 1 \mid q^n - 1$ implies $b = 0$, as asserted.

Assume by way of contradiction that $D \neq Z(D)$ and choose $\alpha \in D - Z(D)$. If C is the centralizer of α, then C is a proper division subring of D, and hence is of order q^d for some $d < n$. Hence the group C^* has order $q^d - 1$ and since it is a subgroup of D^*, we must have $q^d - 1 \mid q^n - 1$. By the preceding paragraph, we conclude that $d \mid n$. Now the conjugacy class of α in D^* consists of $(q^n - 1)/(q^d - 1)$ elements, by virtue of the fact that the number of elements in a conjugacy class of a group is the index of the centralizer of any element in this conjugacy class. These conjugacy classes exhaust all elements of D^* outside the centre, while its centre has order $q - 1$. Since D^* is a disjoint union of conjugacy classes, we obtain the equation

$$q^n - 1 = q - 1 + \sum \frac{q^n - 1}{q^d - 1}, \tag{1}$$

where the sum on the right is over various proper divisors d of n (not necessarily distinct). Consider the cyclotomic polynomial $\phi_n(X)$; it is a factor of $(X^n - 1)/(X^d - 1)$ for every proper factor d of n. Therefore the integer $r = \phi_n(q)$ divides each term in the sum on the right of (1), as well as the left-hand side. Thus

$$r \mid q - 1. \tag{2}$$

However, $r = \phi_n(q) = \prod (q - \varepsilon)$, where ε runs over the primitive nth roots of unity. Taking these roots to be in $\mathbb{C}$, it follows that for $n > 1$, $|q - \varepsilon| > q - 1$ and hence $r > q - 1$, contrary to (2). ∎

As an easy application of Theorem 1.1, we prove

5.1.2 Corollary *Let D be a division ring of non-zero characteristic. Then all finite subgroups of D^* are cyclic.*

Proof. Let F be the prime subfield of $Z(D)$. Given a finite subgroup of G of D^*, put $D_1 = \{\sum \lambda_i g_i \mid \lambda_i \in F, g_i \in G\}$. Then D_1 is obviously a finite subring of D. Hence D_1 is a finite division ring and, by Theorem 1.1, D_1 is a field. Since $G \subseteq D_1$ and D_1^* is cyclic, the result follows. ∎

For future application, we next record the following two subsidiary results.

5.1.3 **Lemma** *Let D be a division ring of characteristic $p \neq 0$. Assume that $\lambda \in D - Z(D)$ is such that $\lambda^{p^s} = \lambda$ for some $s \geqslant 1$. Then*

$$w\lambda w^{-1} = \lambda^n \neq \lambda$$

for some $w \in D$ and some integer n.

Proof.   Consider the map $f: D \to D$ defined by

$$f(x) = x\lambda - \lambda x \qquad (x \in D).$$

Since char $D = p \neq 0$, an elementary calculation shows that

$$f^{p^k}(x) = x\lambda^{p^k} - \lambda^{p^k}x, \qquad \text{for all } k \geqslant 0.$$

Let F be the prime subfield of $Z(D)$. By hypothesis, the element λ is algebraic over F and hence the field $F(\lambda)$ is finite, say of p^m elements. then $\lambda^{p^m} = \lambda$, and so

$$f^{p^m}(x) = x\lambda^{p^m} - \lambda^{p^m}x = x\lambda - \lambda x = f(x) \qquad \text{for all } x \in D,$$

proving that $f^{p^m} = f$.

If $\mu \in F(\lambda)$, then the equality $\mu\lambda = \lambda\mu$ ensures that

$$f(\mu x) = (\mu x)\lambda - \lambda(\mu x) = \mu(x\lambda - \lambda x) = \mu f(x), \qquad \text{for all } x \in D.$$

Hence, if the map $\mu I: D \to D$ is defined by $(\mu I)(x) = \mu x$, then μI and f commute for any choice of $\mu \in F(\lambda)$. Now the polynomial $X^{p^m} - X$ splits in $F(\lambda)$ into linear factors:

$$X^{p^m} - X = \prod_{\mu \in F(\lambda)} (X - \mu).$$

Since μI and f commute, we deduce that

$$0 = f^{p^m} - f = \prod_{\mu \in F(\lambda)} (f - \mu I).$$

Note also that $f \neq 0$ since $\lambda \notin Z(D)$. Let k be the smallest number for which there exist $\mu_1, \ldots, \mu_k$ in $F(\lambda)$ with $(f - \mu_k I)\ldots(f - \mu_1 I)f = 0$. By the foregoing, such k exists and $k \geqslant 1$, since $f \neq 0$. With this choice of $\mu_1, \ldots, \mu_k$ in $F(\lambda)$, there exists $r \in D$ such that

$$(f - \mu_{k-1}I)\ldots(f - \mu_1 I)f(r) = w \neq 0,$$

whereas $(f - \mu_k I)(w) = 0$, i.e. $w\lambda - \lambda w = \mu_k w$. But $w \neq 0$, and hence $w\lambda w^{-1} = \lambda + \mu_k \in F(\lambda)$ and $w\lambda w^{-1} \neq \lambda$.

Finally, the elements $w\lambda w^{-1}$ and λ belong to the multiplicative group of the finite field $F(\lambda)$. Since the latter is cyclic and the elements $w\lambda w^{-1}$, λ have the same order, we conclude that $w\lambda w^{-1}$ is a power of λ, as asserted. ■

A group G is said to be *periodic* (or *torsion*) if each element of G is of finite order. Applying Lemma 1.3, we now provide the following generalization of Theorem 1.1.

5.1.4 Theorem *If D is a division ring such that each element in D^* of the form $xy - yx$ is of finite order, then D^* is abelian. In particular, if D^* is periodic, then D^* is abelian.*

Proof. Assume by way of contradiction that D^* is not abelian. Then we may choose a, b in D such that $c = ab - ba \neq 0$. By hypothesis, c is of finite order, and hence $c^m = c$ for some $m > 1$. Let λ be a non-zero element of $Z(D)$. Then $\lambda c = (\lambda a)b - b(\lambda a)$ and so, by hypothesis, $(\lambda c)^n = \lambda c$ for some $n > 1$. Setting

$$q = (m - 1)(n - 1) + 1,$$

it follows that $c^q = c$ and $(\lambda c)^q = \lambda c$. Hence $(\lambda^q - \lambda)c = 0$ and therefore $\lambda^q - \lambda = 0$. The conclusion is that for each $\lambda \in Z(D)$ there exists $q > 1$ with $\lambda^q = \lambda$. The latter obviously implies that char $Z(D) = p \neq 0$. In what follows we denote by F the prime subfield of $Z(D)$.

We may assume that the chosen element $c = ab - ba \neq 0$ does not belong to $Z(D)$. Indeed, otherwise all elements $xy - yx$ are in $Z(D)$, in particular $c \in Z(D)$, and

$$ac = a(ab) - (ab)a \in Z(D),$$

whence $a \in Z(D)$, contrary to $c \neq 0$. Since c is algebraic over the field F, the field $F(c)$ is finite, say of p^s elements. Then $c^{p^s} = c$ and hence, by Lemma 1.3,

$$wcw^{-1} = c^n \neq c$$

for some $w \in D$ and some integer n. It follows that $wc = c^n w \neq cw$, whence $d = wc - cw \neq 0$, and

$$dc = (wc)c - c(wc) = c^n(wc - cw) = c^n d.$$

By hypothesis, d is of finite order and, by the above, $dcd^{-1} = c^n \neq c$. Hence $\langle c, d \rangle$ is a finite non-abelian subgroup of D^*, contrary to Corollary 1.2. ∎

Our next aim is to apply Theorem 1.4 for the proof of Kaplansky's theorem (Theorem 1.7), which asserts that if $D^*/Z(D^*)$ is periodic, then D^* is abelian.

5.1.5 Theorem *Let K/F be a field extension with $K \neq F$ and such that K^*/F^* is a torsion group. Then at least one of the following holds:*

(i) *K is purely inseparable over F;*

(ii) *char $K = p \neq 0$ and K is algebraic over its prime subfield.*

Proof. By hypothesis, for each $\lambda \in K$ there exists $n \geq 1$ such that $\lambda^n \in F$. Hence the extension K/F is algebraic. Assume that (i) does not hold. Then there is a separable element $\lambda \in K - F$ with $\lambda^n \in F$ for some $n \geq 1$. Hence we may embed $F(\lambda)$ into a finite normal extension L of F. The normality of L implies the existence of $\psi \in \mathrm{Gal}\,(L/F)$, such that $\mu = \psi(\lambda) \neq \lambda$. Since

$$\mu^n = \psi(\lambda)^n = \psi(\lambda^n) \qquad \text{and} \qquad \lambda^n \in F,$$

we have $\mu^n = \lambda^n$, whence $\mu = \varepsilon\lambda$ for some root of unity ε distinct from 1. Taking into account that $\mu + 1 = \psi(\lambda + 1)$ and that $(\lambda + 1)^m \in F$ for some $m > 0$, we obtain

$$\mu + 1 = \sigma(\lambda + 1)$$

where σ is an mth root of unity. We claim that $\sigma \neq \varepsilon$. Indeed, otherwise we would have $\mu + 1 = \varepsilon\lambda + 1 = \varepsilon(\lambda + 1)$, whence $\varepsilon = 1$, a contradiction. Since $\varepsilon\lambda + 1 = \sigma(\lambda + 1)$, we deduce that $\lambda = (1 - \sigma)/(\sigma - \varepsilon)$. Hence, since ε, σ are algebraic over the prime subfield P of F, we see that λ is algebraic over P.

Let x be an arbitrary element of F. Then $\lambda + x \in K - F$ and $\lambda + x$ is separable over F. Hence the element $\lambda + x$ is algebraic over P, which implies that x is also algebraic over P. We have therefore shown that the extension F/P is algebraic. Since K/F is also algebraic, we conclude that K/P is algebraic.

By the foregoing, we are left to verify that char $P = p \neq 0$. To this end, let L_0 denote a finite normal extension of P containing λ and μ. Because L is normal over F, we may assume that $L_0 \subseteq L$. Repeating preceding arguments, we obtain for every integer k the equality $\mu + k = \sigma_k(\lambda + k)$, where σ_k is a root of unity in L. If char $P = 0$, then all the σ_k are distinct and, furthermore, $\sigma_k = (\mu + k)/(\lambda + k) \in L_0$. But this is impossible, since L_0/P is a finite extension. So the lemma is true. ∎

Let A be an algebra over a field F. An element $a \in A$ is said to be *algebraic* over F if $f(a) = 0$ for some $0 \neq f(X) \in F[X]$. The algebra A is called *algebraic* over F if all elements of A are algebraic over F.

5.1.6 Lemma *Let a non-commutative division ring D be an algebraic algebra over $Z(D)$. Then D contains an element separable over $Z(D)$ and not contained in $Z(D)$.*

Proof. If char $D = 0$, then all elements of D are separable over $Z(D)$. So assume that char $D = p \neq 0$. We argue by contradiction, i.e. we assume that D is purely inseparable over $Z(D)$. Then, for any $x \in D$, there exists an integer $n(x) \geq 0$ such that $x^{p^{n(x)}} \in Z(D)$. Hence we may find $\lambda \in D - Z(D)$ with $\lambda^p \in Z(D)$. Define the map $f : D \to D$ by $f(x) = x\lambda - \lambda x$. Then $f^p(x) = x\lambda^p - \lambda^p x$ and $f \neq 0$ since $\lambda \notin Z(D)$. Choose $y \in D$

with $f(y) \neq 0$. Then there exists $k > 1$ with $f^k(y) = 0$, but $f^{k-1}(y) \neq 0$. Setting $x = f^{k-1}(y)$, it follows from $k > 1$ that x can be written in the form

$$x = f(w) = w\lambda - \lambda w.$$

On the other hand, $f(x) = 0$ and so $x\lambda = \lambda x$. Write $x = \lambda u$ for some $u \in D$. Since $x\lambda = \lambda x$, we have $u\lambda = \lambda u$. Applying equality $\lambda u = w\lambda - \lambda w$, we find that

$$\lambda = (w\lambda - \lambda w)u^{-1} = (wu^{-1})\lambda - \lambda(wu^{-1}) = c\lambda - \lambda c,$$

where $c = wu^{-1}$. Hence $c = 1 + \lambda c\lambda^{-1}$ and, by hypothesis, $c^{p^t} \in Z(D)$ for some $t \geq 0$. But then

$$c^{p^t} = (1 + \lambda c\lambda^{-1})^{p^t} = 1 + (\lambda c\lambda^{-1})^{p^t} = 1 + \lambda c^{p^t}\lambda^{-1} = 1 + c^{p^t},$$

which implies that $0 = 1$, a contradiction. ∎

We are at last in a position to attain our main objective which is to prove the following result.

5.1.7 Theorem (Kaplansky 1951) *Let D be a division ring such that the group $D^*/Z(D^*)$ is periodic. Then D^* is abelian.*

Proof. We first observe that $Z(D^*) = Z(D)^*$. Hence, for any $a \in D$ there exists an integer $n(a) > 0$ such that $a^{n(a)} \in Z(D)$. In particular, D is algebraic over $Z(D)$. Therefore, by Lemma 1.6, either D is commutative or there exists $\lambda \in D - Z(D)$ such that λ is separable over $Z(D)$. In the latter case, the field $Z(D)(\lambda)$ is not purely inseparable over $Z(D)$ and satisfies the hypothesis of Lemma 1.5. It follows that $Z(D)(\lambda)$, and hence $Z(D)$, has prime characteristic $p \neq 0$ and is algebraic over the prime subfield P of $Z(D)$. If μ is an arbitrary element of D, then μ is algebraic over $Z(D)$ and hence over P. But then $P(\mu)$ is a finite field and therefore $\mu^{m(\mu)} = \mu$ for some integer $m(\mu) > 1$. The desired conclusion now follows by appealing to Theorem 1.4. ∎

5.2 Subnormal subgroups: preliminary results

A subgroup H of a group G is said to be *n-subnormal* if there exists a chain of $n + 1$ subgroups

$$H = H_0 \subseteq H_1 \subseteq \ldots \subseteq H_n = G,$$

in which H_i is normal in H_{i+1} for all $i \in \{0, 1, \ldots, n-1\}$. In case H is *n-subnormal* for some $n \geq 0$, we say that H is a subnormal subgroup of G. Our aim is to investigate subnormal subgroups of D^*, where D is a division ring. In order not to interrupt future discussions at an awkward stage, we present here a number of preliminary results.

5.2.1 Theorem   (Brauer 1949, Cartan 1947, Hua 1949)   *Let $D_1 \subseteq D$ be division rings. If $D_1^* \lhd D^*$, then $D_1 = D$ or $D_1 \subseteq Z(D)$.*

Proof.   By hypothesis, every inner automorphism of D maps D_1 onto itself. Hence, if $a \in D$, $b \in D_1$, then

$$ba = ab_1 \tag{1}$$

for some $b_1 \in D_1$ (for $a = 0$ this is true with $b_1 = b$). Also

$$b(1 + a) = (1 + a)b_2 \tag{2}$$

for some $b_2 \in D_1$. Subtracting (1) from (2), we find

$$b - b_2 = a(b_2 - b_1). \tag{3}$$

Assume that $a \in D - D_1$. Then (3) implies $b_2 = b_1$, and hence $b = b_2$. Then $b_1 = b$, that is, $ba = ab$. Thus every element of $D - D_1$ commutes with every element of D_1.

Suppose that $D_1 \nsubseteq Z(D)$. Then there exists an element b of D_1 and an element c of D such that

$$bc \neq cb. \tag{4}$$

By the foregoing, we also have $c \in D_1$. If $D_1 \neq D$, there exists an $a \in D - D_1$, in which case $a + c \in D - D_1$. Hence $a + c$ and a both commute with $b \in D_1$,

$$b(a + c) = (a + c)b, \qquad ba = ab.$$

Since these two equalities contradict (4), the result follows. ∎

5.2.2 Corollary   *Let D be a division ring. Then every abelian normal subgroup of D^* is contained in $Z(D)$.*

Proof.   Let A be an abelian normal subgroup of D^* and let D_1 be the division ring generated by A. Since $A \lhd D^*$ we have $D_1^* \lhd D^*$ and hence, by Theorem 2.1, $D_1 = D$ or $D_1 = Z(D)$. But D_1 is commutative, since A is abelian; hence the result. ∎

For future application, we next record

5.2.3 Lemma   *Let R be an arbitrary ring and let $f(X)$, $g(X) \in R[X]$ be of degrees m and n, respectively. Let $k = \max(m - n + 1, 0)$ and let a be the leading coefficient of $g(X)$. Then there exist polynomials $q(X)$ and $r(X)$ such that*

$$a^k f(X) = q(X)g(X) + r(X),$$

where $r(X) = 0$ or $\deg r(X) < n$. Furthermore, if a is not a zero divisor in R, then $q(X)$ and $r(X)$ are uniquely determined.

Proof. If $m < n$, then $k = 0$, and we may take $q(X) = 0$, $r(X) = f(X)$. In case $m \geq n - 1$, we have $k = m - n + 1$. To prove the first part of the lemma, we argue by induction on m, observing it to be true for $m = n - 1$. So assume that $m \geq n$. Then $af(X) - bX^{m-n}g(X)$ has degree $\leq m - 1$, where b is the leading coefficient of $f(X)$. Applying induction hypothesis, there exist polynomials $q_1(X)$ and $r_1(X)$ such that

$$a^{(m-1)-n+1}(af(X) - bX^{m-n}g(X)) = q_1(X)g(X) + r_1(X),$$

where

$$\deg r_1(X) < n \qquad \text{or} \qquad r_1(X) = 0.$$

Taking $q(X) = ba^{m-n}X^{m-n} + q_1(X)$, $r(X) = r_1(X)$, the first assertion follows.

Assume that a is not a zero divisor and that $a^k f = q'g + r'$, $\deg r' < n$. Then $(q - q')g = r' - r$. If $q - q' \neq 0$, then the left-hand side has degree at least n, since the leading coefficient of $g(X)$ is not a zero divisor. However, this is impossible since $\deg(r' - r) < n$. Thus $q - q' = 0$ and $r' - r = 0$, as required. ∎

The next two lemmas are contained in a work of Huzurbazar (1960). In what follows, for any $a, b \in D^*$, we write

$$[a, b] = aba^{-1}b^{-1}.$$

5.2.4 Lemma   *Let D be a division ring and let $x, y \in D^*$ be such that $\lambda = [y, x] \neq 1$ and λ commutes with both x and y. Define R to be $\mathbb{Z}$ if char $D = 0$ and to be the prime subfield of $Z(D)$ if char $D \neq 0$. If x is algebraic over $R[\lambda]$, then λ is a root of unity. Moreover, if y is also algebraic over $R[\lambda]$, then char $D = 0$.*

Proof.   Let $f(X) = \alpha_0 + \alpha_1 X + \ldots + \alpha_r X^r$, $\alpha_r \neq 0$, $\alpha_i \in R[\lambda]$, be a polynomial of least degree such that x is a root $f(X)$. Using the relation $yxy^{-1} = \lambda x$, we obtain

$$yf(x)y^{-1} = \alpha_0 + \alpha_1 \lambda x + \ldots + \alpha_r \lambda^r x^r = 0,$$

whence

$$\alpha_1(\lambda - 1) + \alpha_2(\lambda^2 - 1)x + \ldots + \alpha_r(\lambda^r - 1)x^{r-1} = 0.$$

If $\lambda^r \neq 1$, then x satisfies an equation of degree less than r, which is impossible. Thus λ is a root of unity.

Assume that y is also algebraic over $R[\lambda]$ and that char $D \neq 0$. Since λ is a root of unity, $R[\lambda]$ is a finite field. Moreover, the set T of all finite sums of the form $\sum a_{ij}x^i y^j$, $a_{ij} \in R[\lambda]$ is a finite division ring and hence a field. Since $x, y \in T$, we have $yx = xy$, contrary to the assumption that $\lambda \neq 1$. ∎

5.2.5 Lemma *Let x, y, λ and R be as in Lemma 2.4. Define*

$$x_1 = [y, 1 + x], \quad x_{i+1} = [y, x_i], \qquad i = 1, 2, \ldots .$$

Similarly, given r ∈ R, define

$$(rx)_1 = [y, 1 + rx], \quad (rx)_{i+1} = [y, (rx)_i], \qquad i = 1, 2, \ldots .$$

(i) *If $x_n = 1$ for some $n \geq 1$, then x is algebraic over $R[\lambda]$.*
(ii) *If for every $r \in R$ there exists $n = n(r)$ such that $(rx)_n = 1$, then char $D \neq 0$.*

Proof.
(i) Since $yxy^{-1} = \lambda x$ and λ commutes with x, we have $yx^n = \lambda^n x^n y$ for all $n \geq 1$. Hence

$$yf(x) = f(\lambda x)y, \tag{5}$$

where $f(x) = p(x)q(x)^{-1}$ and $p(x)$, $q(x)$ are polynomials with coefficients in R. If $f(x) \neq 0$, then by (5)

$$[y, f(x)] = f(\lambda x)f(x)^{-1}.$$

Therefore

$$x_1 = [y, 1 + x] = (1 + \lambda x)(1 + x)^{-1},$$
$$x_2 = [y, x_1] = (1 + \lambda^2 x)(1 + \lambda x)^{-2}(1 + x).$$

In general,

$$x_n = [y, x_{n-1}] = \prod_{i=0}^{n} (1 + \lambda^i x)^{\binom{n}{i}(-1)^{n-i}}.$$

If now $x_n = 1$, then

$$(\lambda - 1)^n + a_1 x + a_2 x^2 + \ldots + a_r x^r = 0 \tag{6}$$

for some $a_i \in R[\lambda]$ and $r = 2^{n-1} - 2$. Hence x is algebraic over $R[\lambda]$.

(ii) Assume that for every $r \in R$, there exists $n = n(r)$ with $(rx)_n = 1$. Then, by (6), rx is a root of a polynomial

$$f(X) = (\lambda - 1)^n + a_1 X + a_2 X^2 + \ldots + a_k X^k,$$

where $a_i \in R[\lambda]$, $k = 2^{n-1} - 2$. Let $g(X) = b_0 + b_1 X + \ldots + b_q X^q$, $b_i \in R[\lambda]$, $b_q \neq 0$, be a polynomial of least degree having root x. Then q is also the least degree of a polynomial over $R[\lambda]$ having root rx. Indeed, for such a polynomial we can choose

$$h(X) = b_0 r^q + b_1 r^{q-1} X + b_2 r^{q-2} X^2 + \ldots + b_q X^q.$$

Applying Lemma 2.3, there exists a polynomial $s(X)$ with coefficients in $R[\lambda]$ such that

$$b_q^m f(X) = s(X)h(X), \qquad \text{where } m = k - q + 1.$$

Comparing the constant terms we obtain

$$b_q^m(\lambda - 1)^n = c_0 b_0 r^q,$$

where c_0 is the constant term of $s(X)$. Thus for every $r \in R$, there exist natural numbers $n = n(r)$, $m = m(r)$ such that

$$b_q^m(\lambda - 1)^n \qquad \text{is divisible by } r. \tag{7}$$

Assume by way of contradiction that char $D = 0$ (hence that $R = \mathbb{Z}$). Since λ is a root of unity, $\mathbb{Z}[\lambda]$ is the ring of integers of $\mathbb{Q}(\lambda)$. Hence, if $N = N_{\mathbb{Q}(\lambda)/\mathbb{Q}}$, then for any $f(\lambda) \in \mathbb{Z}[\lambda]$, $N(f(\lambda)) \in \mathbb{Z}$. Since $N(b_q)N(\lambda - 1)$ has only finitely many prime factors, there is a prime $r \in \mathbb{Z}$ that does not divide $N(b_q)N(\lambda - 1)$. But then r does not divide $b_q^m(\lambda - 1)^n$, contrary to (7). This completes the proof of the lemma. ■

5.2.6 Corollary *If* x, $y \in D$, $[y, x]$ *commutes with both* x *and* y, $[y, x] \neq 1$, *and*

$$[y, [y, \ldots [y, 1 + x] \ldots]] = 1$$

then x *is algebraic over* $Z(D)$.

Proof. We keep the notation of Lemma 2.4. Then $R \subseteq Z(D)$ and, by Lemma 2.5, x is algebraic over $R[\lambda]$. Hence, by Lemma 2.4, λ is a root of unity. Therefore $R[\lambda]$ is algebraic over R and thus x is algebraic over R. ■

5.2.7 Lemma *Let* $G_1, \ldots, G_n$ *be subgroups of a group* G *with*

$$G_1 \lhd G_2 \lhd \ldots \lhd G_n = G.$$

Given $x \in G_1$ *and* $y \in G$, *put* $y_1 = [x, y]$, $y_{i+1} = [x, y_i]$, $i \geq 1$. *Then* $y_{n-1} \in G_1$.

Proof. We have $y_1 \in G_{n-1}$ since $x \in G_1 \subseteq G_{n-1} \lhd G$. It follows by induction that $y_i \in G_{n-i}$ and hence that $y_{n-1} \in G_1$. ■

A group is said to be *weakly nilpotent* if any two of its elements generate a nilpotent subgroup. It is clear that every nilpotent group is weakly nilpotent. The next result is essentially due to Huzurbazar (1960).

5.2.8 Theorem *Let* D *be a division ring. Then:*

(i) *Every weakly nilpotent normal subgroup of* D^* *is contained in* $Z(D^*)$. *In particular,* $D^*/Z(D^*)$ *has no non-trivial weakly nilpotent normal subgroups.*

(ii) *Neither* D^* *nor* $D^*/Z(D^*)$ *has any weakly nilpotent non-abelian subnormal subgroups.*

Proof. The second assertion of (ii) is a consequence of the first. Note also that (i) is a consequence of (ii). Indeed, assume that G is a weakly nilpotent normal subgroup of D^*. If $G \not\subseteq Z(D^*)$, then G is non-abelian, by virtue of Corollary 2.2. This, however, contradicts (ii). To prove (ii), assume, by way of contradiction, that G is a weakly nilpotent non-abelian subnormal subgroup of D^*. The assumption on G ensures that there exist $x, y \in G$ such that $[x, y] = xyx^{-1}y^{-1} = \lambda \neq 1$, and λ commutes with both x and y. In what follows we use the notation of Lemma 2.4.

Let r be an arbitrary element of R and let $(rx)_i$, $i \geq 1$, be defined as in Lemma 2.5. Then by Lemma 2.7,

$$(rx)_m \in G, \qquad \text{for some } m \geq 1.$$

Hence the subgroup $\langle y, (rx)_m \rangle$ of G is nilpotent and so $(rx)_k = 1$ for some $k \geq 1$. It follows, from Lemma 2.5(ii), that char $D \neq 0$. On the other hand, taking $r = 1$, we see that x is algebraic over $R[\lambda]$, by Lemma 2.5(i). A similar argument shows that y is algebraic over $R[\lambda]$ which implies, by virtue of Lemma 2.4, that char $D = 0$, a contradiction. ∎

Let H, S be subgroups of a group G. We say that S *normalizes* H if $s^{-1}Hs = H$ for all $s \in S$. In what follows, for any subset X of a division ring D, $C(X)$ denotes the centralizer of X in D and $\bar{X}$ the subdivision ring generated by X.

5.2.9 Theorem (Herstein and Scott 1963) *Assume that if K is a division ring, then for any subdivision ring K_1 with $K_1 \neq K$ and $K_1 \not\subseteq Z(K)$, no non-central $(n - 1)$-subnormal subgroup of K^* normalizes K_1^*. Then, for any subfield F of a division ring D with $F \not\subseteq Z(D)$, no non-central n-subnormal subgroup of D^* normalizes F^*.*

Proof. Assume by way of contradiction that F is a subfield of D with $F \not\subseteq Z(D)$ and such that F^* is normalized by G_0, where

$$G_0 \lhd G_1 \lhd G_2 \lhd \ldots \lhd G_n = D^* \qquad \text{and} \qquad G_0 \not\subseteq Z(D).$$

We first observe that

$$G_0 \not\subseteq C(F).$$

Indeed, if $G_0 \subseteq C(F)$, then $\bar{G}_0 \subseteq C(F)$. Hence $\bar{G}_0$ is a subdivision ring normalized by G_1. But then, by hypothesis, $\bar{G}_0 = D$ so $C(F) = D$ and $F \subseteq Z(D)$, a contradiction. For the sake of clarity, we divide the rest of the proof into two steps.

Step 1: assume that there is an $x \in F$ such that $x \notin Z(D)$ and x is algebraic over $Z(D)$. Let $x_1, \ldots, x_r$ be the conjugates of x in F. Then $Z(D)(x_1, \ldots, x_r)$ is a field invariant under G_0 and not contained in $Z(D)$. Hence $D, F, G_0, G_1, \ldots, G_{n-1}$ may be assumed to be such that $[F:(F \cap Z(D))]$ is finite and as small as possible. Assume that there are

$y \in G_0$, $y \notin C(F)$, and $a \in F$, $a \notin Z(D)$, such that $[y, a] = 1$. Then the minimality of $(F : (F \cap Z(D)))$ is contradicted, for $C(a)$ is a division ring, F a subfield invariant under $G_0 \cap C(a)$, $G_i \cap C(a)$ is a normal subgroup of $G_{i+1} \cap C(a)$, $G_1 \cap C(a) \not\subseteq Z(C(a))$, and since $a \in Z(C(a)) \cap F$,

$$1 < [F : (F \cap Z(C(a)))] < [F : (F \cap Z(D))].$$

Hence $G_0/(G_0 \cap C(F))$ is isomorphic to a non-trivial group of auto-morphisms of F over $F \cap Z(D)$ such that the fixed field of any automorphism $\neq 1$ is $F \cap Z(D)$. It follows that $G_0/(G_0 \cap C(F))$ and each of its non-trivial subgroups is the full Galois group of $F/(F \cap Z(D))$. Therefore $G_0/(G_0 \cap C(F))$ is of prime order. Thus the commutator subgroup Q of G_0 is in $C(F)$. But Q is normal in G_1, so $\bar{Q}$ is invariant under G_1. By hypothesis, either $\bar{Q} = D$ or $Q \subseteq Z(D)$. If $\bar{Q} = D$, then $C(F) = D$, which is impossible. Hence $Q \subseteq Z(D)$ and therefore G_0 is nilpotent. By Theorem 2.8(ii), G_0 is abelian. Consequently, G_0 is a field invariant under G_1. Since $\bar{G}_0 \neq D$, this contradicts our initial assumption.

Step 2:   assume that if $x \in F$ and $x \notin Z(D)$, then x is transcendental over $Z(D)$. To derive a contradiction, first suppose that $F \cap G_0 \subseteq Z(D)$. Since $G_0 \not\subseteq C(F)$, there are $x \in G_0$ and $y \in F$ such that $[x, y] = a \neq 1$. Put $y_0 = [x, y]$, $y_{i+1} = [x, y_i]$, $i \geq 0$. Then, by Lemma 2.7, $y_n \in G_0$ and it is clear that each $y_i \in F$ since F is invariant under x. Therefore

$$y_n \in G_0 \cap F \subseteq Z(D) \qquad \text{and} \qquad y_{n+1} = 1.$$

It follows that there is $u \in F$ (y or an appropriate y_i) such that

$$[x, u] = b \neq 1, \qquad [x, b] = 1 \qquad \text{and} \qquad [u, b] = 1$$

(this last because both u and b are in F). Clearly $(1 + u)_{n+1} = 1$ also. By Corollary 2.6, u is algebraic over $Z(D)$, a contradiction. Thus $F \cap G_0 \not\subseteq Z(D)$. If $u^{-1}(F \cap G_0)u \subseteq C(F)$ for all $u \in G_1$, then the division ring L generated by all $u^{-1}(F \cap G_0)u$ with $u \in G_1$ is invariant under G_1, contradicting the initial hypothesis. Hence, for some $u \in G_1$, $u^{-1}(F \cap G_0)u \not\subseteq C(F)$. Let $y \in u^{-1}(F \cap G_0)u$, $y \notin C(F)$. For some $v \in F$, $[y, v] \neq 1$. Then $v_n \in F \cap G_0$, by Lemma 2.7, so $v_{n+1} \in F \cap G_0 \cap (u^{-1}Fu)$, since $u^{-1}Fu$ is invariant under $u^{-1}G_0u = G_0$. Therefore $v_{n+2} = 1$, since $u^{-1}Fu$ is commutative. As in the preceding paragraph, this leads to a contradiction. ∎

5.3   Subnormal subgroups: main theorems

Let D be a division ring. Our aim is to prove that for any subdivision ring D_1 with $D_1 \neq D$ and $D_1 \not\subseteq Z(D)$, no non-central subnormal subgroup of D^* normalizes D_1^*. The above result will have a number of important applications. For example, it will be shown that every subnormal solvable subgroup of D^* is contained in $Z(D^*)$.

Given a subset S of D, we write $C_D(S)$ for the centralizer of S in D and $\bar{S}$ for the division ring generated by S. If G is a group and $x, y \in G$, then we write

$$[x, y] = xyx^{-1}y^{-1} \quad \text{and} \quad x^y = yxy^{-1}.$$

Further, $Cl(G, x)$ will denote the group generated by the conjugacy class of x in G. Thus $Cl(G, x)$ is the smallest normal subgroup of G containing x.

5.3.1 Lemma *Let* $G_0, \ldots, G_m$ *be groups such that* $G_j \triangleleft G_{j-1}$ *for* $j \in \{1, \ldots, m\}$, *and let* $g \in G_m$. *Put* $G(0, 0) = G_0$ *and* $G(i, 0) = Cl(G(i - 1, 0), g)$ *for* $i \in \{1, \ldots, m\}$. *Then* $G(m, 0)$ *is* m-*subnormal in* G_m.

Proof. Let $G(i, i) = G_i$ for $i = 1, 2, \ldots, m$. Assume inductively that for all $i < k$

$$G(i - 1, 0) \triangleleft \ldots \triangleleft G(i - 1, i - 1),$$

where

$$G(i - 1, j - 1) = Cl(G(i - 2, j - 1), g), \qquad j = 1, \ldots, i - 1.$$

Let $G(k, j) = Cl(G(k - 1, j), g)$ for $j = 0, 1, \ldots, k - 1$. Then

$$G(k, j) \triangleleft G(k - 1, j - 1), \qquad j = 1, \ldots, k - 1$$

and, by hypothesis,

$$G(k, k) \triangleleft G(k - 1, k - 1).$$

Hence $G(k, j - 1) \triangleleft G(k, j)$ for $j = 1, \ldots, k$ as required. ■

5.3.2 Lemma *Let* g *be an element of a group* G *and let*

$$M_0 = G, \quad M_j = Cl(M_{j-1}, g) \qquad (1 \leqslant j \leqslant m).$$

Then, for all $j \in \{0, \ldots, m\}$, M_j *is normalized by* $C_G(g)$.

Proof. We obviously have $M_m \triangleleft \ldots \triangleleft M_0$. Assume inductively that M_{i-1} is normalized by $C_G(g)$ for some $i - 1 < m$. Given $y \in C_G(g)$, $t \in M_{i-1}$, we have $g^t \in M_i$ and hence

$$(g^t)^y = t^y g (t^{-1})^y \in M_i.$$

Thus $x^y \in M_i$ for all $x \in M_i$. Hence M_i is normalized by $C_G(g)$. ■

Let us fix the following assumptions:

(A)　If K is a division ring, then for any subdivision ring K_1 with $K_1 \neq K$ and $K_1 \not\subseteq Z(K)$, no non-central $(n - 1)$-subnormal subgroup of K^* normalizes K_1^*.

(B) D is a division ring with a subdivision ring $D_1 \neq D$, $D_1 \not\subseteq Z(D)$ and G_n is a non-central n-subnormal subgroup of D^*, $G_n \lhd G_{n-1} \lhd \ldots \lhd G_1 \lhd G_0 = D^*$, which normalizes D_1^*.

The proof of the main result (Theorem 3.10) will be based on showing that (A) and (B) lead to a contradiction.

5.3.3 Lemma   *Assume that* (A) *holds and let G be a non-central n-subnormal subgroup of D^*. Then, for each non-central $x \in D$, there exists $g \in G$ such that $[g, x]$ is not central in D.*

Proof.   Assume that there is some non-central $x \in D$ with $[g, x] \in Z(D)$ for all $g \in G$. Since $x^g = [g, x]x$, it follows that

$$C_D(x^g) = C_D(x), \qquad \text{for all } g \in G.$$

Hence x and its G-conjugates commute and they generate a non-central subfield F with F^* normalized by G, contrary to Theorem 2.9.   ∎

In what follows, until Theorem 3.10, it will be assumed that (A) and (B) hold.

5.3.4 Lemma   *Let M be a non-central subgroup of D^* which is normalized by G_n. Then $M \cap G_n \not\subseteq Z(D)$.*

Proof.   Assume inductively that there exists non-central $x \in M \cap G_{i-1}$ for some $i - 1 < n$. By Lemma 3.3, there exists $g \in G_n$ such that $[x, g] \notin Z(D)$. Since $[x, g] \in M \cap G_i$, we have $M \cap G_i \not\subseteq Z(D)$, as required.   ∎

The above lemma implies that $D_1 \cap G_n \not\subseteq Z(D)$, since, by hypothesis, D_1^* is a non-central subgroup of D^* which is normalized by G_n.

5.3.5 Lemma   *Let $x \in D_1 \cap G_n$ and $x \notin Z(D)$. Then there exists $g \in G_n - D_1$ such that $[g, x] \notin Z(D)$.*

Proof.   If $G_n \subseteq D_1$, then by (A), $D = \bar{G}_n \subseteq D_1$ contrary to (B). Therefore $G_n \not\subseteq D_1$, and there exists $y \in G_n - D_1$.
   Assume that $[g, x] \in Z(D)$ for all $g \in G_n - D_1$. Let $w \in D_1 \cap G_n$. Then $wy \in G_n - D_1$ and

$$[wy, x] = w[y, x]xw^{-1}x^{-1} = [y, x][w, x].$$

Since $[wy, x] \in Z(D)$ and $[y, x] \in Z(D)$, then $[w, x] \in Z(D)$. Therefore $[g, x] \in Z(D)$ for all $g \in G_n$, contrary to Lemma 3.3.   ∎

5.3.6 Lemma

(i) $C_D(a) \subseteq D_1$ *for all* $a \in (D_1 \cap G_n) - Z(D)$.
(ii) $C_D(D_1) = Z(D) = Z(D_1)$.

Proof. Since G_n normalizes D_1^*, then $Z(D_1)$ is a field invariant under G_n. By Theorem 2.9, $Z(D_1) \subseteq Z(D)$. If (i) is true, then $C_D(D_1) \subseteq D_1$, so that

$$Z(D) \subseteq C_D(D_1) = Z(D_1) \subseteq Z(D).$$

Hence it suffices to prove (i).

Owing to Lemma 3.4, there exists $x \in (D_1 \cap G_n) - Z(D)$. Assume that $C_D(x) \not\subseteq D_1$. Then there exists $y \in C_D(x) - D_1$. Let $M_0 = D^*$ and $M_i = Cl(M_{i-1}, x)$ for $i = 1, \ldots, n$. The group M_n is non-central and n-subnormal in D^*. By Lemma 3.1, $M_n \subseteq G_n$ and, by Lemma 3.2, M_n is normalized by $C_D(x)^*$.

Given $g \in M_n$, we have $x^g \in D_1$ and $g^y \in M_n$. Hence $g^y x (g^{-1})^y = ygxg^{-1}y^{-1} = c$ is in D_1. Let $z = 1 - y$. Because $z \in C_D(x) - D_1$, then $zgxg^{-1}z^{-1} = d$ is in D_1. Therefore

$$yx^g = cy \qquad \text{and} \qquad zx^g = dz.$$

Adding, it follows that

$$x^g - d = (c - d)y.$$

If $c - d \neq 0$, then $y \in D_1$, a contradiction. Hence

$$d - c = 0 = x^g - d,$$

so that $x^g = c$ and therefore $y \in C_D(x^g)$. If w is an element of $C_D(x) \cap D_1$, then both y and $y - w$ are in $C_D(x) - D_1 \subseteq C_D(x^g)$, so that $w \in C_D(x^g)$. Hence

$$C_D(x) \subseteq C_D(x^g) \qquad \text{for all } g \in M_n.$$

If $b \in C_D(x)$ and $t = u^{-1} \in M_n$, then $b \in C_D(x^t)$, and thus $b^u \in C_D(x)$. Therefore $C_D(x)$ is invariant under M_n. But then $Z(C_D(x))$ is invariant under M_n and is a non-central field in D, contrary to Theorem 2.9. So the lemma is true. ∎

5.3.7 Lemma *Let g be an element of the normalizer of D_1 in D, $g \notin D_1$, and let $k = 1 + g$. Then:*

(i) $D_1 \cap D_1^k = C_D(g) \cap D_1$;
(ii) $G_n \cap D_1 \cap D_1^k \subseteq Z(D)$.

Proof.
(i) If $x \in C_D(g) \cap D_1$, then $x = x^k$, so $x \in D_1 \cap D_1^k$. Hence

$$C_D(g) \cap D_1 \subseteq D_1 \cap D_1^k.$$

Let $h \in D_1 \cap D_1^k$. There exists $h_1 \in D_1$ such that $h = h_1^k$. By hypothesis, there exists $h_2 \in D_1$ such that $h = h_2^g$. Then $hk = kh_1$ and $hg = gh_2$. Subtracting, it follows that $h - h_1 = g(h_1 - h_2)$. If $h_1 - h_2 \neq 0$, then $g \in D_1$, a contradiction. Therefore $h_1 - h_2 = 0 = h - h_1$ and $h = h_2$. Then $h = h^g$, so $h \in C_D(g)$. Hence

$$D_1 \cap D_1^k \subseteq C_D(g) \cap D_1$$

proving (i).

(ii) Applying (i), we have

$$G_n \cap D_1 \cap D_1^k = G_n \cap D_1 \cap C_D(g).$$

If $h \in (G_n \cap D_1 \cap C_D(g)) - Z(D)$, then $g \in C_D(h) \subseteq D_1$ by Lemma 3.6, a contradiction. ∎

5.3.8 Lemma *There is no element in $(D_1 \cap G_n) - Z(D)$ which is algebraic over $Z(D)$.*

Proof. Assume by way of contradiction that $h \in (D_1 \cap G_n) - Z(D)$ is algebraic over $Z(D)$. By Lemma 3.5, there exists $g \in G_n - D_1$. Let $Z(D)(h)$ be the field generated by adjoining h to $Z(D)$. Now h and h^g have the same minimal equation so there exists a $Z(D)$-isomorphism $Z(D) \to Z(D)(h^g)$, induced by $h \mapsto g^g$. By Jacobson (1956; Corollary 2, p. 162), there exists $x \in D_1$ such that x induces the same inner automorphism as does g. Hence $h^g = h^x$ and therefore $x^{-1}g \in C_D(h)$. Owing to Lemma 3.6, $C_D(h) \subseteq D_1$ so that $x^{-1}g \in D_1$. Therefore, $g \in D_1$, a contradiction. ∎

The following lemma is a final step in the preparation for the proof of the main result.

5.3.9 Lemma *There exist $g \in D_n - D_1$ and $b \in (D_1 \cap G_n) - Z(D)$ such that*

$$b^{1+g} \in (D_1^{1+g} \cap G_n) - Z(D)$$

and

$$b^x \in (D_1^x \cap G_n) - Z(D),$$

where $x = (1 + g)^{-1}$.

Proof. Owing to Lemma 3.4, there exists $h_1 \in (D_1 \cap G_n) - Z(D)$. By Lemma 3.5, there exists $g \in G_n - D_1$ such that $[g, h_1] \in (D_1 \cap G_n) - Z(D)$. Let

$$g_1 = (h_1)^{1+g}, \quad g_i = [g, g_{i-1}], \quad h_i = [g, h_{i-1}] \quad (2 \leqslant i \leqslant n).$$

By induction, $h_i \in D_1 \cap G_n$ for $i = 1, \ldots, n$.

Now $g_1 = (h_1)^{1+g} \in D_1^{1+g} \cap G_1$ since $h_1 \in G_n \subseteq G_1$. Assume that for some $j - 1 < n$

$$g_{j-1} = (h_{j-1})^{1+g} \in D_1^{1+g} \cap G_{j-1}.$$

Then

$$g_j = [g, g_{j-1}] = [g, h_{j-1}^{1+g}] = [g, h_{j-1}]^{1+g} = h_j^{1+g} \in D_1^{1+g}.$$

Now $g \in G_n \subseteq G_j$ and $h_{j-1}^{1+g} \in G_{j-1}$ so that $[g, h_{j-1}]^{1+g} \in G_j$. Hence $g_j \in D_1^{1+g} \cap G_j$ and, by induction,

$$g_i = h_i^{1+g} \in D_1^{1+g} \cap G_i \qquad (1 \leqslant i \leqslant n).$$

We now distinguish two cases. First assume that $h_n \notin Z(D)$. Then

$$g_n \in (D_1^{1+g} \cap G_n) - Z(D)$$

and, by the argument above (with $1 + g$ replaced by x) we have

$$h_n^x \in (D_1^{1+g} \cap G_n) - Z(D).$$

Letting $b = h_n$, the lemma follows.

Now assume that $h_n \in Z(D)$. Then there exists an integer $k - 1 < n$ such that $h_{k-1} \in (D_1 \cap G_n) - Z(D)$ and $h_k \in Z(D)$. If $h_k = [g, h_{k-1}] = 1$ then, by Lemma 3.6,

$$g \in C_D(h_{k-1}) \subseteq D_1,$$

which is impossible. Thus $h_k \neq 1$ and h_k commutes with h_{k-1} and g. Therefore $g_k \neq 1$ and g_k commutes with g_{k-1} and g.

Let $t_1 = [g, g_{k-1} + 1]$, $b_1 = [g, h_{k-1} + 1]$, $t_i = [g, t_{i-1}]$, $b_i = [g, b_{i-1}]$, $2 \leqslant i \leqslant n + 1$. By induction, $b_i \in D_1 \cap G_i$ for $i = 1, \ldots, n$. Then, by induction,

$$t_i = b_i^{1+g} \in D_1^{1+g} \cap G_n \qquad (1 \leqslant i \leqslant n).$$

If $t_m \in Z(D)$ for some $m < n + 1$, then $t_{m+1} = 1$, and by Corollary 2.6, $g_{k-1} = h_{k-1}^{1+g}$ is algebraic over $Z(D)$. Hence h_{k-1} is algebraic over $Z(D)$, contrary to Lemma 3.8. Thus t_i is not in $Z(D)$ for $i = 1, 2, \ldots, n$. Then

$$b_n^{1+g} = t_n \in (D_1^{1+g} \cap G_n) - Z(D).$$

Because h_k commutes with h_{k-1} and g, then h_k^x commutes with h_{k-1}^x and g. The argument of this paragraph with g_{k-1} replaced by h_{k-1}^x and $1 + g$ replaced by x shows that

$$b_n^x \in (D_1^x \cap G_n) - Z(D).$$

Letting $b = b_n$, the result follows. ∎

We have done most of the work to demonstrate

5.3.10 Theorem (Stuth 1964) *Let $D_1 \subseteq D$ be division rings with $D_1 \neq$*

D and $D_1 \not\subseteq Z(D)$. Then no non-central subnormal subgroup of D^* normalizes D_1^*.

Proof.   Let us say that D has property P_n, $n \geq 0$, if for any subdivision ring D_1 with $D_1 \neq D$ and $D_1 \not\subseteq Z(D)$, no non-central n-subnormal subgroup of D^* normalizes D_1^*. Then Theorem 2.1 states that every division ring has property P_0, and the statement of the theorem amounts to saying that all division rings have property P_n. Assume inductively that every division ring has property P_{n-1}, but D is a division ring without property P_n. This means that previously introduced assumptions (A) and (B) hold. Therefore we may apply Lemmas 3.3–3.9, proved under the assumptions (A) and (B).

Owing to Lemma 3.9, there exist $g \in G_n - D_1$, $b \in (D_1 \cap G_n) - Z(D)$ such that

$$b^{1+g} \in D_1^{1+g} \cap G_n \qquad \text{and} \qquad b^x \in D_1^* \cap G_n$$

where $x = (1 + g)^{-1}$. Let $b^x = c$. Then

$$d = [b, b^{1+g}] = [c, b]^{1+g} \in D_1^{1+g}.$$

Because $b^{1+g} \in G_n$ and $b \in D_1 \cap G_n$, we have $d \in D_1 \cap G_n$. Therefore $d \in Z(D)$ by Lemma 3.7(ii).

If $d = 1$, then $[c, b] = 1$ and $c \in C_D(b) \subseteq D_1$ by Lemma 3.6(i). Hence, by Lemma 3.7(ii),

$$b = c^{1+g} \in D_1 \cap G_n \cap D_1^{1+g} \subseteq Z(D),$$

a contradiction. Thus $d \neq 1$. Put

$$a_1 = [b, b^{1+g} + 1], \qquad d_1 = [c, b + 1]$$

and

$$a_i = [b, a_{i-1}], \quad d_i = [c, d_{i-1}] \qquad (2 \leq i \leq n + 1).$$

Also let $G_{n+1} = G_n$. By induction,

$$a_i = d_i^{1+g} \in D_1^{1+g} \cap G_i \qquad (1 \leq i \leq n + 1).$$

Now $a_{n+1} = [b, a_n] \in D_1$, so that

$$a_{n+1} \in D_1^{1+g} \cap G_n \cap D_1 \subseteq Z(D).$$

Therefore $[b, a_{n+1}] = 1$. Because $d \in Z(D)$, d commutes with b and b^{1+g}. Hence, by Corollary 2.6, b^{1+g} is algebraic over $Z(D)$, and thus b is algebraic over $Z(D)$, contrary to Lemma 3.8. ■

The rest of this section will be devoted to some applications of Theorem 3.10. First of all, we have

5.3.11 Corollary (Stuth 1964) *Let D be a division ring, let G be a*

non-central subnormal subgroup of D^* and let M be the conjugacy class of $x \in D - Z(D)$ in G. Then:

(i)   *for each non-central $d \in D$, there exists $g \in G$ such that $[g, d] \notin Z(D)$;*

(ii) $C_D(G) = Z(D)$;

(iii) *the division ring generated by M is D.*

Proof. Property (i) is a consequence of Lemma 3.3 and Theorem 3.10, while (ii) follows from (i). Finally, (iii) is a consequence of Theorem 3.10. ∎

5.3.12 Corollary (Stuth 1964)   *Let D be a division ring. Then every subnormal solvable subgroup of D^* is contained in $Z(D^*)$. In particular, every non-abelian subnormal subgroup of D^* is not solvable.*

Proof. Assume, by way of contradiction, that G is a subnormal solvable subgroup of D^* not contained in $Z(D^*)$. Since G is solvable, there exist groups $G_1, \ldots, G_n$ such that

$$1 = G_n \lhd G_{n-1} \lhd \ldots \lhd G_1 \lhd G_0 = G,$$

where G_i is the commutator subgroup of G_{i-1} for $i = 1, \ldots, n$. Now G_i is a subnormal subgroup of D^*, $0 \leq i \leq n$, since G is a subnormal subgroup of D^*. But G is non-central, hence repeated application of Corollary 3.11(i) shows that each G_i, $1 \leq i \leq n$, is non-central, contrary to the fact that $G_n = 1$. ∎

5.3.13 Theorem (Stuth 1964)   *Let D be a division ring and let A and B be non-central subnormal subgroups of D^*. Then $A \cap B$ is a non-central subnormal subgroup of D^*.*

Proof. We may assume that both A and B are n-subnormal in D^*. Let

$$A = A_{2n} = \ldots = A_n \lhd \ldots \lhd A_1 \lhd A_0 = D^*$$

and

$$B = B_{2n} = \ldots = B_n \lhd \ldots \lhd B_1 \lhd B_0 = D^*.$$

Assume inductively that for some $i < 2n$, $A_j \cap B_k$ is a non-central i-subnormal subgroup of D^*, for all non-negative integers j and k with $j + k = i$. Consider $A_s \cap B_t$ for positive integers s and t, with $s + t = i + 1$. Since $s + (t - 1) = (s - 1) + t = i$, there exists $x \in A_s \cap B_{t-1}$ which is not central. By Corollary 3.11(i), there exists an element $y \in A_{s-1} \cap B_t$ such that $[x, y]$ is not central. Since $A_s \cap B_t$ is normal in both $A_s \cap B_{t-1}$ and $A_{s-1} \cap B_t$, we have $[x, y] \in A_s \cap B_t$. Hence $A_s \cap B_t$ is a non-central $(i + 1)$-subnormal subgroup of D^*. By assumption, $A_{i+1} \cap B_0 = A_{i+1}$ and $A_0 \cap B_{i+1} = B_{i+1}$ are non-central $(i + 1)$-subnormal subgroups of D^*.

Hence, by induction, $A \cap B = A_n \cap B_n$ is a non-central $2n$-subnormal subgroup of D^*. ∎

5.4 Periodic multiplicative commutators

Let D be a division ring. Given $x, y \in D^*$, we write

$$[x, y] = xyx^{-1}y^{-1}.$$

Herstein (1978) made the following conjecture:

Conjecture. If for all $x, y \in D^*$, $[x, y]Z(D^*)$ is of finite order in $D^*/Z(D^*)$, then D is commutative.

One of our results verifies this conjecture in the special case where D is finite-dimensional over its centre $Z(D)$. We also obtain a number of related results.

5.4.1 Theorem (Herstein 1978) *Let D be a division ring in which for all $x, y \in D^*$, $[x, y]$ is of finite order. Then D is commutative.*

Proof. For any $a \in D$, let $Z(D)(a)$ be the subfield of D obtained by adjoining a to $Z(D)$. If all $[x, y]$ are in $Z(D)$, then

$$xZ(D)(y)x^{-1} \subseteq Z(D)(y), \qquad \text{for all } y \in D, \ x \in D^*.$$

Thus, for a fixed y in D, $Z(D)(y)$ is invariant under D^*. Hence, by Theorem 2.1, $Z(D)(y) = Z(D)$ and thus $y \in Z(D)$. This shows that D is commutative.

Suppose then that $a = [x, y] \notin Z(D)$ for some $x, y \in D^*$. By hypothesis, $a^n = 1$ for some $n > 1$, so $Z(D)(a)$ is a normal extension of $Z(D)$. Let $\psi \neq 1$ be an automorphism of $Z(D)(a)$ over $Z(D)$. Owing to the Skolem–Noether theorem,

$$\psi(a) = bab^{-1} = a^i \neq a$$

for some $i > 1$ and some $b \in D$. Since $Z(D)(a)$ is a finite extension of $Z(D)$, $\psi^k = 1$ for some $k > 1$ and therefore $b^k a = ab^k$.

Put $D_1 = C_D(b^k)$. Bearing in mind that $a, b \in D_1$ and

$$a^n = 1, \qquad b^k \in Z(D_1), \qquad ba = a^i b$$

we see that a and b generate a subdivision algebra D_2 of D_1 which is finite-dimensional over $Z(D_1)$. Put $\lambda = b^k$. Then

$$c = b(a + \lambda)b^{-1}(a + \lambda)^{-1} = (a^i + \lambda)/(a + \lambda) \neq 1$$

is a commutator in D_2, hence by assumption $c^m = 1$ for some $m > 1$. Because

$$ac = ca \qquad \text{and} \qquad \lambda = (a^i + ac)/(c - 1)$$

and a, c are roots of unity, it follows that λ is obtained from the prime field P by adjunction of roots of unity. Thus λ is algebraic over P.

If char $D \neq 0$, this last statement ensures that $P(\lambda)$ is a finite field, so $1 = \lambda^t = b^{kt}$ for some $t \geq 1$ and thus b is of finite order. Because a is also of finite order and $ba = a^i b$, we see that a and b generate a finite division ring E over P. By Theorem 1.1, E is a field and since $a, b \in E$, we have $ab = ba$, a contradiction.

Thus we may assume that char $D = 0$, in which case λ is algebraic over $\mathbb{Q}$ and so b is algebraic over $\mathbb{Q}$. This fact and the relations $a^n = 1$ and $ba = a^i b$ ensure that a and b generate a division algebra D_3 over $\mathbb{Q}$ which is finite-dimensional over $\mathbb{Q}$. Thus the roots of unity in D_3 are of bounded order. Therefore there exists an integer $s > 0$ such that

$$[u, v]^s = 1, \qquad \text{for all } u, v \in D_3^*.$$

By a result of Amitsur (1966, Theorem 19), D_3 must be commutative. But $a, b \in D_3$, so $ab = ba$, in contradiction to $ab \neq ba$. Thus D is commutative and the result follows. ∎

The following simple observation will enable us to take full advantage of the preceding result.

5.4.2 Theorem *Let D be a division ring which is finite-dimensional over $Z(D)$. If $z \in D$ is a product of multiplicative commutators $[x, y]$, $x, y \in D^*$, and if $z^n \in Z(D)$ for some $n \geq 1$, then z is of finite order.*

Proof. Put $\lambda = z^n$ and observe that, if N is the norm on D to $Z(D)$, then $N([x, y]) = 1$ and therefore $N(z) = 1$. On the other hand, $N(\lambda) = \lambda^t$ where $t = (D : Z(D))$. Since $z^n = \lambda$, we derive

$$1 = N(z^n) = N(\lambda) = \lambda^t$$

as required. ∎

5.4.3 Theorem (Herstein 1978) *Let D be a division ring which is finite-dimensional over its centre $Z(D)$. If for all $x, y \in D^*$, $[x, y]Z(D^*)$ is of finite order in $D^*/Z(D^*)$, then D is commutative.*

Proof. By hypothesis, given $x, y \in D^*$ there exists $n = n(x, y)$ such that $[x, y]^n \in Z(D)$. Hence, by Lemma 4.2, each $[x, y]$ is of finite order. Now apply Theorem 4.1. ∎

The rest of this section will be devoted to the proof of some results related to Theorem 4.3.

5.4.4 Theorem (Herstein 1978) *Let D be a division ring in which for all $x, y \in D^*$, $[x, y]Z(D^*)$ is of finite order in $D^*/Z(D^*)$. If $a \in D - Z(D)$ is algebraic over $Z(D)$, then $Z(D)(a)$ admits no non-trivial automorphism over $Z(D)$.*

Proof.   Assume by way of contradiction that $\psi \neq 1$ is an automorphism of $Z(D)(a)$ over $Z(D)$. Because a is algebraic over $Z(D)$, we have $\psi^k = 1$ for some $k > 1$.

Owing to the Skolem–Noether theorem, there exists an $x \in D^*$ such that $\psi(a) = xax^{-1}$ which implies $x^k a = ax^k$. Observe that both a and x are in $C_D(x^k)$ and both are algebraic over $Z(C_D(x^k))$. Since $xax^{-1} = q(a) \neq a$, where $q(a) \in Z(D)(a)$, we see that x and a generate a finite-dimensional division algebra D_1 over $Z(C_D(x^k))$. Applying Theorem 4.3, we deduce that D_1 is commutative. But $x, a \in D_1$ and $xa \neq ax$, a contradiction.   ∎

Applying Theorem 4.4, we derive

5.4.5 Theorem   (Herstein 1978)   *Let D be a division ring in which for all $x, y \in D^*$, $[x, y]Z(D^*)$ is of finite order in $D^*/Z(D^*)$. Then $D^*/Z(D^*)$ does not have elements of even order. In particular, all $[x, y]Z(D^*)$ are of odd order.*

Proof.   Suppose that $a \in D^* - Z(D^*)$ is such that $a^2 \in Z(D^*)$. Since $a \notin Z(D^*)$, there exists $x \in D^*$ with $u = [x, a] \neq 1$. Because $a^2 \in Z(D^*)$, we have

$$(ua)^2 = xa^2x^{-1} = a^2$$

whence $uau = a$ and so $aua^{-1} = u^{-1}$. By hypothesis, $u^n \in Z(D^*)$ for some $n \geq 1$, so u is algebraic over $Z(D)$. If $u^{-1} = u$ then $u^2 = 1$ and, since $u \neq 1$, we have $u = -1$. But then $xax^{-1} = -a \neq a$ and so $Z(D)(a)$ admits the non-trivial automorphism, induced by conjugation by x, over $Z(D)$. This, however, contradicts Theorem 4.4 and thus $u^{-1} \neq u$. Because $u^{-1} = aua^{-1}$, $Z(D)(u)$ admits the non-trivial automorphism over $Z(D)$ induced by conjugation by a. Theorem 4.4, this is impossible.   ∎

We close this section by providing a local version of Theorem 4.5.

5.4.6 Theorem   (Herstein 1978)   *Let a be a non-zero element of the division ring D such that for all $x \in D^*$, $[a, x]Z(D^*)$ is of finite order in $D^*/Z(D^*)$. If $a^2 \in Z(D)$, then $a \in Z(D)$.*

Proof.   Suppose that $a \notin Z(D)$. Then, by Theorem 2.1, we have that $u = [a, x] \notin Z(D)$ for some $x \in D^*$. As in the proof of Theorem 4.5, we obtain $ua^{-1}u = a^{-1}$ and thus $aua^{-1} = u^{-1} \neq u$. Consider the field $Z(D)(u)$. The element a induces an automorphism of order 2 on $Z(D)(u)$ by conjugation. Therefore the fixed field F of this automorphism is of codimension 2 in $Z(D)(u)$ which implies that

$$Z(D)(u) = \{\alpha + \beta u \mid \alpha, \beta \in F\}.$$

If $\alpha, \beta \in F$, then α, β commute with α, so

$$y = a(\alpha + \beta u)a^{-1}(\alpha + \beta u)^{-1} = (\alpha + \beta u^{-1})/(\alpha + \beta u) = \gamma(\alpha + \beta u)^{-2},$$

where

$$\gamma = (\alpha + \beta u^{-1})(\alpha + \beta u) \in Z(D).$$

By hypothesis, $y^n \in Z(D) \subseteq F$, for some $n \geqslant 1$, and hence

$$(\alpha + \beta u)^{2n} \in F, \qquad \text{for all } \alpha, \beta \in F.$$

If $t \in Z(D)(u)$, then $t = \alpha + \beta u$ for some $\alpha, \beta \in F$, so $t^{m(t)} \in F \neq Z(D)(u)$ for some $m(t) \geqslant 1$. By Lemma 1.5, char $F = p \neq 0$ and either $Z(D)(u)$ is purely inseparable over F or $Z(D)(u)$ is algebraic over the prime field having p elements.

Put $\lambda = u + u^{-1} \in F$ and observe that $u^2 - \lambda u + 1 = 0$. Hence, if char $F \neq 2$, u is separable over F. In case char $F = 2$, u is inseparable over F only if $\lambda = 0$, and so, only if $u^2 = 1$. But this would force $u = 1$, contrary to $u \neq 1$. The conclusion is that $Z(D)(u)$, and hence $Z(D)$, must be algebraic over the prime field. Because $a^2 \in Z(D)$, a^2 is algebraic over $\mathbb{F}_p$ and so $a^m = 1$ for some $m \geqslant 1$. Since $u \in Z(D)(u)$, u is algebraic over $\mathbb{F}_p$, so $u^n = 1$ for some $n \geqslant 1$. Taking into account $aua^{-1} = u^{-1}$, we therefore deduce that a and u generate a finite division ring. It follows, from Theorem 1.1, that $au = ua$, contrary to $au \neq ua$. ∎

5.5 Periodic subnormal subgroups

Let D be a division ring. Our aim is to prove that if N is a periodic subnormal subgroup of D^*, then $N \subseteq Z(D^*)$. Let us fix some notation throughout:

$$[x, y] = xyx^{-1}y^{-1}, \qquad (x, y) = xy - yx;$$

$$(x, y)^0 = x \qquad (x, y)^{(k)} = ((x, y)^{(k-1)}, y).$$

We denote by $c_X(p)$ the coefficient of X in p, where p is a polynomial in the indeterminate X.

The following preliminary results are extracted from a work of Goncalves and Mandel (1986).

5.5.1 Lemma *Let $a, u \in D^*$, and suppose that $u^n = \alpha \in Z(D)$ for some $n \geqslant 1$. Consider the sequence of polynomials*

$$p_i = p_i(a, u; X), \quad \bar{p}_i \in D(X) \quad \text{and} \quad q_i \in Z(D)[X^n] \qquad (i = 0, 1, \ldots)$$

defined by

$$p_0 = (1 - uX)(1 - \alpha X^n), \quad \bar{p}_0 = 1 + uX + u^2 X^2 + \ldots + u^{n-1} X^{n-1},$$

$$q_0 = 1 - \alpha X^n,$$

$$p_i = p_{i-1} a \bar{p}_{i-1} a^{-1}, \quad \bar{p}_i = a p_{i-1} a^{-1} \bar{p}_{i-1}, \quad q_i = q_{i-1}^2 \qquad (i = 1, 2, \ldots)$$

Then:

(i) $p_i(0) = \bar{p}_i(0) = q_i(0) = 1$;
(ii) $c_X(p_i) = -(u, a)^{(i)}a^{-1} = -c_X(\bar{p}_i)$.

Further, suppose that $N_k \lhd N_{k-1} \lhd \ldots \lhd N_0 = D^*$ *is a subnormal series, that* $a \in N_k$, *and that* $\lambda \in Z(D)$ *is such that* $\lambda^n \alpha \neq 1$, *and define elements* $d_i = d_i(\lambda) = d_i(a, u; \lambda) = p_i(\lambda)/q_i(\lambda)$. *Then*:

(iii) $d_i(\lambda) \in N_i$, $i = 0, 1, \ldots, k$;
(iv) $d_i(\lambda)^{-1} = \bar{p}_i(\lambda)/q_i(\lambda)$.

Proof.   The case $i = 0$ is trivial for (i), (ii), (iii), while (iv) follows from the identity

$$1 - \lambda^n \alpha = (1 - \lambda)(1 + \lambda u + \ldots + \lambda^{n-1} u^{n-1}).$$

We now argue by induction on i. If $i > 0$, then (i) is trivial, and using (i) we find

$$\begin{aligned}
c_X(p_i) &= c_X(p_{i-1}) + a c_X(\bar{p}_{i-1})a^{-1} \\
&= -(u, a)^{(i-1)}a^{-(i-1)} + a(u, a)^{(i-1)}a^{-(i-1)}a^{-1} \\
&= -(u, a)^{(i)}a^{-i},
\end{aligned}$$

and a similar computation yields $c_X(\bar{p}_c)$. Upon noting that $q_i(\lambda) \in Z(D^*)$, it follows from the inductive hypothesis that

$$\begin{aligned}
d_i(\lambda) &= (p_{i-1}(\lambda)/q_{i-1}(\lambda))a(\bar{p}_{i-1}(\lambda)/q_{i-1}(\lambda))a^{-1} \\
&= d_{i-1}a d_{i-1}^{-1}a^{-1} \in N_i
\end{aligned}$$

and

$$d_i(\lambda)^{-1} = [d_{i-1}, a]^{-1} = [a, d_{i-1}] = \bar{p}_i(\lambda)/q_i(\lambda),$$

thus completing the proof. ∎

5.5.2 Theorem   *If* $a, u \in D$ *are such that* $(u, a) \neq 0$ *and there exists a* $k \geq 1$ *such that* $(u, a)^{(k)} = 0$, *then there exists an* $x \in D$ *such that*

$$a^{-1}xa = x + 1.$$

Proof.   By hypothesis, we may choose $r \geq 1$ such that

$$(u, a)^{(r)} \neq 0 \qquad \text{and} \qquad (u, a)^{(r+1)} = 0.$$

Setting $x = a(u, a)^{(r-1)}((u, a)^{(r)})^{-1}$, we have

$$\begin{aligned}
(x, a) &= a(u, a)^{(r-1)}((u, a)^{(r)})^{-1}a - a^2(u, a)^{(r-1)}((u, a)^{(r)})^{-1} \\
&= a[(u, a)^{(r-1)}a - a(u, a)^{(r-1)}]((u, a)^{(r)})^{-1} \\
&= a,
\end{aligned}$$

which proves the result. ∎

5.5.3 Theorem *Suppose that the polynomial* $p \in D(X)$ *maps infinitely many elements of* $Z(D)$ *to* $Z(D)$. *Then* $p \in Z(D)[X]$.

Proof. Write $p = a_0 + a_1 X + \ldots + a_n X^n$ with $a_i \in D$. Let $\lambda_0, \lambda_1, \ldots, \lambda_n$ be distinct elements of $Z(D)$ and let $p(\lambda_i) = z_i \in Z(D)$. Then

$$(a_0, a_1, \ldots, a_n) = (z_0, z_1, \ldots, z_n)((\lambda_i^j))^{-1},$$

where $((\lambda_i^j))$ is the Vandermonde matrix. Hence each a_i is in $Z(D)$ as required. ■

5.5.4 Lemma *Let N be a subnormal subgroup of D^*. Then, for every $a \in N - Z(D)$, there is an N-conjugate u of a such that $[u, a] \neq 1$.*

Proof. Deny the statement. Let N_1 be the group generated by the N-conjugates of a. Then $a \in Z(N_1)$, and N_1 is a subnormal subgroup of D^*. Hence $Z(N_1)$ is an abelian subnormal subgroup of D^*. Invoking Corollary 3.12, we infer that $Z(N_1) \subseteq Z(D)$ and hence that $a \in Z(D)$, a contradiction. ■

5.5.5 Lemma *Let F be a subfield of a division ring D and suppose that $x \in D^*$ induces by conjugation an automorphism of F of finite order n. Then the subdivision ring $F(x)$ generated by F and x is finite-dimensional over its centre.*

Proof. Put $F_1 = C_F(x)$ and observe that $F_1(x^n)$ is in the centre of $F(x)$. on the other hand, F_1 is the fixed field of an automorphism of order n, and thus $(F : F_1) = n$. It follows that $(F(x) : F_1(x^n)) \leqslant n^2$, as required. ■

In order to make further progress, we need a classical result of Schur which asserts that a periodic linear group is locally finite. First of all, let us recall the following terminology.

Let V be a finite-dimensional vector space over a field F. We denote by $GL(V)$ the unit group of the ring $\operatorname{End}_F V$ of all linear transformations of V. As is customary, we identify $GL(V)$ with the group $GL(n, F)$ of all $n \times n$ non-singular matrices over F, where $n = \dim_F V$. By a *linear group* we mean any subgroup of $GL(n, F)$, for some $n \geqslant 1$ and some field F.

The linear transformation $x \in \operatorname{End}_F V$ is called *unipotent* if all its eigenvalues are 1 (or, equivalently, if $(x - 1)^n = 0$). A subgroup G of $GL(n, F)$ is called *unipotent* if each of its elements is unipotent. We leave the task of verifying that a unipotent group is necessarily nilpotent to the reader.

5.5.6 Lemma *Extensions of locally finite groups by locally finite groups are locally finite groups. In particular, periodic solvable groups are locally finite.*

Proof. The second assertion is a consequence of the first, since periodic abelian groups are locally finite. Let N be a normal subgroup of the group G such that both N and G/N are locally finite groups. Let X be any finite subset of G. We have to show that the group $\langle X \rangle$ is finite. If Y is any subset of the group G and n is any positive integer, put

$$Y^{[n]} = \{y_1, y_2 \ldots y_n \mid y_i \in Y, 1 \leq i \leq n\}.$$

Since the index $|\langle X \rangle N : N|$ is finite, there exists a finite set T of elements of $\langle X \rangle$ such that X, and some transversal for N in $\langle X \rangle N$ is contained in T. Then, clearly, there exists a finite subset S of N such that $T^{[2]} \subseteq TS$. By induction, we immediately derive $T^{[n+1]} \subseteq TS^{[n]}$ for all $n \geq 1$. However, G is periodic, and so

$$\langle X \rangle = \langle T \rangle = \bigcup_{n=1}^{\infty} T^{[n]} \subseteq T \langle S \rangle.$$

Because the normal subgroup N is locally finite, the subgroup $\langle S \rangle$ of N is finite. Therefore the set $T \langle S \rangle$ is finite and thus $\langle X \rangle$ is finite. $\blacksquare$

We are now ready to prove

5.5.7 Theorem (Schur 1911) *Any periodic linear group is locally finite.*

Proof. Let F be a field and let G be a finitely generated periodic subgroup of $GL(n, F)$. To prove that G is finite, we may harmlessly assume that F is algebraically closed. Denote by R the algebraic closure in F of the prime subfield of F, and by A the R-subalgebra of F generated by the entries of the matrices belonging to G. Because G is finitely generated, A is a finitely generated (commutative) R-algebra. If M is a maximal ideal of A, then A/M is a finite-dimensional R-algebra. Hence A/M is a finite algebraic extension of R. Since R is algebraically closed, $R \cong A/M$.

Let $f : A \to A/M$ be the natural homomorphism. Then f clearly induces a homomorphism of the free A-module M of rank n onto the vector space V of dimension n over R, and an A-algebra homomorphism of $\text{End}_A M$ onto $\text{End}_R V$. Thus f induces a group homomorphism

$$\bar{f} : G \to GL(V).$$

Because G is a torsion group, the eigenvalues of each $g \in G$ all lie in R, and they are preserved under this homomorphism $\bar{f}$. However, the kernel K of $\bar{f}$ is a unipotent (hence nilpotent) group. If $\bar{f}(G)$ is finite, then by Lemma 5.6, G is locally finite and hence finite.

Put $H = \bar{f}(G)$ and denote by S the subring of R generated by all entries of the matrices belonging to H. Let E be the quotient field of S. Because

H is finitely generated, S is a finitely generated ring and E is a finite algebraic extension of the prime subfield of R. If char $R \neq 0$, then E is finite; hence $GL(n, E)$ is finite, which implies that H is finite. We may therefore assume that char $R = 0$.

For any $h \in H$, the eigenvalues of h are roots of unity and also roots of the characteristic polynomial of h. Note that this polynomial has degree n and its coefficients lie in E. Hence the eigenvalues of h have multiplicative order dividing

$$t = l.c.m. \{r \in \mathbb{N} \mid \phi(r) \leq n(E:\mathbb{Q})\},$$

where ϕ is the Euler function. Thus t is finite.

The trace $tr\, h$ of h can take at most t^n distinct values. Denote by k the product of all distinct non-zero values of $(tr\, h) - n$, taken as h ranges over H. Now the element $k \in S$ is non-zero, and hence $k \notin N$ for some ideal N of S, since the Jacobson radical of a finitely generated integral domain is zero. The factor ring $T = S/N$ is a finite field, and the natural homomorphism $\mu : S \to T$ induces a homomorphism $\bar{\mu}$ of H into the finite group $GL(n, T)$. If for $h \in H$, $tr\, h \neq n$, then since $k \neq 0$,

$$tr\, \bar{\mu}(h) = \mu(tr\, h) \neq \mu(tr\, 1) = \mu(n).$$

Hence $tr\, h = n$ for all elements of Ker $\bar{\mu}$. But since the eigenvalues of h are roots of unity, the equation $tr\, h = n$ implies that all the eigenvalues of h are actually 1. The conclusion is that Ker $\bar{\mu}$ is unipotent and hence nilpotent. But then, by Lemma 4.6, H is finite and the result follows. ∎

We are now ready to prove

5.5.8 Theorem (Herstein 1978, Monastyrnyi 1973) *Let D be a division ring. Then all periodic subnormal subgroups of D^* are contained in the centre of D.*

Proof. (Goncalves and Mandel 1986). Assume by way of contradiction that N is a periodic subnormal subgroup of D^* which contains a non-central element a. By hypothesis, $a^n = 1$ for some $n > 2$ (as $a^2 = 1$ implies $a = \pm 1$). Therefore we may choose $\mu > 1$ such that the map $a \mapsto a^\mu$ induces a non-trivial automorphism of the field F, where $F = Z(D)(a)$. By the Skolem–Noether theorem, there is an $x \in D^*$ such that $x^{-1}ax = a^\mu$. Setting $D_1 = F(x)$, it follows from Lemma 5.5 that D_1 is finite-dimensional over its centre. Now $N \cap D_1^*$ is subnormal in D_1^*, periodic, and contains the element $a \notin Z(D_1)$. Thus we may harmlessly assume that D is finite-dimensional over its centre $Z(D)$.

By Lemma 5.4, there exists $u \in N$ such that $[u, a] \neq 1$. Let $G = \langle u, a \rangle$. Since D is finite-dimensional over $Z(D)$, G is a linear group which is finitely generated and periodic. Hence, by Theorem 5.6, G is finite. Let P

be the prime subfield of $Z(D)$ and let A be the linear span of G over P. Then A is a finite-dimensional division algebra over P, which is non-commutative since $[u, a] \neq 1$. Hence, by Theorem 1.1, $P = \mathbb{Q}$. Thus we may assume that D is a finite-dimensional division algebra over $\mathbb{Q}$.

Let $m = (D : \mathbb{Q})$ so that D^* can be identified with a subgroup of $GL(m, \mathbb{Q})$. If $x \in D^*$ is of finite order n, the degree $\phi(n)$ of the cyclotomic polynomial of index n must be at most m. As $\lim_{r \to \infty} \phi(r) = \infty$, there exists an $n \geqslant 1$ such that every $y \in D^*$ which is of finite order satisfies $y^n = 1$.

We now apply Lemma 5.1 to a, u. Then there exists $k \geqslant 0$ such that for each $\lambda \in \mathbb{Q} - \{\pm 1\}$, $p_k(\lambda)/q_k(\lambda) \in N$, so that $p_k(\lambda)^n/q_k(\lambda)^n = 1$. Invoking Lemma 5.3, we therefore deduce that $p_k^n \in Z(D)[X]$. As $p_k(0) = 1$, we obtain

$$c_X(p_k^n) = nc_X(p_k) = -n(u, a)^{(k)}a^{-k} \in Z(D),$$

whence $(u, a)^{(c+1)} = 0$. By Lemma 5.2, there is an x such that $a^{-1}xa = x + 1$; therefore

$$x = a^{-n}xa^n = x + n,$$

which is impossible since $n \geqslant 1$. ∎

5.6 Free subgroups

We are concerned with a problem posed by Lichtman (1977) which we formulate as:

Conjecture 1. The multiplicative group of a non-commutative division ring contains a non-cyclic free subgroup.

The following stronger conjecture made by Goncalves and Mandel (1986) will actually be approached:

Conjecture 2. Any non-central subnormal subgroup of the multiplicative group of a division ring contains a non-cyclic free subgroup.

Let G be an arbitrary group. If H is a subgroup of G, then the *core* of H in G, written $\text{core}_G H$, is the largest normal subgroup of G contained in H.

5.6.1 Lemma Let H be a subgroup of G. Then

$$\text{core}_G H = \bigcap_{g \in G} H^g.$$

Moreover, if $(G : H) = n$, then $(G : \text{core}_G H) \leqslant n!$

Proof. Let $N \triangleleft G$ with $N \subseteq H$. Then, for all $x \in G$, we have $N = N^x \subseteq$

H^x, and therefore $N \subseteq \cap_{x \in G} H^x$. Since $\cap_{x \in G} H^x$ is clearly normal in G, this first fact is proved.

Assume that $(G:H) = n < \infty$. Then G permutes the n right cosets of H by right multiplication, and we therefore obtain a homomorphism $G \to S_n$ with kernel N. Clearly, $N \triangleleft G$, and $(G:N) \le |S_n| = n!$. Moreover, because N fixes the coset H, we have $HN = H$. Thus $N \subseteq H$ and the result follows. ∎

We next quote the following fundamental result, the proof of which is beyond the scope of this book.

5.6.2 Theorem *Any matrix group over a field is either solvable-by-locally finite or contains a non-cyclic free subgroup.*

Proof. The proof is a direct consequence of Lemma 6.1 and Theorems 1 and 2 in Tits (1972). ∎

The following theorem is an easy application of Theorem 6.2 and some preceding results.

5.6.3 Theorem (Goncalves 1984) *Let D be a division ring, finite-dimensional over its centre $Z(D)$, and let G be a subnormal subgroup of D^*, not contained in $Z(D)$. Then G contains a non-cyclic free subgroup.*

Proof. Deny the statement. Then, by Theorem 6.2, G contains a normal solvable subgroup L such that G/L is locally finite. Hence, by Corollary 3.12, $L \subseteq Z(D)$.

Let N be the norm on D to $Z(D)$. Since every element x of G' is a product of commutators $g_1 g_2 g_1^{-1} g_2^{-1}$, g_1, $g_2 \in G$, and, for some m, $x^m = \lambda \in Z(D^*)$, it follows that

$$1 = N(x) = N(x^m) = N(\lambda) = \lambda^r,$$

where $r = (D:Z(D))$. Thus G' is a periodic subnormal subgroup of D^*. Hence, by Theorem 5.7, $G' \subseteq Z(D^*)$ and therefore G is a solvable subnormal subgroup of D^*. Applying Corollary 3.12, we deduce that $G \subseteq Z(D^*)$, a contradiction. ∎

5.6.4 Corollary *Suppose that D_1 is a subdivision ring of D and that D_1 is finite-dimensional over its centre $Z(D_1)$. If G is a subnormal subgroup of D^* and $G \cap D_1 \not\subseteq Z(D_1)$, then G contains a non-cyclic free subgroup.*

Proof. Put $H = G \cap D_1$. Then H is a subnormal subgroup of D_1^* not contained in $Z(D_1)$. Now apply Theorem 6.3. ∎

The following results provide circumstances under which Conjecture 2 holds.

5.6.5 Theorem (Goncalves and Mandel 1986) *Let D be a division ring and let G be a subnormal subgroup of D^*. Assume that at least one of the following conditions holds:*

(i)   *G contains a non-central element a which is algebraic over $Z(D)$, and the field $Z(D)(a)$ admits a non-trivial $Z(D)$-automorphism;*
(ii) *G contains a non-central torsion element;*
(iii) *G contains a non-central element a and $a^p \in Z(D)$, where $p = 2$ or $p = \operatorname{char} D$;*
(iv) *$Z(D^*) \subseteq G$ and $G/Z(D^*)$ contains a non-abelian finite subgroup.*

Then G contains a non-cyclic free subgroup.

Proof.
(i)   By the Skolem–Noether theorem, there exists an element $b \in D^*$ such that the given automorphism is induced by conjugation by b. The assumption on the non-triviality of the automorphism ensures that $[b, a] \neq 1$. Because a is algebraic over $Z(D)$, the automorphism has finite order. Hence, by Lemma 5.5, the subdivision ring $Z(D)(a, b)$ is finite-dimensional over its centre. Since $a \in G \cap Z(D)(a, b)$ and $[b, a] \neq 1$, a is a non-central element of $Z(D)(a, b)$. The desired conclusion is therefore a consequence of Corollary 6.4.

(ii) If $a \in G - Z(D)$ and $a^n = 1$, then $n > 2$ and for some μ coprime with n, $a \mapsto a^\mu$ induces a non-trivial automorphism of $Z(D)(a)$. Now apply (i).

(iii) First assume that $\operatorname{char} D = p > 0$ and that G contains a non-central element a such that $a^p \in Z(D)$. Choose b such that $(b, a) = ba - ab \neq 0$ and define $(b, a)^{(k)}$ as in Section 5. An easy calculation shows that $(b, a)^{(p)} = (b, a^p)$ and hence $(b, a)^{(p)} = 0$. Applying Lemma 5.2, we deduce that $a^{-1}xa = x + 1$ for some $x \in D$. Hence a induces by conjugation a non-trivial automorphism of $Z(D)(x)$ of finite order. It follows, from Lemma 5.5, that the subdivision ring $Z(D)(x, a)$ is finite-dimensional over its centre. Hence the required assertion follows by virtue of Corollary 6.4.

Now assume that $a \in G - Z(D)$ with $a^2 \in Z(D)$. If a is separable over $Z(D)$, then clearly $Z(D)(a)$ has a non-trivial $Z(D)$-automorphism, and the result follows from (i). Otherwise, we are under the hypothesis of the preceding paragraph with $p = 2$.

(iv) Let $H/Z(D^*)$ be a non-abelian finite subgroup of $G/Z(D^*)$. Then the $Z(D)$-linear span of H is a finite-dimensional $Z(D)$-algebra, spanned by a transversal for H in $Z(D^*)$. Thus it is a subdivision ring of D, which allows the application of Corollary 6.4. ∎

We next provide an application of Theorem 6.5.

5.6.6 Theorem (Goncalves and Mandel 1986) *Let D be a division ring and let G be a subnormal subgroup of D^*. Assume that there exists an integer $n \geq 1$ such that $g^n \in Z(D)$ for all $g \in G$. Then $G \subseteq Z(D^*)$.*

Proof. The assumption on G guarantees that it does not have non-cyclic free subgroups. Suppose that $G \not\subseteq Z(D^*)$. By Theorem 6.5(ii), G does not have a non-central torsion element. In particular, $Z(D)$ cannot be finite. Further, we choose n minimal such that the hypothesis is satisfied, and then, if char $D = p > 0$, Theorem 6.5(iii) implies that $p \nmid n$.

Choose $a, u \in G$ such that $[a, u] \neq 1$. Applying Lemma 5.1, we obtain polynomials

$$p = p_k(a, u, X) \in D[X], \qquad q \in Z(D)[X^n]$$

such that for infinitely many $\lambda \in Z(D)$, $p(\lambda)/q(\lambda) \in G$. Then, for infinitely many $\lambda \in Z(D)$, $p''(\lambda) \in Z(D)$. Hence, by Lemma 5.3,

$$c_X(p^n) = nc_X(p) \in Z(D)$$

and as $n \neq 0$,

$$c_X(p) = -(u, a)^{(k)}a^{-k} \in Z(D)$$

by Lemma 5.1(ii). It follows that $(u, a)^{(k+1)} = 0$ and so, by Lemma 5.2, we have

$$a^{-1}wa = w + 1$$

for some $w \in D$. As $a^n \in Z(D)$, $w = a^{-n}wa^n = w + n$, whence $n = 0$ in D, a contradiction. ∎

Our next aim is to verify Conjecture 2 in the case where the subnormal subgroup contains a non-abelian nilpotent subgroup. To achieve this, we need the following preliminary results.

5.6.7 Lemma *Let D be a division ring which is finite-dimensional over its centre $Z(D)$ and let G be a subnormal subgroup of D^*. If G is solvable-by-periodic, then $G \subseteq Z(D^*)$.*

Proof. By hypothesis, there is a solvable normal subgroup N of G such that G/N is periodic. If H is a subgroup of G, then $N \cap H$ is a normal subgroup of H such that $H/(N \cap H) \cong NH/N$ is periodic. Thus G cannot contain a non-cyclic free subgroup. Now apply Theorem 6.3. ∎

5.6.8 Lemma *Let D be a division ring and x, y two elements of D^* such that $1 \neq z = y^{-1}x^{-1}yx$, $zx = xz$ and $zy = yz$. Let Q be the subring of D which is generated by the elements $1, z, x, y$ and let L be the subring of*

D which is generated by the elements 1, *z. Then*:

(i)   *For any* $q \in Q$, *there exists a representation*

$$q = \sum \lambda_{kl} x^k y^l,$$ (1)

    where $\lambda_{kl} \in L$ *and* k, l *are non-negative integers.*

(ii)   *The division subring* D_1 *generated by the elements* x, y *is finite-dimensional over its centre* $Z(D_1)$, *provided that either*

$$\sum \alpha_{ij} x^i y^j = 0$$ (2)

for some non-zero $\alpha_{ij} \in L$, *or*

$$z^n = 1$$ (3)

    for some natural n.

Proof

  (i)   A direct consequence of the relation $yx = zxy$.

 (ii)   First assume that (2) holds. Then we can write

$$\sum_j \beta_j y^j = 0,$$ (4)

where

$$\beta_j = \sum_i \alpha_{ij} x^i.$$ (5)

Denote by P the prime subfield of D, and consider the subfield $P(z, x)$ of D_1. If in (4) some $\beta_j \neq 0$, then by (1), D_1 is finite-dimensional over $P(z, x)$. By Theorem 1, Chapter 7, §9 of Jacobson (1956), we conclude that D_1 is finite-dimensional over $Z(D_1)$. Similarly, if in (5) all the β_j are zero, we obtain from this that D_1 is finite-dimensional over $P(z, y)$ and hence over $Z(D_1)$.

    Now assume that $z^n = 1$ for some $n \geq 1$. Then $x^n y = yx^n$ and, by the foregoing, D_1 is finite-dimensional over the subfield $P(z, x^n, y)$ and hence over $Z(D_1)$. ∎

    The next two theorems were originally proved for the case where G is a normal subgroup of D^*. However, as was observed by Goncalves and Mandel (1986), the word 'normal' can be replaced by 'subnormal'.

5.6.9 Theorem (Lichtman 1978) *Let D be a division ring and let G be a subnormal subgroup of D^*. If G contains a non-abelian nilpotent subgroup, then G contains a non-cyclic free subgroup.*

Proof. Let H be a non-abelian nilpotent subgroup of G. Then we can find two elements $x, y \in H$ such that $1 \neq z = y^{-1}x^{-1}yx$ and $zx = xz$, $zy = yz$. Let D_1 and Q be as in Lemma 6.8, let F be the prime subfield of D, and let $C = \mathbb{Z}$ if char $D = 0$ and $C = F$ if char $D \neq 0$.

Suppose first that one of the conditions (2) or (3) of Lemma 6.8 holds. Then, by Corollary 6.4, G contains a non-cyclic free subgroup. Thus we may assume that neither of the conditions (2) and (3) holds, which implies the uniqueness of the representation (1).

By the foregoing, we may assume that the elements $x^k y^l$, k, $l = 0, 1, \ldots$, are linearly independent over L. It can also be assumed that the subring L is an infinite noetherian domain. Thus we have the following three possibilities:

(a) $F = \mathbb{Q}$, z is algebraic over $\mathbb{Q}$ and hence L is isomorphic to some subring $\mathbb{Z}[\theta]$ of an algebraic number field.
(b) $F = \mathbb{Q}$, z is transcendental over $\mathbb{Q}$ and L is the polynomial ring $\mathbb{Z}[z]$.
(c) F is a finite field, z is transcendental over F, and L is the polynomial ring $F[z]$.

It will next be shown that there exists some P-adic valuation v of the quotient field $F(z)$ of L such that the following conditions hold:

(i)   the valuation ring $\bar{L}$ of v contains L;
(ii) $v(z) = v(1 - z) = 0$;
(iii) if $\bar{A}$ is the maximal ideal of $\bar{L}$, then the quotient field $\bar{L}/\bar{A}$ is finite.

Indeed, in cases (a) and (c), we take any P-adic valuation v of $F(z)$ which satisfies (ii). It then follows immediately that v also satisfies (i) and (iii). In case (b), we choose any maximal ideal A of L with $z(1 - z) \notin A$. Then A obviously defines a P-adic valuation of L and of $F(z)$. Once more, we see that (i)–(iii) hold.

Denote by $\bar{Q}$ the subring of D_1 generated by Q and $\bar{L}$. The elements $x^k y^l$ remain linearly independent over $\bar{L}$, i.e. $\bar{Q} = \bar{L} \otimes_L Q$. The subring $\bar{L}[x]$ of Q is a polynomial ring over the noetherian domain $\bar{L}$ and $\bar{Q}$ is a skew polynomial ring over $\bar{L}[x]$. Thus $\bar{Q}$ is a noetherian ring.

Now put

$$\bar{A} = \{\bar{q} \in \bar{Q} \mid \bar{q} = \sum a_{kl} x^k y^l, \ a_{kl} \in \bar{A}\}.$$

Then $\bar{A}$ is an ideal of $\bar{Q}$ and we claim that $\bar{A}$ is prime. Indeed, assume that there exist $\bar{q}_1, \bar{q}_2 \in \bar{Q}$ such that $\bar{q}_1 \bar{q}_2 \in \bar{A}$ and $\bar{q}_i \notin \bar{A}$, $i = 1, 2$. We order lexicographically the products $x^k y^l$ and write $\bar{q}_i = q_i + r_i$, $i = 1, 2$, where the leading term of r_i is $a_i x^{k_i} y^{l_i}$, $a_i \notin \bar{A}$ and all terms in q_i have coefficients in $\bar{A}$. Hence $\bar{q}_1 \bar{q}_2 = q' + r_1 r_2$, where $q' \in \bar{A}$ and the leading term of $r_1 r_2$ is

$$a_1 a_2 z^{k_2 l_1} x^{k_1 + k_2} y^{l_1 + l_2}.$$

However, $\bar{q}_1\bar{q}_2 \in \bar{A}$ implies $a_1 a_2 z^{k_2 l_1} \in \bar{A}$. Because A is maximal and $z \notin \bar{A}$, we derive $a_1 \in \bar{A}$ or $a_2 \in \bar{A}$, a contradiction.

Let M be the complement of $\bar{A}$ in $\tilde{Q}$. We wish to show that the set M is a right denominator set in $\tilde{Q}$, i.e. M is multiplicatively closed (this is obvious, since $\bar{A}$ is prime), for any two elements $m \in M$, $r \in \tilde{Q}$, there exists $m_1 \in M$, $r_1 \in \tilde{Q}$ such that

$$rm_1 = mr_1 \tag{6}$$

and $rm = 0$ or $mr = 0$ implies $r = 0$. Indeed, $\tilde{Q}$ is a noetherian ring without zero divisors and therefore has a right ring of quotients (see P. M. Cohn 1971). Hence there exist two elements m_1, $r \in \tilde{Q}$ such that (6) holds. It suffices to prove that m_1 can be chosen in M. Let us suppose that in (6) $m_1 \in \bar{A}$, i.e.

$$m_1 = \sum \mu_{kl} x^k y^l \qquad (\mu_{kl} \in \bar{A}).$$

Because $\bar{A}$ is prime and $m \notin \bar{A}$, we obtain $r_1 \in \bar{A}$, i.e.

$$r_1 = \sum \gamma_{rs} x^r y^s \qquad (\gamma_{rs} \in \bar{A}).$$

Let π denote any element of the ideal $\bar{A}$ for which $v(\pi) = 1$, and let π^n $(n \geqslant 1)$ be the greatest common divisor of the elements μ_{kl}, i.e. $m_1 = \pi^n m'$, where $m' \in M$. Then, by (6), π^n is a common divisor of the elements γ_{rs}. Therefore, cancelling π^n from both sides of (6), we obtain

$$rm' = mr', \qquad m' \in M, r' \in \tilde{Q},$$

which proves that M is a right denominator set in $\tilde{Q}$.

Now consider the right quotient ring $\tilde{Q}_M$ of $\tilde{Q}$ with respect to M. Then any element $h \in \tilde{Q}_M$ is of the form $h = (\tilde{q}, m)$ with $\tilde{q} \in \tilde{Q}$, $m \in M$, and we see that h is a unit in $\tilde{Q}_M$ if and only if $\tilde{q} \in M$. This shows that the set B of non-units in $\tilde{Q}_M$ is an ideal, and $B \cap \tilde{Q} = \bar{A}$. Furthermore, $D_1 = \tilde{Q}_M/B$ is a division ring.

Put $\bar{G} = G \cap \tilde{Q}_M$. Then $\bar{G}$ is a subnormal subgroup of the unit group of $\tilde{Q}_M$. Let $\phi : \tilde{Q}_M \to D_1$ be the natural homomorphism and let $\bar{G} = \phi(\bar{G})$. Since $\tilde{Q}_M$ is a local ring, we see that $\bar{G}$ is a subnormal subgroup of D_1^*.

We finally claim that $\bar{G}$ contains a non-cyclic free subgroup; if sustained it will follow that $\bar{G}$ contains a non-cyclic free subgroup, and hence the result.

Put $\bar{x} = \phi(x)$, $\bar{y} = \phi(y)$ and let D_2 be the division subring of D_1 generated by the element $\bar{x}$, $\bar{y}$. Then we have

$$\bar{y}\bar{x} = \bar{z}\bar{x}\bar{y}, \qquad \bar{z} = \phi(z).$$

On the other hand, the linear independence of the elements $x^k y^l$ over L

implies that $\bar{A} \cap \bar{L} = \bar{A}$, and so

$$B \cap \bar{L} = (B \cap \bar{Q}) \cap \bar{L} = \bar{A} \cap \bar{L} = \bar{A}.$$

Our choice of the valuation v now ensures that $\bar{z}$ is a non-identity element of finite order. Hence, by Lemma 6.8, D_2 is finite-dimensional over its centre. But $\bar{G} \cap D_2^*$ is a subnormal subgroup of D_2^* and $\bar{x}, \bar{y}$ are non-commuting elements of $\bar{G} \cap D_2^*$. Therefore, by Theorem 6.3, $\bar{G} \cap D_2^*$ contains a non-cyclic free subgroup, thus completing the proof. ■

We close this section by proving the following result.

5.6.10   Theorem   (Lichtman 1978)   *Let D be a division ring and let G be a subnormal subgroup of D^*. If G has a non-abelian nilpotent-by-finite subgroup, then G contains a non-cyclic free subgroup.*

Proof.   Let H be a non-abelian subgroup of G for which there exists a nilpotent normal subgroup N such that H/N is finite. If N is nonabelian, then the result follows by virtue of Theorem 6.9. We may therefore assume that N is abelian.

Consider the subdivision ring D_1 of D which is generated by $Z(D)$ and H. If F_1 denotes the subfield generated by N and $Z(D)$, then D_1 is finite-dimensional over F_1. Hence, by Theorem 1, Chapter 7, §9 of Jacobson (1956), D_1 is finite-dimensional over its centre. The desired conclusion is therefore a consequence of Corollary 6.4. ■

6
Rings with cyclic unit groups

The object of this chapter is to examine rings with a specified unit group. In the general setting, this problem is extremely complicated. It is therefore natural to choose, as a specified group, a group with the simplest possible structure. For this reason, our main attention is devoted to rings, the unit group of which is a finite cyclic group. We carry out our programme in two steps. As a first step, we examine finite commutative rings with cyclic unit groups. Here the result is as favourable as one could wish. As a second step, using information obtained in the preceding step, we obtain necessary and sufficient conditions on a ring R which guarantee that $U(R)$ is a finite cyclic group. These conditions are given in terms of the subring of R generated by $U(R)$.

6.1 Finite commutative rings with a cyclic group of units

Our aim here is to determine all finite commutative rings R such that $U(R)$ is cyclic.

We begin by exhibiting a relationship between idempotents in a (not necessarily commutative) ring R and idempotents in a factor ring R/N, where N is a nil ideal.

6.1.1 Lemma *Let N be a nil ideal of an arbitrary ring R and, for each $r \in R$, let $\bar{r}$ be the image of r in $\bar{R} = R/N$. Then each idempotent $\varepsilon \in \bar{R}$ can be lifted to an idempotent $e \in R$; that is, $\bar{e} = \varepsilon$.*

Proof. Fix an idempotent $\varepsilon \in \bar{R}$ and choose $u \in R$ with $\bar{u} = \varepsilon$. Then $u - u^2 \in N$ and so $(u - u^2)^m = 0$ for some $m > 1$. We have

$$1 = [u + (1 - u)]^{2m} = \sum_{i=0}^{2m} \binom{2m}{i} u^{2m-i}(1 - u)^i.$$

Observe that, on the right, each term after the first m is divisible by $(1 - u)^m$, while the first m terms are divisible by u^m. Thus, if e denotes the sum of the first m terms, then $1 = e + (1 - u)^m g$, where g is a polynomial in u. Now $u(1 - u) \in N$, and so

$$e \equiv u^{2m} + 2mu^{2m-1}(1 - u) + \cdots \equiv u \pmod{N};$$

that is, $\bar{e} = \bar{u} = \varepsilon$. Since

$$e(1 - e) = e(1 - u)^m g = 0,$$

e is a required idempotent.   ∎

Assume that R is a commutative ring. We say that R is *primary* if it has a unique prime ideal. For example, if R is local and artinian, then R is primary. This is so since any artinian integral domain is a field, and hence any prime ideal of an artinian commutative ring is maximal. In particular, all finite commutative local rings are primary.

6.1.2 Lemma   *Assume that R is artinian commutative ring. Then R is a finite direct product of primary rings.*

Proof.   We may write $R = R_1 \times \cdots \times R_n$, where each R_i is indecomposable and artinian. Hence, by Lemma 1.1, each $R_i/J(R_i)$ is an artinian semisimple ring with no non-trivial idempotents. But then $R_i/J(R_i)$ is a field, and hence R_i is an artinian local ring and therefore is primary.   ∎

6.1.3 Corollary   *Let R be a finite commutative ring. Then R is a direct product of primary rings $R_1, \ldots, R_n$ and $U(R)$ is a direct product of $U(R_1), \ldots, U(R_n)$. Moreover, $U(R)$ is cyclic if and only if each $U(R_i)$ is cyclic and the orders of $U(R_i)$ and $U(R_j)$, $1 \leq i < j \leq n$, are coprime.*

Proof.   The first assertion is a consequence of Lemma 1.2, while the second follows from elementary group theory.   ∎

The above observation reduces the problem of finding all finite commutative rings R such that $U(R)$ is cyclic to the problem of determining all finite primary commutative rings with this property.

Let N be a finite nil ideal in a ring R. If p is a prime divisor of $|N|$, we put

$$N(p) = \{a \in N \mid pa = 0\}.$$

Then $N(p)$ is an ideal of R and so

$$1 + N(p) = \{1 + x \mid x \in N(p)\}$$

is a subgroup of $U(R)$.

6.1.4 Lemma   *Let N be a finite nil ideal in a ring R, let p be a prime dividing $|N|$ and assume that $1 + N(p)$ is cyclic generated by $1 + a$, $a \in N(p)$. If $|N(p)| = p^r$ and n is the least positive integer with $a^n = 0$, then*

(i) $n = p^{r-1} + 1$;
(ii) $p^{r-1} \leq 2$.

Proof.

(i) Because $1 \neq (1+a)^{p^{r-1}} = 1 + a^{p^{r-1}}$, we have $p^{r-1} < n$. For each $1 \leqslant i \leqslant n$,

$$a^i = (1+a)^{s(i)} - 1, \qquad \text{for some } 1 \leqslant s(i) \leqslant p^r.$$

However, if $2 \leqslant i \leqslant n$,

$$0 = a^{n-2+i} = a^{n-2}[(1+a)^{s(i)} - 1]$$
$$= a^{n-2}[s(i)a + a^2 b] = s(i)a^{n-1}$$

and so $p \mid s(i)$. Hence $i \mapsto s(i)$ is an injection of $\{1 \leqslant i \leqslant n\}$ into

$$\{1 \leqslant s \leqslant p^r \mid s = 1 \quad \text{or} \quad p \mid s\},$$

which gives $n \leqslant p^{r-1} + 1$.

(ii) Assume by way of contradiction that $m = p^{r-1} - 1 \geqslant 2$ and let j be such that $(j-1)p < m < jp$. Then, if $s(m) = pt$,

$$a^m = (1+a)^{pt} - 1 = (1+a^p)^t - 1 = \sum_{k=0}^{t} z_k a^{pk},$$

where the z_k are binomial coefficients. If we multiply successively by a^{n-ip-1} for $1 \leqslant i < j$ we obtain $z_i a^{n-1} = 0$. Hence $p \mid z_i$ and so $z_i a^i = 0$. It follows that the sum above runs from j to t. However, if we now multiply by a^{n-m+1}, we get $a^{n-1} = 0$, which is impossible. ■

6.1.5 Lemma *Let N be a finite nil ideal in a ring R. If $|N|$ is odd, then N^+ is cyclic if and only if $1 + N$ is cyclic.*

Proof. Assume that $1 + N$ is cyclic. Then, for any prime $p \mid |N|$, $1 + N(p)$ is a subgroup of $1 + N$, and hence is cyclic. Owing to Lemma 1.4(ii), $|N(p)| \leqslant 2p$ and, since p is odd, $|N(p)| = p$. This obviously implies that N^+ is cyclic.

Conversely, assume that N^+ is cyclic. Given $a \in N$ with $(1+a)^p = 1$ for some prime $p \mid |N|$, it suffices to show that $pa = 0$. Let b be a generator of N^+. Then $ba = nb$ for some integer n and so, if $a = mb$ with $m \in \mathbb{Z}$,

$$a^2 = (mb)a = m(ba) = m(nb) = n(mb) = na.$$

Then, if k is the additive order of a, we can find an integer t with $1 \leqslant t \leqslant k$ and $a^2 = ta$. Because $a^{s+1} = 0$ for some s, we have $0 = t^s a$, whence $k \mid t^s$. Hence every prime which divides k also divides t. However,

$$0 = (1+a)^p - 1 = \sum_{j=1}^{p} z_j a^j = \left(\sum z_j t^{j-1} \right) a,$$

so k divides $\sum z_j t^{j-1}$. In particular, every prime dividing k divides both the sum and t, and hence divides the term at $j = 1$, namely p. Therefore k is a power of p. But the only power of p dividing $\sum z_j t^{j-1}$ is p itself, and hence $k = p$ as required. ∎

We are now in a position to prove

6.1.6 Theorem (Gilmer 1963) *Let R be a finite commutative primary ring such that $U(R)$ is cyclic. Let N and R_0 be the nilradical and the prime subring of R, respectively. Then $R = R_0[N]$ and R is isomorphic to exactly one of the following rings:*

(i)   *the Galois field $GF(p^n)$, where p is a prime and $n \geqslant 1$;*
(ii) $\mathbb{Z}/p^n\mathbb{Z}$, *where p is an odd prime and $n > 1$;*
(iii) $\mathbb{Z}/4\mathbb{Z}$;
(iv) $(\mathbb{Z}/p\mathbb{Z})[X]/(X^2)$, *where p is a prime;*
(v) $(\mathbb{Z}/2\mathbb{Z})[X]/(X^3)$;
(vi) $(\mathbb{Z}/4\mathbb{Z})[X]/(2X, X^2 - 2)$.

Conversely, each of the rings above has a cyclic unit group.

Proof. (Pearson and Schneider 1970). If $N = 0$, then R is a finite field and hence R is of type (i). We may therefore assume that $N \neq 0$. Now R_0 is also a primary ring and so $R_0 \cong \mathbb{Z}/p^s\mathbb{Z}$ for some $s \geqslant 1$ and some prime p. Because $U(R_0)$ is also cyclic, we have the following possibilities:

(a) p is odd;
(b) $p^s = 2$;
(c) $p^s = 4$.

Put $S = R_0[N]$ and $N_0 = N \cap R_0$. Our plan is to determine the structure of S in all possible cases and then to prove that $R = S$.

Assume that (a) holds. By Lemma 1.5, N^+ is cyclic. Then, since char $R = p^s$, we have

$$p^s \geqslant |N| \geqslant |N_0| = p^{s-1}.$$

If $|N| = p^{s-1}$, then $N = N_0$ and $S = R_0 \cong \mathbb{Z}/p^s\mathbb{Z}$. Assume $|N| = p^s$ and let b be a generator of N^+. Because N_0^+ is the unique subgroup of N^+ of order p^{s-1}, we have $pb \in N_0^+$. Write $pb = pt$ with $1 \leqslant t \leqslant p^{s-1}$. Then, since b has order p^s, $(p, t) = 1$. But now $b^n = 0$ for some $n \geqslant 1$, so $0 = pb^n = pt^n$. Hence $s = 1$ since t is a unit. It follows that $t = 1$, so $pb = p$ and $b^2(p - 1) = 0$. Thus $b^2 = 0$ and therefore $S \cong (\mathbb{Z}/p\mathbb{Z})[X]/(X^2)$.

Assume that (b) holds. Then char $R = 2$ and so $N = N(2)$. By Lemma 1.4, $2^r = 2$ and $n = 2$, or $2^r = 4$ and $n = 3$. In the former case, N is a two element ring with trivial multiplication and so $S \cong (\mathbb{Z}/2\mathbb{Z})[X]/(X^2)$. In the latter case, N^+ is the four group, and $a^3 = 0$ where $1 + a$ generates $1 + N$. Hence S is isomorphic to $(\mathbb{Z}/2\mathbb{Z})[X]/(X^3)$.

Finally, assume that (c) holds. Then $R_0 \cong \mathbb{Z}/4\mathbb{Z}$. Suppose that $N_0 \neq N$. Then, applying Lemma 1.4 to $N(2)$, we have $2^r = 2$ and $n = 2$, or $2^r = 4$ and $n = 3$. In the former case N^+ is cyclic, say $N = \{0, b, 2b, 3b\}$. Then $N_0 = \{0, 2b\}$ and, in particular, $0 \neq 2b = 2$. But then $2b^k = 2$, for any $k > 0$, and the nilpotency of b leads to a contradiction. Hence $2^r = 4$ and $n = 3$. Now $N(2)^+$ is the four group and N^+ is the product of two cyclic 2-groups of orders 2^s and 2^t. If a and b are generators of these groups, then 2^{s-1} and $2^{t-1}b$ are generators of $N(2)^+$. Because $1 + N(2)$ is cyclic of order 4, it has two generators and these yield (by Lemma 1.4(i)) two distinct elements of $N(2)$ whose square is non-zero but whose cube is zero $(n = 3)$ By symmetry, we may assume that $(2^{t-1}b)^2 = 0$, which implies $t = 1$. Since char $R = 4$, we have $s \leqslant 2$. Assume that $s = 2$. Then $4a = 0$, $2a \neq 0$, and $2b = 0$. Because $(2a)^2 = 0$, we have $2a = 2$ (the squares of the other non-zero elements of $N(2)$ are non-zero). But then, since a is nilpotent, we derive $2 = 0$, a contradiction. Thus $s = 1$ and therefore $N = N(2)$. Now $N_0 = \{0, 2\}$ and N_0^+ is a direct summand of N^+. Let $N = \{0, 2, d, d + 2\}$. Then $d^3 = 0$ and $0 = (d + 2)^3 = 2d^2$. One verifies that $d^2 = 2$ and we have shown that $S \cong (\mathbb{Z}/4\mathbb{Z})[X]/(2X, X^2 - 2)$.

It will next be shown that $S = R$. Assume, for example, that $S \cong (\mathbb{Z}/p\mathbb{Z})[X]/(X^2)$. Then if $b = \bar{X}$, multiplication by b is a homomorphism from R^+ onto N^+, the kernel of which contains no units and so is contained in N. But $b^2 = 0$, $bN = 0$, and so N is the kernel. Hence $|R| = |N|^2 = p^2 = |S|$ and $R = S$. Similarly, in the cases where

$$S \cong \mathbb{Z}/p^n\mathbb{Z}, (\mathbb{Z}/2\mathbb{Z})[X]/(X^3), \qquad (\mathbb{Z}/4\mathbb{Z})[X](2X, X^2 - 2)$$

if we consider multiplication by p, $\bar{X}$, $\bar{X}$, respectively, we obtain $R = S$.

To prove the converse, it suffices to treat the cases (iv), (v), and (vi). If $R = (\mathbb{Z}/2\mathbb{Z})[X]/(X^3)$, then $|U(R)| = 4$ and $\overline{1 + X}$ has order 4. If $R = (\mathbb{Z}/p\mathbb{Z})[X]/(X^2)$, then $\overline{1 + X}$ has order p and for some element t of $\mathbb{Z}/p\mathbb{Z}$, t has order $p - 1$. Hence $t(1 + X)$ has order $p(p - 1) = |U(R)|$. We leave the task of verifying that $U(R)$ is cyclic in case $R \cong (\mathbb{Z}/4\mathbb{Z})[X]/(2X, X^2 - 2)$ to the reader. So the theorem is true. ∎

6.2   Rings with a cyclic group of units: the general case

Let R be an arbitrary ring and let R_0 be the prime subring of R. If $U(R)$ is cyclic, say generated by g, then the subring $R_1 = R_0[g]$ has the same unit group and is commutative. Furthermore, if g is of finite order, then R_1 is a finitely generated commutative ring which is integral over its prime subring R_0. Our aim is to provide necessary and sufficient conditions on R which guarantee that $U(R)$ is a finite cyclic group. These conditions will be given in terms of the subring $R_0[U(R)]$. The latter ring, as we have seen above, is a finitely generated commutative ring which is integral over its prime subring. The following direct decomposition of such rings provides an important reduction step.

6.2.1 Lemma *Let S be a finitely generated commutative ring which is integral over its prime subring. Then $S = S_1 \oplus S_2$, where S_1 is a finite ring and every torsion element of S_2^+ is nilpotent.*

Proof. Let N denote the nilradical of S and let T be the torsion ideal of S, i.e. the ideal consisting of all torsion elements of S^+. By hypothesis, S must be a finitely generated module over the prime subring. Hence T is finite. In particular, T enjoys the minimum condition and has no non-zero nilpotent elements. Hence $T = Se$ for some idempotent e of S. But then

$$S = Se \oplus S(1 - e) = T \oplus S(1 - e),$$

which proves the assertion by taking $S_1 = T$ and $S_2 = S(1 - e)$.

Turning to the general case, we observe that S/N satisfies the hypotheses of the lemma, and we apply the argument of the preceding paragraph to produce an idempotent splitting off the torsion ideal of S/N. Owing to Lemma 1.1, we may lift this idempotent to an idempotent e in S. Setting $S_1 = Se$ and $S_2 = S(1 - e)$, we obtain $S = S_1 \oplus S_2$. By construction, S_1 is a torsion ring modulo N and hence its prime subring is finite. This shows that S_1 is also finite. Because S_2^+ is torsion-free modulo N^+, every torsion element of S_2^+ is nilpotent. ∎

6.2.2 Lemma *Let R be a torsion-free ring in which $U(R)$ is finite cyclic and $R = R_0[U(R)]$. Then R is isomorphic to $\mathbb{Z}, \mathbb{Z}[X]/(X^2 + 1)$, $\mathbb{Z}[X]/(X^2 - X + 1)$ or $\mathbb{Z}[X]/(X^3 + 1)$.*

Proof. Owing to Corollary 1.2.34, $|U(R)| = 2, 4$ or 6. If $|U(R)| = 2$, then $R = R_0 \cong \mathbb{Z}$. If $|U(R)| = 4$, then let $u \in U(R)$ be of order 4. Then $u^2 = -1$ so $R = R_0[u]$ is a homomorphic image of $\mathbb{Z}[X]/(X^2 + 1)$. But this ring has only finite rings as proper homomorphic images, so $R \cong \mathbb{Z}[X]/(X^2 + 1)$.

Now assume that $|U(R)| = 6$ and let u be a generator of $U(R)$. Then $u^3 + 1 = 0$ and $R = R_0[u]$ is a homomorphic image of $\mathbb{Z}[X]/(X^3 + 1)$. Because R is torsion-free, the homomorphism from $\mathbb{Z}[X]/(X^3 + 1)$ onto R can be extended to a homomorphism from $\mathbb{Q}[X]/(X^3 + 1)$ onto the quotient ring of R with respect to the non-zero elements of R_0. However,

$$\mathbb{Q}[X]/(X^3 + 1) \cong \mathbb{Q}[X]/(X + 1) \oplus \mathbb{Q}[X]/(X^2 - X + 1)$$

has only three proper ideals. Hence $R \cong \mathbb{Z}[X]/(X + 1)$, $\mathbb{Z}[X]/(X^2 - X + 1)$ or $\mathbb{Z}[X]/(X^3 + 1)$ and only the latter two have six units. ∎

We have now accumulated all the information necessary to prove our main result.

6.2.3 Theorem (Pearson and Schneider 1970) *Let R be a ring with unit group $U(R)$ and prime subring R_0. Then $U(R)$ is finite cyclic if and only if $R_0[U(R)]$ is isomorphic to a direct product of rings $R_1, \ldots, R_n$, where each R_i is one of the rings in Theorem 1.6 or one of the rings in (i)–(v) below, and $(|U(R_i)|, |U(R_j)|) = 1$ for $i \neq j$, $1 \leq i, j \leq n$:*

(i) $\mathbb{Z}[X]/(mX, X^2 - tX)$;
(ii) $\mathbb{Z}[X, Y]/(mX, X^2 - tX, Y^2 + 1, XY - vX)$;
(iii) $\mathbb{Z}[X, Y]/(mX, X^2 - tX, Y^3 + 1, XY - vX)$;
(iv) $\mathbb{Z}[X, Y]/(mX, X^2 - tX, Y^2 - Y + 1, XY - vX)$;
(v) $\mathbb{Z}[X, Y]/(mX, X^2 - tX, Y^3 + X + 1, XY - vX)$.

Here, in all cases, m is odd, $t \mid m$, every prime dividing m divides t, $0 \leq n < m$, $3 \nmid m$ in cases (iii) and (iv), $3 \mid m$ in case (v), and m divides $v^2 + 1$, $v^3 + 1$, $v^2 - v + 1$, and $v^3 + t + 1$ in cases (ii), (iii), (iv), and (v), respectively.

Proof. We may harmlessly assume that $R = R_0[U(R)]$. Suppose that $U(R)$ is finite cyclic. Then R satisfies the hypotheses of Lemma 2.1 and so $U(R) = U(R_1) \times U(R_2)$, where R_1 is a finite commutative ring and every torsion element of R_2^+ is nilpotent. Owing to Theorem 1.6, we may therefore assume that R is a commutative ring such that every torsion element of R^+ is nilpotent. In particular, we must have char $R = 0$ and $R_0 \cong \mathbb{Z}$. Let N denote the nilradical of R. Then $1 + N$ is a subgroup of $U(R)$, which is finite, so N is also finite. In particular, each element of N is a torsion element. Therefore N is also the torsion ideal of R. Thus the ring R/N is torsion-free. Now the natural homomorphism $R \to R/N$ induces a surjective homomorphism $U(R) \to U(R/N)$. Hence the structure of R/N is given by Lemma 2.2.

The sequence

$$1 \to 1 + N \to U(R_0[N]) \to U(R_0) \to 1$$

splits and so $|1 + N| = |N|$ is odd and, by Lemma 1.5, N^+ is cyclic. Put $m = |N|$ and denote by b a generator of N^+. Then $b^2 = tb$ with $0 < t \leq m$ and we put $S = R_0[N]$. Because b is nilpotent, we have $t^s b = 0$ for some $s \geq 1$. Hence every prime which divides m also divides t. Without loss of generality we may assume that $t \mid m$. Indeed, let $t' = (t, m)$ and $s(t/t') \equiv 1 \pmod{m}$. Choose r with $r \equiv s \pmod{m/t'}$ and $r \equiv 1 \pmod{q}$ for all q dividing m but not m/t'. Then $(r, m) = 1$, $tr \equiv t' \pmod{m}$, ra has order m, and $(ra)^2 = t'(ra)$. We have therefore shown that

$$S \cong \mathbb{Z}[X]/(mX, X^2 - tX).$$

If $|U(R/N)| = 2$ or 4, then the sequence

$$1 \to 1 + N \to U(R) \to U(R/N) \to 1$$

splits. Hence $U(R)$ is the product of $1 + N$ and a group of order 2 or 4. In the former case, this factor must be $\{\pm 1\}$, so

$$R = R_0[U(R)] = R_0[N] = S.$$

In the latter case, denote by α an element of order 4. Then $\alpha^2 + 1 = 0$ and

$$R_0[\alpha] \cong \mathbb{Z}[X]/(X^2 + 1).$$

Now $R = R_0[b, \alpha]$. Write $\alpha b = vb$ for some $0 \leqslant v < m$. Then $-b = \alpha^2 b = v^2 b$ implies that $m \mid (v^2 + 1)$. The conclusion is that in this case

$$R \cong \mathbb{Z}[X, Y]/(mX, X^2 - tX, Y^2 + 1, XY - vX).$$

Now suppose that $|U(R/N)| = 6$, so that $|U(R)| = 6m$. We first investigate the case where $3 \nmid m$. The group $U(S)$ has order $2m$ and therefore contains no elements of order 6. Choose $u \in U(R)$ with $u^3 = -1$. Then $R = S[u] = R_0[b, u]$ and we put $T = R_0[u]$. If $\bar{u}$ is the image of u in R/N, then $\bar{u}^3 = -1$ and

$$\bar{u}^2 = 1 \Rightarrow u^2 \in 1 + N \Rightarrow u = (-u^2)^{-1} \in -1 + N \subseteq S \Rightarrow R = S.$$

In particular, $U(R/N)$ is generated by $\bar{u}$.

In case $R/N \cong \mathbb{Z}[X]/(X^3 + 1)$, the elements $1, \bar{u}, \bar{u}^2$ are $\mathbb{Z}$-linearly independent and therefore $1, u, u^2$ are R_0-linearly independent. Hence

$$T \cong \mathbb{Z}[X]/(X^3 + 1).$$

Let $ub = vb$ with $0 \leqslant v < m$. Then $-b = u^3 b = v^3 b$ implies that $m \mid (v^3 + 1)$. Thus

$$R \cong \mathbb{Z}[X, Y]/(mX, X^2 - tX, Y^3 + 1, XY - vX).$$

If $R/N \cong \mathbb{Z}[X]/(X^2 - X + 1)$, then $\bar{u}^2 - \bar{u} + 1 = 0$ and $u^2 - u + 1 \in N$. However, for $t \geqslant 1$,

$$(u^2 - u + 1)^t = 3^{t-1}(u^2 - u + 1),$$

so $u^2 - u + 1 \in N$ and $(3, m) = 1$ ensure that $u^2 - u + 1 = 0$. Write $ub = vb$. Since $u^2 - u + 1 = 0$, we have

$$(v^2 - v + 1)b = 0 \qquad \text{and} \qquad m \mid (v^2 - v + 1).$$

We therefore deduce that

$$R \cong \mathbb{Z}[X, Y]/(mX, X^2 - tX, Y^2 - Y + 1, XY - vX).$$

Now assume that $3 \mid m$. Because $-1 - b$ generates $U(S)$, there exists $u \in U(R)$ such that $u^3 = -1 - b$, in which case $R = R_0[b, u]$. Let $ub = vb$. Then

$$-v^3 b = (1 + b)b = (1 + t)b,$$

whence

$$-v^3 \equiv 1 + t \pmod{m}.$$

The latter implies that $R/N \cong \mathbb{Z}[X]/(X^3 + 1)$. Otherwise, letting $\bar{u} = u + N \in R/N$,

$$u^3 = -1 - b \Rightarrow \bar{u}^3 = -1 \Rightarrow \bar{u}^2 - \bar{u} + 1 = 0 \Rightarrow u^2 - u + 1 = kb$$

for some $0 \leqslant k < m$. Then

$$-b = (u + 1)(u^2 - u + 1) = (u + 1)kb = (v + 1)kb$$

and

$$-1 \equiv (v + 1)k \pmod{m}.$$

Hence $-1 \equiv (v + 1)k \pmod 3$ and $-v^3 \equiv 1 + t \pmod 3$ since $3 \mid m$. However, $3 \mid t$ also (since $3 \mid m$) and therefore

$$-v^3 \equiv 1 \pmod 3 \Rightarrow v \equiv 2 \pmod 3 \Rightarrow (v + 1)k \equiv 0 \pmod 3,$$

which is impossible. Thus

$$R \cong \mathbb{Z}[X, Y]/(mX, X^2 - tX, Y^3 + 1 + X, XY - vX).$$

To complete the proof, we notice that the unit groups of rings (i)–(v) are cyclic of orders $2m$, $4m$, $6m$, $6m$, and $6m$, respectively. The actual verification of this routine task is left as an exercise for the reader.   ■

7
Finite generation of unit groups

Our aim in this chapter is to provide circumstances under which the unit group of a ring is finitely generated. After proving some general results, we concentrate on the case where R is a finitely generated commutative ring. It is shown that $U(R)$ is finitely generated if and only if the additive group of the nilradical of R is finitely generated. This is derived as a consequence of a general result pertaining to unit groups of Krull domains. We also show that if R is a commutative ring such that $U(R)$ is finitely generated, then $J(R)$ is nilpotent and the additive group of $J(R)$ is finitely generated. In the second part we introduce some important facts of algebraic K-theory and show that the knowledge of $K_1(R)$ is fundamental in studying finite generation of the general linear group $GL_n(R)$. As an application, we finally prove that if $R = \mathbb{Z}$ or $\mathbb{F}_q[X]$, then:

(i) $GL_n(R)$ is a finitely generated group for all $n \geq 3$;
(ii) $GL_n(R[X_1, \ldots, X_d])$ is a finitely generated group for all $n \geq d + 3$.

7.1 General results

Our aim here is to provide a number of general observations pertaining to unit groups of rings. Throughout, $J(R)$ denotes the Jacobson radical of a ring R. We write $GL_n(R)$ for the unit group of the ring $M_n(R)$ of all $n \times n$ matrices with entries in R. Of course, in case $n = 1$, $GL_n(R)$ coincides with $U(R)$. Given a left R-module M, we denote by $\text{ann}_R(M)$ the annihilator of M, i.e. $\text{ann}_R(M) = \{r \in R \mid rM = 0\}$. If G is a group and $x, y \in G$, then the symbol $[x, y]$ denotes the commutator $xyx^{-1}y^{-1}$. Finally, given subgroups H_1 and H_2 of G, we write $[H_1, H_2]$ for the subgroup of G generated by all commutators $[h_1, h_2]$, $h_i \in H_i$, $i = 1, 2$.

7.1.1 Lemma *Let I be a right ideal of a ring R. Then $I \subseteq J(R)$ if and only if $1 + I$ is a subgroup of $U(R)$.*

Proof. Suppose that $I \subseteq J(R)$. If $x \in I$, put $u = 1 + x$. Then $R = I + uR$, so $R = uR$ by Nakayama's lemma. Choose v so that $uv = 1$. Since $1 = uv = v + xv$, we have $v = 1 - xv \in 1 + I$ also, so v itself has a right inverse. Hence u is invertible and $v = u^{-1} \in 1 + I$.

Conversely, assume that $1 + I$ is a subgroup of $U(R)$. It suffices to show that I is contained in every maximal right ideal K. If not, then $I + K = R$, so $1 = x + y$ with $x \in I$, $y \in K$. Then $y = 1 - x$ is invertible, by hypothesis, so $K = R$, a contradiction.   ■

7.1.2   Lemma   *Let I be an ideal of a ring R with $I \subseteq J(R)$. Then, for all $n \geqslant 1$, the natural homomorphism*

$$GL_n(R) \to GL_n(R/I)$$

is surjective. In fact, $GL_n(R)$ is the inverse image in $M_n(R)$ of $GL_n(R/I)$.

Proof.   Assume that $u \in R$ is such that $u + I \in U(R/I)$. Then

$$uv \equiv vu \equiv 1 (\text{mod } I)$$

for some $v \in R$. It follows that uv, $vu \in 1 + I \subseteq U(R)$, by virtue of Lemma 1.1, and hence $u \in U(R)$. Thus $U(R)$ is the inverse image in R of $U(R/I)$.

Turning to the general case, consider the natural homomorphism

$$M_n(R) \to M_n(R/I) = M_n(R)/M_n(I).$$

Since $J(M_n(R)) = M_n(J(R))$, we have $M_n(I) \subseteq J(M_n(R))$. The required assertion now follows by applying the case $n = 1$.   ■

7.1.3   Lemma   *Let V be a finite-dimensional right vector space over a division ring D and let $R = \text{End}_D(V)$. Given a right ideal I of R, put*

$$IV = \{\phi(v) \mid \phi \in I, \; v \in V\}.$$

Then the map $I \mapsto IV$ is an inclusion preserving bijection between the right ideals of R and subspaces of V.

Proof.   We may write $I = fR$ for some idempotent f of R. Then $IV = f(V)$ is obviously a subspace of V. It is clear that for two right ideals I_1 and I_2 of R, $I_1 \subseteq I_2$ implies $I_1 V \subseteq I_2 V$. Conversely, assume $I_1 V \subseteq I_2 V$ and let $I_i = f_i R$ for a suitable idempotent f_i of I_i, $i = 1, 2$. Given $\phi \in I_1$ and $v \in V$, we have $\phi(v) = f_2(v_1)$ for some $v_1 \in V$. Hence $(f_2 \phi)(v) = f_2(v_1) = \phi(v)$, proving that $\phi = f_2 \phi \in I_2$. Thus $I_1 \subseteq I_2$ and hence $I_1 V = I_2 V$ implies $I_1 = I_2$. Finally, let W be any subspace of V. Then $V = W \oplus W'$ for some subspace W' of V. Hence, if $f : V \to W$ is the projection map, then $W = (fR)V$ as required.   ■

7.1.4   Lemma   *Let I be a right ideal in a ring R such that $R/J(R)$ is artinian. If $\lambda \in R$ satisfies $I + \lambda R = R$, then $I + \lambda$ contains a unit of R.*

Proof.   Owing to Lemma 1.2, an element of R is a unit if and only if it is a unit modulo $J(R)$. Hence, after passing to $R/J(R)$, we may harmlessly assume that $J(R) = 0$. Then R is a finite direct product of full matrix rings over division rings. Therefore, we can further assume that $R = \text{End}_D(V)$, where V is a finite-dimensional right vector space over a division ring D. In this case, it follows from Lemma 1.3 that

$$I = \{\phi \in R \mid \phi(V) \subseteq IV\}.$$

Since $I + \lambda R = R$, we have $IV + \lambda(V) = V$. Choose subspaces $V_0 \subseteq IV$ and $U \subseteq V$ such that

$$V = V_0 + \lambda(V) = \text{Ker } \lambda \oplus U.$$

Then λ induces an isomorphism from U to $\lambda(V)$, so $\text{Ker } \lambda \cong V_0$. Choose $r \in R$ so that $r(U) = 0$ and r induces an isomorphism from $\text{Ker } \lambda$ to V_0. Then $r(V) = V_0 \subseteq IV$, so $r \in I$. Moreover, $r + \lambda \in I + \lambda$ is clearly an automorphism of V. ∎

With the aid of the above lemma, we now derive the following useful result.

7.1.5   Corollary   *Let R be a ring such that $R/J(R)$ is artinian. If I is any ideal in R, then the natural homomorphism*

$$GL_n(R) \rightarrow GL_n(R/I)$$

is surjective for all $n \geqslant 1$.

Proof.   Since $M_n(R)/J(M_n(R)) \cong M_n(R/J(R))$ is artinian, it suffices to treat the case $n = 1$. So assume that $\lambda \in R$ is invertible modulo I. Then $I + \lambda R = R$ and so, by Lemma 1.4, $I + \lambda$ contains a unit of R. Thus $U(R) \rightarrow U(R/I)$ is surjective as required. ∎

Let I be an ideal of a ring R. We put

$$G_n(I) = \text{Ker}(GL_n(R) \rightarrow GL_n(R/I)) = GL_n(R) \cap (1 + M_n(I)).$$

Following Milnor (1966), we refer to $G_n(I)$ as the *congruence subgroup corresponding to I*. The following results are extracted from Bass (1974).

7.1.6   Theorem   *Let I be an ideal of a ring R and let $n \geqslant 1$.*

(i)   *If $I \subseteq J(R)$, then*

$$G_n(I) = 1 + M_n(I) \quad and \quad GL_n(R)/G_n(I) \cong GL_n(R/I).$$

(ii)   *If J is another ideal of R, then*

$$[G_n(I), G_n(J)] \subseteq G_n(IJ + JI).$$

In particular, $[G_n(I^i), G_n(I^j)] \subseteq G_n(I^{i+j})$.

(iii)   *If I is nilpotent, then $G_n(I)$ is a nilpotent group such that $G_n(I^i)/G_n(I^{i+1})$ is isomorphic to the additive group of $M_n(I^i/I^{i+1})$ for each $i \geqslant 1$.*

(iv)   *If I is nilpotent, then $G_n(I)$ is finitely generated if and only if the additive group of I is finitely generated.*

Proof.

(i)   Owing to Lemma 1.1, we have
$$1 + M_n(I) \subseteq 1 + M_n(J(R)) = 1 + J(M_n(R)) \subseteq GL_n(R)$$
and thus $G_n(I) = 1 + M_n(I)$. The second assertion is a consequence of Lemma 1.2.

(ii)   Fix $x \in G_n(I)$, $y \in G_n(J)$ and write $x = 1 + u$, $y = 1 + v$ with $u \in M_n(I)$, $v \in M_n(J)$. Let $u' \in M_n(I)$, $v' \in M_n(J)$ be such that $(1 + u)^{-1} = 1 + u'$, $(1 + v)^{-1} = 1 + v'$. Then
$$u + u' + uu' = v + v' + vv' = 0$$
and so
$$[x, y] = (1 + u)(1 + v)(1 + u')(1 + v')$$
$$= 1 + u'v' + uv' + uu'v' + vu' + vu'v' + uv + uvu'$$
$$+ uvv' + uvu'v'$$
$$\equiv 1 (\mathrm{mod}\, M_n(IJ + JI)).$$
Hence $[x, y] \in G_n(IJ + JI)$, as required.

(iii)   The fact that $G_n(I)$ is a nilpotent group is a consequence of (ii). If $I^2 = 0$ then $x \mapsto 1 + x$ is clearly an isomorphism from the additive group of $M_n(I)$ to the multiplicative group $1 + M_n(I) = G_n(I)$. The required assertion follows from this, applied to the ideal I^i/I^{i+1} in R/I^{i+1}.

(iv)   The additive group of I is finitely generated if and only if so are the additive groups I^i/I^{i+1}. On the other hand, I^i/I^{i+1} is finitely generated if and only if so is $M_n(I^i/I^{i+1})$. But, by (ii) and (iii), all the $M_n(I^i/I^{i+1})$ are successive quotients in a finite filtration of the nilpotent group $G_n(I)$, hence the result.   ∎

7.1.7 Lemma   *Let M be a left R-module. Then M is a finitely generated $\mathbb{Z}$-module if and only if M is a finitely generated R-module and $R/\mathrm{ann}_R(M)$ is a finitely generated $\mathbb{Z}$-module.*

Proof.   We may regard $B = R/\mathrm{ann}_R(M)$ as a subring of $\mathrm{End}_{\mathbb{Z}}(M)$ and M as a B-module. Assume that M is a finitely generated $\mathbb{Z}$-module. Then M is a finitely generated R-module and $\mathrm{End}_{\mathbb{Z}}(M)$ is a finitely generated $\mathbb{Z}$-module. The latter implies that B is a finitely generated $\mathbb{Z}$-module. Conversely, assume that M is a finitely generated R-module and B is a finitely generated $\mathbb{Z}$-module. Since M is a finitely generated B-module, we may assume that $M = F/N$ for some finitely generated free B-module

F and some submodule N of F. Then F is a finitely generated $\mathbb{Z}$-module and hence so is M, as required.   ∎

7.1.8 Corollary *Let I be a nilpotent ideal of R which is a finitely generated left R-module. Then the following conditions are equivalent*:

(i) $G_n(I)$ *is a finitely generated group*;
(ii) I *is a finitely generated $\mathbb{Z}$-module*;
(iii) $R/\mathrm{ann}_R(I)$ *is a finitely generated $\mathbb{Z}$-module*.

Moreover, if $GL_n(R/I)$ is finitely generated then these conditions imply that $GL_n(R)$ is finitely generated.

Proof.   The first assertion is a consequence of Theorem 1.6(iv) and Lemma 1.7, applied to $M = I$. The second assertion follows from Lemma 1.2.   ∎

7.1.9 Corollary *Let R be a commutative noetherian ring with nilradical $N(R)$. Suppose that $U(R/N(R))$ is finitely generated. Then the following conditions are equivalent*:

(i) $U(R)$ *is finitely generated*;
(ii) $N(R)$ *is a finitely generated $\mathbb{Z}$-module*;
(iii) $R/\mathrm{ann}_R(N(R))$ *is a finitely generated $\mathbb{Z}$-module*.

Proof.   Applying Theorem 1.6(i) for $n = 1$ and $I = N(R)$, we see that (i) is equivalent to $G_1(I)$ being finitely generated. On the other hand, since R is noetherian, I is a nilpotent ideal which is a finitely generated R-module. Now apply Corollary 1.8.   ∎

The following result is known as the Artin–Rees lemma.

7.1.10 Lemma *Let R be a commutative noetherian ring and let I be an ideal of R. If M is a finitely generated R-module and if N is a submodule, then there exists an integer m with*

$$I^n M \cap N = I^{n-m}(I^m N \cap N)$$

for all $n \geqslant m$.

Proof.   Let $R(I)$ denote the subring of $R[X]$ given by

$$R(I) = R \oplus IX \oplus I^2 X^2 \oplus \cdots$$

Since R is noetherian, $I = Rx_1 + \cdots + Rx_t$ for some $x_i \in R$ and some $t \geqslant 1$. Then it follows easily that

$$R(I) = \langle R, x_1 X, x_2 X, \ldots, x_t X \rangle.$$

Because R is noetherian, it follows from the Hilbert basis theorem (Proposition 1.2.2) that $R(I)$ is noetherian.

Form the graded module

$$M' = M \oplus XIM \oplus X^2I^2M \oplus \cdots$$

for $R' = R(I)$. Then M' is, clearly, a finitely generated R'-module, and because R' is noetherian, so is M'. Hence the R'-submodule

$$N' = N \oplus X(IM \cap N) \oplus X^2(I^2M \cap N) \oplus \cdots$$

is finitely generated. Thus there exists an integer m with

$$N' = R'(N \oplus X(IM \cap N) \oplus \cdots \oplus X^m(I^mM \cap N)).$$

But then, by equating coefficients of X^n for all $n \geq m$, we deduce easily that

$$I^nM \cap N = I^{n-m}(I^mM \cap N)$$

as required.   ∎

Applying the Artin–Rees lemma, we now prove

7.1.11 Proposition (Malek 1972) *Let R be a commutative noetherian ring and suppose that the sub-ideals of a given ideal I of R satisfy the descending chain condition. Then the natural homomorphism*

$$U(R) \to U(R/I)$$

is surjective.

Proof. Because the sub-ideals of I satisfy the descending chain condition, some power $J = I^n$, $n \geq 1$, satisfies $J^2 = J$. We first claim that it suffices to prove that the natural homomorphism

$$U(R) \to U(R/J)$$

is surjective. Indeed, suppose that $n > 1$ and let

$$U(R/I^n) \to U(R/I^{n-1})$$

be the natural homomorphism. If $x + I^{n-1} \in U(R/I^{n-1})$ then $xy - 1 \in I^{n-1}$ for some $y \in R$. Hence $(xy - 1)^2 \in I^{2n-2} \subseteq I^n$, so $x(2y - xy^2) - 1 \in I^n$, proving that $x + I^n \in U(R/I^n)$. This substantiates our claim.

Consider the ideal Rx for $x \in J$. By Lemma 1.10, there exists an integer $k > 0$ such that for all $n \geq k$,

$$(J^n \cdot J) \cap Rx = J^{n-k}(J^k \cdot J \cap Rx).$$

Since $J^2 = J$, we obtain $J \cap Rx = J(J \cap Rx)$ and hence $Rx = Jx$. The latter implies that there exists $a \in J$ such that $x = ax = a^2x$ and $Ra = Ra^2 = J$. In particular, $a = ua^2$ for some $u \in U(R)$, $ua = u^2a^2 = e$ is an idempotent and $J = Re$. Let $r + J$ be a unit of R/J. Then $rr' - 1 = ye$ for some $r', y \in R$.

Hence $(e + r(1-e))(e + r'(1-e)) = 1$ and $r + J = e + r(1-e) + J$, as required. ∎

7.2 Finitely generated extensions

Throughout this section, all rings are assumed to be *commutative*. Let $R \subseteq S$ be a finitely generated ring extension, i.e. the ring S is finitely generated over its subring R. Our aim is to examine the factor group $U(S)/U(R)$ and to provide circumstances under which $U(S)/U(R)$ is free abelian of finite rank. We also examine in detail the unit group of Krull domains. To establish our main result (Theorem 2.6), we actually impose another condition on $R \subseteq S$, namely we assume that S is an integral domain and R a Krull domain which is integrally closed in S. As an application, we prove that if R is a finitely generated ring, then $U(R)$ is finitely generated if and only if the additive group of the nilradical $N(R)$ of R is finitely generated.

Our point of departure is the following general result.

7.2.1 Theorem (Krempa 1985) *Let $R \subseteq S$ be integral domains with quotient fields $K \subseteq L$. Suppose that R is integrally closed in S and that there exist elements $s_1, s_2, \ldots, s_n \in S$ such that:*

(i)   *$s_1, s_2, \ldots, s_n$ are algebraically independent over R;*
(ii) *S is integral over $R[s_1, \ldots, s_n]$;*
(iii) *L is a finite extension of $K(s_1, \ldots, s_n)$.*

Then $U(S)/U(R)$ is a free abelian group of finite rank.

Proof. We proceed by induction on n. Assume that $n = 1$ and put $s = s_1$. Owing to Lemma 4.1.19, there exists a unique (up to equivalence) valuation w on $K(s)$ such that w is trivial on K and $w(s) > 1$. Since $L/K(s)$ is a finite field extension, there exist only finitely many, say $v_1, \ldots, v_k$, extensions of w to L. The valuations $v_1, v_2, \ldots, v_k$ are discrete; hence they induce a homomorphism from $U(S)$ into a free abelian group of rank k. Let H be the kernel of this homomorphism. It suffices to show that $H = U(R)$. Since $U(R) \subseteq K$, we obviously have $U(R) \subseteq H$. Conversely, fix $h \in H$ and any valuation v on L which is trivial on K. If v is equivalent to one of the v_i, $1 \le i \le k$, then $v(h) = 1$. In the opposite case, we have $v(s) \le 1$ and so s belongs to the valuation ring O_v of v. Thus $R[s] \subseteq O_v$ and $S \subseteq O_v$. Therefore $v(h) \le 1$ and $v(h^{-1}) \le 1$, which implies $v(h) = 1$. Thus $v(h) = 1$ and therefore h is algebraic over K. By hypothesis, h is integral over $R[s]$, so there exists $m \ge 1$ and $\alpha_i \in R[s]$ such that

$$h^m + \alpha_{m-1} h^{m-1} + \cdots + \alpha_0 = 0.$$

Now s is not algebraic over R, so s is not algebraic over K, and hence over $K(h)$. We can write $\alpha_j = \sum_i a_{ij}s^i$ with $j \in \{0, 1, \ldots, m-1\}$ and $a_{ij} \in R$. Since s is not algebraic over $K(h)$, we deduce that

$$h^m + \sum_{j=0}^{m-1} a_{0j}h^j = 0.$$

Thus h is integral over R and, by assumption, $h \in R$. Similarly, we verify that $h^{-1} \in R$ and thus $h \in U(R)$, proving the case $n = 1$.

Assume that $n > 1$ and let C be the integral closure of $R[s_1, \ldots, s_{n-1}]$ in S. Denote by M the quotient field of C. Then the degree of M over $K(s_1, \ldots, s_{n-1})$ is at most the degree of L over $K(s_1, \ldots, s_n)$. Thus, by induction, $U(C)/U(R)$ is a free abelian group of finite rank. The ring S is integral over $C[s_n]$ and the degree of L over $M(s_n)$ is at most the degree of L over $K(s_1, \ldots, s_n)$ which is finite. Hence the case $n = 1$ applies, and therefore the group $U(S)/U(C)$ is free abelian of finite rank. Thus $U(S)/U(R)$ is free abelian of finite rank, as asserted. ∎

The above theorem becomes all the more useful when combined with the following classical result, which is known as the Noether normalization theorem.

7.2.2 Theorem *Let $R = F[x_1, \ldots, x_n]$ be a finitely generated integral domain over a field F and let r be the transcendence degree of $F(x_1, \ldots, x_n)$ over F. Then there exist r elements $y_1, \ldots, y_r$ in R which are algebraically independent over F and such that R is integral over the ring $F[y_1, y_2, \ldots, y_r]$.*

Proof. If $x_1, \ldots, x_n$ are algebraically independent over F, then $r = n$ and we may choose $y_i = x_i$, $1 \leq i \leq n$. In the contrary case, there is a non-trivial relation

$$\sum a_{(j)}x_1^{j_1} \ldots x_n^{j_n} = 0, \tag{1}$$

with each coefficient $a_{(j)} \in F$ and $a_{(j)} \neq 0$. The sum is taken over a finitely many distinct n-tuples of non-negative integers $(j_1, \ldots, j_n)$. Let $s_2, \ldots, s_n$ be positive integers, and put

$$y_2 = x_2 - x_1^{s_2}, \ldots, y_n = x_n - x_1^{s_n}.$$

We substitute $x_i = y_i + x_1^{s_i}$, $2 \leq i \leq n$, in eqn (1). For brevity put

$$(s) = (1, s_2, \ldots, s_n) \quad \text{and} \quad (j)(s) = j_1 + s_2 j_2 + \cdots + s_n j_n.$$

Expanding the relation after making the above substitution, we obtain

$$\sum c_{(j)}x_1^{(j)(s)} + f(x_1, y_2, \ldots, y_n) = 0, \tag{2}$$

where f is a polynomial in which no pure power of x_1 appears. Let d be an integer greater than any component of a vector (j) for which $c_{(j)} \neq 0$. Setting

$$(s) = (1, d, d^2, \ldots, d^{n-1}),$$

it follows that all $(j)(s)$ in (2) are distinct for those (j) with $c_{(j)} \neq 0$. In this way, we obtain an integral equation for x_1 over $F[y_2, \ldots, y_n]$. Because each x_i, $i > 1$, is contained in $F[x_1, y_2, \ldots, y_n]$, we deduce that R is integral over $F[y_2, \ldots, y_n]$. We can now proceed inductively, invoking the transitivity of integral extensions, to diminish the number of y's until we reach an algebraically independent set of y's. ∎

We are now ready to record our first application of Theorem 2.1.

7.2.3 Corollary (Roquette 1958, 1960) *Let R be an integral domain which is finitely generated over a field F. If F is algebraically closed in R, then $U(R)/U(F)$ is free abelian of finite rank.*

Proof. In view of Theorem 2.2, the integral domains $F \subseteq R$ satisfy the hypotheses of Theorem 2.1. ∎

We next provide a definition of the Krull domain. Let R be an integral domain and let F be the quotient field of R. Then R is said to be a *Krull domain* if there exists a family $\{V_i\} i \in I$ of discrete valuation rings with $V_i \subseteq F$, such that:

(i) $R = \cap_{i \in I} V_i$;
(ii) given a non-zero r in R, there is at most a finite number of $i \in I$ such that r is a non-unit of V_i.

It is a consequence of Proposition 1.2.21 that any Dedekind domain is a Krull domain. Note, however, that in contrast to Dedekind domains, Krull domains need not be noetherian.

7.2.4 Lemma *Let R be a Krull domain and let F be the quotient field of R. Let $\{v_i \mid i \in I\}$ be a family of discrete valuations of F for which the corresponding family $\{V_i \mid i \in I\}$ of discrete valuation rings satisfy conditions in the definition of a Krull domain. Let S be a multiplicative subset of R and let J consist of all $j \in I$ for which $v_j(S) = 1$. Then*

$$R_S = \bigcap_{j \in J} V_j$$

and, in particular, R_S is a Krull domain.

Proof. Put $B = \cap_{j \in J} V_j$. Then $S^{-1} \subseteq B$ and $R \subseteq B$, so $R_S \subseteq B$. Conversely, assume that $x \in B$. Let $K \subseteq I$ denote the finite set of those indices k for which $v_k(x) > 1$. If $k \in K$, then $x \notin V_k$ and hence $k \notin J$. Therefore, there exists an element $s_k \in S$ with $v_k(s_k) < 1$. Let $n(k)$ be a positive integer such that $v_k(s_k^{n(k)} x) \leqslant 1$. Put

$$s = \prod_{k \in K} s_k^{n(k)}.$$

Then $v_i(sx) \leqslant 1$ for all $i \in I$. Hence $sx \in R$ and $x \in R_S$, as required. ∎

7.2.5 Lemma *Let R be a Krull domain and let $0 \neq s \in R$. Then $R[s^{-1}]$ is a Krull domain such that*

$$U(R[s^{-1}])/U(R)$$

is free abelian of finite rank.

Proof. Put $S = \{1, s, s^2, \ldots\}$. Then $R[s^{-1}] = R_S$ and hence, by Lemma 2.4, $R[s^{-1}]$ is a Krull domain. Let F be the quotient field of R and let $\{V_i\} i \in I$ be a family of discrete valuation rings in the definition of a Krull domain. For each $i \in I$, choose a discrete valuation v_i, the valuation ring of which is V_i. By hypothesis, there exist only finitely many, say $v_1, \ldots, v_n$, valuations v_i for which $v_i(s) \neq 1$. Then, by Lemma 2.4,

$$R = R[s^{-1}] \cap V_1 \cap \cdots \cap V_n.$$

Now $\{v_1, \ldots, v_n\}$ determines a homomorphism f from $U(R[s^{-1}])$ to a free abelian group of rank n. Since $R[s^{-1}] \subseteq V_i$ for all $i \notin \{1, \ldots, n\}$, we must have $\mathrm{Ker} f = U(R)$. So the lemma is true. ∎

We have done most of the work to establish our main result.

7.2.6 Theorem (Krempa 1985) *Let $R \subseteq S$ be integral domains such that R is integrally closed in S, and S is finitely generated over R. If R is a Krull domain, then $U(S)/U(R)$ is a free abelian group of finite rank.*

Proof. From the Noether normalization theorem (Theorem 2.2) we obtain an element $0 \neq s \in R$ such that $S[s^{-1}]$ is finitely generated integral extension of a polynomial ring over $R[s^{-1}]$. Obviously, $R[s^{-1}]$ is integrally closed in $S[s^{-1}]$. Invoking Theorem 2.1 and Lemma 2.5, we deduce that

$$U(S[s^{-1}])/U(R[s^{-1}]) \qquad \text{and} \qquad U(R[s^{-1}])/U(R)$$

are free abelian of finite rank. Hence $U(S[s^{-1}])/U(R)$ is free abelian of finite rank. But $U(S)/U(R)$ is a subgroup of $U(S[s^{-1}])/U(R)$; hence the result. ∎

7.2.7 Corollary (Roquette 1960, Samuel 1966) *Let R be a finitely generated integral domain. Then $U(R)$ is a finitely generated group.*

Proof. Let F be the quotient field of R and let K be the algebraic closure in F of the prime subfield F_0 of F. Because F is finitely generated over F_0, we have $(K:F_0)<\infty$. Let S be the integral closure in K of the prime subring of R. Then $U(S)$ is finitely generated. Indeed, if char $R \neq 0$, then S is a finite field. In case char $R = 0$, S is the ring of integers of an algebraic number field, and hence $U(S)$ is finitely generated by the Dirichlet Unit Theorem. Note also that, in both cases, S is a Dedekind domain and hence a Krull domain. Now S is integrally closed in $SR \subseteq F$ and, by Theorem 2.6, we know that $U(SR)$ is finitely generated. Thus $U(R)$ is also finitely generated. ■

As an easy application of the above, we now record the following important result.

7.2.8 Corollary (Bass 1974) *Let R be a finitely generated commutative ring, let $N(R)$ be the nilradical of R and let $X = \operatorname{ann}_R(N(R)) = \{r \in R \mid rN(R) = 0\}$. Then the following conditions are equivalent*:

(i)   *$U(R)$ is a finitely generated group;*
(ii) *$N(R)$ is a finitely generated additive group;*
(iii) *R/X is a finitely generated additive group.*

Proof. Since R is finitely generated, R is noetherian, and hence $R/N(R)$ is a subdirect product of finitely many integral domains. Each of these integral domains is finitely generated and hence, by Corollary 2.7, $U(R/N(R))$ is finitely generated. The desired conclusion is therefore a consequence of Corollary 1.9. ■

7.2.9 Corollary *Let R be a commutative ring such that $U(R)$ is finitely generated. Then $J(R)$ is nilpotent and the additive group of $J(R)$ is finitely generated.*

Proof. Let S be the subring of R generated by a set of generators of $U(R)$ and their inverses. Obviously, $J(R) \subseteq J(S) = N(S)$ and $U(R) = U(S)$. Now apply Corollary 2.8. ■

We close by recording the following result required for subsequent investigations.

7.2.10 Proposition (May 1976) *Let R be a finitely generated integral domain, and let $\bar{R}$ be the domain obtained by adjoining all roots of unity to R. Then $U(\bar{R})$ is free modulo torsion.*

Proof. Let F be the quotient field of R, and let $\bar{F}$ be the field obtained by adjoining all roots of unity to F. Then F is a finitely generated field and $\bar{F}$ is an (infinite) abelian extension of F. Hence, by Corollary 4.5.22, $\bar{F}^*$ is free modulo torsion. Since $U(\bar{R}) \subseteq \bar{F}^*$, the result follows. ∎

7.3 The Whitehead group and stability theorem

Our aim in this section is to introduce some important facts pertaining to algebraic K-theory. These include the notion of $K_1(R)$ and the stability theorem due to Bass and Vaserstein. We also illustrate the fact that the knowledge of $K_1(R)$ is fundamental in studying finite generation of the general linear group $GL_n(R)$. All the information obtained will be applied in the next section. Throughout, $R[X_1, \ldots, X_n]$ denotes a polynomial ring in n variables.

Let R be an arbitrary ring and let n be a positive integer. Then $M_n(R)$ is a free R-module with the matrix units e_{ij}, $1 \le i, j \le n$, as a basis. Recall that

$$e_{ij}e_{ks} = \delta_{jk}e_{is}.$$

In particular, if $i \ne j$, then $e_{ij}^2 = 0$. Thus, for any $i \ne j$ and $r \in R$, the matrix

$$e_{ij}(r) = I_n + re_{ij} \qquad (I_n \text{ is the } n \times n \text{ identity matrix})$$

is invertible. Such matrices are called *elementary*. The subgroup of $GL_n(R)$ generated by all elementary matrices will be denoted by $E_n(R)$, and called the *elementary subgroup* of $GL_n(R)$. The elementary matrices satisfy the following easily verified relations:

$$e_{ij}(r)e_{ij}(s) = e_{ij}(r+s), \tag{1}$$

$$[e_{ij}(r), e_{jk}(s)] = e_{ik}(rs) \qquad \text{if } i, j, k \text{ are distinct,} \tag{2}$$

$$[e_{ij}(r), e_{hk}(s)] = 1 \qquad \text{if } j \ne h \text{ and } i \ne k, \tag{3}$$

where $[x, y] = xyx^{-1}y^{-1}$.

The inclusion $M_n(R) \subseteq M_{n+k}(R)$ is given by matrix direct sum with I_k, i.e.

$$M \mapsto M \oplus I_k = \begin{bmatrix} M & 0 \\ 0 & I_k \end{bmatrix}.$$

In particular, identifying each $M \in GL_n(R)$ with the matrix

$$\begin{bmatrix} M & 0 \\ 0 & 1 \end{bmatrix} \in GL_{n+1}(R)$$

we obtain inclusions

$$GL_1(R) \subset GL_2(R) \subset \cdots \subset GL_n(R) \subset \ldots ;$$
$$E_1(R) \subset E_2(R) \subset \cdots \subset E_n(R) \subset \ldots .$$

The union

$$GL(R) = \bigcup_{n=1}^{\infty} GL_n(R)$$

is called the infinite *general linear group*. Similarly, we put

$$E(R) = \bigcup_{n=1}^{\infty} E_n(R).$$

For future use, we now record the following elementary observations.

7.3.1 Lemma

(i) $E_n(R) = [E_n(R), E_n(R)]$ *for all* $n \geq 3$.
(ii) *If* $R \to S$ *is a surjective ring homomorphism, then the induced homomorphism* $E_n(R) \to E_n(S)$ *is also surjective.*
(iii) $E_n(R)$ *is stable under transposition.*
(iv) $E_n(R)$ *contains all matrices of the form*

$$\begin{bmatrix} 1 & & * \\ & \ddots & \\ 0 & & 1 \end{bmatrix}.$$

Proof.

(i)   A direct consequence of (2).
(ii) If $f : R \to S$ is a ring homomorphism, then the induced homomorphism $GL_n(R) \to GL_n(S)$ sends $e_{ij}(r)$ to $e_{ij}(f(r))$.
(iii) This is a direct consequence of the fact that the transpose of $e_{ij}(r)$ is $e_{ji}(r)$.
(iv) If $x = (r_2, \ldots, r_n)$, then

$$\begin{bmatrix} 1 & x \\ 0 & I_{n-1} \end{bmatrix} = e_{12}(r_2) \cdots e_{in}(r_n) \in E_n(R).$$

If $y \in E_{n-1}(R)$, then

$$\begin{bmatrix} 1 & x \\ 0 & y \end{bmatrix} = \begin{bmatrix} 1 & 0 \\ 0 & y \end{bmatrix} \begin{bmatrix} 1 & x \\ 0 & I_{n-1} \end{bmatrix} \in E_n(R)$$

also. The required assertion now follows by induction on n. ∎

We now look at the relationship of $E_n(R)$ to $GL_n(R)$. Note that if our ring R is already a ring of matrices $M_k(R)$, then we may equate $M_n(M_k(R))$ with $M_{nk}(R)$, so long as we make no claims concerning compatibility with inclusions $M_n(R) \subseteq M_{n+1}(R)$.

7.3.2 Lemma   We have $E_2(M_n(R)) \subseteq E_{2n}(R)$.

Proof. The group $E_2(M_n(R))$ is generated by $e_{12}(\alpha)$ and its transpose

$e_{21}(\alpha)$, where α ranges over all matrices $\alpha = (a_{ij})$ of $M_n(R)$. If we write

$$\alpha = \left[\begin{array}{c|c} \alpha' & \begin{matrix} a_{1n} \\ \vdots \\ a_{nn} \end{matrix} \end{array}\right],$$

then

$$e_{12}(\alpha) = \left[\begin{array}{c|c|c} I_n & \alpha' & \begin{matrix} a_{1n} \\ \vdots \\ a_{nn} \end{matrix} \\ \hline 0 & I_n & \end{array}\right]$$

$$= e_{1,2n}(a_{1n}) \cdots e_{n,2n}(a_{nn}) \left[\begin{array}{c|c|c} I_n & \alpha' & \begin{matrix} 0 \\ \vdots \\ 0 \end{matrix} \\ \hline 0 & I_{n-1} & 0 \\ \hline 0 \cdots 0 & & 1 \end{array}\right].$$

Thus, by iterating, we obtain

$$e_{12}(\alpha) = \left(\prod_{i=1}^n e_{i,2n}(a_{in})\right) \cdots \left(\prod_{i=1}^n e_{i,n+j}(a_{ij})\right) \cdots \left(\prod_{i=1}^n e_{i,n+1}(a_{i1})\right)$$

as required. ■

7.3.3 Lemma   *If $\alpha \in GL_n(R)$, then:*

(i)   *α is conjugate in $GL_{2n}(R)$ to*

$$I_n \oplus \alpha = \begin{bmatrix} I_n & 0 \\ 0 & \alpha \end{bmatrix}$$

by an element of E_{2n};

(ii) *$\alpha \oplus \alpha^{-1} = \begin{bmatrix} \alpha & 0 \\ 0 & \alpha^{-1} \end{bmatrix} \in E_{2n}(R).$*

Proof.   Given $a \in U(R)$, we have

$$\begin{bmatrix} 0 & a \\ -a^{-1} & 0 \end{bmatrix} = e_{12}(a)e_{21}(-a^{-1})e_{12}(a) \in E_2(R)$$

and hence, by Lemma 3.2,

$$\begin{bmatrix} 0 & \alpha \\ -\alpha^{-1} & 0 \end{bmatrix} \in E_{2n}(R).$$

Invoking the equalities

$$\begin{bmatrix} \alpha & 0 \\ 0 & I_n \end{bmatrix} = \begin{bmatrix} 0 & I_n \\ -I_n & 0 \end{bmatrix}\begin{bmatrix} I_n & 0 \\ 0 & \alpha \end{bmatrix}\begin{bmatrix} 0 & -I_n \\ I_n & 0 \end{bmatrix},$$

$$\begin{bmatrix} \alpha & 0 \\ 0 & \alpha^{-1} \end{bmatrix} = \begin{bmatrix} 0 & \alpha \\ -\alpha^{-1} & 0 \end{bmatrix}\begin{bmatrix} 0 & -I_n \\ I_n & 0 \end{bmatrix}$$

the result follows. ■

As an easy application of the above, we prove the following fundamental property.

7.3.4 Proposition   (Whitehead lemma)   *We have*

$$E(R) = [GL(R), GL(R)] = [E(R), E(R)].$$

Proof.   That $[E(R), E(R)] = E(R)$ is a consequence of Lemma 3.1(i). In particular,

$$E(R) \subseteq [GL(R), GL(R)].$$

Conversely, suppose $\alpha_1, \alpha_2 \in GL_n(R)$. Then

$$[\alpha_1 \oplus I_n, \alpha_2 \oplus I_n] = [\alpha_1, \alpha_2] \oplus I_n$$
$$= (\alpha_1 \oplus \alpha_1^{-1})(\alpha_2 \oplus \alpha_2^{-1})(\alpha_1^{-1}\alpha_2^{-1} \oplus \alpha_2\alpha_1)$$

belongs to $E_{2n}(R)$ by Lemma 3.3. Thus

$$[GL_n(R), GL_n(R)] \subseteq E_{2n}(R)$$

and hence $[GL(R), GL(R)] \subseteq E(R)$, as required.   ∎

It follows from Proposition 3.4 that $E(R)$ is a normal subgroup of $GL(R)$ with abelian quotient group. The quotient group will be called the *Whitehead group* of R and denoted by $K_1(R)$:

$$K_1(R) = GL(R)/E(R).$$

We will usually think of $K_1(R)$ as an additive group. Clearly K_1 is a covariant functor: that is, any ring homomorphism $R \to R'$ gives rise to a ring homomorphism $K_1(R) \to K_1(R')$.

Note that $E_n(R)$ need not be a normal subgroup of $GL_n(R)$. In what follows we write $GL_n(R)/E_n(R)$ for the set of left cosets modulo $E_n(R)$.

The following lemma explains why the knowledge of $K_1(R)$ is fundamental in studying finite generation of the group $GL_n(R)$.

7.3.5 Lemma   *Define the map*

$$s_n : GL_n(R)/E_n(R) \to K_1(R)$$

to be induced by the inclusion $GL_n(R) \subseteq GL(R)$.

(i)   *If s_n is injective, then $E_n(R) \triangleleft GL_n(R)$ and s_n is a homomorphism.*

(ii)   *If $n \geq 3$ and R is a finitely generated ring, then $E_n(R)$ is a finitely generated group.*

(iii)   *If $n \geq 3$, s_n is injective and both R and $K_1(R)$ are finitely generated, then $GL_n(R)$ is finitely generated.*

Proof.

(i) If s_n is injective, then $E_n(R) = E(R) \cap GL_n(R)$. Hence, if $x \in GL_n(R)$, then

$$x^{-1} E_n(R) x \subseteq E(R) \cap GL_n(R) = E_n(R)$$

as required.

(ii) Let S be a finite subset of R containing 1 that generates R as a ring. Put

$$G = \langle e_{ij}(s) \mid 1 \leqslant i, j \leqslant n, i \neq j, s \in S \rangle$$

and

$$R' = \{r \in R \mid e_{ij}(r) \in G \quad \text{for all} \quad i \neq j\}.$$

We claim that $R' = R$; if sustained, it will follow that $G = E_n(R)$, as required.

By definition, $S \subseteq R'$. Relations (1) imply that R' is an additive group. Then the assumption $n \geqslant 3$ and relations (2) ensure that R' is a subring of R. Hence $R = R'$ as claimed.

(iii) A direct consequence of (i) and (ii). ∎

If the ring R happens to be commutative, then we can also consider the special linear group $SL_n(R)$, consisting of all matrices in $GL_n(R)$ with determinant 1. We put

$$SL(R) = \bigcup_{n=1}^{\infty} SL_n(R).$$

The quotient $SL(R)/E(R)$ will be denoted by $SK_1(R)$, and called the *special Whitehead group.* Note that the inclusion

$$GL_1(R) = U(R) \subseteq GL(R)$$

splits the determinant homomorphism

$$\det : GL(R) \to U(R)$$

and so induces a decomposition

$$K_1(R) = U(R) \oplus SK_1(R).$$

By an *elementary row operation on a matrix* we mean left multiplication by an elementary matrix. Note that $SK_1(R) = 0$ if and only if every matrix in $GL(R)$ with determinant 1 can be reduced to the identity matrix by elementary row operations. For many important rings $SK_1(R) = 0$. For example, if R is a field or $R = \mathbb{Z}$, then $SK_1(R) = 0$. More generally, as we shall see below, if R is any euclidean ring, then $SK_1(R) = 0$.

A ring R is said to be a *euclidean ring* if there is a function

$$d : R - \{0\} \to \mathbb{N}$$

such that for any two non-zero elements a, b in R there is a third, say q, with either $a = qb$ or else $d(a - qb) < d(b)$.

Typical examples of euclidean rings are $\mathbb{Z}$ (with $d(a) = |a|$) and the polynomial ring $F[X]$ over a field F (with $d(a) = 1 + \deg(a)$).

7.3.6 Lemma   *Let R be a commutative euclidean ring. Then $SK_1(R) = 0$ and hence*

$$K_1(R) \cong U(R).$$

Proof.   Fix $\alpha = (a_{ij}) \in SL_n(R)$ and denote by n_α the number of $i \in \{1, \ldots, n\}$ for which $a_{in} = 0$. Then $n_\alpha \leq n - 1$ since α is a unit. It follows that the integer $\bar{d}(\alpha)$ given by

$$\bar{d}(\alpha) = (n - 1 - n_\alpha)\min\{d(a_{in}) \mid a_{in} \neq 0, 1 \leq i \leq n\}$$

is non-negative, attaining the value 0 only when the final column of α contains a single non-zero entry. Choose $t \in \{1, \ldots, n\}$ such that $d(a_{tn}) = \min\{d(a_{in}) \mid a_{in} \neq 0\}$. For each $a_{in} \neq 0$ $(i \neq t)$, there exists $q_i \in R$ with $a_{in} - q_i a_{tn} = 0$ or with $d(a_{in} - q_i a_{tn}) < d(a_{tn})$. If $a_{in} = 0$ or $i = t$, then put $q_i = 0$. It follows that the matrix $\alpha' = (\prod_{i=1}^{n} e_{it}(-q_i))\alpha$ has its last column composed of elements $a'_{in} = a_{in} - q_i a_{tn}$. Therefore $\bar{d}(\alpha') < \bar{d}(\alpha)$. Iteration finally yields a coset representative modulo $E_n(R)$ for which there is only one non-zero entry in the final column. Should this be in the ith row, where $i \neq n$, then premultiplication by $e_{in}(-1)e_{ni}(1)$ converts it to the (n, n)-position. Again, this element emerges as a factor in the expansion of the determinant and so must be a unit, say $a \in U(R)$. Applying Lemma 3.3, we may premultiply by the matrix $I_{n-2} \oplus a \oplus a^{-1} \in E_n(R)$, to obtain a coset representative $\tilde{\alpha}$ with the last column comprising $\overline{a_{in}} = \delta_{in}$. Hence

$$\tilde{\alpha} = \begin{bmatrix} \beta & 0 \\ \bar{a} & 1 \end{bmatrix} = \begin{bmatrix} I_{n-1} & 0 \\ \bar{a}\beta^{-1} & 1 \end{bmatrix}\begin{bmatrix} \beta & 0 \\ 0 & 1 \end{bmatrix}.$$

Bearing in mind that

$$\begin{bmatrix} I_{n-1} & 0 \\ \bar{a}\beta^{-1} & 1 \end{bmatrix} = \prod_{j<n} e_{nj}(c_j)$$

for suitable $c_j \in R$, we are finally left with a coset representative of the form $\beta \oplus I_1 \in SL_{n-1}(R)$. We may therefore repeat the whole procedure to derive a coset representative in $SL_{n-2}(R)$ and so on, until ultimately in $SL_1(R)$, where it can only be the identity matrix. Thus $SK_1(R) = 0$ as asserted. ∎

Let R be an arbitrary ring. If M is a (right) R-module, the *content* of an element $x \in M$ is defined to be the left ideal

$$\mathrm{cont}(x) = \{f(x) \mid f \in \mathrm{Hom}_R(M, R)\}.$$

Following Bass (1974), we call x *unimodular* in M if $\text{cont}(x) = R$; this is equivalent to x being a basis for a free direct summand of M. Note that if M is the free module R^n and $x = (a_1, \ldots, a_n)$, then

$$\text{cont}(x) = Ra_1 + \cdots + Ra_n.$$

For example, if $\alpha \in GL_n(R)$ then the equation $\alpha^{-1}\alpha = I$ shows that the columns of α are unimodular in R^n.

In view of Lemma 3.5(iii), it is important to discover circumstances under which the map

$$s_n : GL_n(R)/E_n(R) \to K_1(R)$$

is injective. For an integer $d \geq 0$, we consider the following condition to which we refer as $(S)_d$:

For all $n > d$, and all unimodular $x = (a_0, a_1, \ldots, a_n)$ in R^{n+1}, there exist $b_1, \ldots, b_n \in R$ such that $(a_1', \ldots, a_n')$ is unimodular in R^n, where $a_i' = a_i + b_i a_0$, $1 \leq i \leq n$.

7.3.7 Theorem (Bass 1974, Vaserstein 1969) (Stability Theorem) *Assume that the condition $(S)_d$ holds. Then*

$$s_n : GL_n(R)/E_n(R) \to K_1(R)$$

is

(a)   *surjective for all $n \geq d + 1$, and*
(b)   *injective for all $n \geq d + 2$.*

Proof.   The proof of property (b), due to Vaserstein, is too long and technical and therefore will be omitted.

To prove (a), we first show that condition $(S)_d$ implies the following condition $(E)_d$: For all $n \geq d + 1$, $E_{n+1}(R)$ acts transitively on the set of unimodular elements in R^{n+1}.

We need only verify that if $x = (a_0, a_1, \ldots, a_n)$ is unimodular in R^{n+1}, then

$$yx = (1, 0, \ldots, 0) \qquad \text{for some } y \in E_{n+1}(R).$$

Observe that the effect of $e_{ij}(b)$ on x is to add ba_j to the coordinate a_i. Assume first that $(a_1, \ldots, a_n)$ is unimodular. Then we can solve

$$1 - a_0 = b_1 a_1 + \cdots + b_n a_n$$

and therefore transform x by an element of $E_{n+1}(R)$ to replace a_0 by 1. Hence, subtracting $a_i \cdot 1$ from the coordinate a_i, $1 \leq i \leq n$, we obtain the desired element $(1, 0, \ldots, 0)$. Therefore, in the general case, it suffices to observe that we can add multiplies $b_i a_0$ of a_0 to a_i, $1 \leq i \leq n$, so that, if $a_i' = a_i + b_i a_0$, then $(a_1', \ldots, a_n')$ is unimodular. The existence of such elements b_i is guaranteed by condition $(S)_d$.

By the foregoing, it suffices to verify that condition $(E)_d$ ensures that the natural map

$$GL_n(R)/E_n(R) \to GL_{n+1}(R)/E_{n+1}(R)$$

is surjective for all $n \geqslant d + 1$. So assume $n \geqslant d + 1$. Given $\alpha \in GL_{n+1}(R)$, we must exhibit $x \in E_{n+1}(R)$ such that

$$x\alpha = \begin{bmatrix} \alpha_1 & 0 \\ 0 & 1 \end{bmatrix} \qquad \text{for some } \alpha_1 \in GL_n(R). \tag{4}$$

Applying condition $(E)_d$ to the last column of α, which is unimodular in R^{n+1}, we can find $x_1 \in E_{n+1}(R)$ such that

$$x_1\alpha = \begin{bmatrix} \alpha_1 & 0 \\ \beta & 1 \end{bmatrix} = \begin{bmatrix} I_n & 0 \\ \beta\alpha_1^{-1} & 1 \end{bmatrix} \begin{bmatrix} \alpha_1 & 0 \\ 0 & 1 \end{bmatrix}$$

for some $\alpha_1 \in GL_n(R)$ and some $\beta \in R^n$. Because

$$x_2 = \begin{bmatrix} I_n & 0 \\ -\beta\alpha_1^{-1} & 1 \end{bmatrix} \in E_{n+1}(R)$$

the element $x = x_2 x_1$ clearly satisfies (4), thus completing the proof. ∎

7.3.8 Corollary *Assume that condition $(S)_d$ holds. If $K_1(R)$ is finitely generated, and if R is a finitely generated ring, then $GL_n(R)$ is finitely generated for all $n \geqslant \max(d + 2, 3)$.*

Proof. The proof is a direct consequence of Theorem 3.7 and Lemma 3.5(iii). ∎

Next we introduce the notion of homological dimension. Let M be a module over an arbitrary ring R. We may choose an epimorphism $\phi_0 : P_0 \to M$, where P_0 is projective. Then choose an epimorphism $\phi_1 : P_1 \to \operatorname{Ker} \phi_0$ with P_1 projective, and so on. This yields a *projective resolution* of M; that is, an exact sequence

$$\cdots \longrightarrow P_2 \xrightarrow{\phi_2} P_1 \xrightarrow{\phi_1} P_0 \xrightarrow{\phi_0} M \longrightarrow 0$$

in which each P_i is projective. If it happens that $\operatorname{Ker} \phi_m$ is projective for some m, then the process of forming such a resolution can be terminated, giving a finite resolution

$$0 \to \operatorname{Ker} \phi_m \to P_m \to P_{m-1} \to \cdots \to P_0 \to M \to 0.$$

Let M be a (right) R-module. Then the *homological dimension* of M, denoted $hd(M)$, is defined to be the least n (possibly ∞) for which there is a projective resolution of M of length n. We agree that $hd(0) = -1$. The *right global dimension* of R is defined to be

$$\text{r.gl.dim}(R) = \sup_M hd(M).$$

We close this section by quoting the following fact, the proof of which is too technical to be presented here.

7.3.9 Proposition *Let R be a right noetherian ring with finite right global dimension. Then*

$$K_1(R) \cong K_1(R[X_1, X_2, \ldots, X_n]) \qquad \text{for all } n \geq 1.$$

Proof. See Bass (1974; Corollary 9.4). ∎

7.4 Finite generation of $GL_n(R)$

Throughout, $\mathbb{F}_q$ denotes a finite field with q elements and $R[X_1, \ldots, X_n]$ a polynomial ring in n variables over a ring R.

Let R be either $\mathbb{Z}$ or $\mathbb{F}_q[X]$. Our main objective is to prove that:

(i) $GL_n(R)$ is a finitely generated group for all $n \geq 3$;
(ii) $GL_n(R[X_1, \ldots, X_d])$ is a finitely generated group for all $n \geq d + 3$.

The situation for *small n,* which our theorem does not cover, is not altogether clear. That some restriction is necessary is shown already by the fact (see Nagao 1959) that $GL_2(\mathbb{F}_q[X])$ is not finitely generated. The fact that $GL_2(\mathbb{Z})$ is finitely generated can be established by elementary methods.

We begin by reviewing some topological notions used in commutative algebra (for details, refer to Bourbaki 1965). Let T be a topological space. Then T is said to be *irreducible* if it is not empty, and is not the union of two proper closed subspaces. One easily verifies that the set of irreducible closed subspaces of T is inductively ordered by inclusion, whence T is the union of its maximal irreducible closed subspaces. We shall refer to those subspaces as the *irreducible components* of T. The *dimension* of T is defined to be the supremum of the lengths n of chains

$$T_0 \subset T_1 \subset \cdots \subset T_n$$

of distinct irreducible closed subspaces of T. We say that T is *noetherian* if its open (respectively, closed) sets satisfy the ascending (respectively, descending) chain condition. In this case every subspace of T is noetherian and T has only a finite number of irreducible components. The latter can be established by reducing to the obvious case where every proper closed subspace of T has only finitely many irreducible components.

Let R be a commutative ring, let $\text{spec}(R)$ be the set of all prime ideals of R, and for each subset S of R, put

$$V(S) = \{P \in \text{spec}(R) \mid P \supseteq S\}.$$

These are easily shown to be the closed sets of a topology on $\text{spec}(R)$. Furthermore, $S \mapsto V(S)$ induces an inclusion-reversing bijection to them from the set of ideals S, which are intersections of prime ideals. In particular, if R is a noetherian ring, then $\text{spec}(R)$ is a noetherian space. We refer to the dimension of $\text{spec}(R)$ as the *Krull dimension* of R, and denote by $\max(R)$ the subspace of $\text{spec}(R)$ formed by the maximal ideals of R. Hence

$$\dim \max(R) \leqslant \text{Krull} \dim(R)$$

and $\max(R)$ is noetherian if $\text{spec}(R)$ is noetherian.

We are now ready to prove

7.4.1 Theorem (Bass 1974) *Let R be a commutative ring such that $\max(R)$ is a noetherian space of dimension $\leqslant d$. Then R satifies the condition $(S)_d$ of Theorem 3.7.*

Proof. Fix $n > d$ and assume that $x = (a_0, a_1, \ldots, a_n)$ is a unimodular element in R^{n+1}. We must prove that there exist $b_1, \ldots, b_n \in R$ such that

$$(a_1', \ldots, a_n') \quad \text{is unimodular in } R^n,$$

where $a_i' = a_i + b_i a_0$, $1 \leqslant i \leqslant n$.

For any $M \in \max(R)$, we can certainly find a linear combination

$$c_0 a_0 + \cdots + c_{n-1} a_{n-1}$$

of $a_0, \ldots, a_{n-1}$ such that

$$a_n'' = c_0 a_0 + \cdots + c_{n-1} a_{n-1} + a_n \not\equiv 0 (\text{mod } M).$$

This is so since not all of the a_i, $0 \leqslant i \leqslant n$, are in M. Applying the Chinese Remainder Theorem, we can even find $c_0, \ldots, c_{n-1}$ which accomplish this simultaneously for all $M \in S$, where S is a given finite subset of $\max(R)$. We do this for an S large enough to meet each of the irreducible components of X (the number of such components is finite, since by hypothesis $\max(R)$ is noetherian).

If $d = 0$ then $\max(R)$ is finite, $S = \max(R)$, and a_n'' is invertible. Hence $(a_1, \ldots, a_{n-1}, a_n + c_0 a_0)$ is unimodular, and therefore the proof is complete in this case (and, more generally, in case a_n'' is invertible).

Now assume that a_n'' is not invertible. We put $\bar{R} = R/Ra_n''$ and claim that

$$\dim \max(\bar{R}) < d.$$

Indeed, $\max(\bar{R})$ can be identified with the closed set of all $M \in \max(R)$ which contain a_n''. Because $a_n'' \notin M$ for all $M \in S$, our choice of S ensures that $\max(\bar{R})$ contains no irreducible component of $\max(R)$. Therefore, a chain of irreducible closed sets in $\max(\bar{R})$ can always be lengthened in $\max(R)$, so $\dim \max(\bar{R}) < \dim \max(R) \leq d$, as claimed.

Consider the unimodular element $\bar{x} = (\bar{a}_0, \ldots, \bar{a}_{n-1})$ in the $\bar{R}$-module $\bar{R}^n$, where $\bar{a}$ denotes the image of a in $\bar{R}$. By the preceding paragraph, we have $\dim \max(\bar{R}) \leq d - 1$, so we can apply induction on d. Thus we obtain $b_1, \ldots, b_{n-1}$ in R such that $(\bar{a}_1 + \bar{b}_1\bar{a}_0, \ldots, \bar{a}_{n-1} + \bar{b}_{n-1}\bar{a}_0)$ is unimodular in the $\bar{R}$-module $\bar{R}^{n-1}$. Therefore $(a_1 + b_1a_0, \ldots, a_{n-1} + b_{n-1}a_0, a_n'')$ is unimodular in R^n. Setting

$$a_i' = a_i + b_ia_0 \qquad (1 \leq i \leq n)$$

we have

$$
\begin{aligned}
a_n'' &= a_n + c_0a_0 + (c_1a_1 + \cdots + c_{n-1}a_{n-1}) \\
&= a_n + c_0a_0 + (c_1a_1' + \cdots + c_{n-1}a_{n-1}') - (c_1b_1a_0 + \cdots + c_{n-1}b_{n-1}a_0) \\
&= a_n + b_na_0 + (c_1a_1' + \cdots + c_{n-1}a_{n-1}'),
\end{aligned}
$$

where

$$b_n = c_0 - (c_1b_1 + \cdots + c_{n-1}b_{n-1}).$$

But then we see that

$$(a_1 + b_1a_0, \ldots, a_n + b_na_0) = (a_1', \ldots, a_{n-1}', a_n + b_na_0)$$

is unimodular, as desired.   ■

It is now an easy matter to prove the following result.

7.4.2   Theorem   (Bass 1974)   *Let R be either $\mathbb{Z}$ or $\mathbb{F}_q[X]$. Then*:

(i) $GL_n(R)$ *is a finitely generated group for all $n \geq 3$;*

(ii) $GL_n(R[X_1, \ldots, X_d])$ *is a finitely generated group for all $n \geq d + 3$.*

Proof.   It is a standard fact of commutative ring theory (see Matsumura 1970) that if A is a noetherian commutative ring, then

$$\dim A[X] = 1 + \dim A = \dim \max(A[X]).$$

In particular, $\dim R = \dim \max(R) = 1$, and hence

$$\dim \max(R[X_1, \ldots, X_d]) = d + 1.$$

Thus, by Theorem 4.1, R satisfies $(S)_1$ and $R[X_1, \ldots, X_d]$ satisfies $(S)_{d+1}$.

By Lemma 3.6, $K_1(R) \cong U(R)$ and the latter is a finite group. Since the global dimension of R is obviously finite, Proposition 3.9 tells us that

$$K_1(R) \cong K_1(R[X_1, \ldots, X_d]).$$

Thus, by Corollary 3.8, $GL_n(R)$ is finitely generated for all $n \geq 3$ and

$GL_n(R[X_1, \ldots, X_d])$ is finitely generated for all $n \geqslant \max(d + 3, 3) = d + 3$. ∎

We close by remarking that Theorem 4.2(i) in case $R = \mathbb{Z}$ is a very special case of a result of Borel and Harish–Chandra (1962), asserting that all S-arithmetic groups are finitely generated (and even finitely presented).

8
Unit groups of group rings

This chapter provides a detailed analysis of the structure of unit groups of group rings. After presenting some preliminary results pertaining to the general theory of group rings, we exhibit a class of rings R for which RG has only trivial central units of finite order. We then prove a classical result of Higman, which surveys all torsion groups G for which $U(\mathbb{Z}G)$ has only trivial units. Turning our attention to unique product groups, we survey all commutative rings R for which RG has only trivial units. Our next topic is the investigation of conjugacy classes of group bases of RG, where G is finite and R is the ring of integers of an algebraic number field. The main result asserts that there are only finitely many such classes. We then describe $U(\mathbb{Z}G)$ in case G is dihedral of order $2p$. Our next result deals with existence of torsion-free complements for G in $V(\mathbb{Z}G)$. It is shown that if G is a finite group with an abelian normal subgroup A such that either G/A is abelian of exponent dividing 4 or 6, or G/A is of odd order, then G has a torsion-free complement in $V(\mathbb{Z}G)$. Next we provide necessary and sufficient conditions under which $U(\mathbb{Z}G)$ is solvable. As an application, we prove that if G is neither abelian nor a Hamiltonian 2-group, then $U(\mathbb{Z}G)$ contains a free subgroup of rank 2. Our final section is devoted to a detailed investigation of units in commutative group rings.

8.1 Definitions and elementary properties

Let R be a commutative ring and let G be a group, possible infinite. The *group ring* RG of G over R is the free R-module on the elements of G, with multiplication induced by that in G. More explicitly, RG consists of all formal linear combinations $\sum x_g \cdot g$, $x_g \in R$, $g \in G$, with finitely many $x_g \neq 0$ subject to:

(i) $\sum x_g \cdot g = \sum y_g \cdot g$ if and only if $x_g = y_g$ for all $g \in G$;
(ii) $\sum x_g \cdot g + \sum y_g \cdot g = \sum (x_g + y_g) \cdot g$;
(iii) $(\sum x_g \cdot g)(\sum y_h \cdot h) = \sum z_t \cdot t$ where $z_t = \sum_{gh=t} x_g y_h$;
(iv) $r(\sum x_g \cdot g) = \sum (r x_g) \cdot g$ for all $r \in R$.

It is straightforward to verify that these operations define RG as an associative R-algebra with $1 = 1_R \cdot 1_G$, where 1_R and 1_G are identity elements of R and G, respectively. For this reason, it is also common to refer to RG as the *group algebra* of G over R.

With the aid of the injective homomorphisms

$$\begin{cases} R \to RG \\ x \mapsto r \cdot 1_G \end{cases} \qquad \begin{cases} G \mapsto RG \\ g \mapsto 1_R \cdot g \end{cases}$$

we shall in the future identify R and G with their images in RG. With these identifications, the formal sums and products become ordinary sums and products. For this reason, from now on we drop the dot in $x_g \cdot g$. We shall also adopt the convention that $RG \cong RH$ means an isomorphism of R-algebras.

It follows from Lemma 4.1.3 that

$$|Rg| = \begin{cases} |R|^{|G|} & \text{if both } |R| \text{ and } |G| \text{ are finite,} \\ \max\{|R|, |G|\} & \text{otherwise.} \end{cases}$$

In particular, RG is countable if and only if both $|R|$ and $|G|$ are countable.

We now proceed to develop our vocabulary. Let $x = \sum x_g g \in RG$. Then the *support* of x, written $\operatorname{Supp} x$, is defined by

$$\operatorname{Supp} x = \{g \in G \mid x_g \neq 0\}.$$

It is plain that $\operatorname{Supp} x$ is a finite subset of G that is empty if and only if $x = 0$. We shall say that $x \in RG$ is a *monomial* if $|\operatorname{Supp} x| = 1$.

The *supporting subgroup* $\langle \operatorname{Supp} x \rangle$ of a non-zero element $x \in RG$ is the subgroup of G generated by $\operatorname{Supp} x$. Thus $\langle \operatorname{Supp} x \rangle$ is a finitely generated subgroup of G.

Suppose that H is a subgroup of G. Then the group ring RH can be embedded naturally in RG, allowing us to write

$$RH = \{x \in RG \mid \operatorname{Supp} x \subseteq H\}.$$

It is clear that for a non-zero $x \in RG$, $\langle \operatorname{Supp} x \rangle$ is the smallest subgroup H of G with $x \in RH$.

Now assume that S is a subring of R. Then the group ring SG can be embedded naturally in RG; therefore we may write

$$SG = \left\{ \sum x_g g \in RG \mid x_g \in S \text{ for all } g \in G \right\}.$$

Let $x = \sum x_g g \in RG$ and let $x \neq 0$. Then the *supporting subring* of x is the subring of R generated by

$$\{x_g \in R \mid g \in \operatorname{Supp} x\}.$$

The above definition implies that the supporting subring S of x is the smallest subring L of R with $x \in LG$. Note also that S is finitely generated, and hence noetherian by Corollary 1.2.3. Thus for any $x \in RG$ there exists a noetherian subring S of R and a finitely generated subgroup H of G such that $x \in SH$.

We next exhibit some elementary properties of group rings. Our starting point is the following:

8.1.1 Proposition   *Let A be an R-algebra and let*

$$\psi : G \to U(A)$$

be a homomorphism of G into the unit group of A. Then the mapping $f : RG \to A$ defined by

$$f\left(\sum x_g g\right) = \sum x_g \psi(g)$$

is a homomorphism of R-algebras. In particular, if ψ is injective and if A is R-free with $\psi(G)$ as a basis, then $RG \cong A$.

Proof.   Because RG is R-free with G as a basis, f is a homomorphism of R-modules. Let

$$x = \sum x_g g \qquad \text{and} \qquad y = \sum y_g g$$

be two elements of RG. Then

$$f(xy) = f\left(\sum_{a,b \in G} x_a y_b ab\right) = \sum x_a y_b \psi(a)\psi(b)$$

$$= \sum_{a \in G} x_a \psi(a) \sum_{b \in G} y_b \psi(b)$$

$$= f(x)f(y)$$

as asserted.   ∎

8.1.2 Corollary   *Let V be an R-module and let $f : G \to \mathrm{Aut}_R(V)$ be a group homomorphism. Then V can be regarded as an RG-module by setting*

$$\left(\sum x_g g\right) v = \sum x_g f(g) v \qquad (v \in V).$$

Proof.   Put $A = \mathrm{End}_R(V)$. Then $\mathrm{Aut}_R(V) = U(A)$ and hence, by Proposition 1.1, the map

$$\begin{cases} RG \to \mathrm{End}_R(V) \\ \sum x_g g \mapsto \sum x_g f(g) \end{cases}$$

is a homomorphism of R-algebras.   ∎

8.1.3 Corollary   *The map*

$$\begin{cases} RG \to R \\ \sum x_g g \mapsto \sum x_g \end{cases}$$

is a homomorphism of R-algebras.

Proof. This is a special case of Proposition 1.1 where $A = R$ and $\psi(g) = 1$ for all $g \in G$. ∎

The *augmentation ideal* $I(R, G)$ of RG is defined to be the kernel of the homomorphism of Corollary 1.3. In other words, $I(R, G)$ consists of all $x = \sum x_g g \in RG$ for which

$$\text{aug}(x) = \sum x_g = 0.$$

We shall refer to $\text{aug}(x)$ as the *augmentation* of x and to the homomorphism

$$\text{aug}: RG \to R$$

as the *augmentation map*.
    It follows from the equality

$$\sum x_g g = \sum x_g(g - 1) + \sum x_g$$

that as an R-module, $I(R, G)$ is a free module with the elements $g - 1$, $1 \neq g \in G$, as a basis. In the future we shall often suppress reference to R and simply denote the augmentation ideal of RG by $I(G)$.
    If X is a subset of RG, we write $RG \cdot X$ and $X \cdot RG$ for the left and right ideals of RG, respectively, generated by X, i.e.

$$RG \cdot X = \sum_{x \in X} RGx \quad \text{and} \quad X \cdot RG = \sum_{x \in X} xRG.$$

Of course, if $RGx = xRG$ for all $x \in X$, then $RG \cdot X = X \cdot RG$ is a two-sided ideal of RG. For example, if N is a normal subgroup of G, then the equalities

$$(n - 1)g = g(g^{-1}ng - 1) \quad \text{and} \quad g(n - 1) = (gng^{-1} - 1)g \qquad (n \in N, g \in G)$$

show that $RG \cdot I(N) = I(N) \cdot RG$ is a two-sided ideal of RG. The significance of this ideal comes from the following fact.

8.1.4 Proposition   Let $\psi: G \to H$ be a surjective homomorphism of groups and let $N = \text{Ker } \psi$. Then the mapping $f: RG \to RH$, which is the

R-linear extension of ψ, is a surjective homomorphism of R-algebras, the kernel of which is $RG \cdot I(N)$. In particular,

$$RG/RG \cdot I(N) \cong R(G/N).$$

Proof. That f is a surjective homomorphism of R-algebras is a consequence of Proposition 1.1. It is plain that $RG \cdot I(N) \subseteq \mathrm{Ker}\, f$. Consequently, f induces a homomorphism $\bar{f}: RG/RG \cdot I(N) \to RH$. The restriction of $\bar{f}$,

$$\lambda: [G + RG \cdot I(N)]/RG \cdot I(N) \to H$$

is an isomorphism. Thanks to Proposition 1.1, λ^{-1} can be extended to a homomorphism $RH \to RG/RG \cdot I(N)$ which is inverse to $\bar{f}$. Thus $\mathrm{Ker}\, f = RG \cdot I(N)$, as required. ∎

Let I be an ideal of RG and let

$$G \cap (1 + I) = \{g \in G \mid g - 1 \in I\}.$$

Then $G \cap (1 + I)$ is the multiplicative kernel of the natural map $G \to RG/I$, and hence a normal subgroup of G. In view of this observation, the next corollary arises from Proposition 1.4 by taking $I = RG \cdot I(N)$.

8.1.5 Corollary *Let N be a normal subgroup of G. Then*

$$G \cap (1 + RG \cdot I(N)) = N.$$

Let H be a subgroup of G. Because RH is a subring of RG, we can view RG as a left or right RH-module by way of ordinary multiplication.

8.1.6 Proposition *Let H be a subgroup of G. If T is a right (left) transversal for H in G, then RG is a free left (right) RH-module with T as a basis.*

Proof. Assume that T is a right transversal for H in G. Then, for any $t \in T$, $(RH)t$ is the R-linear span of the coset Ht. Accordingly, for any $t_1, t_2, \ldots, t_n \in T$, $(RH)t_1 + \ldots + (RH)t_n$ is the R-linear span of $\bigcup_{i=1}^{n} Ht_i$. As is apparent from the definition of RG, if X and Y are disjoint subsets of G, their R-linear spans meet at 0. Hence

$$RG = \bigoplus_{t \in T} (RH)t,$$

proving that RG is a free left RH-module with T as a basis. A similar argument proves the case where T is a left transversal. ∎

Given a subgroup H of G, define $\pi_H : RG \to RH$ by

$$\pi_H\left(\sum_{g \in G} x_g g\right) = \sum_{h \in H} x_h h.$$

We shall refer to π_H as the *projection map*.

8.1.7 Corollary *For any subgroup H of G, the projection map*

$$\pi_H : RG \to RH$$

is a homomorphism of left and right RH-modules.

Proof. Let T be a right transversal for H in G containing 1. Thanks to Proposition 1.6, the map $\delta : T \to RH$ defined by $\delta(1) = 1$ and $\delta(t) = 0$ for $t \neq 1$ in T, extends to a homomorphism $\lambda : RG \to RH$ of left RH-modules. Since $\lambda = \pi_H$, we see that π_H is a homomorphism of left RH-modules. A similar argument, by taking T to be a left transversal, shows that π_H is a homomorphism of right RH-modules. ∎

8.1.8 Corollary *Let H be a subgroup of G. Then*

$$RH \cap U(RG) = U(RH).$$

Proof. Let $u \in RH \cap U(RG)$. Then $uv = vu = 1$ for some $v \in RG$. Applying Corollary 1.7, we conclude that $u\pi_H(v) = \pi_H(v)u = 1$. Thus $u \in U(RH)$ and the opposite containment being obvious, the result follows. ∎

8.1.9 Proposition *Let $\{C_i \mid i \in I\}$ be the set of all finite conjugacy classes of G and, for each $i \in I$, let $C_i^+ = \sum_{x \in C_i} x$. Then $Z(RG)$ is the R-linear span of all C_i^+.*

Proof. Since for all $g \in G$, $g^{-1} C_i^+ g = C_i^+$, we have $C_i^+ \in Z(RG)$ for all $i \in I$. On the other hand, let $x = \sum x_g g \in Z(RG)$. Then, for all $t \in G$, $t^{-1}xt = x$ or, equivalently, $x_{t^{-1}gt} = x_g$, for all $t \in G$, $g \in \mathrm{Supp}\, x$. Hence x is an R-linear combination of those C_i^+ for which $C_i \subseteq \mathrm{Supp}\, x$, as required. ∎

8.1.10 Proposition *Let G be a finite group and let F be a field of characteristic $p \geq 0$. Then FG is semisimple if and only if p does not divide the order of G.*

Proof. If $p > 0$ divides $|G|$, then $x = \sum_{g \in G} g$ satisfies $x^2 = |G|\, x = 0$ and, by Proposition 1.9, $x \in Z(FG)$. Hence $FGx \subseteq J(FG) \neq 0$. Conversely, assume that $p \nmid |G|$, and let W be a submodule of an FG-module V. Write $V = W \oplus W'$ for a suitable F-subspace W', and let $\theta : V \to W$ be the projection map. Define $\psi : V \to V$ by

$$\psi(v) = |G|^{-1} \sum_{x \in G} x\theta x^{-1} v \qquad (v \in V).$$

Since for all $v \in V$ and $y \in G$,

$$\psi(yv) = |G|^{-1} \sum_{x \in G} x\theta x^{-1} yv = |G|^{-1} \sum_{x \in G} (yz)\theta(yz)^{-1} yv$$

$$= |G|^{-1} \sum_{g \in G} yz\theta z^{-1} v = y\psi(v),$$

ψ is an FG-homomorphism.

Suppose now that $v \in W$. Then, for any $x \in G$, $x^{-1}v \in W$, so $\theta(x^{-1}v) = x^{-1}v$. Accordingly, $x\theta x^{-1}v = v$ and $\psi(v) = v$. Setting $W'' = \mathrm{Ker}\,\psi$, it follows that W'' is an FG-submodule of V such that $W'' \cap W = 0$. Finally, let $v \in V$. Then, by the above, $v - \psi(v) \in W''$, so $v = \psi(v) + (v - \psi(v)) \in W + W''$. Thus $V = W \oplus W''$ and the result follows. ∎

8.1.11 Proposition *Let F be a field of characteristic $p \geqslant 0$, let N be a normal p'-subgroup of a finite group G and let $e = |N|^{-1} \sum_{x \in N} x$. Then e is a central idempotent of FG such that $FGe \cong F(G/N)$. In particular,*

$$U(FG) \cong U(FG(1-e)) \times U(F(G/N)).$$

Proof. Since N is a union of some conjugacy classes of G, e is central by Proposition 1.9. It is immediate to verify that e is an idempotent. Let $f : FG \to F(G/N)$ be the natural homomorphism. Since $FG = FGe \oplus FG(1-e)$ and, by Proposition 1.4, $\mathrm{Ker}\,f = FG \cdot I(N)$, it suffices to show that $FG \cdot I(N) = FG(1-e)$.

Since $f(e) = 1$, we have $1 - e \in \mathrm{Ker}\,f$ and therefore $FG(1-e) \subseteq FG \cdot I(N)$. Given $n \in N$, we also have $ne = e$, so that $(n-1)e = 0$ and $FG \cdot I(N)e = 0$. Thus $FG \cdot I(N) \subseteq FG(1-e)$ and the result follows. ∎

8.1.12 Proposition *Let F be a field of characteristic $p > 0$ and let N be a normal p-subgroup of a finite group G. Then*

$$FG \cdot I(N) \subseteq J(FG)$$

with equality if N is a normal Sylow p-subgroup of G.

Proof. Fix an element n in N, say of order p^s. Since $(n-1)^{p^s} = n^{p^s} - 1 = 0$, and since $I(N)$ is the F-linear span of all such $n - 1$, $I(N)$ is nilpotent. But $[FG \cdot I(N)]^m = FG \cdot I(N)^m$ for all $m \geqslant 1$. Hence $FG \cdot I(N)$ is nilpotent and therefore $FG \cdot I(N) \subseteq J(FG)$. If N is a normal Sylow p-subgroup of G, then $FG/FG \cdot I(N) \cong F(G/N)$ is semisimple by Proposition 1.10. Hence $J(FG) = FG \cdot I(N)$ as asserted. ∎

8.1.13   Corollary   *Let F be a field of characteristic p > 0 and let N be a normal p-subgroup of a finite group G. Then the sequence*

$$1 \to 1 + FG \cdot I(N) \to U(FG) \to U(F(G/N)) \to 1$$

is exact.

Proof.   Apply Proposition 1.12 and Lemma 7.1.2.   ■

A unit u of RG is called *normalized* if its augmentation is equal to 1. We shall denote by $V(RG)$ the set of all normalized units of RG. A link between $V(RG)$ and $U(RG)$ is provided by

8.1.14   Proposition $V(RG)$ *is a normal subgroup of* $U(RG)$ *such that*

$$U(RG) = U(R) \times V(RG).$$

Proof.   That $V(RG)$ is a normal subgroup of $U(RG)$ follows from the fact that aug: $RG \to R$ is a homomorphism. Evidently, $U(R) \cap V(RG) = 1$. Observe also that for any $u \in U(RG)$, $r = \text{aug}(u)$ is a unit such that aug $(r^{-1}u) = 1$. The equality $u = r(r^{-1}u)$ now shows that $U(RG) = U(R) \cdot V(RG)$, hence the result.   ■

8.1.15   Proposition   *Let N be a normal subgroup of G of finite index. Then, for any commutative ring R, $J(RN) \subseteq J(RG)$.*

Proof.   Fix an irreducible RG-module V and non-zero v in V. Then we have $V = RGv$. Let T be a transversal for N in G. Then

$$V = \sum_{t \in T} (RN)tv$$

so that V is a finitely generated RN-module. Since $N \lhd G$, $J(RN)V$ is obviously an RG-submodule of V and so either $J(RN)V = 0$ or $J(RN)V = V$. The latter equality is impossible, by Nakayama's lemma. Thus $J(RN)$ annihilates V. Since V is an arbitrary irreducible RG-module, we conclude that $J(RN) \subseteq J(RG)$.   ■

As an application of the above property, we prove

8.1.16   Proposition   *Let G be a finite p-group and R a commutative ring such that $R/J(R)$ is a field of characteristic p. Then $u \in U(RG)$ if and only if aug(u) $\in U(R)$. In particular,*

$$U(RG) = (1 + I(G)) \times U(R).$$

Proof.   Consider the homomorphism $\psi: RG \to (R/J(R))G$ induced by the natural homomorphism $R \to R/J(R)$. Indeed Ker $\psi = J(R)G \subseteq J(RG)$, by Proposition 1.15. Hence, by Lemma 7.1.2, $u \in U(RG)$ if and only if $\psi(u) \in U(FG)$, where $F = R/J(R)$. Since $\text{aug}(\psi(u))$ is the image of $\text{aug}(u)$ under the natural homomorphism $R \to R/J(R)$, we see that $\text{aug}(u) \in U(R)$ if and only if $\text{aug}(\psi(u)) \in U(F)$. Thus we may harmlessly assume that $R = F$ is a field of characteristic p. Owing to Proposition 1.14, we need only verify that $1 + I(G) \subseteq U(RG)$. But, by Proposition 1.12, $I(G)$ is nilpotent, hence the result. ∎

8.1.17 Proposition *Let H be a subgroup of an arbitrary group G and let R be a commutative ring. If X is a generating set for H, then*

$$RG \cdot I(H) = \sum_{x \in X} RG(x-1) \quad and \quad I(H) \cdot RG = \sum_{x \in X} (x-1)RG.$$

Proof. We shall only establish the first equality, since the second follows by a similar argument. It is clear that $\sum_{x \in X} RG(x-1) \subseteq RG \cdot I(H)$. Thus we need only show that for any non-identity h in H, $h - 1 \in \sum_{x \in X} RG(x-1)$. Now h is a group word in the $x_i \in X$ and we proceed by induction on the length t of the word. First, it is true for words of length 1:

$$x_i - 1 \in \sum_{x \in X} RG(x-1), \qquad x_i^{-1} - 1 = -x_i^{-1}(x_i - 1) \in \sum_{x \in X} RG(x-1).$$

Suppose it is true for words h of length t. Any word of length $t+1$ is of the form $x_i^{\pm 1}h$ and both $h - 1$ and $x_i^{\pm 1} - 1$ are in $\sum_{x \in X} RG(x-1)$. Hence

$$x_i^{\pm 1}h - 1 = x_i^{\pm 1}(h-1) + (x_i^{\pm 1} - 1) \in \sum_{x \in X} RG(x-1)$$

as desired. ∎

8.1.18 Corollary *Let G be an arbitrary group and let R be a commutative ring. Then $RG \cdot I(G')$ is the smallest ideal I of RG such that RG/I is commutative.*

Proof. Since $RG/RG \cdot I(G') \cong R(G/G')$ (Proposition 1.4), the ring $RG/RG \cdot I(G')$ is certainly commutative. Let I be any ideal of RG such that RG/I is commutative. Then, for all $x, y \in G$, $xy - yx \in I$. But $xy - yx = yx(x^{-1}y^{-1}xy - 1)$, so $[x, y] - 1 \in I$ and therefore, by Proposition 1.16, $RG \cdot I(G') \subseteq I$. ∎

We close this section with the discussion of direct decompositions of group rings.

8.1.19 Proposition *Let $(R_i)_{i \in I}$ be a family of commutative rings, and let*

G be an arbitrary group.

(i) *If R is a subdirect product of $(R_i)_{i \in I}$, then RG is a subdirect product of $(R_iG)_{i \in I}$.*

(ii) *If $R \cong \prod_{i \in I} R_i$, then $RG \cong \prod_{i \in I} (R_iG)$.*

Proof

(i) By hypothesis, for any $i \in I$, there is a surjective homomorphism $f_i : R \to R_i$ such that $\cap \operatorname{Ker} f_i = 0$. Let $\pi_i : RG \to R_iG$ be the homomorphism induced by f_i. Then $\operatorname{Ker} \pi_i = (\operatorname{Ker} f_i)G$ and hence

$$\bigcap_{i \in I} \operatorname{Ker} \pi_i = \bigcap_{i \in I} (\operatorname{Ker} f_i)G = \left(\bigcap_{i \in I} \operatorname{Ker} f_i \right) G = 0,$$

as required.

(ii) It suffices to prove that for any $(y_i)_{i \in I}$ in $\prod_{i \in I} (R_iG)$ there exists $x \in RG$ such that $\pi_\mu(x) = y_\mu$ for any $\mu \in I$. Write

$$y_\mu = \sum r_{g\mu} g \qquad (r_{g\mu} \in R_\mu).$$

The hypothesis on R ensures that for any $g \in G$, there exists $r_g \in R$ such that for any $\mu \in I$, $f_\mu(r_g) = r_{g\mu}$. Therefore, for any $\mu \in I$,

$$\pi_\mu \left(\sum r_g g \right) = \sum r_{g\mu} g = y_\mu$$

as asserted. ∎

The special case of Proposition 1.19(ii) in which R is noetherian (and hence a direct product of finitely many indecomposable rings) is worth noting. This is recorded in

8.1.20 Proposition *Let R be a commutative noetherian ring and let G be an arbitrary group. Then there exist finitely many indecomposable rings $R_1, \ldots, R_n$ such that*

$$RG \cong R_1G \times R_2G \times \ldots \times R_nG.$$

In particular,

$$U(RG) \cong U(R_1G) \times U(R_2G) \times \ldots \times U(R_nG).$$

8.2 Trace of idempotents

Let A be an R-algebra and let $\operatorname{End}_R(A)$ be the algebra of R-endomorphisms of A. For any $a \in A$, let ψ_a denote the map 'left multiplication by a'. The mapping

$$\Gamma : A \to \operatorname{End}_R(A)$$

defined by $\Gamma(a) = \psi_a$ is an injective homomorphism. We shall refer to Γ as the *regular representation* of A.

Suppose now that A is R-free on a finite basis. Then we can define the character χ of Γ by putting $\chi(a) = \text{trace } \Gamma(a)$ for all $a \in A$. The next lemma illustrates how the character of the regular representation of RG can be brought into argument.

8.2.1   Lemma   *Let G be a finite group, let R be a commutative ring and let χ be the character of the regular representation of RG. Then, for any $x = \sum x_g g \in RG$, $\chi(x) = |G| x_1$.*

Proof.   By taking G as an R-basis for RG, we see that $\chi(g)$ is just the number of elements in G fixed under multiplication by g. Hence $\chi(g) = 0$ for $g \neq 1$ and $\chi(1) = |G|$. Since χ is an R-linear map, the result follows.   ■

The above observation implies that $\chi(x)$ is a fixed scalar multiple of x_1, the identity coefficient of x. Following Passman (1977), we define, for an arbitrary G, a map

$$tr : RG \rightarrow R$$

called the trace by

$$tr\left(\sum x_g g\right) = x_1.$$

The trace of an idempotent enjoys an important property that we shall need later. This is recorded in the following.

8.2.2   Lemma   *Let G be a finite group and let F be a field of characteristic 0. If e is an idempotent of FG, then*

$$tr(e) = |G|^{-1} \dim_F FGe.$$

Proof.   Let χ be the character of the regular representation of FG. Bearing in mind that $FG = FGe \oplus FG(1 - e)$, we may choose a basis for FG which is the union of bases for FGe and $FG(1 - e)$.

Now, for any $y \in FGe$ and for any $z \in FG(1 - e)$, we have $ye = y$ and $ze = 0$. Consequently, it may be inferred that $\chi(e) = \dim_F FGe$. Since, by Lemma 2.1, $\chi(e) = |G| tr(e)$, the result follows.   ■

We next generalize the above result by showing that if e is an idempotent of FG, where F is an arbitrary field and G an arbitrary group, then $tr(e)$ belongs to the prime subfield of F. Unfortunately, in the general case, we do not have a precise formula for $tr(e)$. In particular, if $tr(e) = a/b \in \mathbb{Q}$, $(a, b) = 1$, a, $b \in \mathbb{Z}$, it is not known whether, for any prime p dividing b, G has an element of order p.

8.2.3 Lemma *Let G be an arbitrary group and let R be a commutative ring. Then the map* $tr: RG \to R$ *is R-linear and, for all* $x, y \in RG$, $tr(xy) = tr(yx)$.

Proof.   It is obvious that *tr* is *R*-linear. If $x = \sum x_g g$ and $y = \sum y_g g$, then

$$tr(xy) = \sum_g x_g y_{g-1} = \sum_g y_g x_{g-1} = tr(yx),$$

as required.   ∎

Given an arbitrary ring *R*, we denote by $[R, R]$ the additive subgroup of *R* generated by all Lie products $xy - yx$, $x, y \in R$.

8.2.4 Lemma *Let R be a ring of prime characteristic p. If* $x_1, x_2, \ldots, x_m \in R$ *and if* $n \geq 1$ *is a given integer, then*

$$(x_1 + x_2 + \ldots + x_m)^{p^n} \equiv x_1^{p^n} + x_2^{p^n} + \ldots + x_m^{p^n} (\mathrm{mod}[R, R]).$$

Proof.   We may write

$$(x_1 + x_2 + \ldots + x_m)^{p^n} = x_1^{p^n} + x_2^{p^n} + \ldots + x_m^{p^n} + y$$

where *y* is the sum of all words $x_{i_1} x_{i_2} \ldots x_{i_q}$, $q = p^n$, with at least two distinct subscripts occurring. Note that if y_1 and y_2 are cyclic permutations of each other, that is, if

$$y_1 = x_{i_1} x_{i_2} \ldots x_{i_q},$$

$$y_2 = x_{i_j} x_{i_{j+1}} \ldots x_{i_q} x_{i_1} \ldots x_{i_{j-1}},$$

then $y_1 - y_2 = \alpha\beta - \beta\alpha \in [R, R]$, where

$$\alpha = x_{i_1} x_{i_2} \ldots x_{i_{j-1}} \quad \text{and} \quad \beta = x_{i_j} x_{i_{j+1}} \ldots x_{i_q}.$$

Thus, modulo $[R, R]$, all cyclic permutations of a word are equal. Let $\mathbb{Z}_q$ act on the set of these words by performing the cyclic shifts. Then the number of formally distinct permutations of a word λ occurring in *y* is the size of a non-trivial orbit of $\mathbb{Z}_q$, and so is divisible by *p*. Since char $R = p$, the result follows.   ∎

Let *K* and *F* be fields. A *place* is a map

$$\phi : K \to F \cup \{\infty\}$$

such that $R = \phi^{-1}(F)$ is a subring of *K* with $\phi : R \to F$ a homomorphism and such that $\phi(x) = \infty$ if and only if $\phi(x^{-1}) = 0$.

   We now quote the following standard result whose proof is beyond the scope of this book.

8.2.5 Proposition *Let K be a field of characteristic* 0 *and let*

$x_0, x_1, \ldots, x_n$ *be finitely many elements of K, with x_0 not in* $\mathbb{Q}$. *Then, for infinitely many primes p, there exists a place* $\phi_R : K \to G\tilde{F}(p) \cup \{\infty\}$ *with* $\phi_R(x_i) \neq \infty$ *for all i and with* $\phi_R(x_0) \notin GF(p)$. *Here $G\tilde{F}(p)$ is the algebraic closure of the field $GF(p)$ of p elements.*

Proof. See Passman (1977, p. 44). ∎

We are now ready to prove

8.2.6 Theorem (Zalesskii 1972) *Let K be an arbitrary field and let G be a group. If e is an idempotent in KG, then tr(e) is contained in the prime subfield of K.*

Proof. Write $e = \sum e_g g$ and assume first that char $K = p > 0$. Denote by S the set of all p-elements (including 1) in Supp e. Because S is finite, there exists $m \geq 1$ such that $s^{p^m} = 1$ for all $s \in S$. Now let n be any integer with $n \geq m$. Invoking Lemma 2.4, we may find an element y in $[KG, KG]$ with

$$e = e^{p^n} = \sum e_g^{p^n} g^{p^n} + y.$$

Furthermore, by Lemma 2.3, $tr(y) = 0$. Note also that since $n \geq m$ we have $x^{p^n} = 1$ for $x \in \text{Supp}\, e$ if and only if $x \in S$. Thus, by taking traces of the above, we have

$$tr(e) = \sum_{s \in S} (e_s)^{p^n} = \left(\sum_{s \in S} e_s \right)^{p^n}$$

for all $n \geq m$. In particular, by taking $n = m$ and $n = m + 1$, we obtain

$$[tr(e)]^p = \left(\sum_{s \in S} e_s \right)^{p^m p} = \left(\sum_{s \in S} e_s \right)^{p^{m+1}} = tr(e),$$

proving that $tr(e) \in GF(p)$.

Now assume that char $K = 0$. We argue by contradiction by letting $tr(e) = e_1 \notin \mathbb{Q}$. Then, by Proposition 2.5, there exists a prime p and a place $\phi_R : K \to G\tilde{F}(p)$ with $\phi_R(e_g) \neq \infty$ for all $g \in \text{Supp}\, e$ and with $\phi_R(e_1) \notin GF(p)$. Then $e \in RG$, and therefore $\bar{e}$, the image of e in $(R/M)G \subseteq G\tilde{F}(p)G$, is an idempotent. Furthermore,

$$tr(\bar{e}) = \phi_R(tr(e)) \notin GF(p),$$

which is a contradiction. The theorem is therefore established. ∎

8.3   Units of finite order

The aim of this section is to exhibit a class of rings R for which RG has only trivial central units of finite order. It turns out that, for an arbitrary group G, we can choose as R any integral domain of characteristic 0 in which no rational prime is a unit. In case G is finite, we can strengthen the above, by choosing as R any integral domain in which no rational prime dividing $|G|$ is a unit. A number of applications is also provided.

We start by making two preliminary observations concerning certain properties of algebraic integers. In this way we shall avoid becoming submerged in minor details at a critical stage of the discussion.

8.3.1 **Lemma** *Let $\varepsilon_1, \varepsilon_2, \ldots, \varepsilon_t$ be nth roots of unity over $\mathbb{Q}$ and let $a_1, \ldots, a_t$ be positive rational numbers such that $a_1 + \ldots + a_t = 1$. If $\alpha = a_1\varepsilon_1 + \ldots + a_t\varepsilon_t$ is an algebraic integer, then either $\alpha = 0$ or $\varepsilon_1 = \varepsilon_2 = \ldots = \varepsilon_t$.*

Proof. Observe first that $|\alpha| \leqslant 1$, and that $|\alpha| = 1$ if and only if $\varepsilon_1 = \varepsilon_2 = \ldots = \varepsilon_t$. It therefore suffices to show that if $\alpha \neq 0$, then $|\alpha| = 1$. Assume by way of contradiction that $\alpha \neq 0$ and that $|\alpha| < 1$. Let ε be a primitive nth root unity over $\mathbb{Q}$. Then, for any $\sigma \in \mathrm{Gal}(\mathbb{Q}(\varepsilon)/\mathbb{Q})$, $|\sigma(\alpha)| < 1$; therefore $|N_{\mathbb{Q}(\varepsilon)/\mathbb{Q}}(\alpha)| < 1$, contrary to the fact that the norm of α is a non-zero integer. So the lemma is true. ∎

8.3.2 **Lemma** *Let α be an algebraic number and let n be a natural number such that $n\alpha$ is an algebraic integer. If $\{\alpha_1 = \alpha, \alpha_2, \ldots, \alpha_t\}$ is the set of all $\mathbb{Q}$-conjugates of α, then either α is an algebraic integer or, in the ring $\mathbb{Z}[\alpha_1, \alpha_2, \ldots, \alpha_t]$, at least one rational prime divisor of n is a unit.*

Proof. Let E_i $(1 \leqslant i \leqslant t)$ be the elementary symmetric function of t variables and degree i. If

$$f(X) = X^t + a_1 X^{t-1} + \ldots + a_t \qquad (a_i \in \mathbb{Q})$$

is the irreducible monic polynomial satisfied by α, then

$$a_i = (-1)^i E_i(\alpha_1, \alpha_2, \ldots, \alpha_t) \in \mathbb{Z}[\alpha_1, \alpha_2, \ldots, \alpha_t] \qquad (1 \leqslant i \leqslant t).$$

Suppose α is not an algebraic integer. Then there exists $i \in \{1, 2, \ldots, t\}$ such that

$$E_i(\alpha_1, \alpha_2, \ldots, \alpha_t) = a/b$$

for some coprime integers a and b with $b > 1$. If therefore we denote by p a prime divisor of b, then $a/p \in \mathbb{Z}[\alpha_1, \alpha_2, \ldots, \alpha_t]$. Moreover, p divides n since

$$E_i(n\alpha_1, n\alpha_2, \ldots, n\alpha_t) - n^i(a/b) \in \mathbb{Z}.$$

Finally, because $(a, p) = 1$ there exist $c, d \in \mathbb{Z}$ such that $ac + dp = 1$, and hence

$$\frac{1}{p} = \frac{1}{p}(ac + dp) = c \cdot \frac{a}{p} + d \in \mathbb{Z}[\alpha_1, \alpha_2, \ldots, \alpha_t]$$

as asserted. ∎

Let F be a field, let $f(X) \in F[X]$, and let

$$f(X) = f_1(X)^{e_1} \ldots f_m(X)^{e_m}$$

be the canonical decomposition of $f(X)$. We know, from Proposition 1.2.1, that

$$F[X]/(f(X)) \cong F[X]/(f_1(X)^{e_1}) \times \ldots \times F[X]/(f_m(X)^{e_m}).$$

In particular, if $f(X)$ has no multiple roots and if $f(X)$ splits into linear factors, then $F[X]/(f[X])$ is a direct product of m copies of F. This is certainly the case when $F = \mathbb{C}$ and $f(X) \mid X^n - 1$ for some $n \geq 1$. Assume, for the moment, that A is a $\mathbb{C}$-algebra and that $u \in A$ is a unit of finite order n. Then

$$\mathbb{C}[u] \cong \mathbb{C}[X]/(f(X)),$$

where $f(X)$ is the minimal polynomial satisfied by u, and so $f(X) \mid X^n - 1$. It therefore follows, from what we have said above, that $\mathbb{C}[u]$ is a direct product of finitely many copies of $\mathbb{C}$.

We now have at our disposal all the information necessary to prove

8.3.3   Theorem   *Let R be an integral domain of characteristic 0 and let G be a group. Assume that at least one of the following two conditions hold:*

(i)   (Bass 1976) *No rational prime is a unit of R.*
(ii)   (Saksonov 1971) *G is finite and no rational prime dividing $|G|$ is a unit of R.*

If u is a unit of finite order in RG, then either $tr(u) = 0$ or $u \in U(R)$.

Proof.   Let $u \in \sum u_g g \in RG$ be a unit of order m such that $tr(u) \neq 0$. Observe that $u \in FG$, where F is a finitely generated field obtained from $\mathbb{Q}$ by adjoining the non-zero coefficients of u. Since F is embeddable in $\mathbb{C}$, we may assume that $u \in \mathbb{C}G$. We know that $\mathbb{C}[u]$ is isomorphic to a direct product of finitely many, say t, copies of $\mathbb{C}$. Consequently, there exist non-zero mutually orthogonal idempotents $e_i \in \mathbb{C}[u]$ and complex mth roots of unity ε_i such that

$$u = \sum_{i=1}^{t} \varepsilon_i e_i, \qquad 1 = \sum_{i=1}^{t} e_i.$$

This implies at once that the result will follow provided we prove that $\varepsilon_1 = \varepsilon_2 = \ldots = \varepsilon_t$.

To this end, we first apply Theorem 2.6 and the well-known fact that $tr(e_i) > 0$ (Passman (1977)) to deduce that $tr(e_i) = a_i/b_i$ for some coprime positive integers a_i, b_i. Furthermore, by Lemma 2.2, if G is finite, then each b_i divides the order of G. Next we put $n = b_1 b_2 \ldots b_t$ and observe that $tr(u) = \sum_{i=1}^{t} \varepsilon_i(a_i/b_i)$ is an algebraic number such that $ntr(u)$ is an algebraic integer. Moreover, by looking at $tr(u^\mu)$, where $(\mu, m) = 1$, we infer that any $\mathbb{Q}$-conjugate of $tr(u)$ belongs to R. By hypothesis, no prime dividing n is a unit of R. Hence, by Lemma 3.2, $tr(u)$ is an algebraic integer. Finally, bearing in mind that $tr(u) \neq 0$ and that $\sum_{i=1}^{t} a_i/b_i = 1$, the result follows by virtue of Lemma 3.1. ∎

The following result will enable us to take full advantage of Theorem 3.3.

8.3.4 Proposition (Sehgal 1975) *Suppose that R is an integral domain of characteristic 0 in which no rational prime is invertible. If $x = \sum x_g g \in RG$ is a normalized unit of order p^n, p a prime, then $\mathrm{Supp}\, x$ has an element of order p^n.*

Proof. Applying Lemma 2.4, we have

$$1 = x^{p^n} \equiv \sum x_g^{p^n} g^{p^n} + y \; (\mathrm{mod}\, p(RG))$$

for some $y \in [RG, RG]$. Taking traces of both sides and applying Lemma 2.3, we obtain

$$
\begin{aligned}
1 &\equiv \sum_{g^{p^n}=1} x_g^{p^n} (\mathrm{mod}\, pR) \\
&\equiv \sum_{o(g)=p^n} x_g^{p^n} + \sum_{g^{p^{n-1}}=1} x_g^{p^n} (\mathrm{mod}\, pR) \\
&\equiv \sum_{o(g)=p^n} x_g^{p^n} + \left(\sum_{g^{p^{n-1}}=1} x_g^{p^{n-1}} \right)^p (\mathrm{mod}\, pR) \\
&\equiv \sum_{o(g)=p^n} x_g^{p^n} + (tr(x^{p^{n-1}}))^p (\mathrm{mod}\, pR).
\end{aligned}
$$

Now $x^{p^{n-1}} \neq 1$ is a normalized unit of finite order, and hence $tr(x^{p^{n-1}}) = 0$ by Theorem 3.3. We therefore conclude that

$$\sum_{o(g)=p^n} x_g \not\equiv 0 (\mathrm{mod}\, pR).$$

Hence there exists $g \in \mathrm{Supp}\, x$ with $o(g) = p^n$, as required. ∎

A unit u of RG is said to be *trivial* if $u = rg$ for some $r \in U(R)$ and some $g \in G$.

8.3.5   Corollary   *Let R be an integral domain of characteristic 0 and let G be an arbitrary group. Assume that at least one of the following two conditions hold:*

(i)   *no rational prime is a unit of R;*
(ii) *G is finite and no rational prime dividing $|G|$ is a unit of R.*

Then all central units of finite order in RG are trivial.

Proof.   Let u be a central unit of finite order in RG. We claim that Supp u has an element, say g, of finite order; if sustained it will follow that $v = g^{-1}u$ is a unit of finite order with $tr(v) \neq 0$. But then, by Theorem 3.3, $v \in U(R)$ and the result follows.

The case (ii) being trivial, we assume that (i) holds. Since u is a product of central units of prime power orders, we may assume that $o(u) = p^n$ for some prime p and some $n \geq 1$. Replacing u by $aug(u^{-1})u$, if necessary, we may also assume that u is normalized. The desired conclusion is therefore a consequence of Proposition 3.4.   ∎

Note that, in the special case where G is finite and $R = \mathbb{Z}$, Theorem 3.3 and Corollary 3.5 were proved by Berman (1955) and Cohn and Livingstone (1965).

8.3.6   Corollary   *Let R be an integral domain of characteristic 0 in which no rational prime is a unit. Then $V(RG)$ is torsion-free if and only if G is torsion-free.*

Proof.   If $V(RG)$ is torsion-free, then since $G \subseteq V(RG)$, G is also torsion-free. The converse is a consequence of Proposition 3.4.   ∎

8.3.7   Corollary   *Let G be a finite group and let R be an integral domain of characteristic 0 such that no rational prime dividing $|G|$ is a unit.*

(i)   *If H is a torsion subgroup of $V(RG)$, then H is a linearly independent set and, in particular, H is a finite group.*
(ii) *If H is a finite subgroup of $V(RG)$, then $|H|$ divides $|G|$.*

Proof.

(i)   We carry out the proof by contradiction. Let $\sum_{i=1}^{n} \alpha_i h_i = 0$, where $h_i \in H$, $\alpha_i \in R$, $1 \leq i \leq n$, and let $\alpha_j \neq 0$ for some $j \in \{1, 2, \ldots, n\}$. Then

$$\alpha_j \cdot 1 = -\sum_{i \neq j} \alpha_i (h_i h_j^{-1})$$

and so $tr(h_i h_j^{-1}) \neq 0$ for some $i \neq j$. Hence, by Theorem 3.3(ii)

$h_i h_j^{-1} = r$ for some $r \in U(R)$. Since $\text{aug}(h_i) = \text{aug}(h_j) = 1$ it follows that $r = 1$, i.e. $h_i = h_j$, a contradiction.

(ii) Put $e = |H|^{-1} \sum_{h \in H} h$. Then e is obviously an idempotent of FG, where F is the quotient field of R. If $h \neq 1$, then since $\text{aug}(h) = 1$, we have $tr(h) = 0$ by Theorem 3.3(ii). Thus $tr\, e = |H|^{-1}$. On the other hand, by Lemma 2.2, $tr(e) = |G|^{-1} m$ for some $m \geq 1$. Hence $|G| = |H|\, m$ as asserted. ■

Our next consequence, contained in the work of Passman and Smith (1981), indicates the existence of large torsion-free normal subgroups within $U(\mathbb{Z}G)$.

8.3.8 Proposition *For any finite group G, $U(\mathbb{Z}G)$ has a torsion-free normal subgroup of finite index.*

Proof. Let $\mathbb{F}_p$ be the field of p elements, where p is an odd prime. Then the natural homomorphism $\mathbb{Z}G \to \mathbb{F}_p G$ induces a homomorphism

$$U(\mathbb{Z}G) \to U(\mathbb{F}_p G)$$

with kernel

$$U(\mathbb{Z}G) \cap (1 + p(\mathbb{Z}G)).$$

Since $U(\mathbb{F}_p G)$ is a finite group, we are left to verify that any non-identity unit $u \in U(\mathbb{Z}G)$ of the form $u = 1 + px$, $x \in \mathbb{Z}G$ is of infinite order. But $tr(u) \equiv 1 (\text{mod } p)$, so $u \neq \pm 1$ and $tr(u) \neq 0$. The desired conclusion is therefore a consequence of Theorem 3.3(ii). ■

For future use, we next record the following results, in which $I(G)$ denotes the augmentation ideal of $\mathbb{Z}G$.

8.3.9 Proposition *Let G be an arbitrary group. Then*:

(i) $G/G' \cong I(G)/I(G)^2$;
(ii) $\sum x_g(g - 1) \equiv \prod g^{x_g} - 1 (\text{mod } I(G)^2)$ $(x_g \in \mathbb{Z})$;
(iii) $G \cap (1 + I(G)^2) = G'$.

Proof.
(i) Consider the map $G \xrightarrow{\psi} I(G)/I(G)^2$ defined by

$$\psi(g) = g - 1 + I(G)^2.$$

Then the identity

$$ab - 1 = (a - 1)(b - 1) + (a - 1) + (b - 1)   (a, b \in G)$$

ensures that ψ is a homomorphism. Hence ψ induces a homomorphism

$$\begin{cases} G/G' \xrightarrow{\bar{\psi}} I(G)/I(G)^2 \\ gG' \mapsto (g - 1) + I(G)^2. \end{cases}$$

Since $I(G)$ is a free abelian group with basis $\{g-1\,|\,1\neq g\in G\}$, the map $g-1\mapsto gG'$ extends to a homomorphism $\phi:I(G)\to G/G'$. The identity

$$(a-1)(b-1)=(ab-1)-(a-1)-(b-1)\qquad(a,b\in G)$$

shows that $I(G)^2\subseteq\operatorname{Ker}\phi$ and hence ϕ induces a homomorphism

$$\begin{cases}I(G)/I(G)^2\overset{\lambda}{\to}G/G',\\(g-1)+I(G)^2\mapsto gG'.\end{cases}$$

Since λ is the inverse of f, the required assertion follows.

   (ii)   Note that

$$\lambda\!\left(\sum x_g(g-1)+I(G)^2\right)=\lambda\!\left(\prod g^{x_g}-1+I(G)^2\right).$$

Since λ is an isomorphism, (ii) follows.

   (iii)   This is a direct consequence of the fact that the map f is an isomorphism.   ∎

8.3.10 Proposition   *Let G be an arbitrary group and let $u\in V(\mathbb{Z}G)$ be such that the image of u in $\mathbb{Z}(G/G')$ is of finite order. Then $u\equiv g(\operatorname{mod}I(G)I(G'))$ for some $g\in G$.*

Proof.   The image of u in $\mathbb{Z}(G/G')$ is a normalized unit of finite order. Hence, by Corollary 3.5(i), $u\equiv x(\operatorname{mod}\mathbb{Z}G\cdot I(G'))$ for some $x\in G$. Since

$$\mathbb{Z}G=\mathbb{Z}+I(G)\tag{1}$$

we have

$$\mathbb{Z}G\cdot I(G')=I(G')+I(G)\cdot I(G').$$

Hence, by Proposition 3.9(ii), there exists $a\in G'$ such that

$$u\equiv x+(a-1)=(1-x)(a-1)+xa\equiv xa(\operatorname{mod}I(G)\cdot I(G')).$$

Taking $g=xa$, the result follows.   ∎

8.3.11 Proposition   *Let H be a subgroup of a group G. Then*:

(i) $G\cap(1+I(G)\cdot I(H))=H'$;
(ii) $\mathbb{Z}G\cdot I(H)/I(G)\cdot I(H)\cong H/H'$.

Proof.
   (i)   Let T be a transversal for H in G containing 1, and let $g=t_gh_g$ be a typical element of G, $t_g\in T$, $h_g\in H$. Consider the $\mathbb{Z}$-linear map

$$\phi:\mathbb{Z}G\to\mathbb{Z}H$$

which is the $\mathbb{Z}$-linear extension of $g\mapsto h_g$. Then $\phi(x)=x$ for any $x\in\mathbb{Z}H$ and the equality

$$\phi((g-1)(h_1-1))=(h-1)(h_1-1)\qquad(h=h_g,\,h_1\in H)$$

shows that $\phi(I(G) \cdot I(H)) = I(H)^2$. Consequently,

$$I(G) \cdot I(H) \cap I(H) = I(H)^2 \tag{2}$$

Now Corollary 1.5 and Proposition 3.9(iii) may be employed to infer that

$$G \cap (1 + I(G) \cdot I(H)) = H \cap (1 + I(H)^2) = H'.$$

(ii)   By (1), $\mathbb{Z}G \cdot I(H) = I(H) + I(G) \cdot I(H)$. Hence, applying (2) and Proposition 3.9(i), we infer that

$$\begin{aligned}
\mathbb{Z}G \cdot I(H)/I(G) \cdot I(H) &\cong I(H)/(I(H) \cap I(G) \cdot I(H)) \\
&= I(H)/I(H)^2 \\
&\cong H/H',
\end{aligned}$$

as asserted.   ∎

We close this section by providing some applications pertaining to the isomorphism problem:

$$\text{Does } \mathbb{Z}G \cong \mathbb{Z}H \text{ imply } G \cong H?$$

Let R be any commutative ring. A *normalized group basis* of RG is a group basis consisting of normalized units. We shall write

$$RG = RH$$

to mean that H is a normalized group basis of RG. Note that if H_1 is another group basis of RG, then

$$RG = RH \qquad \text{where} \qquad H = \{\text{aug}(t^{-1})t \mid t \in H_1\}$$

and $H \cong H_1$. Therefore, the isomorphism problem may be stated as follows:

$$\text{Does } \mathbb{Z}G = \mathbb{Z}H \text{ imply } G \cong H?$$

Suppose that H is a normalized group basis of RG and let aug′ be the homomorphism from RG to R induced by collapsing H to 1. Then, for any

$$x = \sum_{h \in H} x_h h, \qquad x_h \in R$$

we have

$$\text{aug}(x) = \sum_{h \in H} x_h = \text{aug}'(x)$$

i.e. $\text{aug} = \text{aug}'$. Consequently, $I(R, G) = I(R, H)$, and every unit normalized with respect to G is also normalized with respect to H. Finally, recall that if I is an ideal of RG, then $G \cap (1 + I)$ is the kernel of the multiplicative map $G \to RG/I$, the image of which is $G + I = \{g + I \mid g \in G\}$. In particular,

$$G/(G \cap (1 + I)) \cong G + I.$$

In what follows, $I(G)$ denotes the augmentation ideal of $\mathbb{Z}G$.

8.3.12   Proposition   *Assume that* $\mathbb{Z}G \cong \mathbb{Z}H$. *Then*

$$G/G' \cong H/H' \qquad and \qquad G'/G'' \cong H'/H''.$$

Proof.   We may assume that $\mathbb{Z}G = \mathbb{Z}H$, in which case $I(G) = I(H)$. By Corollary 1.18, we have

$$\mathbb{Z}G \cdot I(G') = \mathbb{Z}H \cdot I(H'). \tag{3}$$

Multiplying by $I(G) = I(H)$, we obtain

$$I(G) \cdot I(G') = I(H) \cdot I(H'). \tag{4}$$

But then

$$G/G' \cong I(G)/I(G)^2 = I(H)/I(H)^2 \cong H/H'$$

by Proposition 3.9 while, by Proposition 3.11(ii),

$$G'/G'' \cong \mathbb{Z}G \cdot I(G')/I(G)I(G') = \mathbb{Z}H \cdot I(H')/I(H) \cdot I(H') \cong H'/H''$$

as required.   ∎

The following result is essentially due to Whitcomb (1968).

8.3.13   Theorem   *Let G be an arbitrary group such that G/G' is torsion. Then $\mathbb{Z}G \cong \mathbb{Z}H$ implies $G/G'' \cong H/H''$.*

Proof.   We may assume that $\mathbb{Z}G = \mathbb{Z}H$. By Proposition 3.12, $G/G' \cong H/H'$ and so H/H' is also torsion. Let $h \in H$ and let $\pi : \mathbb{Z}G \to \mathbb{Z}(G/G')$ be the natural homomorphism. By hypothesis, $h^n \in H'$ for some $n \geq 1$, so by (3),

$$h^n - 1 \subseteq I(H') \subseteq \mathbb{Z}G \cdot I(G') = \text{Ker } \pi.$$

Thus $\pi(h)$ is of finite order and so, by Proposition 3.10,

$$h \equiv g \pmod{I(G) \cdot I(G')}$$

for some $g \in G$. It follows that $H + I(G) \cdot I(G') \subseteq G + I(G) \cdot I(G')$ and, reversing the roles of G and H, we obtain $G + I(H)I(H') \subseteq H + I(H)I(H')$. Applying (4), we conclude that

$$G + I(G) \cdot I(G') = H + I(H) \cdot I(H').$$

Finally, using Proposition 3.11(ii), we derive

$$G/G'' = G/(G \cap (1 + I(G) \cdot I(G'))) \cong G + I(G)I(G')$$
$$= H + I(H)I(H') \cong H/(H \cap (1 + I(H)I(H')))$$
$$= H/H'',$$

as required. ∎

8.3.14 Proposition (Whitcomb) *Let G be a finite metabelian group. Then $\mathbb{Z}G \cong \mathbb{Z}H$ implies $G \cong H$.*

Proof. We have $|G| = |H|$ and, by Theorem 3.13, $G \cong H/H''$. The latter implies $H'' = 1$ and hence $G \cong H$. ∎

8.4 Trivial units

Our aim in this section is twofold: first to prove a classical result of Higman, which surveys all torsion groups G for which $U(\mathbb{Z}G)$ has only trivial units; and second, to provide necessary and sufficient conditions under which $U(RG)$, where G belongs to a class of torsion-free groups, has only trivial units.

8.4.1 Lemma *Let R be an arbitrary commutative ring. Assume that G is a group with cyclic subgroup $\langle a \rangle$ of finite order n which is not normal in G. Then, for any $g \notin N_G \langle a \rangle$, the element*

$$u = a + (1 + a + a^2 + \ldots + a^{n-1})g(1 - a)$$

is a non-trivial unit of order n.

Proof. Since $\text{aug}(u) = 1$, it suffices to show that $u^n = 1$ and $u^{n-1} \notin G$. To this end, put

$$x = (1 + a + a^2 + \ldots + a^{n-1})g(1 - a).$$

Then $x^2 = 0$ and, for all $m \geq 1$, $a^m x = x$, which implies $xa^m x = 0$. Thus

$$u^m = (a + x)^m = a^m + x(1 + a + \ldots + a^{m-1}). \tag{1}$$

Since $x(1 + a + \ldots + a^{n-1}) = 0$, we obtain $u^n = a^n = 1$.

Now assume that $m = n - 1$. Note that $x(1 + a + \ldots + a^{n-1}) = (1 + a + a^2 + \ldots + a^{n-1})g(1 - a^{n-1}) \neq 0$. Indeed, otherwise $(1 + a + \ldots + a^{n-1})g = (1 + a + \ldots + a^{n-1})ga^{n-1}$, and hence

$$1 + a + \ldots + a^{n-1} = (1 + a + \ldots + a^{n-1})ga^{n-1}g^{-1}.$$

But then $ga^{-1}g^{-1} \in \langle a \rangle$, contrary to the assumption that $g \notin N_G \langle a \rangle$. Since a typical element in the support of $(1 + a + \ldots + a^{n-1})g(1 - a^{n-1})$ is $a^i ga^j \neq a^{n-1}$, it follows from (1) (with $m = n - 1$) that $|\text{Supp } u^{n-1}| \geq 2$, as required. ∎

A group G is said to be *Hamiltonian* if G is non-abelian and all subgroups of G are normal. It is a standard fact of group theory (e.g. see Hall 1959, Theorem 12.5.4) that any Hamiltonian group is a direct product of the quaternion group, abelian torsion group with all elements of odd order, and abelian group of exponent ≤ 2.

8.4.2   Corollary   *Let R be a commutative ring and let G be a group such that all units of finite order in RG are trivial. Then all finite cyclic subgroups of G are normal and, in particular, if G is non-abelian torsion, then G is Hamiltonian.*

Proof.   The proof is a direct consequence of Lemma 4.1.   ∎

We are now ready to prove the following classical result.

8.4.3   Theorem   (Higman 1940b)   *Let G be a torsion group. Then all units in $\mathbb{Z}G$ are trivial if and only if G is one of the following types*:

(i)   *G is abelian with $G^4 = 1$ or $G^6 = 1$;*
(ii) *G is the direct product of a quaternion group and abelian group of exponent ≤ 2.*

Proof.   If G is abelian, then the required assertion is a very special case of a general result to be proved later (see Corollary 9.32). For the sake of clarity, we divide the proof into three steps.

   Step 1:   let $G^* = G \times \mathbb{Z}_2$. We prove that if all units in $\mathbb{Z}G$ are trivial, then so are the units in $\mathbb{Z}G^*$. Let the cyclic group of order 2 be generated by g. Then any element in $\mathbb{Z}G^*$ can be written uniquely in the form $\alpha + \beta g$ with $\alpha, \beta \in \mathbb{Z}G$. If

$$(\alpha + \beta g)(\gamma + \delta g) = \alpha\gamma + \beta\delta + (\alpha\delta + \beta\gamma)g = 1,$$

then

$$\alpha\gamma + \beta\delta = 1 \quad \text{and} \quad \alpha\delta + \beta\gamma = 0.$$

Therefore

$$(\alpha + \beta)(\gamma + \delta) = 1 \quad \text{and} \quad (\alpha - \beta)(\gamma - \delta) = 1$$

and, since $\mathbb{Z}G$ has only trivial units, we must have

$$\alpha + \beta = \pm g_1 \quad \text{and} \quad \alpha - \beta = \pm g_2 \quad (g_1, g_2 \in G).$$

Hence $\alpha = (1/2)(\pm g_1 \pm g_2) \in \mathbb{Z}G$, which is possible only if $g_1 = g_2$ so that $\alpha + \beta = \pm(\alpha - \beta)$. If $\alpha + \beta = \alpha - \beta = \pm g_1$, we have $\alpha = \pm g_1$, $\beta = 0$, and so $\alpha + \beta g = \pm g_1$. If $\alpha + \beta = -(\alpha + \beta) = \pm g_1$, we have similarly, $\alpha + \beta = \pm g_1 g$. In all cases $\alpha + \beta g$ is a trivial unit, as asserted.

   Step 2:   let G be a quaternion group. It will be shown that all the units in $\mathbb{Z}G$ are trivial. Indeed, the group G is generated by x, y, z subject to

$$x^2 = y^2 = z^2 = xyz = u.$$

Let $v = a_0 + a_1 x + a_2 y + a_3 z + b_0 u + b_1 xu + b_2 yu + b_3 zu$ be a unit in $\mathbb{Z}G$. The map $x \mapsto i$, $y \mapsto j$, $z \mapsto k$, $u \mapsto -1$ determines a homomorphism of $\mathbb{Z}G$ onto the ring of integral quaternions. In this ring the only units are ± 1, $\pm i$, $\pm j$, $\pm k$, which can be seen by observing that the norm $a^2 + b^2 + c^2 + d^2$ of the unit $a + bi + cj + dk$ must be equal to 1. Hence v is carried into one of these quantities. Hence, for some $i \in \{0, 1, 2, 3\}$,

$$a_i - b_i = \pm 1, \quad a_j - b_j = 0 \qquad (j \neq i, j = 0, 1, 2, 3). \tag{2}$$

Let H be a direct product of two cyclic groups of order two, and let its elements be 1, x', y', z'. Then $x \mapsto x'$, $y \mapsto y'$, $z \mapsto z'$, $v \mapsto 1$ determines a homomorphism of $\mathbb{Z}G$ onto $\mathbb{Z}H$. Since all units of $\mathbb{Z}H$ are trivial, there exists $m \in \{0, 1, 2, 3\}$ with

$$a_m + b_m = \pm 1, \quad a_j + b_j = 0 \qquad (j \neq m, j = 0, 1, 2, 3). \tag{3}$$

Because a_i, $b_i \in \mathbb{Z}$, we have, on comparing (2) and (3), $i = m$. Hence either

$$a_i = \pm 1, \quad b_i = 0, \quad a_j = b_j \qquad (j \neq i)$$

or

$$a_i = 0, \quad b_i = \pm 1, \quad a_j = b_j = 0 \qquad (j \neq i).$$

In either case the unit v is trivial.

Step 3: completion of the proof. If G is abelian then, as has been observed earlier, the result is true. Suppose G satisfies (ii). To prove that all units of $\mathbb{Z}G$ are trivial, we may harmlessly assume that G is finite. But then the required assertion follows from Steps 1 and 2. Conversely, assume that G is non-abelian and that all units in $\mathbb{Z}G$ are trivial. Then, by Corollary 4.2, G is Hamiltonian. Hence $G = H \times A \times B$, where H is the quaternion group, A abelian whose elements are of odd order, and B abelian of exponent ≤ 2. If A has an element of prime order p, then G has an element of order $4p$, which is impossible by the abelian case. Thus $A = 1$ and the result follows. ■

We now turn our attention to torsion-free groups. A group G is said to be a *unique product group* if, given any two non-empty finite subsets A and B of G, there exists at least one element $x \in G$ that has a unique representation in the form $x = ab$ with $a \in A$ and $b \in B$. Note that a unique product group is necessarily torsion-free. For if $g \neq 1$ is an element of order n in G, we may take $A = B = \langle g \rangle$ and the equalities $g^i g^j = g^j g^i$, $i \neq j$, $0 \leq i$, $j \leq n - 1$, $g^i g^i = g^{2i} \cdot 1$, $1 < i \leq n - 1$, $1 = g^i g^{n-i}$, $1 < i \leq n - 1$, show that G fails to be a unique product group.

A group G is said to be a *right-ordered group* if the elements of G are linearly ordered with respect to the relation $<$ and if, for all x, y, $z \in G$,

$$x < y \text{ implies } xz < yz.$$

8.4.4 Lemma *Let G be a right-ordered group. Then, given any two non-empty finite subsets A and B of G with $|A| + |B| > 2$, there exist at least two distinct elements x and y of G that have unique representations in the form $x = ab$, $y = cd$ with a, $c \in A$ and b, $d \in B$. In particular, G is a unique product group.*

Proof. Let $a_{\max}$ and $a_{\min}$ be the largest and the smallest elements in A, respectively. Choose $b' \in B$ so that $a_{\max} b'$ is the largest element in $a_{\max} B$, and let $b'' \in B$ be chosen with $a_{\min} b''$ the smallest element in $a_{\min} B$. If $a \in A$, $b \in B$, then since $a_{\min} \leq a \leq a_{\max}$, we have $a_{\min} b \leq ab \leq a_{\max} b$, and thus

$$a_{\min} b'' \leq a_{\min} b \leq ab \leq a_{\max} b' \leq a_{\max} b'.$$

Moreover, $a_{\min} b'' = ab$ implies $a_{\min} = a$ and then $b'' = b$. Similarly, $ab = a_{\max} b'$ implies $a = a_{\max}$, $b = b'$. Thus, the two products $a_{\max} b'$ and $a_{\min} b''$ are uniquely represented. ∎

8.4.5 Lemma *Let G be a group with $G = T \cup \{1\} \cup T^{-1}$ (disjoint union), where T is a multiplicatively closed subset of G. Then G is a right-ordered group. In particular, if $N \triangleleft G$ and both N and G/N are right-ordered groups, then so is G.*

Proof. Define $x < y$ to mean that $yx^{-1} \in T$. Then $<$ is a linear ordering on G. If $x < y$ and $z \in G$, then $yx^{-1} \in T$, so that

$$(yz)(xz)^{-1} = yx^{-1} \in T$$

and $xz < yz$, proving that G is a right-ordered group.

Let $G \overset{\phi}{\to} G/N$ denote the natural homomorphism. Put $T(N) = \{x \in N \mid 1 < x\}$, $T(G/N) = \{y \in G/N \mid 1 < y\}$, and define

$$T = T(G) = \{x \in G \mid \phi(x) \in T(G/N) \quad \text{or} \quad x \in T(N)\}.$$

Then it is trivial to verify that T is multiplicatively closed and that

$$G = T \cup \{1\} \cup T^{-1}$$

is a disjoint union. ∎

As an application, we have

8.4.6 Lemma *Assume that a group G has a finite subnormal series*

$$1 = G_0 \triangleleft G_1 \triangleleft \ldots \triangleleft G_n = G$$

with quotients G_{i+1}/G_i *which are torsion-free abelian. Then* G *is a right-ordered group.*

Proof. Owing to Lemma 4.5, we may assume that G is torsion-free abelian. Let L be the class of all subsets T of G that satisfy the condition:

$$T \text{ is multiplicatively closed and } 1 \notin T.$$

By Zorn's lemma, L contains a maximal element T. We claim that

$$G = T \cup \{1\} \cup T^{-1}.$$

Since the union $T \cup \{1\} \cup T^{-1}$ is necessarily disjoint, the result will follow by appealing to Lemma 4.5.

Deny the statement. Then there exists a non-identity g in G such that neither g nor g^{-1} belongs to T. With

$$S = T \cup \{tg^n \mid t \in T, n \geq 1\} \cup \{g^n \mid n \geq 1\}$$

it follows that $S \supset T$ and that S is multiplicatively closed. The maximality of T now yields $1 \in S$, and since G is torsion-free we see that $1 = tg^n$ for some $t \in T$ and some $n \geq 1$. Thus $g^{-n} \in T$ and replacing g by g^{-1} in the above argument, we infer that $g^m \in T$ for some $m \geq 1$. This substantiates our claim, as

$$1 = (g^m)^n(g^{-n})^m \in T,$$

contrary to our assumption. ∎

It will next be shown that unique product groups satisfy the property of right-ordered groups exhibited in Lemma 4.4.

8.4.7 Theorem (Strojnowski 1980) *Let G be a group. Then the following conditions are equivalent:*

(i) *For any two non-empty finite subsets A and B of G with $|A| + |B| > 2$, there exist at least two distinct elements x and y of G that have unique representations in the form $x = ab$, $y = cd$ with a, $c \in A$ and b, $d \in B$.*

(ii) *For any non-empty finite subset A of G there exists at least one element in $AA = \{xy \mid x, y \in A\}$ that has a unique representation in the form xy, where $x, y \in A$.*

(iii) *G is a unique product group.*

Proof. The implications (i) $\Rightarrow$ (ii) and (i) $\Rightarrow$ (iii) being obvious, we are left to verify that (ii) implies (iii) and (iii) implies (i).

(ii) $\Rightarrow$ (iii) Assume by way of contradiction that G is not a unique product group. Then there exist two non-empty finite subsets B and C of G such that every element of BC has at least two distinct representations in the form xy, where $x \in B$ and $y \in C$. Put $A = CB$. Then each element in AA can be written in the form $(c_1b_1)(c_2b_2)$, with c_1, $c_2 \in C$ and b_1, $b_2 \in B$. The element $b_1c_2 \in BC$ has another representation in the form b_3c_3 with $b_3 \in B$, $c_3 \in C$ and $b_1 \neq b_3$. We have

$$(c_1b_1)(c_2b_2) = (c_1b_3)(c_3b_2),$$

which shows that each element in AA has at least two distinct representations in the form xy with x, $y \in A$, a contradiction.

(iii) $\Rightarrow$ (i)   Assume that G satisfies (iii) but not (i). Then there exist two non-empty finite subsets A and B of G such that $|A| + |B| > 2$, and in the set AB there exists exactly one element ab which has unique representation in the form xy with $x \in A$, $y \in B$. Put $C = a^{-1}A$ and $D = Bb^{-1}$. In CD only $1 \cdot 1$ has a unique representation. Let $E = D^{-1}C$ and $F = DC^{-1}$. Then each element of EF can be written in the form $(d_1^{-1}c_1)(d_2c_2^{-1})$. If $c_1 \neq 1$ or $d_2 \neq 1$, then there exist $c_3 \in C$ and $d_3 \in D$ with $c_1d_2 = c_3d_3$ and $c_1 \neq c_3$. Therefore the element $(d_1^{-1}c_1)(d_2c_2^{-1})$ has another representation $(d_1^{-1}c_3)(d_3c_2^{-1})$. If $c_2 \neq 1$ or $d_1 \neq 1$, then there exist $c_4 \in C$ and $d_4 \in D$ with $c_2 \neq c_4$ and $c_2d_1 = c_4d_4$. We have $d_1^{-1}c_2^{-1} = d_4^{-1}c_4^{-1}$, and thus the element $(d_1^{-1} \cdot 1)(1 \cdot c_2^{-1})$ has another representation $(d_4^{-1} \cdot 1)(1 \cdot c_4^{-1})$. Now $|C| + |D| > 2$, so in C or in D there exists an element $\neq 1$. Assume, for definiteness, that $1 \neq c \in C$. Then the element $(1 \cdot 1)(1 \cdot 1)$ in EF has another representation $(1 \cdot c)(1 \cdot c^{-1})$. It follows that there is no element in EF which has unique representation in the form xy with $x \in E$ and $y \in F$, a contradiction. ∎

8.4.8 Corollary   *Let G be a unique product group and let R be an integral domain. If x, y are non-zero elements of RG and if xy is a monomial, then x and y are monomials. In particular, RG has only trivial units.*

Proof.   Suppose the contrary, say x is not a monomial. This means that $|\operatorname{Supp} x| \geq 2$ and hence $|\operatorname{Supp} x| + |\operatorname{Supp} y| > 2$. Write $x = \sum x_g g$ and $y = \sum y_g g$. Then

$$(xy)_g = \sum_{ab=g} x_a y_b, \qquad \text{for all } g \in G.$$

By Theorem 4.7(i), there exist distinct elements g_1, $g_2 \in G$ that have unique representation in the form $g_1 = ab$, $g_2 = cd$ with a, $c \in \operatorname{Supp} x$ and b, $d \in \operatorname{Supp} y$. Hence

$$(xy)_{g_1} = x_a y_b \neq 0 \qquad \text{and} \qquad (xy)_{g_2} = x_c y_d \neq 0,$$

a contradiction. ∎

We have now accumulated all the information necessary to prove our second major result.

8.4.9 Theorem (Krempa 1982) *Let G be a unique product group and let R be an arbitrary commutative ring. Then RG has only trivial units if and only if R is indecomposable and reduced.*

Proof. Assume that RG has only trivial units. If $0 \neq r \in R$ with $r^n = 0$ for some $n \geqslant 1$, then for any $1 \neq g \in G$ we have $(rg)^n = 0$, so that $1 + rg$ is a non-trivial unit. If r is a non-trivial idempotent in R, then for any $1 \neq g \in G$ the unit

$$u = r \cdot 1 + (1 - r)g$$

is non-trivial with $u^{-1} = r \cdot 1 + (1 - r)g^{-1}$.

Conversely, assume that R is indecomposable and reduced. Let

$$x = \sum x_g g \qquad \text{and} \qquad y = \sum y_g g$$

be elements in RG such that $xy = 1 = yx$. Then, by taking coefficients of 1 in $xy = 1$, we obtain

$$\sum x_g y_{g^{-1}} = 1 \tag{1}$$

and therefore

$$x_t y_{t^{-1}} \neq 0, \qquad \text{for some } t \in G. \tag{2}$$

We claim that the result will hold provided the following is true:

$$x_g y_h = 0, \qquad \text{whenever } g \neq h. \tag{3}$$

Indeed, if (3) is valid, then multiplication of (1) by x_t yields

$$x_t^2 y_{t^{-1}} = x_t \qquad \text{and} \qquad (x_t y_{t^{-1}})^2 = x_t y_{t^{-1}}.$$

Because of (2) this forces $x_t y_{t^{-1}} = 1$, and multiplication of this equality by x_g ($g \neq t$) yields the desired result.

We are now left to prove (3). Assume by way of contradiction that $x_g x_h \neq 0$ for some g, h in G with $g \neq h$. Since $x_g x_h$ is not a nilpotent element, there exists a prime ideal P in R such that $x_g x_h \notin P$. Let $\bar{s}$ be the image of $s \in RG$ under the canonical homomorphism $RG \to (R/P)G$. Then

$$\bar{x}_g \neq 0, \bar{x}_h \neq 0 \text{ in } \bar{R} = R/P$$

and

$$\bar{x}\bar{y} = 1 = \bar{y}\bar{x} \text{ in } \bar{R}G.$$

Thus $\bar{R}G$ has a non-trivial unit, contrary to Corollary 4.8. ∎

8.4.10 Corollary *Assume that a group G has a finite subnormal series* *with quotients which are torsion-free abelian. Then, for any commutative ring R, RG has only trivial units if and only if R is indecomposable and reduced.*

Proof.   Apply Theorem 4.9, Lemma 4.6 and Lemma 4.4.   ∎

8.5   Conjugacy of group bases

Throughout, $\mathbb{Z}G = \mathbb{Z}H$ means that H is a normalized group basis. Assume that G is a finite group, that $\mathbb{Z}G = \mathbb{Z}H$ and that $G \cong H$. It is natural to ask whether there is a unit u in $\mathbb{Z}G$ such that $H = u^{-1}Gu$. That this is not always the case as was first discovered by Berman and Rossa (1966). The following example can be found in their work.

Let $G = \{a, b \mid a^4 = b^2 = 1, \ b^{-1}ab = a^{-1}\}$, and let $H = \langle a', b' \rangle$, where

$$a' = -a + 2a^3 - b - ab + a^2b + a^3b,$$

$$b' = -a + a^3 - ab + a^2b + a^3b.$$

Then $\mathbb{Z}G = \mathbb{Z}H$, but it can be shown (see Berman and Rossa 1966) that G and H are not conjugate in $U(\mathbb{Z}G)$. Incidentally, G and H are conjugate in $U(\mathbb{Z}_{(2)}G)$, where $\mathbb{Z}_{(2)}$ is the ring of 2-integral rationals. Indeed, let $u = 1 - b + ab$. Then, by Proposition 1.16, u is a unit in $\mathbb{Z}_{(2)}G$ and one immediately verifies that

$$u^{-1}a'u = a, \qquad u^{-1}b'u = b.$$

In view of the above, we are led to ask whether, for an arbitrary finite group G, the number of conjugacy classes of group bases is finite. Our main result in this section asserts that this is always the case even under more general circumstances, namely when $\mathbb{Z}$ is replaced by the ring of algebraic integers of an algebraic number field. We also provide a number of related results.

Let RG be the group ring of an arbitrary group G over a commutative ring R. Denote by $\mathrm{Aut}(RG)$ the group of all R-automorphisms of RG. For each $u \in U(RG)$, let i_u be the inner automorphism of RG, defined by

$$i_u(x) = u^{-1}xu \qquad (x \in RG).$$

The group of *inner automorphisms* of RG is defined by

$$\mathrm{In}(RG) = \{i_u \mid u \in U(RG)\}.$$

It is clear that $\mathrm{In}(RG)$ is a normal subgroup of $\mathrm{Aut}(RG)$. The corresponding factor group

$$\mathrm{Out}(RG) = \mathrm{Aut}(RG)/\mathrm{In}(RG)$$

is said to be the *outer automorphism group* of RG.

Our first goal is to provide a bijective correspondence between the set $\text{Out}(RG)$ and a set consisting of $R(G \times G)$-isomorphism classes of R-free $R(G \times G)$-modules of a certain type. The following preliminary observation will clear our path.

8.5.1 Lemma   *Let G be an arbitrary group, let R be a commutative ring and, for any $f \in \text{Aut}(RG)$, let M_f be the additive group of RG. Then M_f is a (left) R-free $R(G \times G)$-module under the following action of $R(G \times G)$:*

$$\text{for each } t = \sum_{i=1}^{n} \alpha_i(a_i, b_i) \in R(G \times G) \text{ and for each } x \in M_f$$

$$t \circ x = \sum_{i=1}^{n} \alpha_i a_i x f(b_i^{-1}).$$

Proof.   To prove that M_f is an $R(G \times G)$-module, it is enough to check that M_f is a $G \times G$-module under the composition

$$(a, b) \circ x = axf(b^{-1}), \qquad (a, b) \in G \times G, x \in M_f.$$

It is clear that, for any $x_1, x_2, x \in M_f$,

$$(a, b) \circ (x_1 + x_2) = (a, b) \circ x_1 + (a, b) \circ x_2 \qquad \text{and} \qquad (1, 1) \circ x = x.$$

Let $(c, d) \in G \times G$. Then, for any $x \in M_f$,

$$[(a, b)(c, d)] \circ x = acxf(d^{-1})f(b^{-1}) = (a, b) \circ [cxf(d^{-1})]$$

$$= (a, b) \circ [(c, d) \circ x],$$

proving that M_f is an $R(G \times G)$-module. Since the group ring RG is an R-free module and since $r(1, 1) \circ x = rx$ for any $r \in R$, $x \in RG$, it follows that M_f is also R-free, regarded as an $R(G \times G)$-module, as required. ∎

Denote by $P(RG)$ the set consisting of all $R(G \times G)$-isomorphism classes (M_f) of $R(G \times G)$-modules M_f, where f ranges over all elements of $\text{Aut}(RG)$. Let also $M_f = M$ for $f = $ identity automorphism. We are now ready to prove the following result:

8.5.2 Theorem   (Karpilovsky 1980b)   *The mapping*

$$\begin{cases} \text{Out}(RG) \xrightarrow{\lambda} P(RG) \\ f \, \text{In}(RG) \mapsto (M_f) \end{cases}$$

is a bijection. In particular, $f \in \text{In}(RG)$ if and only if the $R(G \times G)$-modules M and M_f are isomorphic.

Proof.   Let $\phi \in \text{Aut}(RG)$, $u \in U(RG)$, and let $\psi = \phi i_u$. Consider the mapping

$$\mu : M_\phi \to M_\psi \text{ defined by } \mu(x) = x\phi(u)$$

for any $x \in M_\phi$. It is obvious that μ is an R-isomorphism of $R(G \times G)$-modules M_ϕ and M_ψ. On the other hand, for any $(a, b) \in G \times G$ and for any $x \in M_\phi$, we have

$$\mu((a, b) \circ x) = \mu(ax\phi(b^{-1})) = ax\phi(b^{-1})\phi(u)$$
$$= ax\phi(u)\phi(u^{-1}b^{-1}u) = a\mu(x)\psi(b^{-1})$$
$$= (a, b) \circ \mu(x).$$

Hence μ is an $R(G \times G)$-isomorphism and so the map λ is well defined. Now $\theta : M_\phi \to M_f$ be an $R(G \times G)$-isomorphism and put $u = \theta(1)$. Then, for any $(a, b) \in G \times G$ and, for any $x \in M_\phi$, the equality $\theta[(a, b) \circ x] = (a, b) \circ \theta(x)$ implies

$$\theta(ax\phi(b^{-1})) = a\theta(x)f(b^{-1}). \tag{1}$$

Taking $x = b = 1$ in (1), we obtain $\theta(a) = au$ for any $a \in G$, and since θ is necessarily an R-isomorphism of $R(G \times G)$-modules M_ϕ and M_f it follows that

$$\theta(x) = xu, \qquad \text{for any } x \in M_\phi. \tag{2}$$

Next choose $a = x = 1$ in (1), whence

$$\theta[\phi(b^{-1})] = uf(b^{-1})$$

for any $b \in G$ and it follows from (2) that $\phi(g)u = uf(g)$ for all $g \in G$. Thus

$$\phi(x)u = uf(x), \qquad \text{for all } x \in M_\phi. \tag{3}$$

Therefore, $RGu = uRG$ and it follows from (2) that $RG = RGu$ and $RG = uRG$, whence $u \in U(RG)$.

Invoking (3), we deduce that, for any $x \in M_\phi$,

$$f(x) = u^{-1}\phi(x)u; \text{ that is, } f = i_u\phi.$$

This shows that the map λ is injective, and since λ is obviously surjective, the proof is complete. ∎

8.5.3 Corollary (Karpilovsky 1980b) *Let G be a finite group and let R be the ring of integers of an algebraic number field. Then there are only finitely many conjugacy classes of group bases in RG.*

Proof.   Since G is finite, each M_f is R-free of finite fixed rank $n = |G|$. It is well known (see Curtis and Reiner 1962, Theorem 79.13) that there are only finitely many of isomorphism classes of such modules. Thus the set $P(RG)$ is finite and, by Theorem 5.2, the group $\text{Aut}(RG)/\text{In}(RG)$ is finite. Let

$$\text{Aut}(RG) = \text{In}(RG) \cup \text{In}(RG)\phi_2 \cup \ldots \cup \text{In}(RG)\phi_t$$

be the coset decomposition $\text{Aut}(RG)$ with respect to $\text{In}(RG)$. Suppose that H is an arbitrary group basis of RG. Since $|H| = |G|$, there exists only a finite number of non-isomorphic group bases in RG, say, $G_1, G_2, \ldots, G_n$. Hence $H \cong G_i$ for some $i \in \{1, 2, \ldots, n\}$, and therefore there exists $f \in \text{Aut}(RG)$ such that $f(G_i) = H$. Since $f = \theta\phi_j$ for some $\theta \in \text{In}(RG)$ and some $j \in \{1, 2, \ldots, t\}$, we have

$$f(G_i) = u^{-1}\phi_j(G_i)u, \qquad \text{for some } u \in U(RG).$$

Thus H is conjugate to $\phi_j(G_i)$, proving the result. ∎

We next provide circumstances under which any two normalized group bases in $\mathbb{Z}G$ are conjugate in $\mathbb{Q}G$. Note, however, that whether this is always the case is not known. We first record the following consequence of Theorem 5.2.

8.5.4 Corollary *Let G be a finite group, let $\mathbb{Z}G = \mathbb{Z}H$ with $H \cong G$, and let f be the automorphism of $\mathbb{Q}G$ which is the extension of the isomorphism $H \to G$ by $\mathbb{Q}$-linearity. Denote by M (respectively, M_f) the $\mathbb{Q}(G \times G)$-module $\mathbb{Q}G$ defined by $(a, b) \circ x = axb^{-1}$, $(a, b) \in G \times G$, $x \in \mathbb{Q}G$ (respectively, the $\mathbb{Q}(G \times G)$-module $\mathbb{Q}G$ defined by $(a, b) \circ x = axf(b^{-1})$). Then H is conjugate to G in $\mathbb{Q}G$ if and only if $\chi = \chi_f$, where χ (respectively, χ_f) is the character of the $\mathbb{Q}(G \times G)$-module afforded by M (respectively, M_f).*

Proof. All we have to is to notice that the $\mathbb{Q}(G \times G)$-modules M and M_f are isomorphic if and only if $\chi = \chi_f$, and apply Theorem 5.2. ∎

The following standard result will be very useful to us. This is a convenient place to record it for future reference.

8.5.5 Proposition *Let A be a finite-dimensional semisimple algebra over a field. If θ is an automorphism of A which is the identity mapping on the centre of A, then θ is an inner automorphism.*

Proof. See Bourbaki (1958). ∎

A typical application of the above result is the following

8.5.6 Lemma *Let G be a finite group and let $\mathbb{Z}G = \mathbb{Z}H$. If G and H are conjugate in $U(\mathbb{C}G)$, then they are conjugate in $U(\mathbb{Q}G)$.*

Proof. By hypothesis, there exists an element $u \in U(\mathbb{C}G)$ such that $u^{-1}Gu = H$, and therefore the mapping $\phi : G \to H$ defined by $\phi(g) = u^{-1}gu$ is an isomorphism of G onto H. Let ψ be the automorphism of $\mathbb{Q}G$ which is the extension of the map ϕ by $\mathbb{Q}$-linearity. Since ψ is the identity mapping on $Z(\mathbb{Q}G)$, it follows from Proposition 5.5 that ψ is an inner automorphism of $\mathbb{Q}G$. Consequently, there exists an element $v \in U(\mathbb{Q}G)$ such that $\psi(x) = v^{-1}xv$ for any $x \in \mathbb{Q}G$. Because $\psi(G) = H$, we have $v^{-1}Gv = H$, as asserted. ∎

8.5.7   Lemma   *Let G be a finite group and let R be an integral domain of characteristic 0 in which no rational prime divisor of $|G|$ is a unit. Let $RG = RH$, let $N \lhd G$ and let $N^* = H \cap (1 + RG \cdot I(N))$. If $\pi : RG \to R\bar{G}$, $\bar{G} = G/N$, is the natural homomorphism, then*

$$R\bar{G} = R\bar{H}, \quad RG \cdot I(N) = RG \cdot I(N^*) \text{ and } |N| = |N^*|.$$

Proof.   By Corollary 3.7(i), $\bar{H}$, the image of H in $R\bar{G}$, is a linearly independent set in $R\bar{G}$, whence $R\bar{G} = R\bar{H}$. Moreover, since π can be regarded as the extension of the epimorphism $H \to \bar{H}$ (the kernel of which is N^*) by R-linearity, we have

$$\operatorname{Ker} \pi = RG \cdot I(N) = RG \cdot I(N^*).$$

Moreover, because $|G/N| = |\bar{G}| = |\bar{H}| = |H/N^*|$, the groups N and N^* are of the same order. ∎

We next recall some basic facts about group representations.

Let G be a finite group and let

$$\Gamma_i : \mathbb{C}G \to M_{n_i}(\mathbb{C}), \qquad 1 \leq i \leq r$$

be the non-equivalent irreducible matrix representations of $\mathbb{C}G$. Denote by $\chi_i(x) = \operatorname{trace} \Gamma_i(x)$, $x \in \mathbb{C}G$. The family (Γ_i), $1 \leq i \leq r$, defines an isomorphism

$$\Gamma : \mathbb{C}G \to \prod_{i=1}^{r} M_{n_i}(\mathbb{C}),$$

where $\Gamma(x) = (\Gamma_1(x), \Gamma_2(x), \ldots, \Gamma_r(x))$, $x \in \mathbb{C}G$. The set $\{\rho_1, \rho_2, \ldots, \rho_r\}$, where $\rho_i(g) = \Gamma_i(g)$ for any $g \in G$, is a full set of non-equivalent irreducible complex representations of G. Using the bar convention for the homomorphic images, consider the natural homomorphism

$$\mathbb{C}G \to \mathbb{C}\bar{G},$$

where $\bar{G} = G/N$ and $N = \operatorname{Ker} \rho_i$. Then

$$\mathbb{C}G \cdot I(N) \subseteq \operatorname{Ker} \Gamma_i$$

and therefore the mapping

$$f : \mathbb{C}\bar{G} \to M_{n_i}(\mathbb{C}) \tag{4}$$

defined by

$$f(\bar{x}) = \Gamma_i(x) \qquad (x \in \mathbb{C}G)$$

is an irreducible matrix representation of $\mathbb{C}\bar{G}$. Suppose that $z \in Z(G)$. Then, by Schur's lemma, $\rho_i(z) = \text{diag}(\varepsilon, \varepsilon, \ldots, \varepsilon)$, where ε is a root of unity in $\mathbb{C}$. Therefore, if $z = [a, b]$ for some $a, b \in G$, then $\varepsilon\chi_i(a) = \chi_i(a)$. This shows that if G is nilpotent of class 2 and ρ_i is faithful (i.e. $\text{Ker } \rho_i = 1$), then

$$\chi_i(g) = 0, \qquad \text{for all } g \notin Z(G). \tag{5}$$

We are now ready to prove the following result, which is essentially a restatement of Theorem 9 in Whitcomb (1968) (see also Sehgal 1978).

8.5.8 Theorem   *Let G be a finite nilpotent group of class 2. Then any normalized group basis H of $\mathbb{Z}G$ is conjugate to G in $U(\mathbb{Q}G)$.*

Proof.   By Propositions 3.10 and 3.11, for any $g \in G$ there exists a unique $\phi(g) \in H$ such that $g \equiv \phi(g)(\text{mod } I(G) \cdot I(G'))$. Furthermore, the map $g \mapsto \phi(g)$ is obviously an isomorphism of G onto H. On the other hand, by Corollary 3.5, $Z(G) = Z(H)$, whence

$$\phi(g) = g, \qquad \text{for all } g \in Z(G). \tag{6}$$

Preserving the previous notation, we now claim that it suffices to show that, for any $i \in \{1, 2, \ldots, r\}$:

$$\chi_i(g) = \chi_i(h) \qquad \text{whenever} \qquad g \equiv h(\text{mod } I(G) \cdot I(G'). \tag{7}$$

Indeed, if this is the case, then the irreducible matrix representations α_1 and α_2 of G defined by $\alpha_1(g) = \Gamma_i(g)$ and $\alpha_2(g) = \Gamma_i(\phi(g))$ are equivalent, that is, there exists a non-singular matrix B_i such that $B_i^{-1}\Gamma_i(g)B_i = \Gamma_i(\phi(g))$ for all $g \in G$. This forces G and H to be conjugate in $U(\mathbb{C}G)$ and, thanks to Lemma 5.6, G and H are conjugate in $U(\mathbb{Q}G)$, as claimed.

To prove (7), suppose that $g \equiv h(\text{mod } I(G) \cdot I(G'))$ with $h \in H$. By passing to the natural homomorphism $\mathbb{Z}G \to \mathbb{Z}\bar{G}$, where $\bar{G} = G/N$ and $N = \text{Ker } \Gamma_i$, we derive

$$\bar{g} \equiv \bar{h}(\text{mod } I(\bar{G}) \cdot I(\bar{G}')).$$

By Lemma 5.7, $\mathbb{Z}\bar{G} = \mathbb{Z}\bar{H}$ and therefore (4) and induction on $|G|$ may be employed to conclude that $\chi_i(g) = \chi_i(h)$ whenever $\text{Ker } \Gamma_i \neq 1$.

Finally, assume that Γ_i is a faithful representation and put $\Gamma_i'(g) = \Gamma_i(\phi(g))$, $g \in G$. If $\text{Ker } \Gamma_i' = 1$, then since $\mathbb{C}G = \mathbb{C}H$, we have $1 \neq N^* = \{h \in H \mid \Gamma_i(h) = 1\}$. Since $\mathbb{C}G \cdot I(N^*) \subseteq \text{Ker } \Gamma_i$, it follows from Lemma 5.7, that $\mathbb{C}G \cdot I(N) \subseteq \text{Ker } \Gamma_i$ for some $N \lhd G$ with $|N| = |N^*|$. But then $N \subseteq \text{Ker } \Gamma_i'$, a contradiction. Thus Γ_i' is a faithful representation. Since by (6), $g = h$ if $g \in Z(G)$, an application of (5) concludes the proof. ∎

We close by remarking that Corollary 5.3 can also be deduced from property (ii) of the result below.

8.5.9 Theorem *Let R be any ring which as a $\mathbb{Z}$-module is free of finite rank. Then*

(i) *$U(R)$ is finitely presented.*
(ii) *$U(R)$ has only finitely many conjugacy classes of finite subgroups.*
(iii) *$U(R)$ is residually finite.*

Proof. See Serre (1979, pp. 107–8). ■

8.6 A description of $U(\mathbb{Z}G)$, G dihedral of order $2p$

Our aim in this section is to describe $U(\mathbb{Z}G)$ in case G is dihedral of order $2p$, where p is an odd prime. The main result will emerge as a consequence of more general considerations which are of independent interest. Namely, we explicitly compute a certain normal subgroup of $V(\mathbb{Z}G)$, where G is dihedral of order $2p^n$. An application of this result to the case where $n = 1$ will lead us naturally to the desired description.

Throughout, all roots of unity are assumed to be taken from $\mathbb{C}$, and $\mathbb{F}_p$ denotes the field of p elements.

Let R be a ring. By an *involution* of R, we understand a ring antiautomorphism of R of order 2.

Let R be a commutative ring with an involution $r \mapsto r'$. Then we define

$$M_2^*(R) = \left\{ \begin{bmatrix} a & b \\ b' & a' \end{bmatrix} \in M_2(R) \right\}.$$

8.6.1 Lemma *Let R be a commutative ring with involution $r \mapsto r'$. Then:*

(i) *$M_2^*(R)$ is a subring of $M_2(R)$.*
(ii) *The usual adjoint map given by*

$$\mathrm{adj} \begin{bmatrix} a & b \\ b' & a' \end{bmatrix} = \begin{bmatrix} a' & -b \\ -b' & a \end{bmatrix}$$

defines an involution on $M_2^*(R)$.

(iii)

$$x = \begin{bmatrix} a & b \\ b' & a' \end{bmatrix}$$

is a unit of $M_2^*(R)$ if and only if $\det x = aa' - bb'$ is a unit of R.

Proof. The verification of (i) and (ii) is straightforward. To prove (iii), note that if x is a unit of $M_2^*(R)$, then it is a unit of $M_2(R)$, so $\det x$ is a unit of R. Conversely, if $\det x$ is a unit of R, then since $(\det x)' = \det x$ we have

$$x^{-1} = (\det x)^{-1}(\text{adj } x) \in M_2^*(R)$$

as required. ∎

The next lemma illustrates how the ring $M_2^*(R)$ can be brought into argument.

8.6.2 Lemma   *Let $G = A\langle t \rangle$ be the semidirect product of the abelian group A by the group $\langle t \rangle$ of order 2. Then $x \mapsto x^t = t^{-1}xt$ is an involution of $\mathbb{Z}A$, and the map*

$$\begin{cases} \mathbb{Z}G \to M_2^*(\mathbb{Z}A), \\ \alpha + \beta t \mapsto \begin{bmatrix} \alpha & \beta \\ \beta^t & \alpha^t \end{bmatrix} \quad (\alpha, \beta \in \mathbb{Z}A) \end{cases}$$

is a ring isomorphism. In particular:

(i) $\alpha + \beta t \in U(\mathbb{Z}G)$ *if and only if* $\alpha\alpha^t - \beta\beta^t \in U(\mathbb{Z}A)$.
(ii) *If $\alpha + \beta t$ is a unit of finite order, then $\alpha\alpha^t - \beta\beta^t = \pm g$ for some $g \in A \cap Z(G)$.*

Proof.   We know that $\mathbb{Z}G$ is a free $\mathbb{Z}A$-module with basis $\{1, t\}$. Since

$$1 \cdot (\alpha + \beta t) = \alpha \cdot 1 + \beta \cdot t,$$

$$t \cdot (\alpha + \beta t) = \beta^t \cdot 1 + \alpha^t \cdot t$$

and since $\mathbb{Z}G$ is a faithful right $\mathbb{Z}G$-module, the required isomorphism follows. Assertion (i) is a consequence of the above and Lemma 6.1(iii). To prove (ii), denote by w the image of $\alpha + \beta t$. Then w is of finite order and, by Lemma 6.1(ii), so is adj w. Since w and adj w commute, we conclude that w adj $w = \det w \cdot 1$ is also a unit of finite order. Hence, by Corollary 3.5, $\det w = \pm g$ for some $g \in A$. Furthermore, since $(\det w)^t = \det w$, we see that $g \in Z(G)$, as required.   ∎

In view of Lemma 6.2(i), we need to get an insight into the behaviour of units of $\mathbb{Z}A$. A detailed account of the structure of $U(\mathbb{Z}A)$ will be provided in our subsequent investigations. Here we merely prove what is needed for our immediate purposes. The following general observation will be useful.

8.6.3 Lemma   *Let R_1, R_2 and R be rings and let $\pi_1 : R_1 \to R$ and $\pi_2 : R_2 \to R$ be homomorphisms. Let $S \subseteq R_1 \times R_2$ be given by*

$$S = \{(r_1, r_2) \mid \pi_1(r_1) = \pi_2(r_2)\}.$$

Then S is a subring of $R_1 \times R_2$ and

$$U(S) = \{(r_1, r_2) \in S \mid r_1 \in U(R_1), r_2 \in U(R_2)\}.$$

Proof. It is obvious that S is a subring of $R_1 \times R_2$. If $(r_1, r_2) \in U(S)$, then certainly $r_i \in U(R_i)$ for $i = 1, 2$. Conversely, let r_1 and r_2 satisfy the given condition. Then $r_i^{-1} \in R_i$ and

$$\pi_1(r_1^{-1}) = \pi_1(r_1)^{-1} = \pi_2(r_2)^{-1} = \pi_2(r_2^{-1})$$

Thus $(r_1^{-1}, r_2^{-1}) \in S$ and (r_1, r_2) is a unit of S.   ∎

8.6.4 Lemma *Let $A = \langle a \rangle$ be a cyclic group of order p^n, p odd prime, and let ε be a primitive p^nth root of unity. Then the map*

$$\mathbb{Z}A \xrightarrow{\sigma} \mathbb{Z}[\varepsilon],$$

$$\sum_{i=0}^{p^n-1} n_i a^i \mapsto \sum_{i=0}^{p^n-1} n_i \varepsilon^i$$

is a surjective homomorphism, the kernel of which is the principal ideal generated by

$$\sum_{i=0}^{p-1} a^{p^{n-1}i}.$$

Proof. The given map is a $\mathbb{Z}$-linear extension of the isomorphism $A \to \langle \varepsilon \rangle$, $a \mapsto \varepsilon$ and therefore is a surjective ring homomorphism. It remains to compute the kernel of σ. Let $m = p^{n-1} - 1$, $\sigma = \varepsilon^{p^{n-1}}$, let $x \in \mathbb{Z}A$ and write

$$x = \sum_{i=0}^{m} x_i a^i, \qquad \text{with } x_i \in \mathbb{Z}\langle a^{p^{n-1}} \rangle.$$

Then

$$\sigma(x) = \sum_{i=0}^{m} \sigma(x_i) \varepsilon^i, \qquad \text{with } \sigma(x_i) \in \mathbb{Z}[\delta].$$

Because $(\mathbb{Q}(\varepsilon) : \mathbb{Q}(\delta)) = p^{n-1}$, the elements ε^i with $0 \leqslant i \leqslant m$ are linearly independent over $\mathbb{Z}[\delta]$ and thus we see that $\sigma(x) = 0$ if and only if $\sigma(x_i) = 0$ for all i. But $x_i = \sum_{j=0}^{p-1} n_j a^{p^{n-1}j}$ with $n_j \in \mathbb{Z}$, so

$$0 = \sigma(x_i) = \sum_{j=0}^{p-1} n_j \varepsilon^{p^{n-1}j}$$

can occur if and only if all n_j are equal. Thus the lemma is proved.   ∎

8.6.5 Lemma *Let $A = \langle a \rangle$ be a cyclic group of order p^{n-1}, $n \geqslant 1$, p odd prime, and let ε be a primitive p^n-th root of unity. Then the map*

$\varepsilon \mapsto a$ *induces a surjective homomorphism* $\mathbb{Z}[\varepsilon] \to \mathbb{F}_p A$, *the kernel of which is the principal ideal generated by* $\varepsilon^{p^{n-1}} - \varepsilon^{-p^{n-1}}$.

Proof. Let $B = \langle b \rangle$ be a cyclic group of order p^n. Then there is a natural epimorphism $\mathbb{Z}B \to \mathbb{F}_p A$ which sends b to a. Since $\sum_{i=0}^{p-1} b^{p^{n-1}i}$ is in the kernel of this homomorphism, it follows from Lemma 6.4 that $\varepsilon \mapsto a$ induces a homomorphism say π, from $\mathbb{Z}[\varepsilon]$ onto $\mathbb{F}_p A$. It remains to compute the kernel of π.

For any $i \geq 0$, we have $i = j + p^{n-1}k$ with $0 \leq j < p^{n-1}$ and so $\varepsilon^i = \varepsilon^j \delta^k$, where $\delta = \varepsilon^{p^{n-1}}$. Setting $m = p^{n-1} - 1$, it follows that

$$\mathbb{Z}[\varepsilon] = \sum_{j=0}^{m} \mathbb{Z}[\delta] \varepsilon^j.$$

Fix $r \in \mathbb{Z}[\varepsilon]$ and write $r = \sum_{j=0}^{m} r_j \varepsilon^j$ with $r_j \in \mathbb{Z}[\delta]$. Since $\pi(r_j) \in \mathbb{F}_p$ and since the elements $\pi(\varepsilon^j)$, being the distinct elements of A, are $\mathbb{F}_p$-linearly independent, we see that $\pi(r) = 0$ if and only if $\pi(r_j) = 0$ for all j. Hence, to find $\operatorname{Ker} \pi$ it suffices to find the kernel of the restricted homomorphism $\bar{\pi} : \mathbb{Z}[\delta] \to \mathbb{F}_p$ given by $\bar{\pi}(\delta) = 1$.

If $r \in \mathbb{Z}[\delta]$, then $r = (1 - \delta)s + t$ for some $s \in \mathbb{Z}[\delta]$, $t \in \mathbb{Z}$. Since $1 - \delta \in \operatorname{Ker} \pi$, we see that $r \in \operatorname{Ker} \bar{\pi}$ if and only if $t \in \operatorname{Ker} \bar{\pi}$ and hence if and only if $p \mid t$. Therefore to show that $\operatorname{Ker} \bar{\pi} = (1 - \delta)\mathbb{Z}[\delta]$, it suffices to verify that $p \in (1 - \delta)\mathbb{Z}[\delta]$. The latter is a consequence of the fact that the cyclotomic polynomial $\phi(X)$ for δ satisfies

$$1 + X + \ldots + X^{p-1} = \phi(X) = \prod_{i=0}^{p-1} (X - \delta^i)$$

which yields

$$p = \phi(1) = \prod_{i=0}^{p-1} (1 - \delta^i).$$

The conclusion is that $\operatorname{Ker} \bar{\pi} = (1 - \delta)\mathbb{Z}[\delta]$ and hence that $\operatorname{Ker} \pi = (1 - \delta)\mathbb{Z}[\varepsilon]$. Since p is odd, $1 + \delta$ is a unit (Lemma 3.1.2) and hence $1 - \delta$ is an associate of $\delta - \delta^{-1} = -\delta^{-1}(1 + \delta)(1 - \delta)$. Thus $\operatorname{Ker} \pi = (\delta - \delta^{-1})\mathbb{Z}[\varepsilon]$, as asserted. ∎

8.6.6 Lemma *Let* $A = \langle a \rangle$ *and* $B = \langle b \rangle$ *be cyclic groups of order* p^n *and* p^{n-1}, *respectively, where* $n \geq 1$ *and* p *is an odd prime, and let* ε *be a primitive* p^n-*th root of unity. Then the map*

$$\begin{cases} \mathbb{Z}A \to \mathbb{Z}[\varepsilon] \times \mathbb{Z}B, \\ \displaystyle\sum_{i=0}^{p^n-1} n_i a^i \mapsto \left(\sum_{i=0}^{p^n-1} n_i \varepsilon^i, \ \sum_{i=0}^{p^n-1} n_i b^i \right) \end{cases}$$

induces an isomorphism of $\mathbb{Z}A$ *with the subring* S *of* $\mathbb{Z}[\varepsilon] \times \mathbb{Z}B$ *given by*

$$S = \{(x, y) \mid \pi(x) = \rho(y)\},$$

where the homomorphism $\pi : \mathbb{Z}[\varepsilon] \to \mathbb{F}_p B$ *is determined by* $\pi(\varepsilon) = b$ *(see Lemma 6.5) and* $\rho : \mathbb{Z}B \to \mathbb{F}_p B$ *is the natural homomorphism. In particular, by Lemma 6.3,*

$$U(\mathbb{Z}A) = \left\{ \sum_{i=0}^{p^n-1} n_i a^i \in \mathbb{Z}A \; \middle| \; \sum_{i=0}^{p^n-1} n_i \varepsilon^i \in U(\mathbb{Z}[\varepsilon]), \; \sum_{i=0}^{p^n-1} n_i b^i \in U(\mathbb{Z}B) \right\}.$$

Proof. Certainly the given map is a homomorphism. We must show that its image is S and that its kernel is zero. Let σ and τ be the projections on the first and the second factors, respectively. Assume that $\sigma(\alpha) = \tau(\alpha) = 0$. Since $\sigma(\alpha) = 0$, we have $\alpha = \beta(\sum_{i=0}^{p-1} a^{p^{n-1}i})$ for some $\beta \in \mathbb{Z}A$, by Lemma 6.4. Moreover,

$$0 = \tau(\alpha) = \tau(\beta)\tau\left(\sum_{i=0}^{p-1} a^{p^{n-1}i}\right) = p\tau(\beta),$$

so $\tau(\beta) = 0$.   Hence $\beta \in \mathrm{Ker}\, \tau = I(A^{p^{n-1}})\mathbb{Z}A$ and   since $(\sum_{i=0}^{p-1} a^{p^{n-1}i})I(A^{p^{n-1}}) = 0$, we have $\alpha = 0$.

Next observe that the combined homomorphisms $\pi\sigma$ and $\rho\tau$ both map $\mathbb{Z}A$ to $\mathbb{F}_p B$ and $\pi\sigma(a) = b = \rho\tau(a)$. Thus $\pi\sigma = \rho\tau$ and, in particular, if $x = \sigma(\alpha)$ and $y = \tau(\alpha)$, then

$$\pi(x) = \pi\sigma(\alpha) = \rho\tau(\alpha) = \rho(y),$$

proving that the image of the given map is contained in S.

Finally, let $(x, y) \in S$ so that $\pi(x) = \rho(y)$. Since σ is onto, choose $\beta \in \mathbb{Z}A$ with $\sigma(\beta) = x$. Then

$$\rho\tau(\beta) = \pi\sigma(\beta) = \pi(x) = \rho(y)$$

and we see that $\tau(\beta) - y \in \mathrm{Ker}\, \rho = p(\mathbb{Z}B)$. Since τ is onto, $\tau(\beta) - y = p\tau(\gamma)$ for some $\gamma \in \mathbb{Z}A$, and we set $\alpha = \beta - (\sum_{i=0}^{p-1} a^{p^{n-1}i})\gamma$. Observe that $\sum_{i=0}^{p-1} a^{p^{n-1}i} \in \mathrm{Ker}\, \sigma$, so $\sigma(\alpha) = \sigma(\beta) = x$. On the other hand, since $\tau(\sum_{i=0}^{p-1} a^{p^{n-1}i}) = p$, we have

$$\tau(\alpha) = \tau(\beta) - p\tau(\gamma) = y.$$

Thus (x, y) is the image of α and the result follows.   ∎

8.6.7   Lemma   *Let* ε *be a primitive nth root of unity,* $n \geqslant 3$. *Then* $\mathbb{Z}[\varepsilon + \varepsilon^{-1}]$ *is a real subring of* $\mathbb{Z}[\varepsilon]$ *such that*

$$\mathbb{Z}[\varepsilon] = \mathbb{Z}[\varepsilon + \varepsilon^{-1}] \oplus \mathbb{Z}[\varepsilon + \varepsilon^{-1}]\varepsilon^{-1}.$$

Proof. The directness of the sum and the fact that $\mathbb{Z}[\varepsilon + \varepsilon^{-1}]$ is a real subring of $\mathbb{Z}[\varepsilon]$ follow immediately since $\varepsilon + \varepsilon^{-1}$ is real but ε^{-1} is not.

It follows by induction on $i \geqslant 1$ from the identity

$$\varepsilon^{i+1} + \varepsilon^{-(i+1)} = (\varepsilon + \varepsilon^{-1})(\varepsilon^i + \varepsilon^{-i}) - (\varepsilon^{i-1} + \varepsilon^{-(i-1)})$$

that $\varepsilon^i + \varepsilon^{-i} \in \mathbb{Z}[\varepsilon + \varepsilon^{-1}]$. Because $(\varepsilon^j - \varepsilon^{-j})/(\varepsilon - \varepsilon^{-1})$ for any $j \geqslant 0$ is easily seen to be a sum of terms of the form $\varepsilon^i + \varepsilon^{-i}$, we infer that $(\varepsilon^j - \varepsilon^{-j})/(\varepsilon - \varepsilon^{-1}) \in \mathbb{Z}[\varepsilon + \varepsilon^{-1}]$. Therefore, for any $c \in \mathbb{Z}[\varepsilon]$,

$$(c - \bar{c})/(\varepsilon - \varepsilon^{-1}) \in \mathbb{Z}[\varepsilon + \varepsilon^{-1}]$$

where $\bar{c}$ denotes the complex conjugate of c. Hence, given $a \in \mathbb{Z}[\varepsilon]$, we have $x = (a\varepsilon - \bar{a}\varepsilon)/(\varepsilon - \varepsilon^{-1})$, $y = (a - \bar{a})/(\varepsilon - \varepsilon^{-1}) \in \mathbb{Z}[\varepsilon + \varepsilon^{-1}]$ and $a = x - y\varepsilon^{-1}$, which completes the proof. ∎

8.6.8 Lemma *Let ε be a primitive p^n-th root of unity, p odd prime, and let $A = \langle a \rangle$ be a cyclic group of order p^{n-1}, $n \geqslant 1$. Then the kernel of the restriction*

$$\pi : \mathbb{Z}[\varepsilon + \varepsilon^{-1}] \to \mathbb{F}_p A$$

of $\mathbb{Z}[\varepsilon] \to \mathbb{F}_p A$, $\varepsilon \mapsto a$, is the principal ideal generated by $(\delta - \delta^{-1})(\varepsilon - \varepsilon^{-1})$, where $\delta = \varepsilon^{p^{n-1}}$.

Proof. Obviously $(\delta - \delta^{-1})(\varepsilon - \varepsilon^{-1})$ is a real element contained in Ker π. Conversely, assume that $a \in$ Ker π. Owing to Lemma 6.5, $a \in (\delta - \delta^{-1})\mathbb{Z}[\varepsilon]$ and, by Lemma 6.7,

$$a = (b + c\varepsilon^{-1})(\delta - \delta^{-1})$$

for some b, $c \in \mathbb{Z}[\varepsilon + \varepsilon^{-1}]$. Because a, b, c are real, conjugating the expression yields

$$a = -(b + c\varepsilon)(\delta - \delta^{-1})$$

and adding these two we derive

$$2a = -c(\varepsilon - \varepsilon^{-1})(\delta - \delta^{-1}).$$

Hence

$$a = -(c/2)(\varepsilon - \varepsilon^{-1})(\delta - \delta^{-1})$$

and therefore it suffices to show that $c/2 \in \mathbb{Z}[\varepsilon + \varepsilon^{-1}]$.

To this end, put $q = p^{n-1}$ and observe that $\delta - \delta^{-1} = \varepsilon^q - \varepsilon^{-q}$ is a multiple of $\varepsilon - \varepsilon^{-1}$.

Hence, by Lemma 6.5,

$$p \in (\delta - \delta^{-1})\mathbb{Z}[\varepsilon] \subseteq (\varepsilon - \varepsilon^{-1})\mathbb{Z}[\varepsilon],$$

whence

$$p^2 = d(\varepsilon - \varepsilon^{-1})(\delta - \delta^{-1}) \quad \text{for some } d \in \mathbb{Z}[\varepsilon].$$

Since d is real, $d \in \mathbb{Z}[\varepsilon + \varepsilon^{-1}]$. Now $(p^2, 2) = 1$ so there is a linear combination of p^2 and 2 which add to 1 and this yields

$$1 = \lambda(\varepsilon - \varepsilon^{-1})(\delta - \delta^{-1}) + 2\mu$$

for some λ, $\mu \in \mathbb{Z}[\varepsilon + \varepsilon^{-1}]$. Multiplying this equation by $c/2$, we deduce that

$$c/2 = \lambda(c/2)(\varepsilon - \varepsilon^{-1})(\delta - \delta^{-1}) + c\mu$$
$$= -\lambda a + \mu c \in \mathbb{Z}[\varepsilon + \varepsilon^{-1}]$$

as desired. ∎

8.6.9 Lemma   *Let ε be a primitive p^nth root of unity, p odd prime, let $A = \langle a \rangle$ be a cyclic group of order p^{n-1}, $n \geq 1$ and let $\pi : \mathbb{Z}[\varepsilon] \to \mathbb{F}_p A$ be the homomorphism which sends ε to a. Let $s \in \mathbb{Z}[\varepsilon]$ and write $s = x - y\varepsilon^{-1}$ with $x, y \in \mathbb{Z}[\varepsilon + \varepsilon^{-1}]$. Then $\pi(s) = 0$ if and only if*

$$\pi(x) = \pi(y) = \lambda \left(\sum_{i=0}^{p^{n-1}-1} a^i \right) \qquad \text{for some } \lambda \in \mathbb{F}_p.$$

Proof.   Put $z = \sum_{i=0}^{p^{n-1}-1} a^i$ and suppose that $\pi(x) = \pi(y) = \lambda z$. Then

$$\pi(s) = \pi(x) - \pi(y)a^{-1} = \lambda z(1 - a^{-1}) = 0.$$

Conversely, assume that $\pi(s) = 0$. Then $\pi(x) = \pi(y)a^{-1}$. Let * denote the involution of $\mathbb{F}_p A$ determined by $a^* = a^{-1}$. Since $\pi(x)$ and $\pi(y)$ are clearly fixed by *, applying * to the equation $\pi(x) = \pi(y)a^{-1}$, we obtain $\pi(x) = \pi(y)a$. Therefore $\pi(y)(1 - a^2) = 0$. Since p is odd, $A = \langle a^2 \rangle$, and we conclude that $\pi(y) = \lambda z$ for some $\lambda \in \mathbb{F}_p$. Then

$$\pi(x) = \pi(y)a^{-1} = \lambda z a^{-1} = \lambda z = \pi(y),$$

as we wished to show. ∎

  After this digression into the properties of $\mathbb{Z}A$, we return to the study of the unit group of $\mathbb{Z}G$, where G is dihedral of order $2p^n$, $n \geq 1$, p odd prime:

$$G = \langle a, t \mid a^{p^n} = 1, t^2 = 1, a^t = a^{-1} \rangle.$$

For the remainder of this section we fix the following notation:

$b = a^{p^{n-1}}$ and $B = \langle b \rangle$

ε is a primitive p^n-th root of unity, $\delta = \varepsilon^{p^{n-1}}$

$\sigma : \mathbb{Z}A \to \mathbb{Z}[\varepsilon]$ is the homomorphism sending a to ε

$\pi : \mathbb{Z}[\varepsilon] \to \mathbb{F}_p(A/B)$ is the homomorphism sending ε to aB (see Lemma 6.5)

$\tau : \mathbb{Z}A \to \mathbb{Z}(A/B)$ is the natural homomorphism

$\rho : \mathbb{Z}(A/B) \to \mathbb{F}_p(A/B)$ is the natural homomorphism

$\lambda : \mathbb{Z}G \to M_2^*(\mathbb{Z}A)$ is the homomorphism given by Lemma 6.2
$V = V(\mathbb{Z}G)$ and $V(B) = V \cap (1 + \mathbb{Z}G \cdot I(B))$

Observe that if R and S are commutative rings with involution and if $\theta : R \to S$ is a homomorphism preserving these involutions, then θ extends to a natural homomorphism

$$\theta^* : M_2^*(R) \to M_2^*(S).$$

In particular, the four homomorphisms σ, τ, π, ρ give rise to homomorphisms

$$\sigma^* : M_2^*(\mathbb{Z}A) \to M_2^*(\mathbb{Z}[\varepsilon])$$

$$\tau^* : M_2^*(\mathbb{Z}A) \to M_2^*(\mathbb{Z}(A/B))$$

$$\pi^* : M_2^*(\mathbb{Z}[\varepsilon]) \to M_2^*(\mathbb{F}_p(A/B))$$

$$\rho^* : M_2^*(\mathbb{Z}(A/B)) \to M_2^*(\mathbb{F}_p(A/B))$$

Here conjugation by t yields an involution on $\mathbb{Z}A$, $\mathbb{Z}(A/B)$ and $\mathbb{F}_p(A/B)$ and complex conjugation is an involution on $\mathbb{Z}[\varepsilon]$ since $\varepsilon^{-1} = \bar{\varepsilon}$.

8.6.10 Lemma   *The map*

$$(\sigma^* \times \tau^*)\lambda : \mathbb{Z}G \to M_2^*(\mathbb{Z}[\varepsilon]) \times M_2^*(\mathbb{Z}(A/B))$$

is an isomorphism of $\mathbb{Z}G$ *with the subring* S *of* $M_2^*(\mathbb{Z}[\varepsilon]) \times M_2^*(\mathbb{Z}(A/B))$ *given by*

$$S = \{(x, y) \mid \pi^*(x) = \rho^*(y)\}.$$

Proof.   The proof is a direct consequence of Lemmas 6.2 and 6.6.   ∎

Turning our attention to the group $V(B)$, we next prove

8.6.11 Lemma   *The map*

$$\sigma^*\lambda : \mathbb{Z}G \to M_2^*(\mathbb{Z}[\varepsilon])$$

induces an isomorphism from $V(B)$ *onto*

$$H = \{x \in M_2^*(\mathbb{Z}[\varepsilon]) \mid \det x \in U(\mathbb{Z}[\varepsilon]) \text{ and } \pi^*(x) = 1\}.$$

Proof.   Let $\gamma = \alpha + \beta t \in V(B)$ and let $x = \sigma^*\lambda(\gamma)$. Since $\gamma \in 1 + \mathbb{Z}G \cdot I(B)$, we have $\alpha \in 1 + \mathbb{Z}A \cdot I(B)$ and $\beta \in \mathbb{Z}A \cdot I(B)$. Hence $\tau^*\lambda(\gamma) = 1$ and therefore $(\sigma^* \times \tau^*)\lambda(\gamma) = (x, 1)$. By Lemma 6.10, $(\sigma^* \times \tau^*)\lambda$ is injective and thus the restriction of $\sigma^*\lambda$ to $V(B)$ is also injective. Furthermore, since $(x, 1) \in U(S)$ we see that x is a unit of $M_2^*(\mathbb{Z}[\varepsilon])$ and $\pi^*(x) = \rho^*(1) = 1$. Owing to Lemma 6.1(iii), x is a unit in $M_2^*(\mathbb{Z}[\varepsilon])$ if and only if $\det x$ is a unit of $\mathbb{Z}[\varepsilon]$. Thus $\sigma^*\lambda$ induces an injective homomorphism of $V(B)$ into H.

To prove that the image of $V(B)$ is H, fix $x \in H$. Because $\pi^*(x) = 1 = \rho^*(1)$, it follows from Lemma 6.3 that $(x, 1)$ is a unit of S. Therefore, by Lemma 6.10, there exists $\gamma = \alpha + \beta t \in U(\mathbb{Z}G)$ with $(\sigma^* \times \tau^*)\lambda(\gamma) = (x, 1)$. However, $\tau^*\lambda(\gamma) = 1$ implies that $\tau(\alpha) = 1$ and $\tau(\beta) = 0$. Hence $\alpha \in 1 + \mathbb{Z}A \cdot I(B)$ and $\beta \in \mathbb{Z}A \cdot I(B)$, which shows that $\gamma = \alpha + \beta t \in 1 + \mathbb{Z}G \cdot I(B)$. Consequently, $\gamma \in V(B)$ and since $\sigma^*\lambda(\gamma) = x$, the result follows. ∎

As a final preliminary observation, we introduce another map. Given

$$x = \begin{bmatrix} \varepsilon^{-1} & 1 \\ -\varepsilon & -1 \end{bmatrix} \in M_2(\mathbb{Q}[\varepsilon]),$$

we have

$$x^{-1} = (\varepsilon - \varepsilon^{-1})^{-1} \begin{bmatrix} -1 & -1 \\ \varepsilon & \varepsilon^{-1} \end{bmatrix} \in M_2(\mathbb{Q}[\varepsilon]).$$

If

$$\eta : M_2[\mathbb{Q}[\varepsilon]) \to M_2(\mathbb{Q}[\varepsilon])$$

denotes conjugation by x, then η is a ring isomorphism.

Note also that the map $\pi : \mathbb{Z}[\varepsilon] \to \mathbb{F}_p(A/B)$ induces a homomorphism

$$\bar{\pi} : GL_2(\mathbb{Z}[\varepsilon + \varepsilon^{-1}]) \to GL_2(\mathbb{F}_p(A/B)).$$

Finally, it what follows, we write $(A/B)^+$ for the sum of all elements in A/B. With this notation, we now provide the following description of $V(B)$.

8.6.12   Theorem   (Passman and Smith 1981)   *The map*

$$\eta\sigma^*\lambda : \mathbb{Z}G \to M_2(\mathbb{Q}[\varepsilon])$$

induces an isomorphism of $V(B)$ onto

$$W = \{w \in GL_2(\mathbb{Z}[\varepsilon + \varepsilon^{-1}]) \mid \bar{\pi}(w)$$

$$= \begin{bmatrix} 1 & 0 \\ 0 & 1 \end{bmatrix} + \mu(A/B)^+ \begin{bmatrix} 1 & 1 \\ -1 & -1 \end{bmatrix} \quad \text{for some } \mu \in \mathbb{F}_p \}.$$

Proof.   Applying Lemma 6.10 and the fact that η is an isomorphism, it suffices to show that $x^{-1}Hx = \eta(H) = W$.

Let

$$h = \begin{bmatrix} c & d \\ \bar{d} & \bar{c} \end{bmatrix} \in H$$

so that $\det h \in U(\mathbb{Z}[\varepsilon])$ and $\pi^*(h) = 1$. Owing to Lemma 6.7, we can write $c = c_1 - c_2\varepsilon^{-1}$ and $d = d_1 - d_2\varepsilon^{-1}$ with $c_i, d_i \in \mathbb{Z}[\varepsilon + \varepsilon^{-1}]$, $i = 1,2$.

Setting $\lambda = \varepsilon - \varepsilon^{-1}$, it then follows that

$$(c - \bar{c})/\lambda = c_2$$
$$(c\varepsilon - \bar{c}\varepsilon^{-1})/\lambda = c_1$$
$$(c\varepsilon^{-1} - c\bar{\varepsilon})/\lambda = -c_1 + c_2(\varepsilon + \varepsilon^{-1})$$

and, similarly,

$$(d - \bar{d})/\lambda = d_2$$
$$(d\varepsilon - \bar{d}\varepsilon^{-1})/\lambda = d_1$$
$$(d\varepsilon^2 - \bar{d}\varepsilon^{-2})/\lambda = -d_2 + d_1(\varepsilon + \varepsilon^{-1}).$$

Now if

$$x^{-1}hx = w = \begin{bmatrix} a_1 & a_2 \\ a_3 & a_4 \end{bmatrix}$$

then it follows from the above six equations that

$$a_1 = c_1 - c_2(\varepsilon + \varepsilon^{-1}) + d_1$$
$$a_2 = -c_2 + d_2$$
$$a_3 = c_2 - d_1(\varepsilon + \varepsilon^{-1}) + d_2$$
$$a_4 = c_1 - d_1$$

and note that all these entries are contained in $\mathbb{Z}[\varepsilon + \varepsilon^{-1}]$. Observe that $\det w = \det h$ is a real unit of $\mathbb{Z}[\varepsilon]$ and so, by Lemma 6.7, $\det w$ is a unit of $\mathbb{Z}[\varepsilon + \varepsilon^{-1}]$. This shows that $w \in GL_2(\mathbb{Z}[\varepsilon + \varepsilon^{-1}])$.

Since $h \in H$, $\pi^*(h) = 1$ so $\pi(c - 1) = \pi(d) = 0$. Applying Lemma 6.9, we can find $r, s \in \mathbb{F}_p$ such that

$$\pi(c_1) - 1 = \pi(c_2) = r(A/B)^+,$$
$$\pi(d_1) = \pi(d_2) = s(A/B)^+.$$

Therefore, by applying π to the above equations for a_i, $1 \leq i \leq 4$, we obtain

$$\bar{\pi}(w) = \begin{bmatrix} 1 & 0 \\ 0 & 1 \end{bmatrix} + (s - r)(A/B)^+ \begin{bmatrix} 1 & 1 \\ -1 & -1 \end{bmatrix}.$$

Thus $w \in W$ and $\eta(H) \subseteq W$.

Conversely, let

$$w = \begin{bmatrix} a_1 & a_2 \\ a_3 & a_4 \end{bmatrix} \in W$$

and note that

$$xwx^{-1} = \begin{bmatrix} c & d \\ \bar{d} & \bar{c} \end{bmatrix} = h,$$

say, where

$$c = (-a_1\varepsilon^{-1} - a_3 + a_2 + a_4\varepsilon)/\lambda,$$
$$d = (-a_1\varepsilon^{-1} - a_3 + a_2\varepsilon^{-2} + a_4\varepsilon^{-1})/\lambda.$$

Since

$$\bar{\pi}(w) = \begin{bmatrix} 1 & 0 \\ 0 & 1 \end{bmatrix} + \mu(A/B)^+ \begin{bmatrix} 1 & 1 \\ -1 & -1 \end{bmatrix} \qquad (\mu \in \mathbb{F}_p)$$

we have $\pi(a_2) = \mu(A/B)^+$ and

$$0 = \pi(a_1) - 1 - \pi(a_2) = \pi(a_3) + \pi(a_2) = \pi(a_4) - 1 + \pi(a_2).$$

Now, by Lemma 6.8, the kernel of π restricted to $\mathbb{Z}[\varepsilon + \varepsilon^{-1}]$ is the principal ideal generated by $(\varepsilon - \varepsilon^{-1})(\delta - \delta^{-1}) = \lambda(\delta - \delta^{-1})$, so we have

$$a_1 = \lambda(\delta - \delta^{-1})\bar{a}_1 + a_2 + 1$$
$$a_3 = \lambda(\delta - \delta^{-1})\bar{a}_3 - a_2$$
$$a_4 = \lambda(\delta - \delta^{-1})\bar{a}_4 - a_2 + 1$$

for suitable $\bar{a}_i \in \mathbb{Z}[\varepsilon + \varepsilon^{-1}]$, $i = 1, 3, 4$. Substituting these in the above expressions for c and d we obtain

$$c = (\delta - \delta^{-1})\bar{c} + a_2(1 - \varepsilon)(1 - \varepsilon^{-1})/\lambda + 1,$$
$$d = (\delta - \delta^{-1})\bar{d} + a_2(1 - \varepsilon^{-1})^2/\lambda$$

for some $\bar{c}, \bar{d} \in \mathbb{Z}[\varepsilon]$. Since p is odd, $1 + \varepsilon^{-1}$ is a unit (Lemma 3.1.2), and hence $\lambda = \varepsilon - \varepsilon^{-1} = \varepsilon(1 + \varepsilon^{-1})(1 - \varepsilon^{-1})$ is an associate of $1 - \varepsilon^{-1}$ in $\mathbb{Z}[\varepsilon]$. Thus $(1 - \varepsilon^{-1})/\lambda \in \mathbb{Z}[\varepsilon]$ and therefore $c, d \in \mathbb{Z}[\varepsilon]$. Furthermore, since $\pi(a_2) = \mu(A/B)^+$, we have

$$\pi(a_2(1 - \varepsilon)) = 0 = \pi(a_2(1 - \varepsilon^{-1}))$$

whence $\pi(c) = 1$, $\pi(d) = 0$ and $\pi^*(h) = 1$. Finally, $\det h = \det w$ is a unit of $\mathbb{Z}[\varepsilon + \varepsilon^{-1}] \subseteq \mathbb{Z}[\varepsilon]$, which shows that $h \in H$. Hence $\eta(H) \supseteq W$ and the result follows. ∎

We have now come to the demonstration for which this section has been developed.

8.6.13 Theorem (Passman and Smith 1981) *Let G be a dihedral group of order $2p$, p odd prime, and let ε be a primitive pth root of unity. Then $U(\mathbb{Z}G)$ is isomorphic to the subgroup of $GL_2(\mathbb{Z}[\varepsilon + \varepsilon^{-1}])$ given by all*

$$w = \begin{bmatrix} x & y \\ z & v \end{bmatrix}$$

such that

$$\det w \equiv \pm 1(\operatorname{mod}(\varepsilon - \varepsilon^{-1})), \qquad x + z \equiv y + v \equiv \pm 1(\operatorname{mod}(\varepsilon - \varepsilon^{-1}))$$

where $(\varepsilon - \varepsilon^{-1})$ is the principal ideal of $\mathbb{Z}[\varepsilon]$ generated by $\varepsilon - \varepsilon^{-1}$.

Proof. Let $\pi : \mathbb{Z}[\varepsilon] \to \mathbb{F}_p$ be the homomorphism with kernel $(\varepsilon - \varepsilon^{-1})$. This is obtained from Lemma 6.5 by setting $n = 1$.

Setting $n = 1$ in Theorem 6.11, we have $A = B$, $A/B = 1$ and therefore the map $\eta\sigma^*\lambda$ yields an isomorphism

$$V(A) = V(B) \cong \left\{ w \in GL_2(\mathbb{Z}[\varepsilon + \varepsilon^{-1}]) \mid \bar{\pi}(w) \right.$$

$$\left. = \begin{bmatrix} 1 + \mu & \mu \\ -\mu & 1 - \mu \end{bmatrix} \text{ for some } \mu \in \mathbb{F}_p \right\}.$$

Now consider the natural homomorphism $V(\mathbb{Z}G) \to V(\mathbb{Z}(G/A))$. Then the kernel of this homomorphism is $V(A)$. Since $\mathbb{Z}(G/A) \cong \mathbb{Z}\langle t \rangle$ is known to have only trivial units (Theorem 4.3), we see that $\{\pm 1, \pm t\} \subseteq \pm G$ maps onto $U(\mathbb{Z}(G/A))$, and therefore

$$U(\mathbb{Z}G) = V(A) \cdot (\pm \langle t \rangle).$$

An easy calculation also shows that

$$\eta\sigma^*\lambda(-1) = \begin{bmatrix} -1 & 0 \\ 0 & -1 \end{bmatrix}, \qquad \eta\sigma^*\lambda(t) = \begin{bmatrix} 1 & 0 \\ -(\varepsilon + \varepsilon^{-1}) & -1 \end{bmatrix}.$$

The conclusion is that

$$\eta\sigma^*\lambda : U(\mathbb{Z}G) \to GL_2(\mathbb{Z}[\varepsilon + \varepsilon^{-1}])$$

is a homomorphism. It remains to prove that $\eta\sigma^*\lambda$ is injective and to compute the image.

By the foregoing, if $w \in GL_2(\mathbb{Z}[\varepsilon + \varepsilon^{-1}])$ is in the image of $U(\mathbb{Z}G)$, then

$$\bar{\pi}(w) = \pm \begin{bmatrix} 1 + \mu & \mu \\ -\mu & 1 - \mu \end{bmatrix} \quad \text{or} \quad \pm \begin{bmatrix} 1 + \mu & \mu \\ -\mu & 1 - \mu \end{bmatrix} \begin{bmatrix} 1 & 0 \\ -2 & -1 \end{bmatrix}$$

corresponding to the four cosets of $V(A)$ in $U(\mathbb{Z}G)$. However, $\bar{\pi}(w) = 1$ implies that w is in the image of $V(A)$, so we deduce immediately that $\eta\sigma^*\lambda$ is injective on $U(\mathbb{Z}G)$.

Finally, by varying $\mu \in \mathbb{F}_p$, we obtain precisely $4p$ distinct possibilities for

$$\bar{\pi}(w) = \begin{bmatrix} r & s \\ e & f \end{bmatrix},$$

all of which satisfy the equations

$$\det \bar{\pi}(w) = \pm 1, \qquad r + e = s + f = \pm 1.$$

A straightforward verification also shows that there are at most $4p$ solutions of these equations over $\mathbb{F}_p$. Therefore the image of $U(\mathbb{Z}G)$ is precisely all those

$$w = \begin{bmatrix} x & y \\ z & v \end{bmatrix} \in GL_2(\mathbb{Z}[\varepsilon + \varepsilon^{-1}])$$

satisfying

$$\det w \equiv \pm 1(\mathrm{mod}(\varepsilon - \varepsilon^{-1})), \qquad x + z \equiv y + v \equiv \pm 1(\mathrm{mod}(\varepsilon - \varepsilon^{-1}))$$

as required. ∎

8.7   Torsion-free complements

Let $V(\mathbb{Z}G)$ be the group of normalized units of $\mathbb{Z}G$, where G is a finite group. We consider two questions:

(i)   Does the inclusion $G \to V(\mathbb{Z}G)$ split? (Dennis 1977)
(ii) If a splitting exists, is its kernel torsion-free? (Cliff *et al.* 1981)

Our goal is to provide a positive answer to both questions in case G has an abelian normal subgroup A satisfying one of the following properties:

(a) G/A is abelian of exponent dividing 4 or 6;
(b) G/A is abelian of odd order.

Note that condition (b) is indispensable. More precisely, Roggenkamp and Scott (1983) showed that there exists an extension of $\mathbb{Z}_{73}$ by $\mathbb{Z}_8$ for which (i) does not hold.

8.7A   *Preliminary results*

Our aim here is to collect together a number of subsidiary results which will be required in the proof of the main theorem. All the recorded properties are extracted from Cliff *et al.* (1981). Throughout, G denotes a finite group. Given an ideal I of $\mathbb{Z}G$, we write $U(1 + I)$ for the set of units of $\mathbb{Z}G$ congruent to 1 modulo I.

8.7.1 Lemma   *Let A be an abelian normal subgroup of G and let T be a transversal for A in G. Then each element of $\mathbb{Z}G \cdot I(A)$ can be uniquely written in the form*

$$\sum_{t \in T} t(a_t - 1) + \alpha \qquad (a_t \in A,\ \alpha \in \mathbb{Z}G \cdot I(A)^2).$$

Proof. We know, from Proposition 1.6, that $\mathbb{Z}G = \bigoplus_{t \in T} t(\mathbb{Z}A)$ and therefore

$$\mathbb{Z}G \cdot I(A) = \bigoplus_{t \in T} t \cdot I(A).$$

Hence, given $x \in \mathbb{Z}G \cdot I(A)$, we may write x uniquely in the form $x = \sum_{t \in T} t\alpha_t$ with $\alpha_t \in I(A)$. Furthermore, by Proposition 3.9(ii), $\alpha_t \equiv a_t - 1 \pmod{I(A)^2}$ for some $a_t \in A$. Setting $\alpha = \sum_{t \in T} t(\alpha_t - (a_t - 1))$, we have $\alpha \in \mathbb{Z}G \cdot I(A)^2$, and so

$$x = \sum_{t \in T} t(a_t - 1) + \alpha$$

is of the required form. If also $x = \sum t(b_t - 1) + \beta$ with $b_t \in A$ and $\beta \in \mathbb{Z}G \cdot I(A)^2$, then

$$\sum_{t \in T} t(a_t - b_t) \equiv 0 (\bmod \, \mathbb{Z}G \cdot I(A)^2)$$

and therefore $a_t \equiv b_t (\bmod \, I(A)^2)$, for each $t \in T$. Applying Proposition 3.9(iii), we deduce that $a_t = b_t$ for all $t \in T$, as desired. ∎

8.7.2 Lemma *Let A be an abelian normal subgroup of G. If $x \in \mathbb{Z}G \cdot I(A)$ and $(1 + x)^p = 1$ for some prime p not dividing $|A|$, then $x = 0$.*

Proof. If $|A| = 1$, then $x = 0$ and there is nothing to prove. We now argue by induction on $|A|$. Let q be a prime dividing $|A|$. We may assume that A is a q-group. Indeed, if not, let B be a Sylow q-subgroup of A. If $\bar{x}$ is the image of x in $\mathbb{Z}(G/B)$, then $(1 + \bar{x})^p = 1$, and $\bar{x} = 0$ by induction. Then $x \in \mathbb{Z}G \cdot I(B)$ and $x = 0$ by induction.

Since the image of $\mathbb{Z}G \cdot I(A)$ in $(\mathbb{Z}/q\mathbb{Z})G$ is a nilpotent ideal, we have

$$(1 + x)^{q^n} \equiv 1 (\bmod \, q(\mathbb{Z}G)) \qquad \text{for some integer } n.$$

Since $(1 + x)^p = 1$ and $p \neq q$, we must have $1 + x \equiv 1 (\bmod \, q(\mathbb{Z}G))$, and hence $x \in q(\mathbb{Z}G)$. Suppose that $x \in q^i(\mathbb{Z}G)$, where $i > 0$. Because $(1 + x)^p = 1$, we have

$$0 = px + \binom{p}{2}x^2 + \ldots + x^p \equiv px (\bmod \, q^{2i}(\mathbb{Z}G))$$

and thus $x \in q^{2i}(\mathbb{Z}G)$. This shows that $x \in q^j(\mathbb{Z}G)$ for all j and hence that $x = 0$. ∎

8.7.3 Lemma *Let A be an abelian normal subgroup of G and let B be a normal subgroup of G contained in A. Then*

$$\mathbb{Z}G \cdot I(A)^2 \cap \mathbb{Z}G \cdot I(B) = \mathbb{Z}G \cdot I(A) \cdot I(B).$$

Proof. It suffices to prove that the left side is contained in the right, since the opposite inclusion is obvious. Fix $x \in \mathbb{Z}G \cdot I(A)^2 \cap \mathbb{Z}G \cdot I(B)$ and choose a transversal $\{t_1, \ldots, t_r\}$ for A in G. Then we may write x uniquely in the form

$$x = \sum_{i=1}^{r} t_i \alpha_i \qquad (\alpha_i \in I(A)^2, 1 \leq i \leq r).$$

Now choose a transversal $\{y_1, \ldots, y_m\}$ for B in A and write

$$\alpha_i = \sum_{j=1}^{m} y_j \beta_{ij} \qquad (\beta_{ij} \in \mathbb{Z}B, 1 \leq j \leq m).$$

Since $x \in \mathbb{Z}G \cdot I(B)$, we have $\beta_{ij} \in I(B)$ for all i, j. Therefore

$$\alpha_i \in I(A)^2 \cap \mathbb{Z}A \cdot I(B) \qquad (1 \leq i \leq n)$$

and we are left to verify that

$$I(A)^2 \cap \mathbb{Z}A \cdot I(B) \subseteq I(A)I(B).$$

To this end, fix $\alpha \in I(A)^2 \cap \mathbb{Z}A \cdot I(B)$ and write $\alpha = \sum_{j=1}^{m} y_j \gamma_j$ with $\gamma_j \in \mathbb{Z}B$. Since $\alpha \in \mathbb{Z}A \cdot I(B)$, we have $\gamma_j \in I(B)$, $1 \leq j \leq m$. Now

$$\alpha = \sum_{j=1}^{m} (y_j - 1)\gamma_j + \beta,$$

where $\beta = \sum_{j=1}^{m} \gamma_j$. Therefore, since each $(y_j - 1)\gamma_j \in I(A)I(B)$, we have $\alpha \equiv \beta (\mod I(A)I(B))$. Since $\beta \equiv b - 1 (\mod I(B)^2)$ for some $b \in B$ (Proposition 3.9(ii)), we have

$$\alpha \equiv b - 1 (\mod I(A)^2).$$

But $\alpha \in I(A)^2$, so $b \equiv 1 (\mod I(A)^2)$ and, by Proposition 3.9(iii), $b = 1$. Therefore $\alpha \in I(A)I(B)$, as required. ∎

8.7.4 Lemma *Let A be an abelian normal subgroup of G. If u is a unit of $\mathbb{Z}G$ of finite order and $u \equiv 1 (\mod \mathbb{Z}G \cdot I(A)^2)$, then $u = 1$.*

Proof. The case $|A| = 1$ being trivial, we argue by induction on $|A|$. We may harmlessly assume that $u^p = 1$ for some prime p. Suppose that $|A| > 1$ and write $u = 1 + x$ with $x \in \mathbb{Z}G \cdot I(A)^2$. In view of Lemma 7.2, we may assume that p divides $|A|$. If $q \neq p$ is another prime dividing $|A|$, let Q be a Sylow q-subgroup of A and let $-: \mathbb{Z}G \rightarrow \mathbb{Z}(G/Q)$ be the natural homomorphism. Then

$$\bar{u} \equiv 1 (\mod \mathbb{Z}\bar{G} \cdot I(\bar{A})^2)$$

and $\bar{u}^p = 1$, so $\bar{u} = 1$ by induction. Hence $u - 1 \in \mathbb{Z}G \cdot I(Q)$ and, since p does not divide $|Q|$, $u = 1$ by Lemma 7.2. We may therefore assume that A is a p-group.

Put $B = A[p] = \{a \in A \mid a^p = 1\}$. Then $x \in \mathbb{Z}G \cdot I(B)$ by induction, so

$$x \in \mathbb{Z}G \cdot I(A)^2 \cap \mathbb{Z}G \cdot I(B) = \mathbb{Z}G \cdot I(A)I(B)$$

by Lemma 7.3. We now claim that $\Gamma(x) = 0$ for every representation Γ of $\mathbb{C}G$ whose restriction to G is induced from a one-dimensional representation of A. If sustained it will follow, from Frobenius reciprocity, that $\gamma(x) = 0$ for any irreducible representation γ of $\mathbb{C}G$, and hence $x = 0$.

Let p^n be the exponent of A and let ε be a primitive p^nth root of unity in $\mathbb{C}$. Let v be the exponential valuation on $\mathbb{Z}[\varepsilon]$, with respect to the prime ideal $(\varepsilon - 1)$. It is well known (see Janusz 1973, Theorem 1.9.1) that

$$v(p) = p^{n-1}(p - 1), \tag{1}$$
$$v(\varepsilon^{p^{n-1}} - 1) = p^{n-1}. \tag{2}$$

If $r = (G:A)$, then Γ maps $\mathbb{Z}G$ to $M_r(\mathbb{Z}[\varepsilon])$. Extend v to $M_r(\mathbb{Z}[\varepsilon])$ by setting $v(\lambda) = i$ if $\lambda \in (\varepsilon - 1)^i M_r(\mathbb{Z}[\varepsilon]) - (\varepsilon - 1)^{i+1} M_r(\mathbb{Z}[\varepsilon])$. For all $b \in B$, we have $b^p = 1$, so $\Gamma(b)$ is a diagonal matrix, the diagonal entries of which are pth roots of unity, and hence powers of $\varepsilon^{p^{n-1}}$. Therefore $v(\Gamma(\beta)) \geqslant p^{n-1}$ for all $\beta \in I(B)$, by (2). Clearly $v(\Gamma(\alpha)) \geqslant 1$ for all $\alpha \in I(A)$. Assume, by way of contradiction, that $\Gamma(x) \neq 0$. Then $v(\Gamma(x)) = p^{n-1} + k$ for some $k \geqslant 1$. Because $(1 + x)^p = 1$, we have

$$px + \binom{p}{2}x^2 + \ldots + x^p = 0.$$

Since $v(\binom{p}{j}) > 0$ for $1 < j < p$, we have $v(p\Gamma(x)) \geqslant v(\Gamma(x^p))$. Applying (1), we deduce that

$$p^{n-1}(p - 1) + p^{n-1} + k \geqslant p(p^{n-1} + k)$$

whence $k \geqslant pk$, a contradiction. ∎

8.7.5 Lemma *Let B be an elementary abelian p-group for some prime p. Then*

$$pI(B) \subseteq I(B)^p,$$

with equality if $|B| = p$.

Proof. Given $1 \neq b \in B$, choose $k \geqslant 1$ such that $p(b - 1) \in I(B)^k - I(B)^{k+1}$. Since

$$0 = (1 + (b - 1))^p - 1 = p(b - 1) + \binom{p}{2}(b - 1)^2 + \ldots + (b - 1)^p$$

and $\binom{p}{j}(b - 1)^j \in I(B)^{k+1}$, $2 \leqslant j < p$, we have

$$p(b - 1) \equiv -(b - 1)^p (\bmod\, I(B)^{k+1}).$$

If $k \leqslant p - 1$, then $(b - 1)^p \in I(B)^{k+1}$, so $p(b - 1) \equiv 0 (\bmod\, I(B)^{k+1})$, which is impossible. Thus $k \geqslant p$ and $p(b - 1) \in I(B)^p$, proving that $pI(B) \subseteq I(B)^p$.

Assume that $B = \langle b \rangle$ is cyclic of order p. Then $I(B)^p = \mathbb{Z}B(b - 1)^p$ and

$$(b - 1)^p \equiv 0 (\bmod\, pI(B)),$$

proving that $I(B)^p = pI(B)$. ∎

8.7.6 Lemma   *Let A be an abelian normal subgroup of G and let B be an elementary abelian normal p-subgroup of G contained in A.*

(i)   *If $x \in \mathbb{Z}G \cdot I(B)$ and $(1 + x)^p \equiv 1(\mathrm{mod}\ \mathbb{Z}G \cdot I(B)^{p+1})$, then*

$$px + x^p \equiv 0(\mathrm{mod}\ \mathbb{Z}G \cdot I(B)^{p+1}).$$

(ii)   *If $\alpha \in \mathbb{Z}G \cdot I(A) \cdot I(B)$ and $\beta \in \mathbb{Z}G \cdot I(B)$, then*

$$(1 + \alpha + \beta)^p \equiv (1 + \beta)^p(\mathrm{mod}\ \mathbb{Z}G \cdot I(A)I(B)^p).$$

(iii)   *If $G = AH$, $A \cap H = 1$, and $G_1 = BH$, then*

$$\mathbb{Z}G \cdot I(A)I(B)^n \cap \mathbb{Z}G_1 = \mathbb{Z}G_1 \cdot I(B)^{n+1}, \qquad \textit{for all } n \geqslant 1.$$

Proof.
(i)   We have

$$px + \binom{p}{2}x^2 + \ldots + x^p \equiv 0(\mathrm{mod}\ \mathbb{Z}G \cdot I(B)^{p+1}).$$

For $1 < i < p$, $\binom{p}{i}x^i \in p\mathbb{Z}G \cdot I(B)^2$ and, by Lemma 7.5, $p\mathbb{Z}G \cdot I(B) \subseteq \mathbb{Z}G \cdot I(B)^p$. Therefore

$$\binom{p}{i}x^i \in \mathbb{Z}G \cdot I(B)^{p+1}, \qquad \text{for all } 1 < i < p,$$

proving (i).
(ii)   We have

$$(1 + \alpha + \beta)^p = 1 + \sum_{i=1}^{p} \binom{p}{i}(\alpha + \beta)^i.$$

Because $\alpha \in \mathbb{Z}G \cdot I(A) \cdot I(B)$, then for $1 \leqslant i \leqslant p$,

$$\binom{p}{i}(\alpha + \beta)^i \equiv \binom{p}{i}\beta^i(\mathrm{mod}\ p\mathbb{Z}G \cdot I(A) \cdot I(B))$$

and hence, by Lemma 7.5.

$$\binom{p}{i}(\alpha + \beta)^i \equiv \binom{p}{i}\beta^i(\mathrm{mod}\ \mathbb{Z}G \cdot I(A) \cdot I(B)^p)$$

for $1 \leqslant i < p$. Note also that $(\alpha + \beta)^p = \beta^p +$ other terms, each of which has p α's or β's, and at least one α, so

$$(\alpha + \beta)^p \equiv \beta^p(\mathrm{mod}\ \mathbb{Z}G \cdot I(A) \cdot I(B)^p).$$

It follows that

$$(1 + \alpha + \beta)^p \equiv 1 + \sum_{i=1}^{p} \binom{p}{i}\beta^i(\mathrm{mod}\ \mathbb{Z}G \cdot I(A) \cdot I(B)^p)$$

and since

$$1 + \sum_{i=1}^{p} \binom{p}{i} \beta^i = (1 + \beta)^p,$$

the required assertion is proved.

(iii)   Given $\alpha \in \mathbb{Z}G \cdot I(A) \cdot I(B)^n \cap \mathbb{Z}G_1$, we may write $\alpha = \sum_{h \in H} h\alpha_h$, with $\alpha_h \in I(A)I(B)^n \cap \mathbb{Z}B$. It therefore suffices to prove that

$$I(A)I(B)^n \cap \mathbb{Z}B = I(B)^{n+1}.$$

The inclusion $I(B)^{n+1} \subseteq I(A) \cdot I(B)^n \cap \mathbb{Z}B$ being obvious, fix $\beta \in I(A)I(B)^n \cap \mathbb{Z}B$ and write

$$\beta = \sum n_{ij}(a_i - 1)\beta_j \qquad (a_i \in A, \ \beta_j \in I(B)^n, \ n_{ij} \in \mathbb{Z}).$$

Let $\{1, g_2, \ldots, g_m\}$ be a transversal for B in A and define $f : A \to B$ by $f(g_i b) = b$. Extend f linearly to an additive homomorphism $f^* : \mathbb{Z}A \to \mathbb{Z}B$. Because $\beta \in \mathbb{Z}B$, we have $f^*(\beta) = \beta$. If $a = g_k b$, then

$$f^*((a-1)\beta_j) = f^*(g_k b \beta_j) - f^*(\beta_j) = bb_j - \beta_j$$
$$= (b-1)\beta_j \in I(B)^{n+1}.$$

Hence $\beta = f^*(\beta) \in I(B)^{n+1}$, as required.   ∎

8.7.7 Lemma   *Let N and M be normal subgroups of G such that $N \cap M = 1$. Then*

$$\mathbb{Z}G \cdot I(N) \cap \mathbb{Z}G \cdot I(M) = \mathbb{Z}G \cdot I(N) \cdot I(M).$$

Proof.   Let $\{t_1, \ldots, t_s\}$ be a transversal for NM in G. Fix $\alpha \in \mathbb{Z}G \cdot I(N) \cap \mathbb{Z}G \cdot I(M)$, and write $\alpha = \sum_{i=1}^{s} \sum_{n \in N} t_i n \beta_{in}$ with $\beta_{in} \in I(M)$. Since $n = (n-1) + 1$, we have

$$\alpha = \sum_{i=1}^{s} \sum_{n \in N} t_i(n-1)\beta_{in} + \sum_{i=1}^{s} \sum_{n \in N} t_i \beta_{in},$$

whence

$$\alpha \equiv \sum_{i=1}^{s} \sum_{n \in N} t_i \beta_{in} \pmod{\mathbb{Z}G \cdot I(N) \cdot I(M)}.$$

Since $\alpha \in \mathbb{Z}G \cdot I(N)$ and $N \cap M = 1$, then for each i we have

$$\sum_{n \in N} \beta_{in} \in \mathbb{Z}G \cdot I(N) \cap I(M) = 0,$$

proving the lemma.   ∎

Assume that G has an abelian normal subgroup A such that G/A is abelian. If T is a transversal for A in G, then $\mathbb{Z}G \cdot I(A)$ is a free $\mathbb{Z}$-module with basis $\{t(a-1) \mid t \in T, 1 \neq a \in A\}$. Hence, for any $k \in \mathbb{Z}$, the map $t(a-1) \mapsto a^{t^k}$, where $a^{t^k} = t^{-k}at^k$, determines a $\mathbb{Z}$-homomorphism

$$\phi_k : \mathbb{Z}G \cdot I(A) \rightarrow A.$$

We then define

$$I_k = I_k(G, A) = \operatorname{Ker} \phi_k.$$

Recall that the map

$$\begin{cases} \mathbb{Z}G \rightarrow \mathbb{Z}G, \\ \sum x_g g \mapsto (\sum x_g g)^* = \sum x_g g^{-1} \end{cases}$$

is an antiautomorphism of $\mathbb{Z}G$.

8.7.8 Lemma

(i)   *For all $g \in G$, $a \in A$, we have $\phi_k(g(a-1)) = a^{g^k}$.*

(ii) *I_k is a two-sided ideal of $\mathbb{Z}G$ with $\mathbb{Z}G \cdot I(A)^2 \subseteq I_k \subseteq \mathbb{Z}G \cdot I(A)$. Moreover,*

$$I_0 = I(G)I(A).$$

(iii) *For each $x \in \mathbb{Z}G \cdot I(A)$, there exists $a \in A$ such that $x \equiv a - 1 (\operatorname{mod} I_k)$.*

(iv) *$G \cap (1 + I_k) = 1$.*

(v)   *$I_k^* = I_{-k-1}$.*

(vi) *If $(G:A)$ is odd and $s = (1/2)((G:A) - 1)$, then $I_s^* = I_s$ and*

$$g^{-1}(h + h^{-1})g \equiv h + h^{-1} (\operatorname{mod} I_s) \qquad (g, h \in G).$$

Proof

(i)   Write $g = tb$ with $t \in T$, $b \in A$. Then, for all $a \in A$,

$$g(a-1) = tb(a-1) = t(ba-1) - t(b-1)$$

and so

$$\phi_k(g(a-1)) = (ba)^{t^k}(b^{-1})^{t^k} = a^{t^k} = a^{(tb)^k} = a^{g^k}$$

as desired.

(ii) The fact that I_k is an ideal will follow from the identities

$$\phi_k(gx) = \phi_k(x)^{g^k}, \quad \phi_k(xg) = \phi_k(x)^{g^{k+1}} \qquad (g \in G, x \in \mathbb{Z}G \cdot I(A)),$$

which we need verify only for $x = h(a-1)$, $h \in G$, $a \in A$. But, using the fact that G/A is abelian, we have

$$\phi_k(gh(a-1)) = a^{(gh)^k} = a^{h^k g^k} = \phi_k(h(a-1))^{g^k}$$

and, similarly,

$$\phi_k(h(a-1)g) = \phi_k(hg(a^g - 1)) = (a^g)^{(hg)^k} = a^{h^k g^{k+1}} = \phi_k(h(a-1))^{g^{k+1}}.$$

Since I_k is an ideal, the fact that $\mathbb{Z}G \cdot I(A)^2 \subseteq I_k$ follows from

$$\phi_k((a-1)(b-1)) = \phi_k((ab-1) - (a-1) - (b-1)) = aba^{-1}b^{-1} = 1$$

$$(a, b \in A).$$

By definition, $I_k \subseteq \mathbb{Z}G \cdot I(A)$ and therefore $\mathbb{Z}G \cdot I(A)^2 \subseteq I_k \subseteq \mathbb{Z}G \cdot I(A)$.
To prove that $I(G)I(A) \subseteq I_0$, it suffices to verify that

$$\phi_0((g-1)(a-1)) = 1 \qquad (g \in G, a \in A)$$

which is the case since

$$\phi_0((g-1)(a-1)) = \phi_0(g(a-1) - (a-1)) = a^{g^0} a^{-1} = 1.$$

Conversely, assume that $x \in I_0$. Owing to Lemma 7.1,

$$x = \sum_{t \in T} t(a_t - 1) + \alpha \qquad (\alpha \in \mathbb{Z}G \cdot I(A)^2 \subseteq I_0)$$

so

$$\sum_{t \in T} t(a_t - 1) \in I_0$$

and therefore $\prod_{t \in T} a_t = 1$. Now

$$\sum_{t \in T} t(a_t - 1) = \sum_{t \in T} (t-1)(a_t - 1) + \sum_{t \in T} (a_t - 1)$$

and, by Proposition 3.9(ii), $\sum (a_t - 1) \equiv \prod a_t - 1 \pmod{I(A)^2}$. Because $\prod a_t = 1$, we have $x \equiv \sum (t-1)(a_t - 1) \pmod{\mathbb{Z}G \cdot I(A)^2}$, which shows that $x \in I(G)I(A)$.

(iii) Given $x \in \mathbb{Z}G \cdot I(A)$, we have $\phi_k(x) \in A$, so $\phi_k(\phi_k(x) - 1) = \phi_k(x)$. Thus, if $a = \phi_k(x)$, then $x \equiv a - 1 \pmod{I_k}$.

(iv) Assume that $g \in G \cap (1 + I_k)$. Then $g = 1 + x$ with $x \in I_k$ and $g - 1 \in \mathbb{Z}G \cdot I(A)$, so $g = a \in A$. Then $x = a - 1$, and $1 = \phi_k(x) = \phi_k(a - 1) = a$.

(v) It suffices to verify that $\phi_k(x^*) = \phi_{-k-1}(x)^*$ for all $x \in \mathbb{Z}G \cdot I(A)$. We may harmlessly assume that $x = g(a - 1)$ with $g \in G$, $a \in A$. Since

$$(g(a-1))^* = g^{-1}((a^{-1})^{g^{-1}} - 1)$$

we have

$$\phi_k((g(a-1))^*) = ((a^{-1})^{g^{-1}})^{g^{-k}} = (a^{-1})^{g^{-k-1}} = (a^{g^{-k-1}})^*$$
$$= (\phi_{-k-1}(g(a-1)))^*$$

as required.

(vi) By hypothesis, $s \equiv s - 1(\text{mod}\,(G : A))$, so $I_s^* = I_s$, by (v). Now

$$g^{-1}(h + h^{-1})g - (h + h^{-1}) = (h^g - h) + (h^{-g} - h^{-1})$$
$$= h(h^{-1}h^g - 1) + h^{-1}(hh^{-g} - 1)$$

which ϕ_s maps to

$$[h, g]^{h^s}(hh^{-g})^{h^{-s}} = h^{-s}[h, g]h^s h^s(hh^{-g})h^{-s}$$
$$= h^{-s}[h, g]^{2s+1}h^{-g}h^{-s}.$$

Because G/A is abelian, $G' \subseteq A$ and since $(G : A) = 2s + 1$, h^{2s+1} commutes with $[h, g]$, giving

$$h^{-s}h^{2s+1}[h, g]h^{-g}h^{-s} = h^{s+1}(h^{-1}h^g)h^{-g}h^{-s} = 1$$

and the proof is complete. ∎

As a final preparatory result, we prove

8.7.9 Lemma   *Let A be a finite abelian group.*

(i) $V(\mathbb{Z}A) = A \times U(1 + I(A)^2)$.
(ii) $V(\mathbb{Z}A)$ *is finitely generated.*
(iii) $U(1 + I(A)^2)$ *is a free abelian group.*
(iv) *If* $u \in U(1 + I(A)^2)$, *then* $u^* = u$.

Proof.   Properties (ii) and (iii) are consequences of a more general result to be proved later (see Theorem 9.31). By Proposition 3.9(ii), given $u \in V(\mathbb{Z}A)$, we have

$$u - 1 \equiv a - 1(\text{mod}\,I(A)^2)$$

for some $a \in A$. Hence $ua^{-1} = 1 + \alpha$ for some $\alpha \in I(A)^2$ and so (i) is a consequence of Proposition 3.9(iii).

To prove (iv), write $v = u^* u^{-1}$, so $v^* v = 1$ since A is abelian. If $v = \sum_{a \in A} v_a a$, $v_a \in \mathbb{Z}$, then $tr(v^* v) = \sum_a v_a^2$, which is 1 since $v^* v = 1$. Hence $v = \pm a$ for some $a \in A$ and since $\text{aug}(v) = 1$, we have $v = a$ and $u^* = ua$. Now $u \equiv 1(\text{mod}\,I(A)^2)$ and, since $I(A)^* = I(A)$, $u^* \equiv 1(\text{mod}\,I(A)^2)$. Therefore $a \equiv 1(\text{mod}\,I(A)^2)$ and, by Proposition 3.9(iii), $a = 1$. Thus $u^* = u$ as required. ∎

8.7B   Main theorems

We preserve all the notation introduced in Section 8.7A. Our first aim is to prove that if A is an abelian normal subgroup of G such that G/A is abelian, then $U(1 + I_k)$ is torsion-free for all $k \in \mathbb{Z}$. The burden of the proof of this fact is contained in the following result.

8.7.10 Proposition   (Cliff *et al.* 1981)   *Suppose that* $G = AH$, $A \lhd G$,

A being an elementary abelian p-group, and assume that H is abelian and acts faithfully and irreducibly on A. If $x \in I_k(G, A)$ and

$$(1 + x)^p \equiv 1(\text{mod } \mathbb{Z}G \cdot I(A)^{p+1})$$

then $x \in \mathbb{Z}G \cdot I(A)^2$.

Proof. It follows from Huppert (1967, Satz II, 3.10) that if $|A| = p^d$ and K is the field of p^d elements, then there is an isomorphism γ from A to the additive group of K and an injective homomorphism $\phi : H \to K^*$ such that

$$\gamma(a^h) = \phi(h)\gamma(a) \qquad (h \in H, a \in A). \tag{1}$$

Owing to Lemma 7.1, $x = \sum_{h \in H} h(a_h - 1) + \alpha$, where $a_h \in A$, $h \in H$ and $\alpha \in \mathbb{Z}G \cdot I(A)^2$. We have

$$(1 + x)^p = \left(1 + \sum_{h \in H} h(a_h - 1) + \alpha\right)^p$$

$$\equiv \left(1 + \sum_{h \in H} h(a_h - 1)\right)^p (\text{mod } \mathbb{Z}G \cdot I(A)^{p+1})$$

by Lemma 7.6(ii). Thus we may assume that $x = \sum h(a_h - 1)$ and we shall prove that $x = 0$. To prove the latter, we may assume that $|H| = p^d - 1$. Indeed, let H_1 be cyclic of order $p^d - 1$ containing H, and let $\phi_1 : H_1 \to K^*$ be an isomorphism extending ϕ. If G_1 is the semidirect product AH_1, then $x \in I_k(G_1, A)$ and $x = 0$ since $|H_1| = p^d - 1$. From now on, we write $|H| = p^d - 1 = r$.

Let $1 \neq \chi \in \text{Hom}(A, \mathbb{C}^*)$, let $\varepsilon \in \mathbb{C}$ be a primitive pth root of unity and let $R = \mathbb{Z}[\varepsilon]$. Let $\Gamma = \chi^G$ be the induced representation of G. Then we may regard Γ as a representation of RG. We may also consider the representation space of Γ to be RH, with

$$\Gamma(a)y = \chi(a^y)y, \quad \Gamma(h)y = hy \qquad (a \in A, h, y \in H).$$

We now claim that it suffices to verify that

$$\Gamma(x) \equiv 0(\text{mod}(\varepsilon - 1)^2). \tag{2}$$

Indeed, if

$$\Gamma\left(\sum_{h \in H} h(a_h - 1)\right)y \in (\varepsilon - 1)^2 RH \qquad \text{for all } y \in H,$$

then

$$\sum_k (\chi(a_h^y) - 1)hy \in (\varepsilon - 1)^2 RH \qquad (y \in H).$$

Fix $h \in H$ and let $a = a_h$, so

$$\chi(a^y) - 1 \equiv 0(\text{mod}(\varepsilon - 1)^2) \qquad (y \in H).$$

However, $\chi(a^y)$ is a power of ε, so the above congruence ensures that $\chi(a^y) = 1$, $y \in H$. If $a \neq 1$, then $\{a^y \mid y \in H\}$ generates A since A is irreducible. The conclusion is that $\chi = 1$, contrary to our assumption. Thus $a_h = 1$ for all $h \in H$, and $x = 0$ as claimed.

Given $a \in A$, $y \in H$, we have $\Gamma(a - 1)y = (\chi(a^y) - 1)y$, and since $\chi(a^y) - 1 \equiv 0 (\mathrm{mod}(\varepsilon - 1))$, it follows that for all $\alpha \in \mathbb{Z}G \cdot I(A)$, there exists an operator $S(\alpha)$ on RH with

$$\Gamma(\alpha) = (\varepsilon - 1)S(\alpha) \qquad (\alpha \in \mathbb{Z}G \cdot I(A)).$$

Let $\bar{S}(\alpha)$ denote the reduction of $S(\alpha)\mathrm{mod}(\varepsilon - 1)$, so that $\bar{S}(\alpha)$ is an operator on $\mathbb{F}_pH$, and by extension of scalars, on KH. To prove (2), we are left to verify that

$$\bar{S}(x) = 0. \tag{3}$$

Let $\bar{\Gamma}(h)$ be the reduction of $\Gamma(h)$ modulo $(\varepsilon - 1)$. Because $\Gamma(h(a - 1)) = \Gamma(h)\Gamma(a - 1)$, we have

$$\bar{S}(h(a - 1)) = \bar{\Gamma}(h)\bar{S}(a - 1) \qquad (h \in H, a \in A).$$

By hypothesis, $(1 + x)^p \equiv 1 (\mathrm{mod}\ \mathbb{Z}G \cdot I(A)^{p+1})$, so by Lemma 7.6(i),

$$px + x^p \equiv 0 (\mathrm{mod}\ \mathbb{Z}G \cdot I(A)^{p+1}).$$

Therefore

$$p(\varepsilon - 1)S(x) + ((\varepsilon - 1)S(x))^p \equiv 0 (\mathrm{mod}(\varepsilon - 1)^{p+1}),$$

or

$$(p/(\varepsilon - 1)^{p-1})S(x) + S(x)^p \equiv 0 (\mathrm{mod}(\varepsilon - 1)).$$

Our next task is to compute $p/(\varepsilon - 1)^{p-1}\mathrm{mod}(\varepsilon - 1)$. We have

$$p = \prod_{i=1}^{p-1} (1 - \varepsilon^i) = (1 - \varepsilon)^{p-1} \prod_{i=1}^{p-1} (1 - \varepsilon^i)/(1 - \varepsilon)$$

$$= (1 - \varepsilon)^{p-1} \prod_{i=1}^{p-1} (1 + \varepsilon + \ldots + \varepsilon^{i-1})$$

$$\equiv (-1)^{p-1}(\varepsilon - 1)^{p-1}(p - 1)!$$

$$\equiv -(\varepsilon - 1)^{p-1}(\mathrm{mod}(\varepsilon - 1)^p).$$

Hence $p/(\varepsilon - 1)^{p-1} \equiv -1(\mathrm{mod}(\varepsilon - 1))$ and therefore

$$-S(x) + S(x)^p \equiv 0 (\mathrm{mod}(\varepsilon - 1)). \tag{4}$$

Let $tr: K \to \mathbb{F}_p$ be the field trace. Then, taking $\chi(a) = \varepsilon^{tr\gamma(a)}$, $a \in A$, we have

$$(1/(\varepsilon - 1))(\chi(a) - 1) = 1 + \varepsilon + \ldots + \varepsilon^{tr\gamma(a)-1}$$

$$\equiv tr\gamma(a)(\mathrm{mod}(\varepsilon - 1)),$$

whence

$$\bar{S}(a-1)y = (tr\gamma(a^y))y = tr(\phi(y)\gamma(a))y \qquad (a \in A, y \in H).$$

We choose a basis of KH consisting of primitive orthogonal idempotents

$$e_i = (1/r) \sum_{h \in H} \phi(h)^{-i}h,$$

for $i \bmod r$. Then $\bar{\Gamma}(h)e_i = \phi(h)^i e_i$. Given $a \in A$, we have

$$\bar{S}(a-1)e_j = \bar{S}(a-1)(1/r) \sum_h \phi(h)^{-j}h$$

$$= (1/r) \sum_h \phi(h)^{-j}\bar{S}(a-1)h$$

$$= (1/r) \sum_h \phi(h)^{-j}tr(\phi(h)\gamma(a))h$$

$$= (1/r) \sum_h \phi(h)^{-j} \sum_{n=0}^{d-1} \phi(h)^{p^n}\gamma(a)^{p^n}h$$

$$= \sum_{n=0}^{d-1} \gamma(a)^{p^n}(1/r) \sum_h \phi(h)^{-j+p^n}h$$

whence

$$\bar{S}(a-1)e_j = \sum_{n=0}^{d-1} \gamma(a)^{p^n}e_{j-p^n} \qquad (a \in A, j \bmod r).$$

It follows that

$$\bar{S}(x)e_j = \sum_h \bar{\Gamma}(h)\bar{S}(a_h-1)e_j = \sum_h \sum_n \gamma(a_h)^{p^n}\phi(h)^{j-p^n}e_{j-p^n}$$

$$= \sum_n \sum_h (\gamma(a_h)\phi(h)^{p^{-n}j-1})^{p^n}e_{j-p^n}.$$

Setting

$$x_i = \sum_h \phi(h)^i\gamma(a_h) \qquad (i \bmod r),$$

we have

$$\bar{S}(x)e_j = \sum_{n=0}^{d-1} (x_{p^{-n}j-1})^{p^n}e_{j-p^n} \qquad (j \bmod r). \tag{5}$$

Now write $\bar{S}(x)e_j = \sum_{i \bmod r} s(i, j)e_i$. Then, by (5),

$$s(i, j) = 0 \qquad \text{unless } j - i \equiv p^n(\bmod r) \text{ for some } n \tag{6}$$

and, taking $n = 0$ in (5), we obtain

$$s(j, j+1) = x_j \qquad (j \bmod r). \tag{7}$$

By (4), $\bar{S}(x)^p = \bar{S}(x)$, and hence

$$s(i, j) = \sum_{m_1, \ldots, m_{p-1}} s(i, m_1)s(m_1, m_2) \ldots s(m_{p-1}, j). \tag{8}$$

Invoking (6), we see that the sum collapses to those $m_1, \ldots, m_{p-1}$ for which $m_1 - i, \; m_2 - m_1, \ldots, m_{p-1} - j$ are all powers of p mod r. Assume that $j - i \equiv p^n (\bmod r)$. Then

$$(m_1 - i) + (m_2 - m_1) + \ldots + (j - m_{p-1}) = j - i \equiv p^n (\bmod r)$$

giving

$$p^{n_1} + p^{n_2} + \ldots + p^{n_p} \equiv p^n (\bmod r).$$

It will be shown that

$$n_1 \equiv n_2 \equiv \ldots \equiv n_p \equiv n - 1 (\bmod d).$$

Indeed, since $p^d \equiv 1 (\bmod(p^d - 1))$, we may replace n_i by any of its residues mod d, and therefore may assume that $n \le n_i$ for all i. Then, dividing by p^n, we may assume $n = 0$. Now replace each n_i by its least non-negative residue mod d. We have

$$p \le p^{n_1} + p^{n_2} + \ldots + p^{n_p} \le p \cdot p^{d-1} = p^d.$$

But then $p^{n_1} + \ldots + p^{n_p} \equiv 1 (\bmod(p^d - 1))$ can only happen if $p^{n_1} + \ldots + p^{n_p} = p^d$, so the upper bound is attained, and each $p^{n_i} = p^{d-1}$. This shows that $n_1 = \ldots = n_p = d - 1$, as required.

By the foregoing, the sum in (8) collapses to a single term

$$s(i, i + p^n) = \prod_{m=0}^{p-1} s(i + mp^{n-1}, i + (m+1)p^{n-1}). \tag{9}$$

With $n = 1$, (9) becomes:

$$s(i, i + p) = \prod_{m_0} s(i + m_0, i + m_0 + 1)$$

while with $n = 2$, (9) becomes:

$$s(i, i + p^2) = \prod_{m_1} s(i + m_1 p, i + (m_1 + 1)p),$$

giving

$$s(i, i + p^2) = \prod_{m_1} \prod_{m_0} s(i + m_1 p + m_0, i + m_1 p + m_0 + 1).$$

Then substitute this equation into (9) with $n = 3$. Continuing in this manner, we derive

$$s(i, i+1) = s(i, i+p^d) = \prod_{m_{d-1}} \ldots \prod_{m_1} \prod_{m_0} s(i + m_{d-1}p^{d-1} + \ldots + m_1 p + m_0,$$

$$i + m_{d-1}p^{d-1} + \ldots + m_1 p + m_0 + 1).$$

Bearing in mind that $m_{d-1}p^{d-1} + \ldots + m_1 p + m_0$ is the expansion of the integers from 0 to r in base p, we have

$$s(i, i+1) = s(i, i+1)s(i+1, i+2) \ldots s(i-1, i)s(i, i+1) \qquad (i \bmod r).$$
(10)

Now $x \in I_k(G, A)$, so $\prod_h (a_h)^{h^k} = 1$ and therefore, by (1), $x_k = \sum_h \phi(h)^k \gamma(a_h) = 0$. By (7), $s(k, k+1) = x_k = 0$ which, by (10), implies that $s(i, i+1) = 0$ for all $i(\bmod r)$. Applying (7), we obtain $x_i = 0$ for all $i(\bmod r)$ and finally from (5), $\bar{S}(x) = 0$. This proves (3) and hence the result. ∎

We now apply the foregoing result to prove the following property, which is decisive for the proof of the main theorems.

8.7.11 Proposition (Cliff *et al.* 1981) *Let* $G = AH$, $A \lhd G$, $A \cap H = 1$, *with both A and H abelian. Assume that A is an elementary abelian p-group for some prime p, and that a Hall p'-subgroup H_0 of H acts faithfully and irreducibly on A. If $x \in I_k(G, A)$ and*

$$(1+x)^p \equiv 1 (\bmod \mathbb{Z}G \cdot I(A)^{p+1})$$

then $x \in \mathbb{Z}G \cdot I(A)^2$.

Proof. If $A = 1$, then $(1+x)^p = 1$ and, by Lemmas 7.2 and 7.8(ii), $x = 0$. Thus we may assume that $A \neq 1$. Let S be the Sylow p-subgroup of H. The case $|S| = 1$ being the content of Proposition 7.10, we argue by induction on $|S|$. Assume that $|S| > 1$ and let $B = C_A(S)$. Then $|B| \neq 1$ and $B \lhd G$, so $B = A$ since H_0 acts irreducibly on A. Thus B is central in G. Now fix $s \in S$ of order p. If $-: \mathbb{Z}G \to \mathbb{Z}(G/\langle s \rangle)$ is the natural homomorphism, then by induction $\bar{x} \in \mathbb{Z}\bar{G} \cdot I(\bar{A})^2$, so $x = \alpha + x_1$, where $\alpha \in \mathbb{Z}G \cdot I(A)^2$ and $x_1 \in \mathbb{Z}G \cdot I\langle s \rangle$. Applying Lemma 7.6(ii), we have

$$(1+x_1)^p = (1+x-\alpha)^p \equiv (1+x)^p (\bmod \mathbb{Z}G \cdot I(A)^{p+1})$$

and so, by hypothesis,

$$(1+x_1)^p \equiv 1 (\bmod \mathbb{Z}G \cdot I(A)^{p+1}).$$

Then, by Lemma 7.6(i),

$$px_1 + x_1^p \equiv 0 (\bmod \mathbb{Z}G \cdot I(A)^{p+1}).$$

By Lemma 7.7, we also have

$$x_1 \in \mathbb{Z}G \cdot I\langle s \rangle \cap \mathbb{Z}G \cdot I(A) = \mathbb{Z}G \cdot I\langle s \rangle I(A).$$

Then, applying the facts that $I\langle s \rangle^p = \mathbb{Z}\langle s \rangle (1-s)^p = \mathbb{Z}\langle s \rangle p(1-s)$ and $p\mathbb{Z}G \cdot I(A) \subseteq \mathbb{Z}G \cdot I(A)^p$ from Lemma 7.5, we derive

$$x_1^p \in (1-s)^p \mathbb{Z}G \cdot I(A)^p = p(1-s)\mathbb{Z}G \cdot I(A)^p \subseteq (1-s)\mathbb{Z}G \cdot I(A)^{p+1}.$$

Because $px_1 + x_1^p \equiv 0 \pmod{\mathbb{Z}G \cdot I(A)^{p+1}}$, we deduce that $px_1 \in \mathbb{Z}G \cdot I(A)^{p+1}$. Now write $x_1 = \sum_{h \in H} h\alpha_h$, $\alpha_h \in \mathbb{Z}A$. Because $px_1 \in \mathbb{Z}G \cdot I(A)^{p+1}$, we have $p\alpha_h \in I(A)^{p+1}$, $h \in H$. We are therefore left to verify that if $\alpha \in \mathbb{Z}A$, $p\alpha \in I(A)^{p+1}$, then $\alpha \in I(A)^2$.

To this end, note that $\alpha \in I(A)$ and, by Proposition 3.9(ii), $\alpha \equiv a - 1 \pmod{I(A)^2}$ for some $a \in A$. Hence we may assume that $\alpha = a - 1$, and must prove that $a = 1$. If $a \neq 1$, let ε be a primitive pth root of unity and let $\chi \in \mathrm{Hom}(A, \mathbb{C}^*)$ be such that $\chi(a) = \varepsilon$. Extend χ linearly to $\mathbb{Z}A$, so that χ maps $\mathbb{Z}A$ to $\mathbb{Z}[\varepsilon]$. Now $\chi(p(a-1)) = p(\varepsilon - 1)$, but $\chi(I(A)^{p+1}) = (\varepsilon - 1)^{p+1}\mathbb{Z}[\varepsilon]$. Since we are assuming that $p(a-1) \in I(A)^{p+1}$, we derive $p(\varepsilon - 1) \in (\varepsilon - 1)^{p+1}\mathbb{Z}[\varepsilon]$, a contradiction. ∎

We are now ready to prove our first major result.

8.7.12 Theorem (Cliff *et al.* 1981) *Let $A \lhd G$ with A and G/A abelian. Let T be a transversal for A in G and, for each $k \in \mathbb{Z}$, let I_k be the kernel of the homomorphism of abelian groups $\mathbb{Z}G \cdot I(A) \to A$ which sends $t(a-1)$ to $t^{-k}at^k$, $t \in T$, $a \in A$. Then, for any $k \in \mathbb{Z}$, I_k is an ideal of $\mathbb{Z}G$ such that $U(1 + I_k)$ is torsion-free. In particular,*

$$U(1 + I(G)I(A))$$

is torsion-free.

Proof. In view of Lemma 7.8(i), we need only verify that $U(1 + I_k)$ is torsion-free. Denote G/A by H. Let $\tilde{G}$ be the wreath product $A \wr H$. It is well known (see Huppert 1967; Satz I, 15.9) that there is an embedding $\phi : G \to \tilde{G}$, and $\tilde{G}$ is a split metabelian group $\bar{A}H$, with $\tilde{G} = \bar{A}\phi(G)$. Then $I_k(G, A) \subseteq I_k(\tilde{G}, \bar{A})$. Thus we may harmlessly assume that $G = A \cdot H$ with $A \cap H = 1$ and H abelian.

We argue by induction on $|G|$. Note first that if $N \lhd G$ and $-: \mathbb{Z}G \to \mathbb{Z}(G/N)$ is the natural homomorphism, then the image of $I_k(G, A)$ in $\mathbb{Z}\tilde{G}$ is contained in $I_k(\tilde{G}, \bar{A})$. For if $\sum ng(a-1) \in I_k(G, A)$, then $\prod a^{ng^k} = 1$, so $\prod \bar{a}^{n\bar{g}^k} = 1$ and $\sum n\bar{g}(\bar{a} - 1) \in I_k(\tilde{G}, \bar{A})$.

Assume that $u \in U(1 + I_k(G, A))$ and suppose that $u^p = 1$ for some prime p. Suppose that N is a central subgroup of G with $1 \neq N \subseteq H$. Then $\tilde{G} = G/N$ is a split metabelian group, $\bar{u} \equiv \bar{1} \pmod{I_k(\tilde{G}, \bar{A})}$, and $\bar{u}$ has finite order, so $\bar{u} = \bar{1}$ by induction. Hence $u \equiv 1 \pmod{\mathbb{Z}G \cdot I(N)}$. If

$\{1, g_2, \ldots, g_n\}$ is a transversal for N in G, then

$$u = 1 + \alpha + \sum_{i=2}^{n} g_i \alpha_i \qquad (\alpha, \alpha_i \in I(H)).$$

If $\alpha = 0$, then $1 \in \text{Supp}\, u$, so $u = 1$ by Theorem 3.3(ii). If $\alpha \neq 0$, then $\text{Supp}\, u$ contains an element $h \in N$ and then Theorem 3.3(ii) applied to uh^{-1} implies that $u = h$. However, by Lemma 7.8(iv), $G \cap (1 + I_k(G, A)) = 1$ and thus $u = 1$. Hence we may assume that H does not contain a non-trivial central subgroup of G.

We now show that we may assume every normal subgroup B of G, with $B \subseteq A$, to be indecomposable, in the sense that if $B = B_1 \times B_2$ with $B_i \triangleleft G$, $i = 1, 2$, then $B_1 = 1$ or $B_2 = 1$. If not, let $-: \mathbb{Z}G \rightarrow \mathbb{Z}(G/B_1)$ be the natural homomorphism, and then $\bar{u} = \bar{1}$ by induction, so $u \equiv 1 (\text{mod } \mathbb{Z}G \cdot I(B_1))$. Similarly, $u \equiv 1 (\text{mod } \mathbb{Z}G \cdot I(B_2))$ and therefore, by Lemma 7.7, $u \equiv 1 (\text{mod } \mathbb{Z}G \cdot I(B_1)I(B_2))$. But

$$\mathbb{Z}G \cdot I(B_1)I(B_2) \subseteq \mathbb{Z}G \cdot I(A)^2$$

and hence $u = 1$ by Lemma 7.4. Thus we may assume that B is indecomposable. In particular, taking $B = A$, we find that A is a q-group for some prime q. Moreover, by Lemma 7.2, we may assume that $p = q$. To avoid trivialities, we assume that $A \neq 1$.

Write $H = H_p \times H_0$, where H_p is a p-group and $p \nmid |H_0|$. Since H contains no non-trivial central subgroup of G, it follows that H acts faithfully on A. Note also that $C_A(H_p) \neq 1$ since AH_p is a p-group. It is well known (see Gorenstein 1968, Theorem 5.3.4) that H_0 acts faithfully on $C_A(H_p)$, and then (see Gorenstein 1968, Theorem 5.2.4), H_0 acts faithfully on $B = \{a \in C_A(H_p) \mid a^p = 1\}$. Because B is indecomposable, H_0 also acts irreducibly on B.

Write $u = 1 + x$ with $x \in I_k(G, A)$. Because $|B| > 1$, $u \equiv 1 (\text{mod } \mathbb{Z}G \cdot I(B))$ by induction, so $x \in \mathbb{Z}G \cdot I(B)$. Let $\{a_1, \ldots, a_n\}$ be a transversal for B in A. Then

$$x = \sum_{h \in H} \sum_i h a_i \beta_{ih} \qquad (\beta_{ih} \in I(B)).$$

Let

$$\alpha = \sum_{h \in H} \sum_i h(a_i - 1)\beta_{ih}, \qquad x_1 = \sum_{h \in H} \sum_i h \beta_{ih}$$

so $x = \alpha + x_1$, and $\alpha \in \mathbb{Z}G \cdot I(A) \cdot I(B)$. Also, since $\alpha \in \mathbb{Z}G \cdot I(A)^2 \subseteq I_k(G, A)$, we have $x_1 \in I_k(G, A)$. Because $x_1 = \sum_{h \in H} \sum_i h \beta_{ih}$, we have $x_1 \in I_k(G_1, B)$, where $G_1 = BH$. Owing to Lemma 7.6(ii),

$$1 = (1 + x)^p = (1 + \alpha + x_1)^p \equiv (1 + x_1)^p (\text{mod } \mathbb{Z}G \cdot I(A) \cdot I(B)^p).$$

Also, $x_1 \in \mathbb{Z}G_1$, so

$$1 - (1 + x_1)^p \in \mathbb{Z}G \cdot I(A) \cdot I(B)^p \cap \mathbb{Z}G_1 = \mathbb{Z}G_1 \cdot I(B)^{p+1}$$

by Lemma 7.6(iii). Thus

$$(1+x_1)^p \equiv 1(\operatorname{mod} \mathbb{Z}G_1 \cdot I(B)^{p+1})$$

and, by Proposition 7.11, $x_1 \in \mathbb{Z}G_1 \cdot I(B)^2$. Hence $x \in \mathbb{Z}G \cdot I(A)^2$ and, by Lemma 7.4, $x = 0$, as required. ∎

As an application of the above, we now prove our second major result.

8.7.13   Theorem   (Cliff *et al.* 1981) *Let G be a finite group with an abelian normal subgroup A such that at least one of the following conditions hold:*

(a)   *G/A is abelian of exponent dividing 4 or 6;*
(b)   *G/A is abelian of odd order.*

Then G has a torsion-free normal complement in $V(\mathbb{Z}G)$.

Proof.   Assume that (a) holds. Then, by Theorem 4.3, $\mathbb{Z}(G/A)$ has only trivial units. It will be shown that $U(1 + I_k)$ is a normal complement to G in $V(\mathbb{Z}G)$. This will prove the case (a), by appealing to Theorem 7.12.

Let $-: \mathbb{Z}G \to \mathbb{Z}(G/A)$ be the natural homomorphism. If $u \in V(\mathbb{Z}G)$, then $\bar{u}$ is a normalized unit of $\mathbb{Z}(G/A)$, and hence $\bar{u} = \bar{g}$ for some $g \in G$. Then $u = g(1 + x)$ for some $x \in \mathbb{Z}G \cdot I(A)$. By Lemma 7.8(iii), $x \equiv a - 1 (\operatorname{mod} I_k)$ for some $a \in A$. Hence

$$1 + x = a(1 + x') \qquad \text{for some } x' \in I_k$$

and we have $u = ga(1 + x')$. Thus $V(\mathbb{Z}G) = GU(1 + I_k)$. By Lemma 7.8(iv), $G \cap (1 + I_k) = 1$ and the proof is complete in case (a).

Now assume that (b) holds and set $s = (1/2)((G:A) - 1)$. It will be shown that G has a normal complement N in $V(\mathbb{Z}G)$, and N is an extension of $U(1 + I_s)$ by a finitely generated free abelian group. This will complete the proof, by applying Theorem 7.12.

Write $-: \mathbb{Z}G \to \mathbb{Z}(G/A)$ for the natural homomorphism. Because $I_s \subseteq \mathbb{Z}G \cdot I(A)$, we have a commutative diagram of rings:

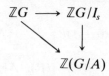

By Lemma 7.8(vi), $I_s^* = I_s$, so $\mathbb{Z}G/I_s$ admits *, which commutes with the ring homomorphisms of the diagram. We denote by $\bar{u}$ the image of $u \in \mathbb{Z}G/I_s$ under the vertical row. Because $I_s \subseteq I(G)$, $\mathbb{Z}G/I_s$ admits the augmentation map, and we write $V(\mathbb{Z}G/I_s)$ for units of augmentation 1. Owing to Lemma 7.8(iv), $G \cap (1 + I_s) = 1$, so we may identify G with its image in $\mathbb{Z}G/I_s$. We now claim that it suffices to prove that G has a normal complement Y in $V(\mathbb{Z}G/I_s)$ and Y is a finitely generated free abelian group. Indeed, given $u \in V(\mathbb{Z}G)$, consider u as an element of $V(\mathbb{Z}G/I_s)$ and write $u = yg$, $y \in Y$, $g \in G$. Define $\pi: V(\mathbb{Z}G) \to G$ by $\pi(u) = g$. Then $\pi(g) = g$, $g \in G$, and Ker π is an extension of $U(1 + I_s)$ by a subgroup of Y, as claimed.

Set $\bar{G} = G/A$ and define

$$V = \{u \in V(\mathbb{Z}G/I_s) \mid u^* = u \quad \text{and} \quad \bar{u} \equiv 1 (\operatorname{mod} I(\bar{G})^2)\}.$$

We first show that V is a subgroup. To this end, put

$$W = \{u \in V(\mathbb{Z}G/I_s) \mid \bar{u} \equiv 1 (\operatorname{mod} I(\bar{G})^2)\}.$$

It suffices to verify that W is abelian. For then $V \subseteq W$, so if $v_1, v_2 \in V$, then $v_1 v_2 = v_2 v_1$ and so $(v_1 v_2)^* = v_1^* v_2^* = v_1 v_2$, which implies that V is a subgroup. To prove that W is abelian, fix $u \in W$. Then $\bar{u}^* = \bar{u}$ by Lemma 7.9(iv), and so

$$\bar{u} = \sum_{x \in G/A} u_x x, \ u_{x^{-1}} = u_x.$$

Because $|G/A|$ is odd, we may choose a transversal T for A in G such that $g^{-1} \in T$ whenever $g \in T$. Then $v = \sum_{g \in T} u_{\bar{g}} g$ in $\mathbb{Z}G/I_s$ satisfies $v^* = v$, and v can be written as a $\mathbb{Z}$-linear combination of 1 and terms of the form $g + g^{-1}$. Hence, by Lemma 7.8(vi), v is central in $\mathbb{Z}G/I_s$. Because $\bar{v} = \bar{u}$, we have $v = u + \alpha$ where $\alpha \in \mathbb{Z}G \cdot I(A)$, so $v = u(1 + \beta)$ in $\mathbb{Z}G/I_s$ with $\beta \in \mathbb{Z}G \cdot I(A)$, $u\beta = \alpha$. Owing to Lemma 7.8(iii), $\beta \equiv b - 1 (\operatorname{mod} I_s)$ for some $b \in A$. Therefore $v = ub$ in $\mathbb{Z}G/I_s$, so $u = vb^{-1}$. But v is central and A is abelian, and hence W is indeed abelian.

It will next be shown that

$$V \lhd V(\mathbb{Z}G/I_s) \quad \text{and} \quad V(\mathbb{Z}G/I_s) = VG.$$

Take $u_1 \in V(\mathbb{Z}G/I_s)$, so $\bar{u}_1 \equiv \bar{g} (\operatorname{mod} I(\bar{G})^2)$ for some $g \in G$, by Lemma 7.9(i). Then $u = u_1 g^{-1}$ satisfies $\bar{u} \equiv 1 (\operatorname{mod} I(\bar{G})^2)$, so $u \in W$. We have seen that $u = vb^{-1}$, where $v^* = v$, and $b \in A$. It follows that $v = ub$ is a unit of $\mathbb{Z}G/I_s$ and $v \in V$. Hence $u_1 = vb^{-1}g$, and $V(\mathbb{Z}G/I_s) = VG$. Normality of V follows from this and the equality $(g^{-1}vg)^* = g^{-1}v^*g$, $g \in G$, $v \in \mathbb{Z}G$.

Put $C = G \cap V$. If $g \in C$, then $g = g^* = g^{-1}$, so $g^2 = 1$ and $g \in A$ since $|G/A|$ is odd. Hence $C = \{a \in A \mid a^2 = 1\}$. Consider the map $f: V \to V(\mathbb{Z}\bar{G})$ defined by $f(v) = \bar{v}$, $v \in V$. If $v \in \operatorname{Ker} f$, then $\bar{v} = \bar{1}$ in $\mathbb{Z}\bar{G}$, so $v = 1 + x$ with $x \in \mathbb{Z}G \cdot I(A)$. By Lemma 7.8(iii), $x \equiv a - 1 (\operatorname{mod} I_s)$ for some $a \in A$, and hence $v = a$ and $\operatorname{Ker} f = A \cap V = C$. Now the image of f is contained in $U(1 + I(\bar{G})^2)$, so V/C is free abelian, by Lemma 7.9(iii). Thus there exists a splitting $\sigma: V \to C$ of the inclusion $C \to V$ of abelian groups.

Because $V \lhd V(\mathbb{Z}G/I_s)$, V is a G-module. We have $V \subseteq W$, an abelian group, and since $A \subseteq W$, A acts trivially on V and thus V is a $\bar{G}$-module. Finally, define $\tau : V \to C$ by

$$\tau(v) = \prod_{x \in \bar{G}} \sigma(v^{x^{-1}})^x.$$

It is clear that τ is a $\bar{G}$-module homomorphism and, since $|G/A|$ is odd,

$$\tau(a) = a^{(G:A)} = a, \qquad \text{for } a \in C.$$

Hence $V = C \times Y$ for some $\bar{G}$-submodule Y of V. Then Y is also a G-module, and because $V(\mathbb{Z}G/I_s) = VG$, we conclude that $Y \lhd V(\mathbb{Z}G/I_s)$ and Y is a normal complement to G in $V(\mathbb{Z}G/I_s)$. Furthermore, $Y \cong V/C$ is isomorphic to a subgroup of $U(1 + I(\bar{G})^2)$, so Y is finitely generated free abelian, by Lemma 7.9. The proof is therefore complete. ∎

8.8   Solvability and nilpotence of $U(\mathbb{Z}G)$

Let G be a finite group. Our aim is to provide necessary and sufficient conditions under which $U(\mathbb{Z}G)$ is solvable. It turns out that the latter is equivalent to $U(\mathbb{Z}G)$ being nilpotent. It fact, we will prove that $U(\mathbb{Z}G)$ is solvable if and only if G is either abelian or a Hamiltonian 2-group. As an easy consequence, it will also be shown that if G is neither abelian nor a Hamiltonian 2-group, then $U(\mathbb{Z}G)$ contains a free subgroup of rank 2. We follow a simplified approach due to Goncalves (1985b).

8.8.1 Lemma   *Let G be a finite group and let e be a central idempotent of $\mathbb{Q}G$. Then $[U(\mathbb{Z}Ge) : U(\mathbb{Z}G)e] < \infty$.*

Proof.   Choose a positive integer m such that $me \in \mathbb{Z}G$. Taking into account that

$$\mathbb{Z}Ge/m\mathbb{Z}Ge \cong (\mathbb{Z}/m\mathbb{Z})Ge$$

we see that $\mathbb{Z}Ge/m\mathbb{Z}Ge$ is finite. Let

$$\mathbb{Z}Ge \to \mathbb{Z}Ge/m\mathbb{Z}Ge$$

be the natural homomorphism. Its restriction to $U(\mathbb{Z}Ge)$ gives a surjective homomorphism

$$f : U(\mathbb{Z}Ge) \to V,$$

where V is the image of f. Since V is finite, we see that

$$H = \operatorname{Ker} f = \{x \in U(\mathbb{Z}Ge) \mid x \equiv e \pmod{m\mathbb{Z}Ge}\}$$

is a normal subgroup of $U(\mathbb{Z}Ge)$ of finite index. It therefore suffices to verify that $H \subseteq U(\mathbb{Z}G)e$.

To this end, fix $h \in H$ and consider the element $x = 1 - e + eh \in \mathbb{Q}G$.

By definition, $h = e + mye$ for some $y \in \mathbb{Z}G$, and therefore

$$x = 1 - e + eh = 1 - e + e(e + mye) = 1 + mye \in \mathbb{Z}G.$$

A similar argument shows that $x' = 1 - e + eh^{-1} \in \mathbb{Z}G$. Since $x' = x^{-1}$ and $xe = eh = h$, the desired conclusion follows. ∎

8.8.2 Lemma   *Let G be a finite group and let e be a central idempotent of $\mathbb{Q}G$ such that $U(\mathbb{Q}Ge) \cong GL_n(R)$, $n > 1$, where R is a ring of characteristic 0. Then, upon identifying $U(\mathbb{Q}Ge)$ with $GL_n(R)$, we have*

$$[GL_n(\mathbb{Z}):(GL_n(\mathbb{Z}) \cap U(\mathbb{Z}G)e)] < \infty.$$

Proof.   Let e_{ij}, $1 \leq i, j \leq n$, be the element of $\mathbb{Q}Ge$ corresponding to the matrix of $M_n(R)$ that has 1 at the position (i, j) and 0 elsewhere. Denote by m a positive integer for which $me_{ij} \in \mathbb{Z}Ge$ for all i, j. Then $\mathbb{Z}Ge$ contains every $n \times n$ matrix over $\mathbb{Z}$ which is congruent to 1 modulo m, and hence $U(\mathbb{Z}Ge)$ contains the subgroup of $GL_n(\mathbb{Z})$ consisting of all such matrices. Thus $[GL_n(\mathbb{Z}):(U(\mathbb{Z}Ge) \cap GL_n(\mathbb{Z}))]$ is finite. On the other hand, by Lemma 8.1,

$$[U(\mathbb{Z}Ge):U(\mathbb{Z}G)e] < \infty$$

and so

$$[(U(\mathbb{Z}Ge) \cap GL_n(\mathbb{Z})):(U(\mathbb{Z}G)e \cap GL_n(\mathbb{Z}))] < \infty,$$

proving the result. ∎

8.8.3 Lemma   *Let G be a group such that $G/Z(G)$ is infinite and contains no non-trivial abelian normal subgroups. Then G is not solvable-by-finite.*

Proof.   Assume that H is a normal solvable subgroup of G of finite index. Since $G/Z(G)$ is infinite, $H \nsubseteq Z(G)$ and so $HZ(G)/Z(G) \neq 1$ is a normal solvable subgroup of $G/Z(G)$ of finite index. But then the last non-trivial term of the derived series of $HZ(G)/Z(G)$ is a non-trivial normal abelian subgroup of $G/Z(G)$, a contradiction. ∎

8.8.4 Lemma   *For all $n > 1$, $GL_n(\mathbb{Z})$ is not solvable-by-finite.*

Proof.   The property of being solvable-by-finite is inherited by subgroups and homomorphic images. Therefore, it suffices to verify that $PSL_2(\mathbb{Z}) = SL_2(\mathbb{Z})/\{\pm I\}$ is not solvable-by-finite. By Lemma 8.3, we need only show that $PSL_2(\mathbb{Z})$ has no non-trivial abelian normal subgroups. Assume by way of contradiction that $A/\{\pm I\}$ is a non-trivial abelian normal subgroup of $PSL_2(\mathbb{Z})$. For each $a \in A$, let $\bar{a}$ be the image of a in $A/\{\pm I\}$. We may choose $a \in A$ with $\bar{a} \neq 1$. Then, given $\gamma \in SL_2(\mathbb{Z})$, we have

$$\overline{(\gamma^{-1}a\gamma)}\bar{a} = \bar{a}\overline{\gamma^{-1}a\gamma}$$

and so, either $(\gamma^{-1}a\gamma)a = a(\gamma^{-1}a\gamma)$ or $(\gamma^{-1}a\gamma)a = -a(\gamma^{-1}a\gamma)$. Now taking

$$\gamma = \begin{bmatrix} 1 & 1 \\ 0 & 1 \end{bmatrix} \quad \text{and} \quad \gamma = \begin{bmatrix} 1 & 0 \\ 1 & 1 \end{bmatrix}$$

and writing

$$a = \begin{bmatrix} x & y \\ z & u \end{bmatrix}$$

with $xu - yz = 1$, we arrive at a contradiction.   ■

8.8.5 Lemma   *Let G be a group with a central subgroup Z of finite index. Then G' is finite and, in particular, the torsion elements of G form a subgroup.*

Proof.   Let us show first that the second statement is a consequence of the first. Indeed, assume that G' is finite. Let $H = \langle h_1, \ldots, h_n \rangle$, where each h_i is of finite order. Then both H/H' and H' are finite and hence H is finite, as required.

To prove that G' is finite, let $[x, y] = x^{-1}y^{-1}xy$ denote a commutator in G and let $x_1, x_2, \ldots, x_n$ be a transversal for Z in G. Setting $c_{ij} = [x_i, x_j]$, we observe that these are, in fact, all the commutators of G. Indeed, if x, $y \in G$, say $x = ux_i$, $y = vx_j$ with $u, v \in Z$, then $[x, y] = [x_i, x_j] = c_{ij}$.

Now fix $x, y \in G$. Since Z is central in G and G/Z has order n, we have $[x, y]^n \in Z$. Hence

$$
\begin{aligned}
[x, y]^{n+1} &= x^{-1}y^{-1}xy[x, y]^n = x^{-1}y^{-1}x[x, y]^n y \\
&= x^{-1}y^{-1}x(x^{-1}y^{-1}xy)[x, y]^{n-1}y \\
&= x^{-1}y^{-2}xy^2y^{-1}[x, y]^{n-1}y \\
&= [x, y^2][y^{-1}xy, y]^{n-1},
\end{aligned}
$$

since conjugation by y being an automorphism of G ensures that

$$y^{-1}[x, y]^{n-1}y = [y^{-1}xy, y^{-1}yy]^{n-1} = [y^{-1}xy, y]^{n-1}.$$

It will next be shown that every element of G' can be written as a product of at most n^3 commutators, and this will prove the result. Assume that $u \in G'$ and $u = c_1 c_2 \ldots c_m$ is a product of m commutators. If $m > n^3$, then since there are at most n^2 distinct c_{ij}, it follows that some c_{ij}, say, $c = [x, y]$, occurs at least $n + 1$ times. We shift $n + 1$ of these successively to the left using

$$
\begin{aligned}
[x_r, x_s][x, y] &= [x, y]c^{-1}[x_r, x_s] \\
&= [x, y][c^{-1}x_r c, c^{-1}x_s c]
\end{aligned}
$$

and derive $u = [x, y]^{n+1}c'_{n+2}c'_{n+3} \ldots c'_m$, where each c'_i is possibly a new commutator. Applying

$$[x, y]^{n+1} = [x, y^2][y^{-1}xy, y]^{n-1}$$

we can write u as a product of $m - 1$ commutators. Hence every element of G' is a product of at most n^3 of the c_{ij}, as required. ∎

Let p be an odd prime, let ε be a primitive pth root of unity in $\mathbb{C}$, and let H be the usual quaternion algebra over $\mathbb{Q}$, i.e.

$$H = \{x + yi + zj + wk \mid i^2 = j^2 = -1,\ ij = -ji = k,\ x, y, z, w \in \mathbb{Q}\}.$$

Put $H_p = \mathbb{Q}(\varepsilon) \otimes_\mathbb{Q} H$, and inside this $\mathbb{Q}$-algebra consider the subring

$$R = \{x + yi + zj + wk \mid x, y, z, w \in \mathbb{Z}[\varepsilon]\}. \tag{1}$$

8.8.6 Lemma With the above notation, $Z(U(R)) = U(\mathbb{Z}[\varepsilon])$ and $U(R)/Z(U(R))$ is infinite.

Proof. Only the second assertion deserves a proof. Deny the statement. Then, by Lemma 8.5, the torsion elements of $U(R)$ form a subgroup. Now since p is an odd prime, the identity

$$X^p + 1 = \prod_{i=0}^{p-1} (X + \varepsilon^i)$$

shows that $1 = (1 + \varepsilon)(1 + \varepsilon^2) \ldots (1 + \varepsilon^{p-1})$. Thus $1 + \varepsilon^2 \in U(\mathbb{Z}[\varepsilon])$ and therefore

$$(1 + \varepsilon i)^{-1} = (1 - \varepsilon i)/(1 + \varepsilon^2) \in U(R).$$

We now show that the product of the torsion units $[(1 - \varepsilon i)/(1 + \varepsilon^2)]j(1 + \varepsilon i)$ and $-j$ has infinite order and this will give a desired contradiction.

To this end, note that

$$[(1 - \varepsilon i)/(1 + \varepsilon^2)]j(1 + \varepsilon i)(-j) = (1 - \varepsilon i)^2/(1 + \varepsilon^2) \in \mathbb{C}$$

and so if this number is a root of unity, then its absolute value is 1. Therefore $|1 - \varepsilon i|^2 = |1 + \varepsilon^2|$. Let us calculate both sides of this equality. Let $\varepsilon = \cos(2\pi/p) + i \sin(2\pi/p)$ and $\bar{\varepsilon} = \cos(2\pi/p) - i \sin(2\pi/p)$. Then

$$|1 - \varepsilon i|^2 = (1 - \varepsilon i)(1 + \bar{\varepsilon} i) = 2(1 + \sin(2\pi/p))$$

and, on the other hand,

$$|1 + \varepsilon^2| = |1 + \cos(4\pi/p) + i \sin(4\pi/p)| = 2\,|\cos(2\pi/p)|.$$

Because $p \geq 3$, the angle $2\pi/p$ belongs to the first or to the second quadrant, and therefore $1 + \sin(2\pi/p) > 1$ and $|\cos(2\pi/p)| < 1$, a contradiction. ∎

As a preliminary to the next lemma, we make the following simple observation. Let L be the subfield of H_p generated by ε and i. Then $H_p = L \oplus Lj$ is a left vector space over L, and the right regular representation of H_p gives an injective homomorphism $\psi : H_p \to M_2(L)$.

$$a \mapsto \psi_a = \begin{bmatrix} x + yi & z + wi \\ -z + wi & x - yi \end{bmatrix},$$

where $a = x + yi + zj + wk$. The determinant of $M_2(L)$ gives us a multiplicative function

$$N(a) = x^2 + y^2 + z^2 + w^2.$$

We are now ready to prove our crucial lemma.

8.8.7 Lemma *Let R be as in eqn (1) above. Then $U(R)$ is not solvable-by-finite.*

Proof. Owing to Lemmas 8.3 and 8.6, it suffices to verify that $U(R)/Z(U(R))$ contains no non-trivial abelian normal subgroups. Assume by way of contradiction that $A/Z(U(R))$ is a non-trivial abelian normal subgroup of $U(R)/Z(U(R))$. Choose $\alpha \in A$ such that its image $\bar{\alpha}$ in $A/Z(U(R))$ is not $\bar{1}$ and let $\gamma \in U(R)$. Then

$$\overline{\gamma^{-1}\alpha\gamma}\,\bar{\alpha} = \bar{\alpha}\overline{\gamma^{-1}\alpha\gamma}, \text{ i.e. } \alpha(\gamma^{-1}\alpha\gamma) = (\gamma^{-1}\alpha\gamma)\alpha\delta,$$

where $\delta \in U(\mathbb{Z}[\varepsilon])$, by Lemma 8.6. Applying the function N to both sides of the last equality, we conclude that $N(\delta) = \delta^2 = 1$, and so $\delta = \pm 1$. It therefore follows that if $\gamma \in U(R)$, then either (a) $\alpha(\gamma^{-1}\alpha\gamma) = (\gamma^{-1}\alpha\gamma)\alpha$ or (b) $\alpha(\gamma^{-1}\alpha\gamma) = -(\gamma^{-1}\alpha\gamma)\alpha$. Let $\alpha = x + yi + zj + wk$ and $\gamma = i$. We shall derive a contradiction in case (a). The proof of the case (b) is similar and therefore will be omitted. Therefore assume (a) holds. We claim that α has one of the following forms:

   (i) $\alpha = x + wk$;
  (ii) $\alpha = x + zj$;
 (iii) $\alpha = zj + wk$;
  (iv) $\alpha = x + yi$.

Indeed, we have $(x + yi)(zj + wk) = (zj + wk)(z + yi)$, or

$$yz = yw = 0.$$

If $y \neq 0$, then $w = z = 0$ and α is of form (iv). Assume that $y = 0$. Conjugating α by j we obtain either

   (a1) $\alpha(j^{-1}\alpha j) = (j^{-1}\alpha j)\alpha$ or   (b1) $\alpha(j^{-1}\alpha j) = -(j^{-1}\alpha j)\alpha$.

Let us assume (a1). Then we obtain $(x + zj)wk = wk(x + zj)$ or $wz = 0$, so either $z = 0$ or $w = 0$. On the other hand, if we assume (b1), then

$$(x + zj)^2 = (wk)^2 \quad \text{or} \quad zx = 0 \quad \text{and} \quad x^2 + w^2 = z^2$$

and from the first equation we obtain that either $x = 0$ or $z = 0$. This substantiates our claim.

Let us assume (i) and let $\gamma = 1 + \varepsilon i$. Then $\gamma^{-1} = (1 - \varepsilon i)/(1 + \varepsilon^2)$ and so either

(a2) $\quad \alpha(\gamma^{-1}\alpha\gamma) = (\gamma^{-1}\alpha\gamma)\alpha \qquad$ or $\qquad$ (b2) $\quad \alpha(\gamma^{-1}\alpha\gamma) = -(\gamma^{-1}\alpha\gamma)\alpha.$

Let us assume (a2). Then we have

$$(1 + \varepsilon^2)(\gamma^{-1}\alpha\gamma)\alpha = [(1 + \varepsilon^2)x^2 - w^2(1 - \varepsilon^2)] + 2\varepsilon w^2 i + 2\varepsilon wxj + 2wxk$$

and

$$(1 + \varepsilon^2)\alpha(\gamma^{-1}\alpha\gamma) = [(1 + \varepsilon^2)x^2 - w^2(1 - \varepsilon^2)] - 2\varepsilon w^2 i + 2\varepsilon xwj + 2wxk.$$

Comparing the coefficients of i in both expressions, we conclude that $4\varepsilon w^2 = 0$ and so $w = 0$. Therefore $\alpha = x \in Z(U(R))$, a contradiction.

Let us assume (b2). Then, comparing the coefficients of 1 and k, we obtain

$$4wx = 0 \qquad \text{and} \qquad (1 + \varepsilon^2)x^2 = w^2(1 - \varepsilon^2).$$

Thus $w = x = 0$ and α is not a unit, a contradiction.

The case (ii) can be treated in entirely similar manner. So let us assume (iii) and let $\gamma = 1 + \varepsilon i$. Let us assume (a2). Then we have

$$(1 + \varepsilon^2)(\gamma^{-1}\alpha\gamma)\alpha = (\varepsilon^2 - 1)(z^2 + w^2) + 2\varepsilon(z^2 + w^2)i$$

and

$$(1 + \varepsilon^2)\alpha(\gamma^{-1}\alpha\gamma) = (\varepsilon^2 - 1)(z^2 + w^2) - 2\varepsilon(z^2 + w^2)i.$$

Comparing the coefficients of i, we obtain $w^2 + z^2 = 0$. If $w \neq 0$, then $(z/w)^2 = -1$ and $\sqrt{-1} \in \mathbb{Q}(\varepsilon)$ and hence $\sqrt{-1} \in \mathbb{Z}[\varepsilon]$, a contradiction. Therefore $w = z = 0$, again a contradiction.

Let us assume (b2). Then, from the equality of the coefficients of 1, we obtain $(\varepsilon^2 - 1)(z^2 + w^2) = 0$, and $z = w = 0$, a contradiction. Finally, note that the case $\alpha = x + yi$ can be handled by conjugation by $\gamma = 1 + \varepsilon j$. So the lemma is true. ∎

We have done most of the work to demonstrate

8.8.8 Theorem (Hartley and Pickel 1980, Sehgal 1978) *Let G be a finite group. Then the following conditions are equivalent*:

(i) $U(\mathbb{Z}G)$ *is solvable*;
(ii) $U(\mathbb{Z}G)$ *is nilpotent*;
(iii) *G is abelian or a direct product of the quaternion group and an abelian group of exponent ≤ 2.*

Proof. If nonabelian G satisfies (iii), then by Theorem 4.3, $U(\mathbb{Z}G) = \pm G$ is a nilpotent group. Thus (iii) implies (ii). Since the implication (ii) $\Rightarrow$ (i) is obvious, we are left to verify that (i) $\Rightarrow$ (iii).

Assume $U(\mathbb{Z}G)$ is solvable and that G is non-abelian. We must show that G is Hamiltonian without elements of order p, where p is any odd prime. Write

$$\mathbb{Q}G = \prod_{i=1}^{r} M_{n_k}(D_i),$$

where each D_i is a division ring of characteristic 0. Suppose that for some k, $1 \leqslant k \leqslant r$, we have $n_k > 1$, and let e be the corresponding central idempotent in the decomposition above. Since $U(\mathbb{Z}G)$ is solvable, it follows that $U(\mathbb{Z}G)e$ is solvable. But then, by Lemmas 8.2 and 5.6.1, $GL_{n_k}(\mathbb{Z})$ is solvable-by-finite contrary to Lemma 8.4. Hence, for any $i \in \{1, \ldots, r\}$, $n_i = 1$ and therefore every idempotent of $\mathbb{Q}G$ is central. Let H be any subgroup of G and let $e = 1/|H| \sum_{h \in H} h$. Then e is central and therefore $H \lhd G$, proving that G is Hamiltonian. Let K be the quaternion group of order 8. It suffices to assume that $G = \langle g \rangle \times K$, where g is of order p, p odd prime, and to derive a contradiction.

To this end, let ε be a primitive pth root of 1 in $\mathbb{C}$. Then

$$\mathbb{Q}(\langle g \rangle \times K) = (\mathbb{Q}\langle g \rangle)K \cong (\mathbb{Q} \times \mathbb{Q}(\varepsilon))K$$
$$\cong \mathbb{Q}K \times \mathbb{Q}(\varepsilon)K$$
$$\cong \mathbb{Q}K \times \mathbb{Q}(\varepsilon)^4 \times (\mathbb{Q}(\varepsilon) \otimes_{\mathbb{Q}} H),$$

where H is the quaternion algebra over $\mathbb{Q}$ and $\mathbb{Q}(\varepsilon)^4$ is the direct product of four copies of $\mathbb{Q}(\varepsilon)$. Let e be the central idempotent of $\mathbb{Q}G$ corresponding to $H_p = \mathbb{Q}(\varepsilon) \otimes_{\mathbb{Q}} H$. Then

$$\mathbb{Z}Ge = \{x + yi + zj + wk \in H_p \mid x, y, z, w \in \mathbb{Z}[\varepsilon]\} = R.$$

Since $U(\mathbb{Z}G)$ is solvable, so is $U(\mathbb{Z}G)e$ and therefore, by Lemmas 8.1 and 5.6.1, $U(R)$ is solvable-by-finite. The latter, however, is contrary to Lemma 8.7 and the result follows. ∎

8.8.9 Corollary (Hartley and Pickel 1980) *Assume that a finite group G is neither abelian nor a direct product of the quaternion group with an abelian group of exponent $\leqslant 2$. Then $U(\mathbb{Z}G)$ contains a free subgroup of rank 2.*

Proof.   First assume that G is Hamiltonian. Then, by hypothesis, G has an element of prime order $p > 2$. Hence, by the proof of Theorem 8.8, $U(\mathbb{Z}G)$ has a homomorphic image that is not solvable-by-finite. Next assume that G is not Hamiltonian. Then there exists a central idempotent e of $\mathbb{Q}G$ such that $\mathbb{Q}Ge \cong M_n(D)$ for some $n > 1$ and some division ring D of characteristic 0. Indeed, otherwise, as we have seen in the proof of Theorem 8.8, G is Hamiltonian. Applying Lemmas 8.2 and 8.4, we see that the homomorphic image $U(\mathbb{Z}G)e$ of $U(\mathbb{Z}G)$ is not solvable-by-finite.

Thus in both cases $U(\mathbb{Z}G)$ is not solvable-by-finite. The desired conclusion is now a consequence of Theorems 5.6.2 and 5.9(i).   ■

8.9   Units in commutative group rings

Throughout this section, all groups are assumed to be *abelian* and all rings to be *commutative*. Given a group G, we write G_0 for the torsion subgroup of G. Our aim is to provide a detailed analysis of the structure of $U(RG)$. In order not to interrupt future discussions at an awkward stage, we first present some general results pertaining to the theory of commutative group rings. Once this has been accomplished, we introduce a class of rings R for which RG has only trivial units of finite order. We then describe $U(RG)$ in case G is torsion-free. Next we investigate conditions under which $U(RG)$ is free modulo torsion, G has a free complement in $V(RG)$, and G has torsion-free complement in $V(RG)$. As an application of the above, we determine the isomorphism class of $U(\mathbb{Z}G)$ for an arbitrary group G. Finally, we discuss the problem of determining precisely the units existing within the ring $\mathbb{Z}G$. For convenience, we split this long section into a number of subsections.

8.9A   General results

Our aim here is to present a number of results pertaining to the general theory of commutative group rings. We begin by describing the nilradical of RG exclusively in terms of R and G.

Throughout, G_p denotes the p-component of G, and supp G stands for the set of all primes p for which $G_p \neq 1$. If S is a subset of RG, we denote by $\langle S \rangle$ the ideal generated by S. In case S is empty $\langle S \rangle$ is to be interpreted as the zero ideal. For any $p \in$ supp G, put

$$N_p(R) = \{r \in R \mid pr \in N(R)\}.$$

8.9.1   Lemma   *Let k and n be positive integers and let p be a prime. Then for any n elements $x_1, x_2, \ldots, x_n$ of a ring R*

$$(x_1 + x_2 + \ldots + x_n)^{p^k} \equiv x_1^{p^k} + x_2^{p^k} + \ldots + x_n^{p^k} \pmod{pR}.$$

Proof.   By an induction argument, we may harmlessly assume that $n = 2$ and $k = 1$. The desired conclusion is now a consequence of the binomial theorem and the fact that for $0 < i < p$, the binomial coefficient

$$\binom{p}{i}$$

is multiple of p.   ■

Observe that if $\alpha = r(g - 1)$, where $r \in N_p(R)$ and $g \in G_p$, then α is a nilpotent element. For let n be such that $g^{p^n} = 1$. By Lemma 9.1, there exists $x \in RG$ such that $\alpha^{p^n} = r^{p^n} p x$. Hence $\alpha^{p^n} \in N(R)G \subseteq N(RG)$, as claimed.

We shall refer to α as a *special nilpotent element*. Our aim is to show that $N(RG)$ is the sum of $N(R)G$ and the ideal generated by the special nilpotent elements. The following preliminary observations will clear our path.

8.9.2 Lemma   *Let R be a ring and let G be a torsion-free group.*

(i)   *If R is reduced, then every element of RG that is integral over R lies in R. In particular, if R is reduced, then so is RG.*

(ii) $N(RG) = N(R)G$.

Proof.

(i)   Suppose that x is a non-zero element in RG such that

$$x^n = r_1 x^{n-1} + r_2 x^{n-2} + \ldots + r_n \qquad (r_i \in R). \qquad (1)$$

We know, from Lemma 4.6, that G is an ordered group. Denote by $g_{\max}$ and $g_{\min}$ the greatest and the smallest elements in Supp x. Because R is reduced, we see that, for any positive integer m, $g_{\max}^m$ is the greatest element in Supp x^m. Suppose that $g_{\max} > 1$. Then, for any $i \in \{0, 1, \ldots, n-1\}$, $g_{\max}^{n-i} > 1$, so that $g_{\max}^n > g_{\max}^i$ and hence

$$g_{\max}^n \in \text{Supp } x^n - \text{Supp } x^i \qquad (0 \leq i < n).$$

This, however, contradicts (1) and so $g_{\max} \leq 1$. A similar argument shows that $g_{\min} \geq 1$. Hence $g_{\max} = g_{\min} = 1$, and so $x \in R$.

(ii)   Owing to (i), the ring $[R/N(R)]G$ is reduced. Hence the natural map

$$RG \rightarrow [R/N(R)]G$$

sends $N(RG)$ to 0. This shows that $N(RG) \subseteq N(R)G$. The opposite inclusion being trivial, the result follows.   ∎

We have now at our disposal all the information necessary to describe the nilradical of RG.

8.9.3 Theorem   (May 1976b)   *Let G be a group and let R be a ring. Then*

$$N(RG) = N(R)G + \langle r(g - 1) \mid r \in N_p(R), g \in G_p \text{ for some } p \in \text{supp } G \rangle.$$

Proof.   For the sake of clarity, we shall divide the proof into two steps.

Step 1:   reduction to the case $N(R) = 0$, R is noetherian, and G finite. Assume the theorem to be true under the above hypothesis. Turning to the general case, suppose $x \in N(RG)$. By looking at the supporting subring of x, we may assume that R (and hence $\bar{R} = R/N(R)$) is noetherian. Let $-: RG \to \bar{R}G$ be the natural map. Then $\bar{x} \in N(\bar{R}G)$, and by looking at $\langle \text{Supp } \bar{x} \rangle$ we have $\bar{x} \in \bar{R}(G_1 \cdot F) = (\bar{R}G_1)F$, where G_1 is finite and F a free subgroup of G. By Lemma 9.2(ii), we may therefore conclude that $\bar{x} \in N(\bar{R}G_1)F$. By assumption, $N(\bar{R}G_1)$ as an ideal is generated by the set of special nilpotent elements. Thus $\bar{x}$ is a $\bar{R}G$-linear combination of special nilpotent elements. Now, if $\bar{r}(g-1)$ is a special nilpotent element of $\bar{R}G$, then so is $r(g-1)$ in RG. Hence $x = y + z$ for some $y \in N(R)G$ and some z belonging to the ideal generated by special nilpotent elements in RG. The reduction step is therefore established.

  Step 2:   proof of the case where $N(R) = 0$, R is noetherian, and G finite. Let x be a non-zero element in $N(RG)$ and let $\{P_1, P_2, \ldots, P_m\}$ be the set of minimal prime ideals of R. Since $N(R) = \cap_{i=1}^{m} P_i = 0$, the natural map

$$\pi: RG \to (R/P_1)G \times \ldots \times (R/P_m)G$$

is an injective homomorphism. We claim that not all the R/P_i have zero characteristic. Suppose false, and let F_i be the quotient field of R/P_i. Then, by Maschke's theorem, F_iG is a direct product of finitely fields; hence $\pi(x) = 0$ and $x = 0$, contrary to our choice of x. Let $p_1, p_2, \ldots, p_t$ be the distinct non-zero characteristics that occur. Then there exist integers $d_1, d_2, \ldots, d_t$ such that $d_1 + d_2 + \ldots + d_t = 1$, and such that for each $i \in \{1, \ldots, m\}$, p_j divides d_i for every j, where $j \neq i$, $1 \leq j \leq m$. Setting $x_i = d_i x$ we have $x = x_1 + x_2 + \ldots + x_t$, where $x_i \in N(RG)$ for each i. By symmetry, it suffices to verify that x_1 belongs to the ideal generated by the special nilpotent elements. To this end, we first note that $p_1 x_1 = 0$. Indeed, $p_1 x_1$ is nilpotent and $p_1, p_2, \ldots, p_t$ divide $p_1 x_1$; hence $\pi(p_1 x_1) = 0$ and $p_1 x_1 = 0$. Write $G = H \cdot K$, where K is a p_1-group and H has trivial p_1-component. Let $\psi: RG \to RH$ be induced by the projection of G onto H along K. Then $\psi(x_1)$ is a nilpotent element whose image in $(R/P_i)G$ is zero provided R/P_i has characteristic zero or p_1. On the other hand, if char R/P_i is p_s $(s \neq 1)$, then $p_s \mid x_1$; hence $p_s \mid \psi(x_1)$, and so $\pi(\psi(x_1)) = 0$. Hence $\psi(x_1) = 0$, and $x_1 \in \text{Ker } \psi = RG \cdot I(K) = RH \cdot RK \cdot I(K) = RH \cdot I(K)$. Write

$$x_i = \sum_{k \in K} y_k(k-1) \qquad (y_k \in RH).$$

It follows from $p_1 x_1 = 0$ that

$$\sum_{k \in K} p_1 y_k(k-1) = 0.$$

This forces, in view of Proposition 1.6, $p_1 y_k = 0$ for all $k \neq 1$. Consequently, p_1 annihilates the coefficients of y_k. From this fact, it is apparent that x_1 is in the ideal generated by the special nilpotent elements. The theorem is therefore established. ∎

As an immediate consequence, we derive the following.

8.9.4 Corollary   *The group ring RG is reduced if and only if the following two conditions hold*:

(i)   *R is reduced*;
(ii) *for all $p \in \text{supp } G$, p is not a zero divisor of R.*

Proof.   Apply Theorem 9.3.   ■

In what follows, we refer to the elements of $\text{Hom}(G, F^*)$ as *characters* of G over F. As a preliminary to the next result, we prove

8.9.5 Lemma   *Let G be a finite group of exponent n, and let F be a field containing a primitive nth root of* 1 *(hence* char $F \nmid |G|$). *For any character χ of G over F, put*

$$e_\chi = |G|^{-1} \sum_{g \in G} \chi(g^{-1})g.$$

Then $\{e_\chi \mid \chi \in \text{Hom}(G, F^*)\}$ is the full system of primitive idempotents of FG, which is also an orthogonal F-basis for FG.

Proof.   Given $g \in G$ and $\chi \in \text{Hom}(G, F^*)$, we have

$$ge_\chi = |G|^{-1} \sum_{h \in G} \chi(h^{-1})hg = \left[|G|^{-1} \sum_{g \in G} \chi(h^{-1}g^{-1})hg\right]\chi(g)$$

$$= \chi(g)e_\chi$$

and therefore

$$e_\chi^2 = \left[|G|^{-1} \sum_{g \in G} \chi(g^{-1})g\right]e_\chi = |G|^{-1} \sum_{g \in G} \chi(g^{-1})(ge_\chi)$$

$$= |G|^{-1} \sum_{g \in G} \chi(g^{-1})\chi(g)e_\chi = e_\chi.$$

Since $ge_\chi = \chi(g)e_\chi$, we also have $FGe_\chi = Fe_\chi \cong F$; hence e_χ is a primitive idempotent. For any $\alpha, \beta \in \text{Hom}(G, F^*)$, $\alpha \neq \beta$, we have $FGe_\alpha \neq FGe_\beta$ since $e_\alpha \neq e_\beta$, and since e_β is the only non-zero idempotent of the field FGe_β. Accordingly,

$$e_\alpha e_\beta \in FGe_\alpha \cap FGe_\beta = 0,$$

which shows that $\{e_\chi \mid \chi \in \text{Hom}(G, F^*)\}$ is an F-linearly independent set. Since F contains a primitive nth root of unity, $\text{Hom}(G, F^*) \cong G$. Hence the number of distinct e_χ is equal to the F-dimension of FG, as required.   ■

In future we shall refer to e_χ as the *primitive idempotent corresponding* to χ. Given a ring R and a natural number n, let R^n denote the direct product of n copies of R. The theorem that follows gives circumstances under which RG is isomorphic to R^n and provides an explicit formula for such an isomorphism.

8.9.6 Theorem (Higman 1940b) *Let G be a finite group of order n and let R be an integral domain. Suppose that R satisfies the following two conditions:*

(a)   *n is a unit of R*
(b)   *R contains a primitive mth root of unity, where m is the exponent G.*

Then there exists an isomorphism $RG \cong R^n$ of R-algebras such that the elements $x = \sum x_g g \in RG$ and $y = (r_1, r_2, \ldots, r_n) \in R^n$ correspond if and only if:

(c) $x_g = (1/n) \sum_{i=1}^n r_i \chi_i(g^{-1})$ *for all $g \in G$;*
(d) $r_i = \sum_{g \in G} x_g \chi_i(g)$ *for all $i \in \{1, 2, \ldots, n\}$.*

Here F is the quotient field of R and $\mathrm{Hom}(G, F^*) = \{\chi_1, \chi_2, \ldots, \chi_n\}$.

Proof. Condition (b) guarantees that $\mathrm{Hom}(G, F^*) \cong G$ and hence $|\mathrm{Hom}(G, F^*)| = n$. We now claim that it suffices to consider the case $R = F$. Indeed, assume that there exists an isomorphism $f : FG \to F^n$ of F-algebras satisfying (c) and (d). Then conditions (c) and (d) together with (a) and (b) ensure that $x \in RG$ if and only if $f(x) \in R^n$. Hence f restricts to an isomorphism $RG \cong R^n$ of R-algebras which also satisfies (c) and (d). This substantiates our claim and we may therefore assume that $R = F$.

Put $e_i = (1/n) \sum_{g \in G} \chi_i(g^{-1}) g$, $1 \leqslant i \leqslant n$. Then, by Lemma 9.5, $\{e_1, \ldots, e_n\}$ is the full system of primitive idempotents of FG, which is also an orthogonal F-basis for FG. Accordingly, the mapping $FG \to F^n$, defined by $x \mapsto (r_1, \ldots, r_n)$ where

$$x = \sum x_g g = \sum_{i=1}^n r_i e_i,$$

is an isomorphism of F-algebras. We are therefore left to verify that x_g and r_i satisfy (c) and (d), respectively.

Bearing in mind that $g e_i = \chi_i(g) e_i$, to deduce (d) we need only multiply both sides in the expression of x by e_i. From the definition of e_i, it also follows that

$$x = \sum x_g g = \sum_{i=1}^n r_i \left[(1/n) \sum_g \chi_i(g^{-1}) g \right]$$

$$= \sum_g \left[(1/n) \sum_{i=1}^n r_i \chi_i(g^{-1}) \right] g$$

and so (c) also holds. ∎

Let G be a finite group and let F be a field whose characteristic does not divide the order of G. Then FG is completely reducible and therefore is isomorphic to a finite direct product of fields of finite degree over F. We wish to describe the fields that occur in the decomposition of FG and to relate their structure to that of G. The precise formulation of this requires a certain amount of notation and the notion of F-conjugacy, which we proceed to develop.

The following notation will be used in the rest of this section:

G a finite group of exponent n

F a field with char $F \nmid |G|$

ε_d a primitive dth root of unity over F

$\hat{F} = F(\varepsilon_n)$

$\hat{G} = \mathrm{Hom}(G, \hat{F}^*)$, the character group of G over $\hat{F}$

$e_\chi = |G|^{-1} \sum_{g \in G} \chi(g^{-1})g$, the primitive idempotent of $\hat{F}G$ corresponding to $\chi \in \hat{G}$

$F(\chi)$ the field obtained from F by adjoining all values of $\chi \in \hat{G}$ or, equivalently,

$$F(\chi) = F(\varepsilon_d), \quad \text{where } d = (G{:}\mathrm{Ker}\,\chi)$$

I_n the multiplicative group consisting of those integers μ, taken modulo n, for which the mapping $\varepsilon_n \mapsto \varepsilon_n^\mu$ determines an automorphism of $\hat{F}$ over F.

Two elements $a, b \in G$ are called *F-conjugate* if $b = a^\mu$ for some $\mu \in I_n$. It is clear that F-conjugacy is an equivalence relation, and so G is the disjoint union of F-conjugacy classes. To clarify the concept, we shall consider the following examples.

Examples

1. Let $F = \mathbb{Q}$. Then any μ that is coprime to n determines an automorphism of $\mathbb{Q}(\varepsilon_n)$ over $\mathbb{Q}$. Thus a and b are $\mathbb{Q}$-conjugate if and only if $\langle a \rangle = \langle b \rangle$.
2. Assume that $F = \mathbb{R}$. Then it is easy to see that a and b are $\mathbb{R}$-conjugate if and only if $b = a$ or $b = a^{-1}$.
3. Assume that F contains a primitive nth root of unity. Then a and b are F-conjugate if and only if $a = b$.

For any $\chi \in \hat{G}$ and $\sigma \in \mathrm{Gal}(\hat{F}/F)$, put $\chi^\sigma = \sigma \circ \chi$. It is clear that $\chi^\sigma \in \hat{G}$. We shall two characters $\alpha, \beta \in \hat{G}$ *F-conjugate* (or simply *conjugate* if no confusion can arise) if

$$\beta = \alpha^\sigma, \quad \text{for some } \sigma \in \mathrm{Gal}(\hat{F}/F).$$

The above definition implies that, if α, $\beta \in \hat{G}$ are F-conjugate, then

$$F(\alpha) = F(\beta) \qquad \text{and} \qquad \text{Ker } \alpha = \text{Ker } \beta.$$

For any $x \in \hat{F}G$ and any automorphism σ of $\hat{F}$, denote by x^σ the image of x under the automorphism of $\hat{F}G$ induced by σ.

Let e_1, e_2 be two primitive idempotents of $\hat{F}G$. We say that e_1 and e_2 are F-*conjugate* if $e_2 = e_1^\sigma$ for some $\sigma \in \text{Gal}(\hat{F}/F)$. It is obvious that F-conjugacy of idempotents and characters are equivalence relations. Note also that for any $\chi \in \hat{G}$ and $\sigma \in \text{Gal}(\hat{F}/F)$,

$$e_\chi^\sigma = e_{\chi^\sigma}.$$

Thus two idempotents e_α, e_β (α, $\beta \in \hat{G}$) are F-conjugate if and only if α and β are F-conjugate.

We have now accumulated all the information necessary to determine the primitive idempotents of FG.

8.9.7 Theorem (Berman 1958) *Let M be a set of all representatives of equivalence classes of F-conjugate characters of G over $\hat{F}$. Given $\chi \in M$, put*

$$\bar{e}_\chi = \sum_{\sigma \in \text{Gal}(F(\chi)/F)} e_\chi^\sigma.$$

Then $\{\bar{e}_\chi \mid \chi \in M\}$ is the full set of primitive idempotents of FG. In particular, $\{FG\bar{e}_\chi \mid \chi \in M\}$ is the full set of non-isomorphic irreducible FG-modules.

Proof. Let $\chi \in M$ and let $d = (G:\text{Ker } \chi)$. Then $\chi(G) = \langle \varepsilon_d \rangle$, $F(\chi) = F(\varepsilon_d)$, and therefore $e_\chi \in F(\varepsilon_d)G$. Now any automorphism of $\hat{F}$ restricts to an automorphism of $F(\varepsilon_d)$ over F. Hence any idempotent which is F-conjugate to e_χ is of the form e_χ^σ for some $\sigma \in \text{Gal}(F(\chi)/F)$. Moreover, all the e_χ^σ are distinct since ε_d is a coefficient of some $g \in \text{Supp } e_\chi$. Hence $\bar{e}_\chi$ is the sum of all idempotents which are F-conjugate to e_χ. Because the set of primitive idempotents of $\hat{F}G$ is a disjoint union of F-conjugate idempotents, it suffices to verify that $\bar{e}_\chi$ is a primitive idempotent of FG.

To this end, we first observe that $\bar{e}_\chi$ is fixed under any automorphism of $F(\chi)G$ induced by that of $F(\chi)$ over F; hence $\bar{e}_\chi \in FG$. Assume next that $\bar{e}_\chi = u + v$ for some idempotents u, v in FG. Since any idempotent of $F(\chi)G$ is a unique sum of primitive idempotents, at least one of the summands of $\bar{e}_\chi$, say u, is such that e_χ occurs in its decomposition. But any F-conjugate to e_χ must also appear in the decomposition of u; hence $u = \bar{e}_\chi$. This shows that $\bar{e}_\chi$ is a primitive idempotent of FG and proves the result. ∎

Let χ be a character of G over $\hat{F}$, i.e. $\chi \in \hat{G}$. Then χ can be viewed as a homomorphism from G to $F(\chi)$. We shall also consider χ as an F-algebra homomorphism

$$\chi : FG \to F(\chi).$$

We are now ready to prove our second major result.

8.9.8 Theorem (Berman 1958, Perlis and Walker 1950)   *Let G be a finite group of exponent n, let F be a field with char $F \nmid |G|$, and let $\hat{F}$ be the nth cyclotomic extension of F. Denote by $\{\chi_1, \chi_2, \ldots, \chi_s\}$ a full set of representatives of equivalence classes of F-conjugate characters of G over $\hat{F}$.*

(i)   *The natural map $FG \to \prod_{i=1}^s F(\chi_i)$ induced by $\chi_i : FG \to F(\chi_i)$ is an isomorphism.*

(ii) *$FG \cong \prod_{d|n} F(\varepsilon_d)^{a_d}$, where $F(\varepsilon_d)^{a_d}$ is the direct product of a_d copies of $F(\varepsilon_d)$, a_d is the number of F-conjugacy classes of G consisting of elements of order d and $(F(\varepsilon_d):F)$ is the number of elements in each of these classes.*

Proof.
  (i)   It follows from Theorem 9.7 that

$$FG = FG\bar{e}_{\chi_1} \oplus \ldots \oplus FG\bar{e}_{\chi_s}.$$

Hence it suffices to show that $y\bar{e}_{\chi_i} \mapsto \chi_i(y)$, $y \in FG$, determines an isomorphism $FG\bar{e}_{\chi_i} \to F(\chi_i)$, $1 \leq i \leq s$. Because $FG\bar{e}_{\chi_i}$ is a field and χ_i a homomorphism, we need only verify that for all $y \in FG$

$$y\bar{e}_{\chi_i} = 0 \text{ implies } \chi_i(y) = 0. \tag{2}$$

Since $ge_{\chi_i} = \chi_i(g)e_{\chi_i}(g \in G)$, it follows that $ye_{\chi_i} = \chi_i(y)e_{\chi_i}(y \in FG)$. Now e_{χ_i} is a primitive idempotent of $\hat{F}G$ which occurs in the decomposition of $\bar{e}_{\chi_i}$ and all such idempotents are F-linearly independent. This proves (2) and therefore (i) is established.

  (ii) Let $d \mid n$ and let m_d be the number of elements in a given F-conjugacy class of G consisting of elements of order d. Then m_d is equal to the number of automorphisms of $F(\varepsilon_d)$ over F induced by those of $F(\varepsilon_n)$ over F. Hence $m_d = (F(\varepsilon_d):F)$. Now put

$$M_d = \{\chi \in \hat{G} \mid \chi(G) = \langle \varepsilon_d \rangle\}, \ \bar{M}_d = \{e_\chi \mid \chi \in M_d\}$$

$$e_d = \sum_{\chi \in M_d} e_\chi.$$

Every element of the group $\mathrm{Gal}(F(\varepsilon_d)/F)$ induces an automorphism of $F(\varepsilon_d)G$ and in this way $\mathrm{Gal}(F(\varepsilon_d)/F)$ acts on the set $\bar{M}_d$ as a permutation group. Note that the size of each orbit under this action is equal to m_d since, for any $\chi \in M_d$, there exists $g \in G$ such that $\chi(g^{-1}) = \varepsilon_d$. Owing to

Theorem 9.7, a typical orbit sum is of the form $\bar{e}_\chi (\chi \in M_d)$. Because

$$FG\bar{e}_\chi \cong F(\chi) = F(\varepsilon_d)$$

we may therefore conclude that $FGe_d \cong F(\varepsilon_d)^{b_d}$, where b_d is the number of orbits. Finally,

$$m_d b_d = |\bar{M}_d| = |M_d| = \phi(d) \times (\text{number of cyclic factor groups of order } d)$$
$$= \phi(d) \times (\text{number of cyclic subgroups of order } d)$$
$$= \text{number of elements of order } d \text{ in } G$$
$$= m_d a_d$$

and so $b_d = a_d$. Since 1 is the sum of pairwise orthogonal e_d's with d ranging over all divisors of n, the result follows. ■

Remark 1:   For any given subring R of F, the isomorphism of (ii) induces an isomorphism

$$RG[\bar{e}_{\chi_1}, \bar{e}_{\chi_2}, \ldots, \bar{e}_{\chi_s}] \rightarrow \prod_{d|n} R[\varepsilon_d]^{a_d}$$

of R-algebras. Therefore, if we identify RG with its image, then $\prod_{d|n} R[\varepsilon_d]^{a_d}$ can be regarded as an integral ring extension of RG. In particular, by taking $F = \mathbb{Q}$, $R = \mathbb{Z}$, and applying the fact that $\mathbb{Z}[\varepsilon_d]$ is integrally closed, we infer that the integral closure of $\mathbb{Z}G$ in $\mathbb{Q}G$ can be identified with

$$\prod_{d|n} \mathbb{Z}[\varepsilon_d]^{a_d}.$$

8.9.9   Corollary   *Let G be a finite group of exponent n. For any $d \mid n$, let a_d be the number of cyclic subgroups of order d in G. Then:*

(i) $\mathbb{Q}G \cong \prod_{d|n} \mathbb{Q}(\varepsilon_d)^{a_d}$.
(ii)   *The integral closure of $\mathbb{Z}G$ in $\mathbb{Q}G$ can be identified with*

$$\prod_{d|n} \mathbb{Z}[\varepsilon_d]^{a_d}.$$

Proof.   Direct consequence of the foregoing remark and Theorem 9.8. ■

8.9.10   Corollary   (Berman 1958, Witt 1952)   *Let F be a field, the characteristic of which does not divide the order of a finite group G. Then the number $n(FG)$ of non-isomorphic irreducible FG-modules is equal to the number of F-conjugacy classes of G. In particular:*

(i) $n(\mathbb{Q}G)$ *is equal to the number of cyclic subgroups in G:*

(ii) $n(\mathbb{R}G)$ *is equal to the number of unordered pairs* $\{g, g^{-1}\}$ *in G or,
equivalently, to* $(1/2)(|G| + t + 1)$, *where t is the number of elements
of order* 2.

Proof.   Note that the number of fields that occur in the decomposition
of FG is the same as the number of non-isomorphic irreducible
FG-modules. The desired conclusion is therefore a consequence of
Theorem 9.8. ∎

8.9B   Units of finite order

Our aim here is to exhibit a class of rings R for which RG has only trivial
units of finite order. We begin by proving the following general fact.

8.9.11   Proposition   (May 1976b)   *Let A be the integral closure of R in
RG. Then* $A = RG_0 + N(RG)$.

Proof.   It is clear that $RG_0 + N(RG) \subseteq A$. To prove the opposite
containment, assume that $\alpha \in A$. By looking at the supporting subgroup
of α, we may assume that G is finitely generated, say $G = H \times F$, where
H is finite and F is free. Setting $S = RH$, it follows from $RG = SF$ that we
may view α as an element of the integral closure of S in SF. Hence the
image of α in $[S/N(S)]F$ is integral over $S/N(S)$. It follows that it lies in
$S/N(S)$, by Lemma 9.2(i). Accordingly, there exists $\beta \in S$ such that
$\alpha - \beta \in N(S)F$,

$$\alpha \in S + N(S)F \subseteq S + N(SF) \subseteq RG_0 + N(RG)$$

as required. ∎

8.9.12   Corollary   *Suppose that R is reduced and that, if G has an
element of prime order p, then p is not a zero divisor in R. Then* RG_0 *is
the integral closure of R in RG. In particular,* RG_0 *is a characteristic
subalgebra of RG.*

Proof.   Direct consequence of Proposition 9.11 and Corollary 9.4. ∎

8.9.13   Corollary   (May 1976b)   *Let G be an arbitrary group and let R
be an integral domain of characteristic* 0. *If G has an element of prime
order p, assume that p is a non-unit of R. Then all units of finite order in
RG are trivial.*

Proof.   Let $u = \sum u_g g \in RG$ be a unit of finite order. Since u is integral
over R, $u \in RG_0$ by virtue of Corollary 9.12. Hence, by looking at the
supporting subgroup of u, we may assume that G is finite. The desired
conclusion is therefore a consequence of Corollary 3.5(ii). ∎

Can the hypotheses of the preceding corollary be weakened without
sacrificing the conclusion? Here are some pertinent examples.

Example 1 (May 1976b):   let us show that the hypothesis that R is an integral domain cannot be replaced directly by the hypothesis that R is indecomposable. Assume that G is a cyclic group of order 3, with generator g, and let ε be a primitive cube root of unity over $\mathbb{Z}$. Let R be $\mathbb{Z}[\varepsilon][X]$ modulo the ideal generated by $(3X - 1)(X - 1)$. Taking into account that $\mathbb{Z}[\varepsilon]$ is a *UFD* and that $3X - 1$ and $X - 1$ are both contained in the proper ideal generated by 2 and $X - 1$, it is not hard to see that R is indecomposable. Furthermore, char $R = 0$ and no rational prime is a unit of R. Let r be the natural image of X in R. Then

$$u = r + (r - 1)\varepsilon^2 g + (r - 1)\varepsilon g^2$$

is a non-trivial unit of order 2.

Example 2 (Saksonov 1971):   here we demonstrate that the condition that the orders of elements of G are non-units of R is indispensable. Let G be a cyclic group of order 4, with generator g, and let $R = \mathbb{Z}[1/2, i]$. Then

$$u = (1/2)[(1 - i)g + (1 + i)g^3]$$

is a non-trivial unit of order 2.

Example 3:   the aim of this example is to illustrate that the requirement char $R = 0$ cannot be dropped. Let G be a cyclic group of order 4, with generator g, and let $R = \mathbb{Z}/2\mathbb{Z}$. Then $U(RG)$ is finite, $(g - g^2)^4 = 0$, and therefore $1 + g - g^2$ is a non-trivial unit of finite order.

8.9C   Torsion-free groups

Let G be a torsion-free group and R an arbitrary ring. The aim of this section is to describe $U(RG)$ exclusively in terms of R and G. We shall also present necessary and sufficient conditions under which $U(RG)$ is finitely generated.

8.9.14 Theorem (Karpilovsky 1983a)   *Let R be a ring and let G be a torsion-free group. For each $g \in G$ and each idempotent $e \in R$, put*

$$u_{e,g} = eg + (1 - e).$$

These elements are units in RG; denote by T the multiplicative group they generate.

(i)   *$U(RG) = U(R) \times T \times [1 + N(R) \cdot I(G)]$.*
(ii)   *An element $u \in RG$ is a unit if and only if u is of the form*

$$u = r(v_1 g_1 + \ldots + v_k g_k)\left[1 + \sum_{g \in G} x_g(g - 1)\right]$$

where $r \in U(R)$, $x_g \in N(R)$, $g_i \in G$, and the v_i are orthogonal idempotents of R with sum 1.

(iii)   *Suppose R is a direct product of finitely many indecomposable rings (e.g. R is noetherian) and let $e_1, e_2, \ldots, e_m$ be the primitive idempotents corresponding to the decomposition of R. Then*

$$T = e_1G + e_2G + \ldots + e_mG \cong G \times G \times \ldots \times G \ (m \ times).$$

Proof. That T is a subgroup of $V(RG)$ is a consequence of the equalities

$$\text{aug}(u_{e,g}) = 1, \qquad u_{e,g}u_{e,g^{-1}} = 1.$$

Let $v_1, v_2, \ldots, v_k$ be orthogonal idempotents of R with sum 1. Then, for any $g_1, g_2, \ldots, g_k \in G$, $v_1g_1 + \ldots + v_kg_k \in T$, since

$$v_1g_1 + v_2g_2 + \ldots + v_kg_k = u_{v_1,g_1}u_{v_2,g_2} \ldots u_{v_k,g_k}.$$

Conversely, a straightforward induction on the number of factors shows that any element of T is of the above form. This proves (iii), by expressing v_i as a sum of primitive idempotents and observing that

$$\begin{cases} G \times G \times \ldots \times G \to e_1G + e_2G + \ldots + e_mG \\ (g_1, g_2, \ldots, g_m) \mapsto e_1g_1 + e_2g_2 + \ldots + e_mg_m \end{cases}$$

is an isomorphism. The above also proves that (ii) is a consequence of (i).

   To prove (i) (and hence the result), we first note that $N(R) \cdot I(G)$ is a nil ideal contained in $I(G)$. Thus $1 + N(R) \cdot I(G)$ is a subgroup of $V(RG)$ and so we need only show that

$$V(RG) = T \times [1 + N(R) \cdot I(G)].$$

To this end, assume that

$$v_1g_1 + \ldots + v_kg_k - 1 \in N(R) \cdot I(G).$$

Then $v_1(g_1 - 1) + \ldots + v_k(g_k - 1) \in N(R) \cdot I(G)$; hence there is a positive integer n such that $v_i(g_i - 1)^n = 0$, $1 \leq i \leq k$. This certainly implies $g_i = 1$, since otherwise

$$g_i^n \in \text{Supp } v_i(g_i - 1)^n.$$

Thus we must have

$$T \cap (1 + N(R) \cdot I(G)) = 1.$$

Suppose u is a unit of RG. Then, by looking at the supporting subrings of u and u^{-1}, we conclude that there is a finitely generated subring R_1 of R such that u and u^{-1} are in R_1G. Therefore, to prove that

$$V(RG) = T \cdot (1 + N(R) \cdot I(G))$$

we may assume that R is noetherian.

   Let $e_1, e_2, \ldots, e_m$ be the primitive idempotents of R, let $x \in V(RG)$, let $\bar{R} = R/N(R)$, and let $-: RG \to \bar{R}G$ be the natural map. Because the idempotents of $\bar{R}$ will lift uniquely to those of R, each of the idempotents $\bar{e}_i$, $1 \leqslant i \leqslant m$, is primitive. Hence $\bar{R}\bar{e}_i$ is both indecomposable and reduced. Now $\overline{e_i x} = \bar{e}_i \bar{x}$ is a unit of $(\bar{R}\bar{e}_i)G$, the augmentation of which is $\bar{e}_i$, the identity element of $\bar{R}\bar{e}_i$. By Corollary 4.8, we may therefore conclude that $\overline{e_i x} = \bar{e}_i \bar{g}_i$ for a suitable $g_i \in G$, $1 \leqslant i \leqslant m$. It follows, from the fact that $N(R) \cdot I(G)$ is the ideal of $N(R)G$ consisting of elements of zero augmentation, that there is a $t_i \in N(R) \cdot I(G)$ such that

$$e_i x = e_i g_i + t_i \qquad (1 \leqslant i \leqslant m).$$

Hence, $x = e_1 g_1 + \ldots + e_m g_m + t$, where $t = t_1 + \ldots + t_m \in N(R) \cdot I(G)$. This shows that

$$x(e_1 g_1 + \ldots + e_m g_m)^{-1} \in 1 + N(R) \cdot I(G)$$

as required. ∎

The rest of this section is devoted to finite generation of unit groups.

8.9.15   Theorem   (Karpilovsky 1983a)   *Let G be a non-identity torsion-free group and let R be an arbitrary ring. Then $U(RG)$ is finitely generated if and only if the following four conditions hold:*

(i)   *$U(R)$ is finitely generated;*
(ii) *G is finitely generated;*
(iii) *R is reduced;*
(iv) *R is a direct product of finitely many indecomposable rings.*

Proof.   Assume that (i)–(iv) hold. Then, by Theorem 9.14.

$$U(RG) = U(R) \times (\text{finitely many copies of } G),$$

so $U(RG)$ is finitely generated.

Conversely, assume that $U(RG)$ is finitely generated. Then (i) and (ii) hold, since $U(R)$ and G are subgroups of $U(RG)$. Note that R is reduced if and only if any finitely generated subring of R is reduced. Because for any subring S of R, $U(SG)$ is finitely generated, to prove (iii) we may assume that R is finitely generated.

Assume by way of contradiction that $N(R) \neq 0$. Since R is noetherian, there exists $s \geqslant 2$ such that $N(R)^s = 0$ but $N(R)^{s-1} \neq 0$. It therefore follows that $[N(R)^{s-1}G]^2 = 0$; hence the mapping $x \mapsto 1 + x$ from the additive group of $N(R)^{s-1} \cdot I(G)$ to $1 + N(R)^{s-1} \cdot I(G)$ is an isomorphism. Now every element in $N(R)^{s-1} \cdot I(G)$ can be uniquely written in the form

$$r(g_1 - 1) + \ldots + r_n(g_n - 1) \qquad (r_i \in N(R)^{s-1}, g_i \in G).$$

Therefore the additive group of $N(R)^{s-1} \cdot I(G)$ is isomorphic to a direct sum of $|G|$ copies of $N(R)^{s-1}$, and hence cannot be finitely generated.

This proves that $U(RG)$ is not finitely generated and provides the desired contradiction. Thus R is reduced.

To prove (iv), we again argue by contradiction. So assume that for any integer $n \geqslant 2$, there is a system $\{e_1, \ldots, e_n\}$ of orthogonal idempotents of r with sum 1. Taking into account that $e_1 G + \ldots + e_n G$ is a subgroup of $U(RG)$ which is isomorphic to a direct product of n copies of G, we conclude that $U(RG)$ has infinite torsion-free rank. This, however, is contrary to our assumption that $U(RG)$ is finitely generated. The theorem is therefore established. ∎

Remark 1: the foregoing proof shows that for an arbitrary $G \neq 1$, conditions (i), (ii) and (iv) are necessary in order that $U(RG)$ be finitely generated. However, condition (iii) is not necessary in case G is finite (e.g. take $R = \mathbb{Z}/4\mathbb{Z}$).

Remark 2: the preceding theorem yields a device for manufacturing finitely generated rings whose unit groups are not finitely generated— simply take $R = \mathbb{Z}/4\mathbb{Z}$ and G a free group of finite rank.

For the rest of this section, we abandon the assumption that G is torsion-free. The following is interesting and apparently difficult:

Problem.   Determine necessary and sufficient conditions for $U(RG)$ to be finitely generated.

The following theorem provides a solution for the case where R is a finitely generated ring.

8.9.16   Theorem   (Karpilovsky 1983a)   *Suppose R is a finitely generated ring and G an arbitrary non-identity group.*

(i)   *G finite: $U(RG)$ is finitely generated if and only if for all $p \in \operatorname{supp} G$, $N_p(R)$ is a finitely generated additive group.*

(ii)   *G infinite: $U(RG)$ is finitely generated if and only if the following three conditions hold;*
   (a)   *G is finitely generated,*
   (b)   *R is reduced, and*
   (c)   *for all $p \in \operatorname{supp} G$, p is not a zero divisor of R.*

Proof.   Suppose momentarily that R and G are arbitrary. Then, by Theorem 9.3,

$$N(RG) = N(R)G + \langle r(g-1) \mid r \in N_p(R), g \in G_p \text{ for some } p \in \operatorname{supp} G \rangle.$$
(3)

Because $N_p(R)G$ is an ideal of RG which contains $N(R)G$, it follows that

$$N(RG) \subseteq \sum_{p \in \operatorname{supp} G} N_p(R)G.$$

Note also that, by Corollary 7.2.8, if R is finitely generated, then $U(R)$ is finitely generated if and only if so is the additive group of $N(R)$.   (4)

Suppose now that G is finite, say $G = \{g_1, g_2, \ldots, g_n\}$, and let P denote the prime subring of R. Because R is finitely generated, there exists $\alpha_1, \alpha_2, \ldots, \alpha_m$ in R such that $R = P[\alpha_1, \ldots, \alpha_m]$. Therefore we have

$$RG = P[\alpha_1, \ldots, \alpha_m, g_1, \ldots, g_n, g_1^{-1}, \ldots, g_n^{-1}]$$

and thus RG is a finitely generated ring. The additive group of $N_p(R)G$, being a direct sum of n copies of $N_p(R)$, is finitely generated if and only if $N_p(R)$ is. Note further that if $1 \neq g \in G_p$, then

$$\{r(g - 1) \mid r \in N_p(R)\}$$

is an additive copy of $N_p(R)$ contained in $N(RG)$. Hence the additive group of $N(RG)$ is finitely generated if and only if $N_p(R)$ is, for all $p \in \operatorname{Supp} G$. This proves the case where G is finite, by appealing to (4).

Now assume that G is infinite. To prove sufficiency, write $G = G_0 \times F$, where F is free of finite rank and G_0 is finite, and put $R_1 = RG_0$. By (b) and (c), if $p \in \operatorname{supp} G_0$, then $N_p(R) = 0$. It follows from (3) that $N(RG) = 0$ (hence R_1 is reduced) and, from the finite case, that $U(R_1)$ is finitely generated. Since R_1 is a finitely generated ring, R_1 is noetherian, and hence a direct product of finitely many indecomposable rings. But $RG = R_1 F$ and hence, by Theorem 9.15, $U(RG)$ is finitely generated.

Conversely, assume that $U(RG)$ is finitely generated. Then so is G, say $G = G_0 \times F$. Because $RG = (RG_0)F$, Theorem 9.15 tells us that $N(RG_0) = 0$, which implies (b) and (c), by virtue of (3). So the theorem is true. ■

8.9D *Splitting unit groups*

Our main concern here is to discover conditions under which the group ring RG satisfies one of the following properties;

(a) G has a torsion-free complement in $V(RG)$;
(b) G has a free complement in $V(RG)$;
(c) $U(RG)$ is free modulo torsion.

That $V(\mathbb{Z}G)$ satisfies (a) is nearly obvious. This follows from the following simple observation, due to Dennis (1976), which shows that

$$M = \left\{ \sum x_g g \in V(\mathbb{Z}G) \,\middle|\, \prod g^{x_g} = 1 \right\}$$

is a torsion-free complement of G in $V(\mathbb{Z}G)$. Indeed, let

$$f : V(\mathbb{Z}G) \to \mathbb{Z}G / I(\mathbb{Z}, G)^2$$

be the restriction of the natural map $\mathbb{Z}G \to \mathbb{Z}G/I(\mathbb{Z}, G)^2$. If $x = \sum x_g g$ is in $V(\mathbb{Z}G)$, then $x - 1 = \sum x_g(g - 1)$, and so $\operatorname{Ker} f = M$, by virtue of Proposition 3.9(ii). Evidently, $G \cap M = 1$. Since, by Proposition 3.9(ii),

$$x \equiv \prod g^{x_g} (\operatorname{mod} I(\mathbb{Z}, G)^2)$$

it follows that

$$f(x) = f\left(\prod g^{x_g}\right) \in f(G),$$

whence the claim.

The above argument, however, does not tell us whether M is free, nor can it be extended to a larger class of rings. In this section we shall develop a different approach to the problem.

Let us first of all illustrate that, in general, even a weaker form of (a) does not hold. Namely, it will be shown that G need not be a direct factor of $V(RG)$. For convenience, let us recall the following piece of information.

Let p be a prime number and A an abelian group. Inductively define A_α for every ordinal α by

$$A_0 = A, \quad A_{\alpha+1} = A_\alpha^p \quad \text{and} \quad A_\alpha = \bigcap_{\beta<\alpha} A_\beta, \quad \text{if } \alpha \text{ is a limit ordinal.}$$

This defines a descending chain of subgroups whose intersection we denote by A_∞. The least ordinal γ for which $A_\gamma = A_\infty$ is called the *p-length* of A. It is clear that A_∞ is the maximal p-divisible subgroup of A. If $a \in A_\infty$, we say that a has p-height ∞. If $a \notin A_\infty$, the p-height of a is the largest ordinal α such that $a \in A_\alpha$.

8.9.17 Proposition (May 1986) *Let $R = \mathbb{Z}[1/p]$, let A be a p-group, and let $\alpha \in V(RA)$ have p-height at least ω in $V(RA)$. Then the p-height of α is in fact ∞.*

Proof. By examining how the coefficients of α are related to the values of characters, we see that the torsion subgroups of RA and $\mathbb{Q}A$ are the same. Hence we may assume that $R = \mathbb{Q}$. The result will follow provided we show that if $\alpha \in V(\mathbb{Q}A)$ has p-height at least ω, then there exists $\beta \in V(\mathbb{Q}A)$ such that $\alpha = \beta^p$, and β has p-height at least ω in $V(\mathbb{Q}A)$. Choose a finite subgroup $B \subseteq A$ such that α and a pth root of α lie in $\mathbb{Q}B$. We know that

$$\mathbb{Q}B \cong F_1 \times F_2 \times \ldots \times F_r$$

for certain fields F_i, $1 \le i \le r$. Regard α as lying in this product, and choose β as follows. If α has ith coordinate 1, let β have ith coordinate 1.

This, of course, assures that we are dealing with normalized units. If the coordinate of α is distinct from 1, then our choice of B ensures that we may take the coordinate of β to be a pth root. It suffices to verify that, given a positive integer k, we can always find a p^kth root of β in $V(\mathbb{Q}A)$. Choose a finite subgroup $C \supseteq B$ such that a p^{k+1}th root of α lies in $\mathbb{Q}C$. Then

$$\mathbb{Q}C \cong L_1 \times L_2 \times \ldots \times L_s$$

for certain fields L_j, $1 \leqslant j \leqslant s$. For every coordinate of β that is 1, we select 1 as the coordinate for a p^kth root. Now consider a coordinate of β which is distinct from 1, say in L_j. This comes from a coordinate of β in some F_i which is distinct from 1. Therefore the coordinate of α in F_i will be distinct from 1, and hence also in L_j. By choice of C, this coordinate of α has a p^{k+1}th root in L_j. Consequently, the coordinate of β must have a p^kth root in L_j. Thus β has a p^kth root. ∎

8.9.18 Proposition (May 1986) *Let R be a ring in which a rational prime p is a unit. Then there exists a p-group A such that A is not a direct factor of $V(RA)$. Moreover, if $R = \mathbb{Q}$, then A is a pure subgroup of $V(\mathbb{Q}A)$.*

Proof. Let A be the abelian group with generators $\{t_i \mid i \geqslant 0\}$ subject to relations

$$t_0^p = 1 \quad \text{and} \quad t_i^{p^i} = t_0 \quad (i \geqslant 1).$$

Then A is a p-group such that t_0 has p-height ω in A. Since p is a unit of R, there is a ring homomorphism $\mathbb{Z}[1/p] \to R$ which induces a homomorphism

$$V(\mathbb{Z}[1/p]A) \to V(RA).$$

Owing to Proposition 9.17, the p-height of t_0 in $V(\mathbb{Z}[1/p]A)$ is ∞, hence also in $V(RA)$, since homomorphisms cannot decrease height. But it is evident that the height of an element in a direct factor is the same as its height in the full group. Hence A cannot be a direct factor of $V(RA)$. The last assertion is a consequence of a general result to be proved below. ∎

8.9.19 Proposition (May 1986) *Any abelian group A is a pure subgroup of $V(\mathbb{Q}A)$.*

Proof. We need only verify that A is p-pure in $V(\mathbb{Q}A)$ for any prime p. Let $a \in A$, and assume that a has a p^k-th root in $V(\mathbb{Q}A)$. Then there is a subgroup of A of form $T \cdot F$ such that T is a finite subgroup, F is free of finite rank, and the p^k-th root of a lies in $\mathbb{Q}(TF)$. We may write $a = tf$ with $t \in T$, $f \in F$. The projection $T \cdot F \to F$ induces $V(\mathbb{Q}(TF)) \to V(\mathbb{Q}F)$.

But $V(\mathbb{Q}F) = F$ by Corollary 4.8, and hence f has a p^k-th root in F. Therefore it suffices to verify that t has a p^k-th root in T. The projection $TF \to T$ shows that t has a p^k-th root in $\mathbb{Q}T$. If p does not divide the order of t, then we can extract a p^k-th root in T. Therefore we may assume that t has finite p-height m in T. Then there is a homomorphism of $\mathbb{Q}T$ onto $\mathbb{Q}(\varepsilon)$, where ε is a root of unity such that $\varepsilon^{p^m} \neq 1$, and t maps to ε^{p^m}. Because the image of t must have a p^k-th root in $\mathbb{Q}(\varepsilon)$, we deduce that $k \leqslant m$. Thus t has a p^k-th root in T and the result follows. ∎

Our next aim is to discover a class of rings R for which the properties (a), (b), and (c) mentioned at the beginning of the section hold. The following two group-theoretic properties will clear our path.

8.9.20 Lemma

(i) *Let B be a pure subgroup of a group A. Then, for all $n > 0$, A/B^n splits over B/B^n.*

(ii) *Let G_0 be bounded and let $G_0 \times H$ be a pure subgroup of G. Then there exists a subgroup L of G containing H and such that $G = G_0 \times L$.*

Proof.

(i) We shall show that if C is a subgroup of B, then B/C is pure in A/C. Since B/B^n is bounded, the required assertion will follow by appealing to Lemma 4.1.9(ii). Let $x = a^m C = bC \in (A/C)^m \cap (B/C)$. Since $a^m \in B$ and since B is pure in A, $a^m = b_1^m$ for some $b_1 \in B$. It follows that $x = (b_1 C)^m \in (B/C)^m$, proving that B/C is pure in A/C.

(ii) By hypothesis, there is an $n \geqslant 1$ such that $G_0^n = 1$. Setting $D = G_0 \times H$, it follows from (i) that there is a subgroup $K \supseteq D^n$ such that

$$G/D^n = D/D^n \times K/D^n.$$

Put $L = H \cdot K$; then obviously $G = G_0 \cdot L$. Let $x = hk \in L \cap G_0$, $h \in H$, $k \in K$. Then $xh^{-1} \in D \cap K = D^n = H^n$, so $x \in H \cap G_0 = 1$ and so $G = G_0 \times L$. ∎

The next result provides an important tool for subsequent investigations.

8.9.21 Theorem (May 1976a) *A group G splits over its torsion subgroup G_0 if and only if there exists a countable ascending sequence*

$$A_1 \subseteq A_2 \subseteq \cdots \subseteq A_n \subseteq \cdots$$

of subgroups of G such that the following three conditions hold:

(i) $G = \bigcup_n A_n$;
(ii) *for every* $n = 1, 2, \ldots$, *the torsion subgroup* $t(A_n)$ *of* A_n *is bounded*;
(iii) *for every* $n = 1, 2, \ldots$, $t(G/A_n) = (G_0 \cdot A_n)/A_n$.

Proof. Assume that there is a subgroup F of G such that $G = G_0 \times F$. Then the groups $A_n = G_0[n!] \times F$, $n = 1, 2, \ldots$, satisfy conditions (i)–(iii). Conversely, suppose that (i)–(iii) hold. Then, by (ii) and Theorem 4.1.10, we infer that there is a subgroup G_1 of A_1 such that $A_1 = t(A_1) \times G_1$. It therefore suffices to exhibit a chain

$$G_1 \subseteq G_2 \subseteq \ldots \subseteq G_n \subseteq \ldots$$

of torsion-free subgroups of G such that $A_i = t(A_i) \times G_i$, $i = 1, 2, \ldots$ (for then $G = G_0 \times B$, where $B = \cup G_n$).

Suppose that, for all $i \leq n$, we have already obtained decompositions $A_i = t(A_i) \times G_i$ such that $G_1 \subseteq G_2 \subseteq \ldots \subseteq G_n$. Set $B_{n+1} = t(A_{n+1}) \times G_n$ and note that

$$(G_0 \cdot A_n) \cap A_{n+1} = B_{n+1}.$$

Therefore, we may apply (iii) to deduce that the group

$$A_{n+1}/B_{n+1} \cong (G_0 \cdot A_{n+1})/(G_0 \cdot A_n)$$
$$\cong [(G_0 \cdot A_{n+1})/A_n]t(G/A_n)$$

is torsion-free. The latter ensures that B_{n+1} is a pure subgroup of A_{n+1}. Applying Lemma 9.20(ii), we conclude that there is a subgroup $G_{n+1} \supseteq G_n$ such that

$$A_{n+1} = t(A_{n+1}) \times G_{n+1}$$

as required. ∎

8.9.22 Lemma *Let G be a finite group of exponent m, and let F be a field containing a primitive mth root of unity. Suppose that H is a subgroup of G of index n and that χ is a character of H over F.*

(i) *There exists exactly n characters $\chi_1, \ldots, \chi_n$ of G extending χ.*
(ii) *If e (respectively e_i) is the idempotent corresponding to χ (respectively, χ_i), then $e = e_1 + e_2 + \ldots + e_n$. Moreover,*

$$FGe_i \cong FHe$$

as FH-algebras.

Proof
(i) The group $\mathrm{Hom}(G/H, F^*)$ can be identified with the subgroup of $\mathrm{Hom}(G, F^*)$ consisting of all characters α for which $H \subseteq \mathrm{Ker}\,\alpha$. Furthermore, any two characters of G restrict to the same character of H

if and only if they belong to the same coset of $\text{Hom}(G/H, F^*)$. This is so since $\alpha \mid H = \beta \mid H$ if and only if $(\alpha\beta^{-1})(h) = 1$ for all $h \in H$. Since $G \cong \text{Hom}(G, F^*)$, it follows that any character of H is a restriction of a character of G and that there are exactly n characters of G restricting to χ.

(ii) Let v be the primitive idempotent corresponding to a given character $\theta \in \text{Hom}(G, F^*)$. If $v \in GFe$, then $ve = v$; therefore for any $h \in H$,

$$\theta(h)v = hv = v(he) = v(\chi(h)e) = \chi(h)v.$$

This means that θ is an extension of χ; hence $\theta = \chi_i$ and $v = e_i$ for some i, $1 \le i \le n$. Bearing in mind that for any $h \in H$, $\chi(h) = \chi_i(h)$ and $he_i = \chi_i(h)e_i$, we also derive

$$ee_i = \left[|H|^{-1} \sum_h \chi(h^{-1})h \right] e_i = |H|^{-1} \sum_h \chi(h^{-1})\chi_i(h)e_i = e_i \in FGe.$$

We have therefore shown that $\{e_1, \ldots, e_n\}$ is the set of all primitive idempotents of FG belonging to FGe; hence $e = e_1 + e_2 + \ldots + e_n$. Finally, note that each element of FGe_i (respectively, FHe) can be uniquely written in the form λe_i (respectively, λe) for some $\lambda \in F$. It is now straightforward to verify that the mapping $\psi : FGe_i \to FHe$ defined by $\psi(\lambda e_i) = \lambda e$ $(\lambda \in F)$ is an isomorphism of FH-algebras. This completes the proof. ∎

For future use, we now record the following useful observation.

8.9.23   Lemma   *Let G be the product of its subgroups G_1 and G_2, and let $H = G_1 \cap G_2$. Then, for any ring R,*

$$RG \cong RG_1 \underset{RH}{\otimes} RG_2$$

as RH-algebras.

Proof.   Let T_i be a transversal for H in G_i containing 1, $i = 1,2$. Then, by Proposition 1.6, RG_i is a free RH-module with T_i as a basis. Consequently, $RG_1 \otimes_{RH} RG_2$ is a free RH-module with $\{\alpha \otimes \beta \mid \alpha \in T_1, \beta \in T_2\}$ as a basis. Because $T_1 T_2$ is a transversal for H in G, it is an RH-basis for RG, again by Proposition 1.6. Therefore, the mapping $f : RG \to RG_1 \otimes_{RH} RG_2$, which is the RH-linear extension of $\alpha\beta \mapsto \alpha \otimes \beta (\alpha \in T_1, \beta \in T_2)$, is an RH-isomorphism, preserving identity elements. Since $f(xy) = f(x)f(y)$ for all $x,y \in T_1 T_2$, f is in fact an isomorphism of RH-algebras, as required. ∎

As a final preparatory result, we prove

8.9.24   Proposition   *Let G be a torsion group, let H be a subgroup of finite index n, and let F be a field. For every prime number p dividing n, and for every positive integer r, assume that F contains a primitive p^r-th root of unity. Then FG and $(FH)^n$ are isomorphic as FH-algebras.*

Proof. Note first that any p-element in G/H is of the form gH for some p-element g in G. We may therefore choose a finite subgroup G_1 of G such that $G = G_1 \cdot H$, and such that $p \mid n$ if $p \mid |G_1|$. Setting $H_1 = H \cap G_1$, it follows that $(G_1 : H_1) = n$. We next argue that FG_1 and $(FH_1)^n$ are isomorphic FH_1-algebras. Indeed, let $m = |H_1|$. Then, by Lemma 9.5, we may write $1 = e_1 + \ldots + e_m$, where the e_i's are primitive pairwise orthogonal idempotents of FH_1. Consequently, Lemma 9.22 may be employed to infer that each e_i is a sum of n primitive idempotents e_{ij} $(1 \leq j \leq n)$ of FG_1, such that for any j, $FG_1 e_{ij} \cong FHe_i$ as FH_1-algebras. Thus, putting $f_j = e_{ij} + \ldots + e_{mj}$ $(1 \leq j \leq n)$, we deduce that

$$(FG_1)f_j \cong FH_1$$

as FH_1-algebras. Because $1 = f_1 + f_2 + \ldots + f_n$ is a decomposition of 1 into the sum of pairwise orthogonal idempotents, it follows that $FG_1 \cong (FH_1)^n$ as FH_1-algebras. Invoking Lemma 9.23, we can now deduce that

$$FG \cong FH \underset{FH_1}{\otimes} FG_1 \cong FH \underset{FH_1}{\otimes} (FH_1)^n \cong (FH)^n.$$

Since the image of $x \in FH$ under the above isomorphism $FG \to (FH)^n$ is $(x, x, \ldots, x)$, the result follows. ∎

After this digression, we shall now confine our attention to the unit group of RG. Our first goal is to find sufficient conditions in order that $U(RG)$ be free modulo torsion.

Let A be an R-algebra. By the *idempotent subalgebra* of A we understand the R-subalgebra of A generated by the idempotent elements. Every element of the idempotent subalgebra is an R-linear combination of idempotent elements in A.

8.9.25 Theorem (May 1976b) *Let R be a finitely generated integral domain, and let G be a group which is free modulo torsion and of finite torsion-free rank. If R has characteristic p, assume that G has no elements of order p. The $U(RG)$ is free modulo torsion.*

Proof. By hypothesis, there exists a free group F of finite rank such that $G = G_0 \times F$. We know that $RG = (RF)G_0$ and that RF is a finitely generated integral domain such that G_0 has no elements of order p if char $RF = p$. Consequently, without restricting generality, we may assume that G is torsion.

Denote by $\bar{R}$ the integral domain obtained by adjoining all roots of unity to R, let K be the quotient field of $\bar{R}$, and let A be the idempotent $\bar{R}$-subalgebra of KG. Assume that $x \in \bar{R}G$. Setting $H = \langle \operatorname{Supp} x \rangle$, it follows that H is finite, say of order m. By looking at the image $(y_1, \ldots, y_m)$ of x under the isomorphism $KH \to K^m$ described in

Theorem 9.6, we infer that $y_i \in \bar{R}$, $1 \leq i \leq m$. This implies at once that $\bar{R}G \subseteq A$, so it suffices to prove that $U(A)$ is free modulo torsion.

Thanks to Proposition 7.2.10, the latter will follow provided we show that there is an index set I such that

$$U(A) \cong \coprod_I U(\bar{R}). \tag{1}$$

The proof of (1) rests on the following observation. Let $H \subset H'$ be subgroups of G such that H has finite index n in H'. Let U' (respectively, U) be the unit group of the idempotent subalgebra of KH' (respectively, KH). Then there exists an isomorphism

$$U' \to U \times U \times \ldots \times U \qquad (n \text{ copies}) \tag{2}$$

whose restriction to U is an isomorphism $U \to U \times \{1\} \times \ldots \times \{1\}$. To prove (2), we first apply Proposition 9.24 to deduce that KH' and $(KH)^n$ are isomorphic KH-algebras. From this we see that $(x_1, x_2, \ldots, x_n) \in (KH)^n$ is a unit of the idempotent subalgebra of $(KH)^n$ if and only if $x_i \in U$, $1 \leq i \leq n$. It follows that U' is isomorphic to $U \times U \times \ldots \times U$ (n copies) by an isomorphism which maps U (as a subgroup of U') to the diagonal in $U \times U \times \ldots \times U$. Because $U \times U \times \ldots \times U$ admits an automorphism which maps its diagonal to $U \times \{1\} \times \ldots \times \{1\}$, (2) is verified.

It is now an easy matter to prove (1). Consider the set X of all ordered pairs $(H_\alpha, \{U_{\alpha,i}\}_{i \in I_\alpha})$ that satisfy the following conditions:

(a) H_α is a subgroup of G;
(b) if U_α is the unit group of the idempotent $\bar{R}$-subalgebra of KH_α, then $\{U_{\alpha,i}\}_{i \in I_\alpha}$ is a family of subgroups of U_α that give an inner direct decomposition of it;
(c) $U_{\alpha,i} \cong U(\bar{R})$ for every $i \in I_\alpha$.

The set X is non-empty: $(\{1\}, \{U(\bar{R})\}) \in X$. Partially order X by specifying that

$$(H_\alpha, \{U_{\alpha,i}\}_{i \in I_\alpha}) \leq (H_\beta, \{U_{\beta,j}\}_{j \in I_\beta})$$

if and only if

$$H_\alpha \subseteq H_\beta \quad \text{and} \quad \{U_{\alpha,i}\}_{i \in I_\alpha} \subseteq \{U_{\beta,j}\}_{j \in I_\beta}.$$

Note that a finite number of elements of KG involve only finitely many elements of G non-trivially and that the order relation respects the inner direct decompositions. Thus X is inductively ordered and so we may choose a maximal element $(H, \{U_i\}_{i \in I})$, say. We now apply (2) to conclude that H cannot be contained properly in a subgroup H' of G such that $(H':H) < \infty$. But G is torsion; hence $H = G$ and the result follows. ∎

The next result gives circumstances under which G has a torsion-free complement in $V(RG)$ and G has a free complement in $V(RG)$.

8.9.26 Theorem (May 1976a,b) *Let G be a group and let R be an integral domain of characteristic 0. If G has an element of prime order p, then assume that p is a non-unit of R.*

(i) *There exists a torsion-free subgroup F of $V(RG_0)$ such that $V(RG_0) = G_0 \times F$. Furthermore, for any such F, $V(RG) = G \times F$.*

(ii) *If R is finitely generated, then F is free.*

Proof.

(i) Owing to Corollary 9.13, the torsion subgroup of $V(RG_0)$ is G_0. Choose an ascending sequence

$$G_1 \subseteq G_2 \subseteq \ldots$$

of subgroups of G_0 such that $G_0 = \cup_i G_i$ and such that for all i, G_i is bounded. Let

$$f_i : V(RG_0) \to V(R(G_0/G_i))$$

be the homomorphism induced by the natural homomorphism

$$- : RG_0 \to R(G_0/G_i)$$

and let $A_i = \operatorname{Ker} f_i$. Our first goal is to prove that the ascending chain

$$A_1 \subseteq A_2 \subseteq \ldots \subseteq A_n \subseteq \ldots$$

satisfies conditions (i)–(iii) laid down in Theorem 9.21. To this end, note that if $x \in V(RG_0)$, then $x \in RG_j$ for some j, and so $x - 1 \in RG_0 \cdot I(G_j)$. This shows that $x \in A_j$, whence $V(RG_0) = \cup_i A_i$. Since G_0 is the torsion subgroup of $V(RG_0)$, it follows that

$$t(A_i) = G_0 \cap [1 + RG_0 \cdot I(G_i)] = G_i;$$

hence $t(A_i)$ is bounded. Assume next that $u \in V(RG_0)$ is such that $u^m \in A_i$ for some $m \in \mathbb{N}$. Then $\bar{u}^m = 1$ in $R\bar{G}_0$, so there is a $g_0 \in G_0$ such that $\bar{u} = \bar{g}_0$, and thus $u = g_0 \cdot a$ for some $a \in A_i$. Therefore

$$t(V(RG_0)/A_i) = G_0 \cdot A_i/A_i = t(V(RG_0)) \cdot A_i/A_i$$

and so, by Theorem 9.21, $V(RG_0) = G_0 \times F$ for some torsion-free subgroup F of $V(RG_0)$. Because $G \cap V(RG_0) = G_0$, we are left to verify that

$$V(RG) = G \cdot V(RG_0). \tag{3}$$

To prove (3), we may harmlessly assume that G is finitely generated, say $G = G_0 \times F_1$. Now RG_0 is indecomposable (see Karpilovsky 1983b, Corollary 3.3.13) and, by Corollary 9.4, RG_0 is reduced. Hence the unit group of $(RG_0)F_1 = RG$ is equal to $U(RG_0) \times F_1$, by virtue of Corollary 4.8. It follows that $V(RG) = V(RG_0) \times F_1$, proving (3) and hence (i).

(ii) By Theorem 9.25, $U(RG_0)$ is free modulo torsion, hence so is a subgroup $V(RG_0)$ of $U(RG_0)$. The desired assertion is now a consequence of (i). ∎

8.9E   *The isomorphism class of $U(\mathbb{Z}G)$*

Throughout this section, $r_p(G)$ denotes the p-rank of G and ε_n a primitive nth root of unity over $\mathbb{Q}$.

8.9.27   Lemma   *Let S be a subring of a ring R. Assume that the additive group R^+ of R is finitely generated and that the torsion-free ranks of S^+ and R^+ are equal. Then the torsion-free ranks of $U(R)$ and $U(S)$ are also equal.*

Proof.   The hypothesis on ranks of S^+ and R^+ ensures that S^+ has finite index, say m, in R^+ and so $mR \subseteq S$. Since R^+ is finitely generated, we also have $(R^+ : mR^+) < \infty$. It remains to show that for any x, $y \in U(R)$,

$$x - y \in mR \qquad \text{implies} \qquad x^{-1}y \in U(S).$$

Because $x - y \in mR$ implies both $1 - x^{-1}y$ and $y^{-1}x - 1 \in mR \subseteq S$, the result follows. ∎

8.9.28   Lemma   *Let S_1, S_2, ..., S_n, ... be non-empty finite sets and let $S_n \xleftarrow{f_n} S_{n+1}$ be maps, $n \geqslant 1$. Then there exists $(s_i) \in \prod_{i \geqslant 1} S_i$ such that*

$$f_i(s_{i+1}) = s_i, \qquad \text{for all } i \geqslant 1.$$

Proof.   Put the discrete topology on each set S_i and define $A_n \subseteq \prod S_i$, $n \geqslant 1$, by

$$A_n = \{(s_i) \mid f_i(s_{i+1}) = s_i \text{ for any } i \leqslant n\}.$$

Then each A_n is a closed subset of $\prod S_i$ and

$$A_{n_1} \cap A_{n_2} \cap \ldots \cap A_{n_k} = A_m,$$

where $m = \max\{n_1, n_2, \ldots, n_k\}$. Furthermore, $A_m \neq \emptyset$. But $\prod S_i$ is compact, hence $\cap_{n \geqslant 1} A \neq \emptyset$, as required. ∎

The above lemma is used in the proof of the following useful observation.

8.9.29   Lemma   *Let G be an infinite p-group which is countable and reduced. Then $r_p(G) = \aleph_0$.*

Proof.   If G is bounded, then the result is true by applying Corollary 4.1.8. Assume that G is unbounded and suppose by way of contradiction that $r_p(G) = n < \infty$. Put

$$S_n = \{\langle x \rangle \mid x \in G, \text{ order } x = p^n\}.$$

Then S_n is finite, since $G[p^n]$ has finite p-rank. Furthermore, each $S_n \neq \emptyset$, since otherwise $G = G[p^n]$ for some $n \geq 1$. Define the map $f_n : S_{n+1} \to S_n$ by $f_n(\langle x \rangle) = \langle x^p \rangle$. Owing to Lemma 9.28, there exist $x_1, x_2, \ldots, \in G$ such that for all $n \geq 1$, the order of x_n is p^n, and such that

$$\langle x_1 \rangle \subseteq \langle x_2 \rangle \subseteq \ldots .$$

It follows that $\bigcup_{n \geq 1} \langle x_n \rangle$ is a divisible subgroup of G, contrary to the assumption that G is reduced. So the lemma is true. ∎

The following group-theoretic result is crucial for the description of $U(\mathbb{Z}G)$.

8.9.30 Lemma   *Let G be an infinite torsion group satisfying the following two conditions:*

(i)   *G is reduced or uncountable;*
(ii)   *$G^4 \neq 1$ and $G^6 \neq 1$.*

Then G contains a subgroup H which is a direct product of $|G|$ non-trivial cyclic groups, at least one of which does not have order dividing 4 or 6.

Proof.   Let us first show that it suffices to consider the case where G is a p-group. Indeed, assume that G is uncountable. Write $G = A \times B \times C$, where A is a direct product of all countable p-components and where B (respectively, C) a direct product of all uncountable p-components satisfying (respectively, violating) (ii). Note that $|G| = |B \times C|$ and that, by Corollary 4.1.8, C is a direct product of cyclic groups. Hence, if $B \neq 1$, then the result for the p-component of B would imply the result for G. In the contrary case, either C or $\langle a \rangle \times C$ (for some $a \in A$) fulfills the role of H. Assume that G is reduced and countable. Then, either there are infinitely many non-trivial p-components, hence the result, or there is a countable reduced p-component and the problem again reduces to the primary case. Thus we may assume that G is a p-group.

Let G be a p-group satisfying (i) and (ii). We know that

$$G[p] \subseteq G[p^2] \subseteq \ldots \subseteq G[p^n] \subseteq \ldots$$

and that, for all $i, j \geq 1$,

$$r_p(G[p^i]) = r_p(G[p^j]).$$

Since $G = \bigcup_{i \geq 1} G[p^i]$, if G is uncountable, then

$$r_p(G[p^i]) = |G|.$$

We take $H = G[p^3]$ and are finished. Hence we may assume that G is countable and reduced. Then, by Lemma 9.29,

$$r_p(G[p^i]) = \aleph_0$$

and so we are again done. Thus the lemma is proved. ∎

It is now possible to determine the isomorphism class of $U(\mathbb{Z}G)$ for an arbitrary group G. The result for the case where G is finite is essentially due to Higman (1940b) (see also Ayoub and Ayoub 1969), while its extension to the general case was furnished by May in a private communication to the author.

8.9.31   Theorem   *Let G be an arbitrary group. Then*

$$U(\mathbb{Z}G) = \pm\, G \times F,$$

where F is a free group whose rank is defined as follows:

$$\text{rank } F = \begin{cases} (1/2)(|G_0| - 2l + m + 1) & \text{if } G_0 \text{ is finite,} \\ 0 & \text{if } G_0^4 = 1 \text{ or } G_0^6 = 1, \\ |G_0| & \text{if } G_0 \text{ is infinite, } G_0^4 \neq 1 \text{ and } G_0^6 \neq 1. \end{cases}$$

Here m (respectively, l) is the number of cyclic subgroups of G_0 of order 2 (respectively, the number of cyclic subgroups of G_0).

Proof.   By Theorem 9.26, there is a free subgroup F of $V(\mathbb{Z}G_0)$ such that

$$U(\mathbb{Z}G_0) = \pm\, G_0 \times F \qquad \text{and} \qquad U(\mathbb{Z}G) = \pm\, G \times F.$$

Therefore, we need only calculate the torsion-free rank of $U(\mathbb{Z}G)$, where G is a torsion group. We distinguish two cases. Suppose first that G is finite. Let a_d be the number of cyclic subgroups of order d in G and let n be the exponent of G. By Corollary 9.9,

$$\mathbb{Z}G \subseteq \prod_{d|n} \mathbb{Z}[\varepsilon_d]^{a_d} = R, \qquad \mathbb{Q}G = \prod_{d|n} \mathbb{Q}(\varepsilon_d)^{a_d}.$$

Because $(\mathbb{Z}[\varepsilon_d] : \mathbb{Z}) = (\mathbb{Q}(\varepsilon_d) : \mathbb{Q})$, it follows that the groups $(\mathbb{Z}G)^+$ and R^+ have the same torsion-free rank, equal to $|G|$. Since R^+ is free, we may apply Lemma 9.27 to deduce that $U(\mathbb{Z}G)$ and $U(R)$ have the same torsion-free rank. Owing to Dirichlet's unit theorem, if $d > 2$ then the torsion-free rank of $U(\mathbb{Z}[\varepsilon_d])$ is equal to $(1/2)\phi(d) - 1$. Therefore

$$\text{rank } F = \sum_{d>2} a_d[\phi(d)/2 - 1] = (1/2)\sum_{d>2} a_d\phi(d) - \sum_{d>2} a_d$$

$$= (1/2)\left[\sum_d a_d\phi(d) - 2\sum_d a_d + a_2 + a_1\right]$$

$$= (1/2)(|G| - 2l + m + 1) \tag{1}$$

as required.

It will next be shown that

$$\text{rank } F = 0 \text{ if and only if } G^4 = 1 \text{ or } G^6 = 1. \tag{2}$$

To do this, we may assume that G is finite, in which case rank $F = 0$ if and only if for all $d > 2$, $(1/2)\phi(d) - 1 = 0$. Because the latter is equivalent to $d \in \{3, 4, 6\}$, (2) is established.

From now on it will be assumed that G is an infinite torsion group such that $G^4 \neq 1$ and $G^6 \neq 1$. Clearly, rank $F \leqslant |\mathbb{Z}G| = |G|$. To prove the reverse inequality, we first consider the case where G is a countable group which has a subgroup D of type p^∞. Because D contains a cyclic subgroup of order p^n for all $n \geqslant 1$, it follows from (1) that $U(\mathbb{Z}D)$ has infinitely many independent units of infinite order. Hence

$$|G| = \aleph_0 = \text{rank(free part of units in } \mathbb{Z}D) \leqslant \text{rank } F$$

and therefore we may assume that G is either uncountable or reduced. Applying Lemma 9.30, we conclude that G contains a subgroup H which is a direct product of $|G|$ non-trivial cyclic groups such that at least one of them does not have order dividing 4 or 6.

Let $\{H_i\}_{i \in I}$ $(H_i = \langle h_i \rangle, |I| = |G|)$ be a family of cyclic subgroups of H that give an inner direct decomposition of it, and let $h_1^4 \neq 1$ and $h_1^6 \neq 1$. Then, by (2), $\mathbb{Z}H_1$ has a unit u_1 of infinite order. Moreover, for any $i \neq 1$,

$$(h_1 h_i)^4 \neq 1, \quad (h_1 h_i)^6 \neq 1 \text{ and } h_1 h_i \notin H_1.$$

Applying formula (1), we see that the torsion-free rank of $U(\mathbb{Z}H_1)$ is strictly less than that of $U(\mathbb{Z}(H_1 \cdot H_i))$. Hence $\mathbb{Z}(H_1 \cdot H_i)$ has a non-torsion unit u_i such that for any non-zero integer m, $u_i^m \notin \mathbb{Z}H_1$.

We now claim that the system $\{u_i\}_{i \in I}$ of non-torsion units of $\mathbb{Z}G$ is independent; if sustained, then rank $F \geqslant |I| = |G|$; hence the result. Assume, for definiteness, that

$$u_1^{n_1} u_2^{n_2} \dots u_k^{n_k} = 1 \qquad (n_i \in \mathbb{Z}, k \neq 1).$$

Then

$$u_k^{-n_k} \in \mathbb{Z}(H_1 \cdot H_k) \cap \mathbb{Z}(H_1 \cdot H_2 \dots H_{k-1}) = \mathbb{Z}(H_1 \cdot H_k \cap H_1 \dots H_{k-1})$$
$$= \mathbb{Z}H_1.$$

This shows that $n_k = 0$ and completes the proof. ∎

8.9.32 Corollary   *Let G be an arbitrary group.*

(i) $U(\mathbb{Z}G) = \pm G$ *if and only if* $G_0^4 = 1$ *or* $G_0^6 = 1$.
(ii) $U(\mathbb{Z}G)$ *is torsion if and only if* $G^4 = 1$ *or* $G^6 = 1$.

Proof. The proof is a direct consequence of Theorem 9.31. ∎

8.9F   *Effective construction of units of $\mathbb{Z}G$*

Let G be a finite group. A number of results of differential topology (see Milnor 1966) suggest the importance of finding out precisely what units exist within the ring $\mathbb{Z}G$. To state the problem more explicitly, let us introduce the following notation:

n = the exponent of G

a_d = the number of cyclic subgroups of G of order d

$$r = \sum_{\substack{d>2 \\ d\mid n}} a_d((1/2)\phi(d) - 1) \quad \text{(equivalently,} \quad r = (1/2)(|G| - 2l + m + 1)$$

in the notation of Theorem 9.31).

We know, from Theorem 9.31, that there exists a system of r units $u_1, \ldots, u_r$ in $\mathbb{Z}G$ such that every unit of $\mathbb{Z}G$ is represented uniquely in the form

$$\pm g u_1^{n_1} u_2^{n_2} \ldots u_r^{n_r} \quad (n_i \in \mathbb{Z}, g \in G).$$

We call $\{u_1, \ldots, u_r\}$ a *fundamental system* of units in $\mathbb{Z}G$. The mentioned problem may be stated in the following manner:

Problem A. Given a finite group G, find a specific fundamental system of units in $\mathbb{Z}G$.

This is an extremely difficult problem. The main difficulty seems to be the fact that effective calculations in $U(\mathbb{Z}G)$ are intimately connected with those in $U(\mathbb{Z}[\varepsilon_n])$. Taking into account that, in general, no specific system of fundamental units of $\mathbb{Z}[\varepsilon_n]$ is known, any optimism in solving Problem A must be guarded.

From the results of the previous section, it seems more appropriate to attack the following related problem:

Problem B. Given a finite group G, find a specific system of r independent units of infinite order in $\mathbb{Z}G$ or, equivalently, a system of independent units of infinite order which generates a subgroup of finite index in $U(\mathbb{Z}G)$.

It is the aim of this section to present a theorem, due to Bass (1966), which solves Problem B for cyclic groups.

Let ε_n be a primitive nth root of unity over $\mathbb{Q}$, and let $\mathbb{Q}_n = \mathbb{Q}(\varepsilon_n)$. As usual, by a character of G we shall understand a homomorphism

$$\chi : G \to \langle \varepsilon_n \rangle.$$

For convenience, we shall identify χ with its extension to the ring homomorphism $\mathbb{Z}G \to \mathbb{Z}[\varepsilon_n]$. In particular, the identity character will be identified with the augmentation map.

8.9.33 Lemma *Let $\chi_1, \chi_2, \ldots, \chi_s$ be a full set of representatives of equivalence classes of $\mathbb{Q}$-conjugate characters of G.*

(i) *If x, $y \in \mathbb{Z}G$, then $x = y$ if and only if $\chi_i(x) = \chi_i(y)$ for all $i \in \{1, 2, \ldots, s\}$.*

(ii) *An element $x \in \mathbb{Z}G$ is a unit if and only if for all $i \in \{1, 2, \ldots, s\}$,
$\chi_i(x)$ is a unit of $\mathbb{Z}[\varepsilon_n]$.*

Proof. Property (i) is a consequence of Theorem 9.8. To prove (ii), we
apply Corollary 9.9(ii) to deduce that x is a unit if and only if for all
$i \in \{1, 2, \ldots, s\}$, $\chi_i(x)$ is a unit of Im χ_i. The required assertion is now a
consequence of the fact that $\mathbb{Z}[\varepsilon_n]$ is integral over Im χ_i. ■

Let us pause for two examples which illustrate the interplay between
the unit group of $\mathbb{Z}G$ and that of $\mathbb{Z}[\varepsilon_n]$.

8.9.34 Example (Kaplansky, unpublished) *Let G be a cyclic group of
order 5 with generator g. Then*

$$U(\mathbb{Z}G) = \pm G \times \langle g^2 + g^3 - 1 \rangle.$$

Proof. We start with a general observation pertaining to a cyclic group
G of odd prime order p. Let $\chi : \mathbb{Z}G \to \mathbb{Z}[\varepsilon_p]$ be the homomorphism which
sends g to ε_p. We claim that χ induces an injective homomorphism of
unit groups and that Ker $\chi = \mathbb{Z}(1 + g + \ldots + g^{p-1})$. Indeed, setting χ_1 to
be the identity character, it follows that $\{\chi_1, \chi\}$ is a full system of
non-conjugate characters of G. Therefore, if $u \in U(\mathbb{Z}G)$ is such that
$\chi(u) = 1$, then

$$\chi(u^2) = \chi_1(u^2) = \text{aug}(u^2) = 1$$

so $u^2 = 1$ by Lemma 9.33(i). Because $\pm G$ is the torsion subgroup of
$U(\mathbb{Z}G)$, the assumption that p is odd forces $u = \pm 1$, whence $u = 1$. The
assertion regarding Ker χ is a direct consequence of the equality

$$1 + \varepsilon_p + \ldots + \varepsilon_p^{p-1} = 0$$

and the fact that $\{1, \varepsilon_p, \ldots, \varepsilon_p^{p-2}\}$ is a $\mathbb{Z}$-basis for $\mathbb{Z}[\varepsilon_p]$.
Returning to the case where G is of order 5, we first note that the
torsion-free rank of $U(\mathbb{Z}[\varepsilon_5])$ is 1 and that $\varepsilon_5 + 1$ is a fundamental unit
(Corollary 3.3.5). If $x \in \mathbb{Z}G$ satisfies $\chi(x) = \varepsilon_5 + 1$, then

$$x = (g + 1) + m(1 + g + g^2 + g^3 + g^4)$$

for some $m \in \mathbb{Z}$. Because $\text{aug}(x) = 2 + 5m \neq \pm 1$, we deduce that there is
no unit in $\mathbb{Z}G$ which is mapped to $\varepsilon_5 + 1$. In particular, this shows that
the image of $U(\mathbb{Z}G)$ is a proper subgroup of $U(\mathbb{Z}[\varepsilon_5])$. On the other
hand, the element

$$u = (g + 1)^2 - (1 + g + g^2 + g^3 + g^4) = -g(g^2 + g^3 - 1)$$

is mapped to $(\varepsilon_5 + 1)^2$ and has augmentation -1. Therefore u is a unit
(Lemma 9.33(ii)) such that the image of $\langle \pm G, u \rangle$ is a subgroup of
$U(\mathbb{Z}[\varepsilon_5])$ of index 2. This certainly implies $U(\mathbb{Z}G) = \langle \pm G, u \rangle$, whence
the result. ■

8.9.35 Example *Let G be a cyclic group of order 8 with generator g.
Then*

$$U(\mathbb{Z}G) = \pm G \times \langle g^6 + 2g^5 + g^4 - g^2 - g - 1 \rangle.$$

Proof. Let $\chi : \mathbb{Z}G \to \mathbb{Z}[\varepsilon_8]$ be the homomorphism which sends g to ε_8.
Note that the torsion-free rank of $U(\mathbb{Z}[\varepsilon_8])$ is 1 and that $1 + \varepsilon_8 + \varepsilon_8^2$ is a
fundamental unit (see Corollary 3.3.5). Invoking the fact that
$\{1, \varepsilon_8, \varepsilon_8^2, \varepsilon_8^3\}$ is a $\mathbb{Z}$-basis for $\mathbb{Z}[\varepsilon_8]$, it is easy to verify that

$$\mathrm{Ker}\,\chi = \{\alpha_0(1 + g^4) + \alpha_1(g + g^5) + \alpha_2(g^2 + g^6) + \alpha_3(g^3 + g^7) \mid \alpha_i \in \mathbb{Z}\}.$$

It will next be shown that there is no unit in $\mathbb{Z}G$ which is mapped to
$1 + \varepsilon_8 + \varepsilon_8^2$. Indeed, let $x \in \mathbb{Z}G$ be such that $\chi(x) = 1 + \varepsilon_8 + \varepsilon_8^2$. Then
there exist $\alpha_i \in \mathbb{Z}$ such that

$$x = 1 + g + g^2 + \alpha_0(1 + g^4) + \alpha_1(g + g^5) + \alpha_2(g^2 + g^6) + \alpha_3(g^3 + g^7).$$

Note that $\{\chi_1 = \text{identity}, \chi_2, \chi_3, \chi\}$, where $\chi_2(g) = -1$ and $\chi_3(g) = i$, is a
full set of non-conjugate characters of G. Hence x is a unit if and only if
$\chi_i(x)$ is a unit for $i = 1, 2, 3$. The latter is equivalent to the following
system of equations:

$$3 + 2\alpha_0 + 2\alpha_1 + 2\alpha_2 + 2\alpha_3 = \pm 1$$

$$1 + 2\alpha_0 - 2\alpha_1 + 2\alpha_2 - 2\alpha_3 = \pm 1$$

$$(2\alpha_0 - 2\alpha_2) + (1 + 2\alpha_1 - 2\alpha_3)i = \pm 1 \quad \text{or} \quad \pm i.$$

It is straightforward to verify that this system does not admit integral
solutions, and so x is a non-unit. Thus the image of $U(\mathbb{Z}G)$ is a proper
subgroup of $U(\mathbb{Z}[\varepsilon_8])$.

Setting

$$u = -(1 + g + g^2)^2 - (g + g^5) - 2(g^2 + g^6) - (g^3 + g^7)$$
$$= -g(g^6 + 2g^5 + g^4 - g^2 - g - 1)$$

it follows that

$$\chi(u) = (1 + \varepsilon_8 + \varepsilon_8^2)^2, \qquad \chi_1(u) = -1,$$

$$\chi_2(u) = -1, \qquad \chi_3(u) = 1.$$

Therefore, u is a unit such that $\chi(\langle \pm G, u \rangle)$ is a subgroup of $U(\mathbb{Z}[\varepsilon_8])$ of
index 2. This implies at once that

$$\chi(\langle \pm G, u \rangle) = \chi(U(\mathbb{Z}G)).$$

Thus the result will follow provided that we show that the kernel of the
induced homomorphism $U(\mathbb{Z}G) \to U(\mathbb{Z}[\varepsilon_8])$ is contained in $\pm G$.
To this end, we invoke Theorem 9.8(i) to conclude that the mapping

$$\begin{cases} \mathbb{Z}G \to \mathbb{Z} \times \mathbb{Z} \times \mathbb{Z}[i] \times \mathbb{Z}[\varepsilon_8] \\ x \mapsto (\chi_1(x), \chi_2(x), \chi_3(x), \chi(x)) \end{cases}$$

is an injective homomorphism. Because the unit group of the first three factors is finite, it follows that any $v \in U(\mathbb{Z}G)$ for which $\chi(v) = 1$ is of finite order. This shows that $v \in \pm G$ and completes the proof. ∎

We turn now to the task of constructing independent units which generate a subgroup of finite index in $U(\mathbb{Z}G)$.

8.9.36 Theorem (Bass 1966) *Let G be a cyclic group of order $n > 2$, let m be a multiple of $\phi(n)$, and for each $d \mid n$, $d > 2$, let g_d be a fixed element of order d in G. Put*

$$S_d = \{g_d^\mu \mid 1 < \mu < d/2 \text{ and } (\mu, d) = 1\}$$

and, for any $s = g_d^\mu \in S_d$, put

$$u_s = (1 + g_d + \ldots + g_d^{\mu-1})^m + \frac{1 - \mu^m}{d}(1 + g_d + \ldots + g_d^{d-1}).$$

Then $\bigcup_{d \mid n, d > 2} \{u_s \mid s \in S_d\}$ is a system of independent units of infinite order which generates a subgroup of finite index in $U(\mathbb{Z}G)$.

Proof. By assumption, $\phi(n) \mid m$, so $\phi(d) \mid m$ for all $d \mid n$. Because $(\mu, d) = 1$, we also have $\mu^{\phi(d)} \equiv 1 \pmod{d}$, so $d \mid 1 - \mu^m$ and therefore $u_s \in \mathbb{Z}G$. Let χ be a character of G. If $\chi(g_d) = 1$, then

$$\chi(u_s) = \mu^m + \frac{1 - \mu^m}{d} d = 1.$$

If $\chi(g_d) = \varepsilon \neq 1$, then

$$\chi(1 + g_d + \ldots + g_d^{d-1}) = 1 + \varepsilon + \ldots + \varepsilon^{d-1}$$
$$= \frac{\varepsilon^d - 1}{\varepsilon - 1} = 0;$$

so

$$\chi(u_s) = (1 + \varepsilon + \ldots + \varepsilon^{\mu-1})^m = \left(\frac{1 - \varepsilon^\mu}{1 - \varepsilon}\right)^m \in U(\mathbb{Z}[\varepsilon]) \qquad (\text{I})$$

since $(1 - \varepsilon^\mu)/(1 - \varepsilon)$ is a (cyclotomic) unit. Therefore, by Lemma 9.33(ii), we have $u_s \in U(\mathbb{Z}G)$. Next, note that since G has exactly one subgroup of order $d \mid n$, the torsion-free rank of $U(\mathbb{Z}G)$ is

$$\sum_{\substack{d \mid n \\ d > 2}} [(1/2)\phi(d) - 1].$$

Since $|S_d| = \phi(d)/2 - 1$, to complete the proof it suffices to show that the units u_s are independent.

To prove independence, suppose we are given a relation

$$\prod_{\substack{d\mid n \\ d>2}} \prod_{s\in S_d} u_s^{a_s} = 1. \tag{II}$$

We must show that all $a_s = 0$. We shall apply Theorem 3.4.3, so we must construct the necessary data. Let $\Phi_d(G)$ be the set of elements of order d in G. If $g \in G$, then $g \in \Phi_d(G)$ for a unique $d \mid n$. We now define the integer b_g as follows. If $d \le 2$ set $b_g = 0$, otherwise exactly one of the following four cases occurs;

(1) $g \in S_d$,   (2) $g^{-1} \in S_d$,   (3) $g = g_d$,   or   (4) $g^{-1} = g_d$.

Define

$$b_g = \begin{cases} ma_g & \text{if } g \in S_d \\ (-m)\sum_{s\in S_d} a_s & \text{if } g = g_d \end{cases}$$

and

$$b_{g^{-1}} = b_g.$$

Using the notation of Section 4 of Chapter 3, choose an isomorphism $\chi_0 : G \to \Psi_n$ and put $b_\varepsilon = b_g$ if $\varepsilon = \chi_0(g)$. We claim that n and the $b_\varepsilon (\varepsilon \in \Psi_n)$ satisfy conditions (3) and (4) of Theorem 3.4.3.

Indeed, condition (3) follows from the construction. Thus we need only prove that if χ is a character of Ψ_n, then

$$\prod_{\varepsilon\in\Psi_n} e(\chi(\varepsilon))^{b_\varepsilon} = 1.$$

Denote by u_s^* the image of u_s under the automorphism of $\mathbb{Z}G$ which sends every element in G to its inverse. Then (II) gives

$$\prod_{\substack{d\mid n \\ d>2}} \prod_{s\in S_d} (u_s^*)^{a_s} = 1$$

and the product of this with (II) implies

$$\prod_{\substack{d\mid n \\ d>2}} \prod_{s\in S_d} (u_s u_s^*)^{a_s} = 1.$$

Applying the character $\theta = \chi \circ \chi_0$ of G to the above, we derive

$$\prod_{\substack{d\mid n \\ d>2}} \prod_{s\in S_d} [\theta(u_s)\theta(u_s^*)]^{a_s} = 1$$

and this, by (I), is

$$\prod_{\substack{d|n \\ d>2}} \prod_{s \in S_d} \left[\frac{e(\theta(s))}{e(\theta(g_d))} \frac{e(\theta(s^{-1}))}{e(\theta(g_d^{-1}))} \right]^{ma_s} = 1$$

$$= \prod_{d|n} \prod_{\varepsilon \in \Phi_d} e(\chi(\varepsilon))^{b_\varepsilon},$$

the last equality being true by the definition of the b_ε. This verifies condition (4) of Theorem 3.4.3, and thus $b_\varepsilon = 0$ for all $\varepsilon \neq 1$. In particular, for any $s \in S_d$,

$$b_{\chi_0(s)} = ma_s = 0$$

so $a_s = 0$. This completes the proof of the theorem. ∎

We conclude this section by deriving the following classical result.

8.9.37 Corollary (Hilbert 1897, Satz 144) *Let d range over all divisors of n, and let Φ_d be the set of primitive d-th roots of unity over $\mathbb{Q}$. Then the cyclotomic units*

$$\frac{1-\varepsilon}{1-\delta} \qquad (\varepsilon, \delta \in \Phi_d)$$

generate a subgroup of finite index in $U(\mathbb{Z}[\varepsilon_n])$.

Proof. Let G and the u_s be as in Theorem 9.36, and let H be the group generated by the u_s. We know, from Theorem 9.8, that if we identify $\mathbb{Z}G$ with its image in $\prod_{d|n} \mathbb{Z}[\varepsilon_d]$, then the coordinate projections $\mathbb{Z}G \to \mathbb{Z}[\varepsilon_d]$ can be identified with certain characters of G. This implies, by virtue of property (I) of theorem 9.36, that the projection H_n of H into $U(\mathbb{Z}[\varepsilon_n])$ is contained in the group described in the corollary. Hence we need only show that H_n has finite index in $U(\mathbb{Z}[\varepsilon_n])$. This being true, by virtue of the fact that H has finite index in $\prod_{d|n} U(\mathbb{Z}[\varepsilon_d])$ (both groups have the same torsion-free rank), the corollary is proved. ∎

Bibliography

Albert, A. A. (1939). Structure of algebras. *Am. Math. Soc. Colloq.* Publ. Vol. 24.

Allen, P. J. and Hobby, C. (1980). A characterization of units in $Z[A_4]$. *J. Algebra* **66**, 534–43.

Amitsur, S. A. (1955). Finite subgroups of division rings. *Trans. Am. Math. Soc.* **80**, 361–86.

—— (1966). Rational identities and applications to algebra and geometry. *J. Algebra* **3**, 304–59.

Apostol, T. M. (1976). *Introduction to analytic number theory.* Springer-Verlag, New York–Heidelberg–Berlin.

Appelgate, H. and Onishi, H. (1982). Periodic expansion of modules and its relation to units. *J. Number Theory* **15**, 283–94.

Armitage, J. V. and Fröhlich, A. (1967). Classnumbers and unit signatures. *Mathematika* **14**, 94–8.

Artin, E. (1967). *Algebraic numbers and algebraic functions.* Gordon and Breach, New York.

—— and Whaples, G. (1945). Axiomatic characterization of fields by the product formula for valuations, *Bull. Am. Math. Soc.* **51**, 469–92.

Arwin, A. (1929). On cubic fields. *Ann. Math.* **30**, 1–11.

Ayoub, C. W. (1968). On the units in certain integral domains. *Arch. Math.* **19**, 43–6.

—— (1969). On finite primary rings and their group of units, *Compositio Math.* **21**, 247–52.

—— (1970). On diagrams for abelian groups, *J. Number Theory* **2**, 442–58.

—— (1972). On the group of units in certain rings. *J. Number Theory* **4**, 383–403.

—— and Ayoub, R. G. (1969). On the group ring of a finite abelian group. *Bull. Austral. Math. Soc.* **1**, 245–61.

Ax, J. (1965). On the units of an algebraic number field. *Illinois J. Math.* **9**, 584–9.

Azuhata, T. (1986). On the calculation of the units of algebraic number fields. *Nagoya Math. J.* **101**, 181–5.

Baer, R. (1934). Der Kern, eine charakteristische Untergruppe. *Compositio Math.* **1**, 254–83.

—— (1936). The subgroup of the elements of finite order of an abelian group. *Ann. Math.* **37**, 766–81.

Bass, H. (1965). A remark on an arithmetic theorem of Chevalley. *Proc. Am. Math. Soc.* **16**, 875–8.

—— (1966). The Dirichlet unit theorem, induced characters, and Whitehead groups of finite groups. *Topology* **4**, 391–410.

—— (1974). *Introduction to some methods of algebraic K-theory.* Regional Conference Series in Mathematics 20.

—— (1976). Euler characteristics and characters of discrete groups. *Invent. Math.* **35**, 155–96.

Bateman, J. M. (1971). On the solvability of unit groups of group algebras. *Trans. Am. Math. Soc.* **157**, 73–86.

—— and Coleman, D. B. (1968). Group algebras with nilpotent unit groups. *Proc. Am. Math. Soc.* **19**, 448–9.

Behr, H. (1967). Über die endliche Definierbarkeit verallgemeinerter Einheitengruppen, II. *Invent. Math.* **4**, 265–74.

Bergmann, G. (1966a). Zur numerischen Bestimmung einer Einheitenbasis. *Math. Ann.* **166**, 103–5.

—— (1966b). Beispiel numerischer Einheitenbestimmung. *Math. Ann.* **167**, 143–68.

Berman, S. D. (1953). On certain properties of integral group rings. *Dokl. Akad. Nauk, SSSR* (N.S.) **91**, 7–9.

—— (1955). On the equation $x^m = 1$ in an integral group ring. *Ukrain. Mat. Ž.* **7**, 253–61.

—— (1958). Characters of linear representations of finite groups over an arbitrary field. *Mat. Sb.* **44**(86), 409–56.

Berman, S. D. and Rossa, A. R. (1966). Integral group rings of finite and periodic groups. *Algebra and mathematical logic: studies in algebra.* Izdat Univ. Kiev, pp. 44–53.

Bernstein, L. (1966). The modified algorithm of Jacobi–Perron. *Mem. Am. Math. Soc.* No. 67, 1–44.

—— (1971a). *The Jacobi–Perron algorithm, its theory and application.* Lecture Notes in Mathematics 207, Springer-Verlag, Berlin–Heidelberg–New York.

—— (1971b). Infinite non-solubility classes of the Delaunay–Nagell diophantine equation $x^3 + my^3 = 1$. *J. Lond. Math. Soc.* (2) 3, 118–20.

—— (1972a). A 3-dimensional periodic Jacobi–Perron algorithm of length 8. *J. Number Theory* **4**, 48–69.

—— (1972b). On units and fundamental units. *J. Reine Angew. Math.* **257**, 129–45.

—— (1974). Fundamental units from the preperiod of generalized Jacobi–Perron algorithm. *J. Reine Angew. Math.* **268/269**, 391–409.

—— (1975a). Units and their norm equation in real algebraic number fields of any degree. *Symp. Math.* **XV**, 307–40.

—— (1975b). Der Hasse–Bernsteinsche Einheitensatz fuer den verallgemeinerten Jacobi–Perronschen Algorithmus. *Abh. Math. Sem.* **43**, 192–202.

—— (1975c). Truncated units in finitely many algebraic number fields of degree $n \geqslant 4$. *Math. Ann.* **213**, 275–9.

—— (1975d). Units and periodic Jacobi–Perron algorithms in real algebraic number fields of degree 3. *Trans. Am. Math. Soc.* **212**, 295–306.

—— (1977). Gaining units from units. *Can. J. Math.* **29**(1), 93–106.

—— and Hasse, H. (1969). An explicit formula for the units of algebraic number field of degree $n \geqslant 2$. *Pacif. J. Math.* **30**(2), 293–365.

Berwick, W. E. H. (1932). Algebraic number fields with two independent units. *Proc. Lond. Math. Soc.* (2) **34**, 360–78.

—— (1934). The classification of ideal numbers in a cubic field. *Proc. Lond. Math. Soc.* (2) **38**, 217–42.

Besicovitch, A. S. (1940). On the linear independence of fractional powers of integers. *J. Lond. Math. Soc.* **15**, 3–6.

Bhandari, A. K. (1985). Some remarks on the unit groups of integral group rings. *Arch. Math.* **44**, 319–22.

—— and Luthar, I. S. (1983a). Torsion units of integral group rings of metacyclic groups. *J. Number Theory* **17**, 270–83.

—— and —— (1983b). Conjugacy classes of torsion units of the integral group ring of D_p. *Comm. Algebra* **11**(14), 1607–27.

Billevich, K. K. (1956). On units of algebraic fields of 3rd and 4th degree. *Mat. Sb.* **40**, 123–36.

—— (1964). A theorem on unit elements of algebraic fields of degree n. *Mat. Sb.* N.S. **64**(106), 145–52.

Borel, A. and Harish-Chandra. (1962). Arithmeric subgroups of algebraic groups. *Ann. Math.* **75**(3), 485–535.

Borevich, Z. I. and Shafarevich, I. R. (1966). *Number theory*. Academic Press, New York.

Bourbaki, N. (1958). *Algébre*. Hermann, Paris, Ch. 8: Modules et anneaux semi-simples.

—— (1959). *Algébre*. Hermann, Paris, Chs 4 and 5.

—— (1961). *Algébre commutative*. Hermann, Paris, Chs 1–4.

—— (1964). *Algébre commutative*. Hermann, Paris, Chs 5 and 6.

—— (1965). *Algébre commutative*. Hermann, Paris, Ch. 7.

—— (1966). *Éléments de mathematique*. Topologie Generale, Hermann, Paris.

—— (1971/1972). *Groupes et algébres de Lie*. Hermann, Paris, Chs 1–3.

Bovdi, A. A. (1960). Group rings of torsion-free groups. *Sibirsk. Mat. Ž.* **1**, 555–8.

—— (1968). Periodic normal divisors of the multiplicative group of a group ring, I. *Siberian Math. J.* **9**, 374–6.

—— (1970). Periodic normal divisors of the multiplicative group of a group ring, II. *Siberian Math. J.* **11**, 374–8.

—— (1974). *Group rings*. Uzhgorod.

—— (1980). On the construction of integral group rings with trivial elements of finite order. *Sibirsk. Mat. Zh.* **21**(4), 28–37.

—— (1984). Unitarity of the multiplicative group of an integral group ring. *Math. USSR Sb.* **47**(2), 377–89.

—— and Hripta, I. I. (1972). Normal subgroups of a multiplicative group of a ring. *Math. USSR Sb.* **16**, 349–62.

—— and —— (1974). Finite dimensional group algebras having solvable unit groups. *Trans. Sci. Conf.* Uzhgorod State Univ., pp. 227–35.

Brandis, R. (1965). Über die multiplikative struktur von korpererweiterungen. *Math. Z.* **87**, 71–3.

Brauer, R. (1949). On a theorem of H. Cartan. *Bull. Am. Math. Soc.* **55**, 619–20.

Brumer, A. (1967). On the units of algebraic number fields. *Mathematika* **14**, 121–4.

—— (1969). On the group of units of an absolutely cyclic number field of prime degree. *J. Math. Soc. Japan* **21**, 357–8.

Buchmann, J. (1985a). A generalization of Voronoi's unit algorithm, I. *J. Number Theory* **20**, 177–91.

—— (1985b). A generalization of Voronoi's unit algorithm, II. *J. Number Theory* **20**, 192–209.

Čarin, V. S. (1954). On automorphism groups of nilpotent groups. *Ukrain. Math. J.* **6**, 295–304.

Carlitz, L. (1955). Note on the class number of quadratic fields. *Duke Math. J.* **22**, 589–93.

Cartan, H. (1947). Les principaux théorémes de la théorie de Galois pour les corps non nécessairement commutatifs. *C.R. Acad. Sci. Paris* **224**, 249–51.

Cassels, J. W. S. (1950). The rational solutions of the diophantine equation $y^2 = x^3 - D$. *Acta Math.* **82**, 243–73.

Chevalley, C. (1940). La theorie du corps de classes. *Ann. Math.* **41**, 394–418.

Chowla, S. (1961a). On the class number of real quadratic fields. *Proc. Nat. Acad. Sci. U.S.A.* **47**, 878.

—— (1961b). A remarkable solution of the Pellian equation $X^2 - pY^2 = -4$ in the case when $p \equiv 1(\mathrm{mod}\ 4)$ and the class number of $R(\sqrt{p})$ is 1. *J. Ind. Math. Soc.* **25**, 43–6.

Cliff, G., Sehgal, S. K., and Weiss, A. (1981). Units of integral group rings of metabelian groups. *J. Algebra* **73**, 167–85.

Cohen, I. (1946). On the structure and ideal theory of complete local rings. *Trans. Amer. Math. Soc.* **59**, 54–106.

Cohn, H. (1954). A periodic algorithm for cubic forms. *Am. J. Math.* **76**, 904–14.

—— (1971). A numerical study of units in composite real quartic and octic fields, *Computers in Number Theory* (eds A. O. L. Atkin and B. J. Birch), 153–65. London and New York, Academic Press.

—— and Gorn, S. (1957). A computation of cyclic cubic units. *J. Res. Nat. Bur. Standards* **59**, 155–68.

Cohn, J. A. and Livingstone, D. (1965). On the structure of group algebras. *Can. J. Math.* **17**, 583–93.

Cohn, J. H. E. (1967). The diophantine equation $x^3 = dy^3 + 1$. *J. Lond. Math. Soc.* **42**, 750–2.

Cohn, P. M. (1962). Eine bemerkung über die multiplikative gruppe eines körpers. *Arch. Math.* **13**, 344–8.

—— (1966). On the structure of the GL_2 of a ring. *Publ. Math. IHES* **30**, 5–53.

—— (1971). *Free rings and their relations*. Academic Press, New York.

—— (1977). *Algebra,* Vol. 2. John Wiley, Chichester–New York–Brisbane–Toronto.

—— (1985). *Free rings and their relations* (2nd edn). Academic Press, New York.

Coleman, D. B. and Passman, D. S. (1970). Units in modular group algebras. *Proc. Am. Math. Soc.* **25**, 510–2.

Coleman, R. F. (1986). On an archimedean characterization of the circular units, *J. Math.* **356**, 161–73.

Connell, I. G. (1963). On the group ring. *Can. J. Math.* **15**, 650–85.

Cornel, G. and Rosen, M. I. (1984). The l-rank of the real class group of cyclotomic fields. *Compositio Math.* **53**, 133–41.

Cornell, G. and Washington, L. C. (1985). Class numbers of cyclotomic fields. *J. Number Theory* **21**, 260–74.

Corner, A. L. S. (1963). Every countable reduced torsion-free ring is an endomorphism ring. *Proc. Lond. Math. Soc.* (3)**13**, 687–710.

Curtis, C. W. and Reiner, I. (1962). *Representation Theory of Finite Groups and Associative Algebras*. Interscience, New York.

Dedekind, R. (1871). *Über die Theorie der ganzen algebraischen Zahlen*. Supplement XI to Dirichlet's *Vorlesungen über Zahlentheorie* (2nd edn).

—— (1900). Über die Anzahl der Idealklassen in reinen kubischen Zahlkörpern. *J. Reine Angew. Math.* **121**, 40–123.

Degert, G. (1958). Über die Bestimmung der Grundeinheit gewisser reell-quadratischer Zahlkörper. *Abh. Math. Sem. Univ. Hamburg* **22**, 92–97.

Delaunay, B. N. (1928). Vollständige Lösung der unbestimmten Gleichung $X^3q + Y^3 = 1$ in ganzen Zahlen. *Math. Z* **28**, 1–9.

—— and Faddeev, D. K. (1964). The theory of irrationals of the third degree. *Am. Math. Soc. Transl. Math. Monogr.,* Vol. 10.

Dennis, R. K. (1976). Units in group rings. *J. Algebra* **43**, 655–64.

—— (1977). *The structure of the unit group of group rings*. Lecture Notes in Pure and Applied Math., 26, Marcel Dekker, New York, pp. 103–30.

de Orozco, M. A. and Vélez, W. Y. (1984). The torsion group of a field defined by radicals. *J. Number Theory* **19**, 283–94.

Dicker, R. M. (1968). A set of independent axioms for a field and a condition for a group to be the multiplicative group of a field. *Proc. Lond. Math. Soc.* (3)**18**, 114–24.

Dirichlet, P. G. Lejeune. (1840). Sur la theorie des nombres. *C.R. Acad. Sci. Paris* **10**, 285–8.

—— (1846). Zur Theorie der complexen Einheiten. *Verhand., Preuss. Acad. Wiss.*, 103–7.

Ditor, S. Z. (1971). On the group of units of a ring. *Am. Math. Monthly* **78**, 522–3.

Ehrlich, G. (1976). Units and one-sided units in regular rings. *Trans. Am. Math. Soc.* **216**, 81–90.

Eldridge, K. E. (1969). On ring structures determined by groups. *Proc. Am. Math. Soc.* **23**, 472–7.

—— (1970). Artinian q-torsion rings with p-groups. *Math. Ann.* **185**, 13–16.

—— (1973). On rings and groups of units. In *Colloq. Math. Soc. Janos Bolyai, 6*; Rings, modules and radicals. North-Holland, Amsterdam, pp. 177–81.

—— and Fischer, I. (1967). D.C.C. rings with a cyclic group of units. *Duke Math. J.* **34**, 243–8.

Ennola, V. (1972). On relations between cyclotomic units. *J. Number Theory* **4**, 236–47.

Epstein, P. (1934). Zur Auflösbarkeit der Gleichung $x^2 - Dy^2 = -1$. *J. Reine Angew. Math.* **171**, 243–52.

Farahat, H. K. (1965). The multiplicative groups of a ring. *Math. Z.* **87**, 378–84.

Feng, K. (1982). The rank of group of cyclotomic units in abelian fields. *J. Number Theory* **14**, 315–26.

Fisher, J. L., Parmenter, M. M., and Sehgal, S. K. (1976). Group rings with solvable n-Engel unit groups. *Proc. Am. Math. Soc.* **59**, 195–200.

Fomin, C. V. (1937). Über periodische Untergruppen der unendlichen abelschen Gruppen. *Mat. Sb.* **2**, 1007–9.

Franz, W. (1935). Über die Torsion einer Uberdeckung. *Crelles J.* **173**, 245–54.

Frei, G. (1982). Fundamental systems of units in biquadratic parametric number fields. *J. Number Theory* **15**, 295–303.

—— and Levesque, C. (1979). Independent systems of units in certain algebraic number fields. *J. Reine Angew. Math.* **311/312**, 116–44.

——, —— (1980). On an independent system of units in the field $K = \mathbb{Q}(\sqrt[n]{D^n \pm d})$ where $d \mid D^n$. *Abh. Math. Sem. Univ. Hamburg* **51**, 160–3.

Fuchs, L. (1958). *Abelian groups*. Publ. house of the Hungar, Acad. Sci., Budapest.

—— (1960). *Abelian groups*. Pergamon Press, London.

—— (1970). *Infinite abelian groups,* Vol. I. Academic Press, New York.

—— (1973). *Infinite abelian groups,* Vol. II. Academic Press, New York.

Furukawa, T. (1986). The group of normalized units of a group ring. *Osaka J. Math.* **23**, 217–21.

Furuta, Y. (1959). Norm of units in quadratic fields. *J. Math. Soc. Japan* **11**, 139–45.

Galovich, S. and Rosen, M. (1982). Units and class groups in cyclotomic function fields. *J. Number Theory* **14**, 156–84.

——, ——, Reiner, I., and Ullom, S. (1972). Class groups for integral representations of metacyclic groups. *Mathematika* **19**, 105–11.

Garbanati, D. (1976). Units with norm-1 and signature of units. *J. reine Angew. Math.* **283/284,** 164–75.

Gauss, C. F. (1832). Theoria residuorum biquadraticorum, *Comm. Sec. Comm. Soc. Reg. Scient. Göttingensis recent.* Werke, Göttingen 1863, II, pp. 93–198.

Gay, D. and Vélez, W. Y. (1981). The torsion group of a radical extension. *Pacif. J. Math.* **92,** 317–27.

Gilbarg, D. (1942). The structure of the group of p-adic 1-units. *Duke Math. J.* **9,** 262–71.

Gilmer, R. (1963). Finite rings having a cyclic multiplicative group of units, *Am. J. Math.* **85,** 447–52.

—— (1972). *Multiplicative ideal theory.* Marcel Dekker, New York.

—— (1984). *Commutative semigroup rings.* Univ. of Chicago Press, Chicago.

—— and Heitman, R. C. (1979). The group of units of a commutative semigroup ring. *Pacif. J. Math.* **85,** 49–64.

—— and Teply, M. L. (1977). Units of semigroup rings. *Comm. Algebra,* **5**(12), 1275–303.

Godwin, H. J. (1960). The determination of units in totally real cubic fields. *Proc. Camb Phil. Soc.* **56,** 318–21.

Goncalves, J. Z. (1984a). Free groups in subnormal subgroups and the residual nilpotence of the group of units of group rings. *Can. Math. Bull.* **27**(3), 365–70.

—— (1984b). Free subgroups of units in group rings, *Can. Math. Bull.* **27**(3), 309–12.

—— (1985a). Free subgroups in the group of units of group rings II. *J. Number Theory* **21**(2), 121–7.

—— (1985b). Integral group rings whose group of units is solvable: an elementary proof. *Bol. Soc. Mat.* **16,** 1–9.

—— (1986a). Free subgroups and the residual nilpotence of the group of units of modular and p-adic group rings. *Can. Math. Bull.* **29**(3), 321–8.

—— (1986b). Group rings with solvable unit groups. *Comm. Algebra* **14,** 1–20.

—— and Mandel, A. (1986). Are there free groups in division rings? *Israel J. Math.* **53,** 69–80.

Gorenstein, D. (1968). *Finite groups.* Harper & Row, New York.

Greenfield, G. R. (1978). A note on subnormal subgroups of division rings. *Can. J. Math.* **30,** 161–3.

—— (1981). Subnormal subgroups of p-adic division algebras. *J. Algebra* **73,** 65–9.

Groenewald, N. J. (1980). Units of the group ring, *Can. Math. Bull.* **23,** 445–8.

Guan, Chi-Wen (1963). On the structure of multiplicative group of discretely valued complete fields. *Sci. Sinica* **12,** 1079–103.

Hafner, P. (1968). Automorphismen von binären quadratischen Formen. *Elem. Math.* **23,** 25–30.

Hall, M., Jr (1959). *The theory of groups.* Macmillan, New York.

Hallett, J. T. and Hirsch, K. A. (1965). Torsion-free groups having finite automorphism groups. I. *J. Algebra* 2, 287–98.

—— and —— (1970). Die Konstruktion von Gruppen mit vorgeschriebenen Automorphismengruppen. *J. Reine Angew. Math.* **241,** 32–46.

Halter-Koch, F. (1975). Unanbhaengige Einheitsysteme fuer eine allgemeine Klasse algebraischer Zahlkörper. *Abh. Math. Sem. Univ. Hamburg* **43,** 85–91.

—— and Stender, H. J. (1974). Unanbhängige Einheiten für die Körper $K = \mathbb{Q}(\sqrt[n]{D^n \pm d})$ mit $d \mid D^n$. *Abh. Math. Sem. Univ. Hamburg* **42,** 33-40.

Harder, G. (1969). Minkowskische Reduktionstheorie über Funktionkörpern. *Invent. Math.* **7,** 33–54.

Hartley, B. and Pickel, P. F. (1980). Free subgroups in the unit groups of integral group rings. *Can. J. Math.* **32**, 1342–52.

Hasse, H. (1948). Die Einheitengruppe in einem total-reellen nicht-zyklichen kubischen Zahlkörper und in zugehörigen bikubischen Normalkörper. *Archiv. Math.* **1**, 42–6.

—— (1965). Über mehrklassige, aber eingeschlcchtige reell-quadratische Zahlkörper. *Elemente der Mathematik* **20**, 49–72.

—— (1980). *Number theory.* Springer-Verlag, Berlin–Heidelberg–New York.

—— and Bernstein, L. (1965). Einheiten berechnung mittels des Jacobi–Perronschen Algorithmus. *J. Reine Angew. Math.* **218**, 51–69.

—— and —— (1969). An explicit formula for the units of algebraic number field of degree *n*. *Pacif. J. Math.* **30**, 293–365.

Hensel, K. (1916). Die multiplikative Darstellung der algebraischen Zahlen fur den bereich eines Beliebigen Primteilers. *J. Reine Angew. Math.* **146**, 189–215.

Herstein, I. N. (1953). Finite multiplicative subgroups in division rings. *Pacif. J. Math.* **1**, 121–6.

—— (1956). Conjugates in division rings. *Proc. Am. Math. Soc.* **7**, 1021–2.

—— (1968). *Noncommutative rings,* The Carus Math. Monographs, 15. Maths Assoc. Am.

—— (1978). Multiplicative commutators in division rings. *Israel J. Math.* **31**, 180–8.

—— (1980). Multiplicative commutators in division rings II. *Rend. Circ. Math. Palermo,* Serie II **XXIX**, 485–9.

—— and Scott, W. R. (1963). Subnormal subgroups of division rings. *Can. J. Math.* **15**, 80–3.

Higman, G. (1940a). Units in group rings. D. Phil. thesis, University of Oxford, Oxford.

—— (1940b). The units of group rings. *Proc. Lond. Math. Soc.* (2)**46**, 231–48.

Hilbert, D. (1897). Die Theorie der algebraischen Zahlkorper. *Ber. Jahres, D.M.* **4**, 175–546.

Hoechsmann, K. (1986). Functors on finite vector spaces and units in abelian group rings. *Can. Math. Bull.* **29**(1), 79–83.

—— and Sehgal, S. K. (1986). Units in regular elementary abelian group rings. *Arch. Math.* **47**, 413–7.

——, ——, and Weiss, A. (1985). Cyclotomic units and the unit group of an elementary abelian group ring. *Arch. Math.* **45**, 5–7.

Holzapfel, R. P. (1967). Eine Bemerkung zur Zahlentheorie von H. Hasse. *Wiss. Z. Humboldt. Univ. Berlin* **16**, 317.

Hua, L. K. (1949). Some properties of *s*-fields. *Proc. Nat. Acad. Sci. U.S.A.* **35**, 533–7.

—— (1950). On the multiplicative group of a *s*-field. *Sci. Record Acad. Sin.* **3**, 1–6.

Hughes, I. and Mollin, R. (1983). Totally positive units and squares. *Proc. Am. Math. Soc.* **87**, 613–6.

—— and Pearson, K. R. (1972). The group of units of the integral group ring $\mathbb{Z}S_3$. *Can. Math. Bull.* **15**, 529–34.

—— and Wei, C. (1972). Group rings with only trivial units of finite order. *Can. J. Math.* **24**, 1137–8.

Humbert, P. (1940). Théorie de la réduction des formes quadratiques définies positive dans un corp algébrique *K* fini. *Comment. Math. Helv.* **12**, 263–306.

Huppert, B. (1967). *Endliche Gruppen I.* Springer-Verlag, Heidelberg.

Hurwitz, A. (1933). Die unimodularen Substitutionen in einem algebraischen Zahlkörper. In *Collected works,* pp. 244–68. Basel.

Huzurbazar, M. Š. (1960). The multiplicative group of a division ring. *Soviet Math. Dokl.* **1**(2), 433–5.

—— (1961). On the theory of multiplicative groups of division rings. *Soviet Math. Dokl.* **2**, 241–3.

Ivory, L. R. (1980). A note on normal complements in mod p envelopes. *Proc. Am. Math. Soc.* **79**, 9–12.

Iwasawa, K. (1956). A note on the group of units of an algebraic number field. *J. Math. Pures Appl.* **35**, 189–92.

Iyanaga, S. (1935). Sur les classes d'idéaux dans les corps quadratiques. Paris.

Jacobi, C. G. J. (1868), Allgemeine Theorie der kettenbruchaehnlichen Algorithmen, in welchen jede Zahl aus drei vorhergehenenden gebildet wird. *J. Reine Angew Math.* **69**, 29–64.

Jacobson, N. (1945). Structure theory for algebraic algebras of bounded degree. *Ann. Math.* **46**, 605-707.

—— (1956). Structure of rings. *Am. Math. Soc. Colloq. Publ.* Vol. 37. Providence, R.I.

—— (1964). *Lectures in abstract algebra, III.* Theory of fields and Galois theory. Springer-Verlag, New York–Heidelberg–Berlin.

Janusz, G. J. (1973). *Algebraic number fields.* Academic Press, New York.

—— (1983). Finite subgroups of units of group algebras, I. *J. Algebra* **82**(2), 508–15.

Jespers, E. (1986). The group of units of a commutative semigroup ring of a torsion-free semigroup. In *Proc. Int. Conf. Group and Semigroup Rings,* (ed. G. Karpilovsky), pp. 35–41. North-Holland Mathematics Studies 126, North-Holland, Amsterdam.

Kaplansky, I. (1951). A theorem on division rings. *Can. J. Math.* **3**, 290–2.

—— (1972). *Fields and rings.* Chicago Lecture in Mathematics. The University of Chicago Press, Chicago.

Karpilovsky, G. (1980a). On certain group ring problems. *Bull. Austral. Math. Soc.* **21**, 329–50.

—— (1980b). On some properties of group rings. *J. Austral. Math. Soc.,* **29**, 385–92.

—— (1982). On units in commutative group rings. *Arch. Math.* **38**(5), 420–2.

—— (1983a). On finite generation of unit groups of commutative group rings. *Arch. Math.* **40**, 503–8.

—— (1983b). *Commutative group algebras.* Marcel Dekker, New York.

—— (1984). On group rings of ordered groups. *Colloq. Math.* **XLIX**, 1–4.

Khripta, I. I. (1971). On the multiplicative group of a group ring. Thesis, Uzhgorod.

—— (1972). The nilpotence of the multiplicative group of a group ring. *Math. Notes* **11**, 119–24.

Kleinert, E. (1981). Einheiten in $\mathbb{Z}[D_{2m}]$. *J. Number Theory* **13**, 541–61.

Kneser, M. (1975). Lineare Abhängigkeit von Wurzeln. *Acta Arith.* **26**, 307–8.

Kramer, K. (1985). Residue properties of certain quadratic units. *J. Number Theory* **21**, 204–13.

Krempa, J. (1977). On semigroup rings. *Bull. Acad. Polon. Sci. Ser. Math. Astron. Phys.* **25**, 225–31.

—— (1982). Homomorphisms of group rings. In *Banach Center Publications,* Vol. 9, pp. 233–55. PWN, Warsaw.

—— (1985). Finitely generated groups of units in group rings. Preprint.

Kronecker, L. (1845). De unitatibus complexis. Diss., Berlin.

—— (1857). Über complexe Einheiten. *J. Reine Angew. Math.* **53**, 176–81.

—— (1882). Grundzüge einer arithmetischer Theorie der algebraischen Grössen. *J. Reine Angew. Math.* **92**, 1–122; Werke II, 237–387.

Krull, W. (1931). Allgemeine Bewertungstheorie. *J. Reine Angew. Math.* **167**, 160–96.

Kubota, T. (1955). A note on units of algebraic number fields. *Nagoya Math. J.* **9**, 115–8.

—— (1957). Unit groups of cyclic extensions. *Nagoya Math. J.* **12**, 221–9.

Kulikov, Ya. (1945). On the theory of abelian groups of arbitrary power. *Mat. Sb.* **16**, 129–62.

Kummer, E. (1851). Mémoire sur la théorie des nombres complexes composés de racines de l'unite et de nombres entiers. *J. Math. Pures Appl.* **16**, 377–498.

—— (1870). Über die aus 31-ten Wurzeln der Einheit gebildeten complexen Zahlen. *Ber. Königl. Akad. Wiss. Berlin,* 755–66.

Kurschak, J. (1913). Uber Limesbildung und allgemeine Körpertheorie. *J. Reine Angew. Math.* **142**, 211–53.

Lagarias, J. C. (1978). Signatures of units and congruences (mod 4) in certain real quadratic fields. *J. Reine Angew. Math.* **301**, 142–6.

—— (1980). Signatures of units and congruences (mod 4) in certain totally real fields. *J. Reine Angew. Math.* **320**, 1–5.

Lang, S. (1965). *Algebra.* Addison-Wesley, Reading, Mass.

—— (1970). *Algebraic number theory.* Addison-Wesley, Reading, Mass.

—— (1978). *Cyclotomic fields.* Springer-Verlag, New York–Heidelberg–Berlin.

—— (1980). *Cyclotomic fields, II.* Springer-Verlag, New York.

—— (1982). Units and class groups in number theory and algebraic geometry. *Bull. Am. Math. Soc.* **6**(3), 253–316.

Lansky, C. (1970a). The group of units of a simple ring I. *J. Algebra* **15**, 554–69.

—— (1970b). The group of units of a simple ring, II. *J. Algebra* **16**, 108–28.

Latimer, C. G. (1934). On the units in a cyclic field. *Am. J. Math.* **56**, 69–74.

Lednev, N. A. (1939). On units of relative cyclic algebraic fields. *Mat. Sb.* **6**, 227–61.

Levesque, C. (1979). A class of fundamental units and some classes of Jacobi–Perron algorithms in pure cubic fields. *Pacif. J. Math.* **81**(2), 447–66.

—— (1985). An independent system of units in certain algebraic number fields. *Can. J. Math.* **XXXVII**(4), 644–63.

Liang, J. (1972). On relations between units of normal algebraic number fields and their subfields. *Acta Arith.* **XX**, 331–44.

Lichtman, A. I. (1963). On the normal subgroups of the multiplicative group of a skewfield. *Soviet Math. Dokl.* **153**, 1424–9.

—— (1977). On subgroups of the multiplicative group of skewfields. *Proc. Am. Math. Soc.* **63**, 15–16.

—— (1978). Free subgroups of normal subgroups of the multiplicative group of skewfields. *Proc. Am. Math. Soc.* **71**(2), 174–8.

—— (1982). On normal subgroups of multiplicative group of skewfields generated by a polycyclic-by-finite group. *J. Algebra* **78**, 548–77.

—— (1987). Free subgroups in linear groups over some skewfields. *J. Algebra* **105**, 1–28.

Losey, G. (1974). A remark on the units of finite order in the group ring of a finite group. *Can. Math. Bull.* **17**, 129–30.

Malek, E. (1972). On the group of units of a finite *R*--algebra. *J. Algebra* **23**, 538–52.

Masley, J. M. (1976). Solution of small class number problems for cyclotomic fields. *Compositio Math.* **33**, 179–86.

—— (1979). Where are number fields with small class numbers? In *Springer Lecture Notes,* No. 751, pp. 221–42. Springer-Verlag, Berlin.

Masuda, K. (1961). Note on characters of the group of units of algebraic number fields. *Tohoka Math. J.* **13**(2), 248–52.

Matsumura, H. (1970). *Commutative algebra.* Benjamin, New York.

May, W. (1969). Commutative group algebras. *Trans. Am. Math. Soc.* **136**, 139–49.

—— (1970). Unit groups of infinite abelian extensions. *Proc. Am. Math. Soc.* **25**(3), 680–3.

—— (1971). Invariants for commutative group algebras. *Illinois J. Math.* **15**(3), 525–31.

—— (1972). Multiplicative groups of fields. *Proc. Lond. Math. Soc.* (3) **24**, 295–306.

—— (1976a). Isomorphism of group algebras. *J. Algebra* **40**(1), 10–18.

—— (1976b). Group algebras over finitely generated rings. *J. Algebra* **39**(2), 483–511.

—— (1979a). Multiplicative groups under field extension. *Can. J. Math.* **XXXI**(2), 436–40.

—— (1979b). Modular group algebras of totally projective *p*-primary groups. *Proc. Am. Math. Soc.* **76**(1), 31–4.

—— (1980). Fields with free multiplicative groups modulo torsion. *Rocky Mountain J. Math.* **10**(3), 599–604.

—— (1986). Unit groups and isomorphism theorems for commutative group algebras. In *Proc. Int. Conf. Group and Semigroup Rings* (ed.) G. Karpilovsky pp. 163–78. North-Holland Mathematics Studies 126. North-Holland.

Milnor, J. (1966). Whitehead torsion. *Bull. Am. Math. Soc.* **3**, 358–426.

—— (1971). Introduction to algebraic *K*-theory. *Ann. Math. Studies,* No. 72. Princeton, N.J.

Minkowski, H. (1900). Zur Theorie der Einheiten in den algebraischen Zahlkörpern. *Nachr. Ges. Wiss. Göttingen,* 90–3.

Mitsuda, T. (1986). On the torsion units of integral dihedral group rings. *Comm. Algebra* **14**(9), 1707–28.

Miyata, T. (1980). On the units of the integral group ring of a dihedral group. *J. Math. Soc. Japan* **32**, 703–8.

Mollov, T. Ž. (1971). On multiplicative groups of modular group algebras of primary abelian groups of any cardinality, I. *Publ. Math. Debrecen* **18**, 9–21.

—— (1972). On multiplicative groups of modular group algebras of primary abelian groups of any cardinality, II. *Publ. Math. Debrecen* **19**, 87–96.

—— (1975). Sylow *p*-subgroups of the unit groups of modular group algebras of abelian *p*-groups. *Serdika* **I**(3–4), 249–60.

—— (1976). Sylow *p*-subgroups of group algebras of countable abelian groups over a field of characteristic *p*. *Serdika* **2**(3), 219–35.

Monastyrnyi, V. I. (1973). Subnormal periodic subgroups of the multiplicative group of a field (in Russian). *Mathematics (Russian),* pp. 13–14, Minsk. Gos. Ped. Inst., Minsk.

Montgomery, M. S. (1969). Left and right inverses in group algebras. *Bull. Am. Math. Soc.* **75**, 539–40.

Mordell, L. J. (1953). On the linear independence of algebraic numbers. *Pacif. J. Math.* **3**, 625–30.

—— (1960). On a Pellian equation conjecture. *Acta Arith.* **6**, 137–44.

—— (1961). On a Pellian equation conjecture II. *J. Lond. Math. Soc.* **36**, 282–88.

Morikawa, R. (1968). On the unit group of an absolutely cyclic number field of degree five. *J. Math. Soc. Japan* **20**(1–2), 263–5.

Morishima, T. (1933). Über die Einheiten und Idealklassen des Galoischen Zahlkorpers und die Theorie der Kreiskörper der l^v-ten Einheitswurzeln. *Japan. J. Math.* **10**, 83–126.

—— (1934). Uber die Theorie der Kreiskörper der l^v-ten Einheitswurzeln. *Japan. J. Math.* **11**, 225–40.

Motose, K. and Ninomiya, Y. (1972). On the solvability of unit groups of group rings. *Math. J. Okayama Univ.* **15**, 209–14.

—— and Tominaga, H. (1969). Group rings with nilpotent unit groups. *Math. J. Okayama Univ.* **14**, 43–6.

——, —— (1971). Group rings with solvable unit groups. *Math. J. Okayama Univ.* **15**, 37–40.

Musson, I. and Weiss, A. (1982). Integral group rings with residually nilpotent unit groups. *Arch. Math.* **38**, 514–30.

Nagao, H. (1959). On $GL(2, K[x])$. *J. Inst. Polytech. Osaka City Univ.* Ser. A10, 117–21.

Nagell, T. (1922). Vollständige Lösung einiger unbestimmer Gleichungen dritten Grades. *Skr. NorskeVid. Akad. Oslo,* I, *Mat.-Naturv. Klasse,* No. 14, 1–13.

—— (1923). Über die Einheiten in reinen kubischen Zahlkörpern. *Skr. NorskeVid. Akad. Oslo,* I. *Mat.-Naturv. Klasse,* No. 11, 1–34.

—— (1925). Solution compléte de quelques équations cubiques á deux indéterminées. *J. Math. Pures Appl.* (9) 4, 204–70.

—— (1926). Über eininge Kubische Gleichungen mit zwei Unbestimmten. *Math. Z.* **24**, 422–47.

—— (1970). Sur un type particulier d'unites algébriques. *Arkiv. Mat.* **8**, 163–84.

Nakagoshi, N. (1979). The structure of the multiplicative group of residue classes modulo P^{N+1}. *Nagoya Math. J.* **73**, 41–60.

Nakamara, T. (1970). On the determination of the fundamental units of certain real quadratic fields. *Mem. Fac. Sci. Kyusku* **A24**, 300–304.

Narkiewicz, W. (1974). *Elementary and analytic theory of algebraic numbers.* PWN, Warsaw.

Newman, M. (1971). Units in cyclotomic number fields, *J. Math.* **250**, 3–11.

Nicholson, W. K. (1973). Semiperfect rings with abelian group of units. *Pacif. J. Math.* **49**(1), 191–8.

O'Meara, O. T. (1965). On the finite generation of linear groups over Hasse domains. *J. Reine Angew. Math.* **217**, 79–108.

Onishi, H. (1986). Conjugacy problem in $GL_2(\mathbb{Z}[\sqrt{-1}])$ and units of quadratic extensions of $\mathbb{Q}(\sqrt{-1})$. *Trans. Am. Math. Soc.* **293**(1), 83–98.

Ostrowski, A. (1913). Über eininge Fragen der allgemeinen Körpertheorie. *J. Reine Angew. Math.* **143**, 255–84.

—— (1917). Über sogenannte perfekte Körper. *J. Reine Angew. Math.* **147**, 191–204.

—— (1918). Über eininge Lösungen der Funktionalgleichung $\phi(x)\phi(y) = \phi(xy)$. *Acta Math.* **41**, 271–84.

—— (1935). Untersuchungen zur arithmetischer Theorie der Körper. *Math. Z.* **39**, 269–320, 321–61, 361–404.

Ozeki, K. (1986). On binomial units in certain cubic fields. *Proc. Am. Math. Soc.* **98**(2), 215–6.

Passman, D. S. (1977). *Algebraic structure of group rings.* Interscience, New York.

—— and Smith, P. F. (1981). Units in integral group rings. *J. Algebra* **69**, 213–39.

Pearson, K. R. (1972). On the units of a modular group ring. *Bull. Austral. Math. Soc.* **7**, 169–82.

—— (1973). On the units of a modular group ring, II. *Bull. Austral. Math. Soc.* **8**, 435–42.

—— and Schneider, J. E. (1970). Rings with a cyclic group of units. *J. Algebra* **16**, 243–51.

Pei, D. Y. and Feng, K. (1980). On independence of the cyclotomic units. *Acta Math. Sinica* **23**, 773–8.

Perlis, S. and Walker, G. L. (1950). Abelian group algebras of finite order. *Trans. Am. Math. Soc.* **68**, 420–6.

Perron, O. (1907). Grundlagen fuer eine Theorie des Jacobischen Kettenbruchalgorithmus. *Math. Ann.* **64**, 1–76.

Perrot, J. (1888). Sur l'equation $t^2 - Du^2 = -1$. *J. Reine Angew. Math.* **102**, 188–223.

Pohst, M. and Zassenhaus, H. (1977). An effective number geometric method of computing the fundamental units of an algebraic number field. *Maths Comput.* **31**, No. 139, 754–70.

—— and —— (1982). On effective computation of fundmental units, I. *Maths Comput.* **38**, No. 157, 275–91.

——, ——, and Weiler, P. (1982). On effective computation of fundamental units, II. *Maths Comput.* **38**, No. 157, 293–329.

Polcino Milies, C. (1972). The units of the integral group ring $\mathbb{Z}D_4$. *Bol. Soc. Brasil, Mat.* **4**, 85–92.

—— (1976). Integral group rings with nilpotent unit groups. *Can. J. Math.* **28**, 954–60.

—— and Sehgal, S. K. (1984). Torsion units in integral group rings of metacyclic groups. *J. Number Theory* **19**, 103–14.

——, ——, and Ritter, J. (1986). On a conjecture of Zassenhaus on torsion units in integral group rings II. *Proc. Am. Math. Soc.* **97**, 201–6.

Pollaczek, F. (1924). Über die irregularen Kreiskörper der l-ten und l^2-ten Einheitswurzeln. *Math. Z.* **21**, 1–38.

Prüfer, H. (1923). Untersuchungen über die Zerlegbarkeit der abzählbaren primaren abelschen Gruppen. *Math. Z.* **17**, 35–61.

Raggi Cárdenas, F. F. (1967). Units in group rings I. *An. Inst. Mat. Univ. Nac. Autonoma Mexico* **7**, 27–35.

—— (1968). Units in group rings II. *An. Inst. Mat. Univ. Nac. Autonoma Mexico* **8**, 91–103.

Ramachandra, K. (1966). On the units of cyclotomic fields. *Acta Arith.* **12**, 165–73.

Redei, L. (1932). Uber die Klassenzahl und Fundamentaleinheit des reellen quadratischen Zahlkörpers. *Mat. Természett. Értes* **48**, 648–82.

—— (1934). Über die Grundeinheit und die durch 8 teilbaren Invariaten der absoluten Klassengruppe in quadratischen Zahlkörpern. *J. Reine Angew. Math.* **171**, 131–48.

—— (1935). Über die Pellsche Gleichung $t^2 - du^2 = -1$. *J. Reine Angew. Math.* **173**, 193–211.

—— (1938). Ein neues Zahlentheoretisches Symbol mit Anwendungen auf die Theorie der quadratischen Zahlkörper. *J. Reine Angew. Math.* **180**, 1–43.

—— (1943). Ubcr den geraden teil der Ringklassengruppe quadratischer Zahlkörper, die Pellsche Gleichung und die Diophantische Gleichung $rx^2 + sy^2 = z^{2n}$. *Math. Naturwiss. Anz. Ungar. Akad. Wiss.* **62**, 13–34, 35–47, 48–62.

—— (1953). Bedingtes Artinsches Symbol mit Anwendung in der Klassenkörpertheorie. *Acta Math. Hungar.* **4**, 1–29.

Rella, T. (1920). Über die multiplikative Darstellung von algebraischen Zahlen eines Galoisschen Zahlkörpers fur den Bereich eines beliebigen Primteilers. *J. Reine Angew. Math.* **150**, 157–74.

Richaud, C. (1866). Sur la resolution des equations $x^2 - Ay^2 = \pm 1$. *Atti Acad. Pontif. Nuovi Lincei,* 177–82.

Riehm, C. (1970). The norm 1 group of p-adic division algebra. *Am. J. Math.* **92**, 499–523.

Ritter, J. and Sehgal, S. K. (1983). On a conjecture of Zassenhaus on torsion units in integral group rings. *Math. Ann.* **264**, 257–70.

Roggenkamp, K. W. (1981). Units in integral metabelian group rings, I. Jackson's unit theorem revisited. *Q. J. Math.* (2), **32**, 209–24.

—— and Scott, L. L. (1983). Units in metabelian group rings: Nonsplitting examples for normalized units. *J. Pure Appl. Algebra* **27**, 299–314.

—— and —— (1985a). Units in group rings: splittings and the isomorphism problem. *J. Algebra* **96**(2), 397–417.

—— and —— (1985b). Isomorphisms of p-adic group rings. Preprint. *Ann. Maths.*

Roquette, P. (1958). Einheiten und Divisorenklassen in endlich erzenglar Körpern. *J. Deutsch. Math. Verein* **60**, 1–27.

—— (1959). Abspaltung des Radikals in vollständigen lokalen Ringen. *Hamb. Abh.* **23**, 75–113.

—— (1960). Some fundamental theorems on abelian function fields. In *Proc. Int. Congr. Math.* 1958, pp. 322–9. Cambridge University Press.

Rosen, M. (1973). S-units and S-class group in algebraic function fields. *J. Algebra* **26**, 98–108.

Rosenberg, A. (1958). The structure of the infinite general linear group. *Ann. Math.* **68**, 278–93.

Rudman, R. J. (1973). On the fundamental unit of a purely cubic field. *Pacif. J. Math.* **46**, 253–6.

Saksonov, A. I. (1971). On group rings of finite groups, I. *Publ. Math. Debrecen* **18**, 187–209.

Samuel, P. (1966). Á propos due théoréme des unités. *Bull. Sci. Math.* (2) **90**, 84–96.

Sandling, R. (1974). Group rings of circle and unit groups. *Math. Z.* **140**, 195–202.

—— (1981). Graham Higman's thesis "Units in group rings". *Springer Lecture Notes,* No. 882, pp. 93–116. Springer-Verlag, Berlin.

—— (1984). Units in the modular group algebra of a finite abelian p-group. *J. Pure Appl. Algebra* **33**, 337–46.

Scarowsky, M. (1984). On units of certain cubic fields and the diophantine equation $x^3 + y^3 + z^3 = 3$. *Proc. Am. Math. Soc.* **91**(3), 351–6.

Schenkman, E. (1958). Some remarks on the multiplicative group of a s-field. *Proc. Am. Math. Soc.* **9**, 231–5.

—— (1964). On the multiplicative group of a field. *Arch. Math.* **15**, 282–5.

—— and Scott, W. R. (1960). A generalization of the Cartan–Brauer–Hua theorem. *Proc. Am. Math. Soc.* **11**, 396–8.

Schmidt, F. K. (1930). Zur Klassenkörpertheorie im Kleinen. *J. Reine Angew. Math.* **162**, 155–68.

Schur, I. (1911). Über Gruppen periodischer substitutionen, *S.-B. Preuss. Akad. Wiss.,* 619–27.

Scott, W. R. (1957). On the multiplicative group of a division ring. *Proc. Am. Math. Soc.* **8**, 303–5.

Sehgal, S. K. (1970). Units in commutative integral group rings. *Math. J. Okayama Univ.* **14**, 135–8.

—— (1975). Certain algebraic elements in group rings. *Arch. Math.* **26**, 139–43.

—— (1978). *Topics in group rings.* Marcel Dekker, New York and Basel.

—— and Weiss, A. (1986). Torsion units in integral group rings of some metabelian groups. *J. Algebra* **103**(2), 490–9.

—— and Zassenhaus, H. J. (1977a). Group rings whose units form an *FC*-group. *Math. Z.* **153**, 29–35.

—— (1977b). Integral group rings with nilpotent unit groups. *Comm. Algebra* **5**, 101–11.

Serre, J.-P. (1979). *Arithmetic Groups, Homological Group Theory,* édité par C. T. C. Wall, LMS Lect. Note Series No. 36, Cambridge Univ. Press, 105–36.

Setzer, B. (1978). Units in totally complex S_3 fields, *J. Number Theory* **10**, 244–9.

Siegel, C. L. (1972). Algebraische Abhängigkert von Wurzeln. *Acta Arith.* **21**, 59–64.

Sinnott, W. (1978). On the Stickelberger ideal and the circular units of a cyclotomic field. *Ann. Math.* **108**, 107–34.

—— (1980). On the Stickelberger ideal and the circular units of an abelian field. *Invent. Math.* **62**, 181–234.

—— (1981). On the Stickelberger ideal and the circular units of an abelian field. *Progr. Math.* **12**, 277–86.

Skolem, T. (1947). On the existence of a multiplicative basis. *Norske Vid. Selsk. Forh.* **2**, 4–7.

Smith, J. H. (1970). On *S*-units almost generated by *S*-units of subfields. *Pacif. J. Math.* **34**(3), 803–805.

Stender, H. J. (1969). Über die Grundeinheit fur spezielle unendliche Klassen reiner Kubischer Zahlkörper. *Abh. Math. Sem. Univ. Hamburg* **33**, 203–215.

—— (1972). Einheiten fur eine allgemeine klasse total reeller algebraischer Zahlkörper. *J. Reine Angew. Math.* **257**, 151–78.

—— (1973). Grundeinheiten fur einige unendliche klassen reiner biquadratischer Zahlkörper mit einer Anwendung auf die diophantische Gleichung $x^4 - ay^4 = \pm c (c = 1, 2, 4$ order 8). *J. Reine Angew. Math.* **264**, 207–20.

—— (1974). Uber die Einheitengruppe der reinen algebraischen Zahlkörper sechsten Grades. *J. Reine Angew. Math.* **268/269**, 78–93.

—— (1975). Eine Formel fur Grundeinheiten in reinen algebraischen Zahlkörpern dritten, vierten und sechsten Grades. *J. Number Theory* **7**, 235–250.

—— (1977). Losbare gleichungen $ax^n - by^n = c$ und Grundeinheiten für einige algebraische Zahlkörper vom Grade n, $n = 3, 4, 6$. *J. Reine Angew. Math.* **290**, 24–62.

Stewart, I. (1972). Finite rings with a specified group of units. *Math. Z.* **126**, 51–8.

Strojnowski, A. (1980). A note on *u.p.*-groups. *Comm. Algebra* **8**(3), 231–4.

Stuth, C. J. (1964). A generalization of the Cartan–Brauer–Hua theorem. *Proc. Am. Math. Soc.* **15**, 211–7.

Suzuki, M. (1982). *Group Theory I.* Springer-Verlag, Berlin–Heidelberg–New York.

Takaku, M. (1971). Units in real quadratic fields. *Nagoya Math. J.* **44**, 51–5.

Tano, F. (1889). Sur quelques théorémes de Dirichlet. *J. Reine Angew. Math.* **105**, 160–9.

Taylor, D. E. (1977). Groups whose modular group rings have soluble unit groups. In *Springer Lecture Notes in Mathematics*, No. 573, pp. 112–7. Springer-Verlag, Berlin.

Taylor, M. (1975). Galous module structure of class groups and units. *Mathematika* **22**, 156–60.

Tits, J. (1972). Free subgroups in linear groups. *J. Algebra* **20**, 250–70.

Trotter, H. F. (1969). On the norm of units in quadratic fields. *Proc. Am. Math. Soc.* **22**, 198–201.

Uspensky, J. V. (1931). A method for finding units in cubic orders of a negative discriminant. *Trans. Am. Math. Soc.* **33**, 1–31.

Van der Linden, F. (1982). Class number computations of real abelian number fields. *Maths Comput.* **39**, 693–707.

Vandiver, H. S. (1929). A theorem of Kummer's concerning the second factor of the class number of a cyclotomic field. *Bull. Am. Math. Soc.* **35**, 333–5.

—— (1930). Some properties of a certain system of independent units in a cyclotomic field. *Ann. Math.* **31**, 123–5.

—— (1934). A note on units in super-cyclic fields. *Bull. Am. Math. Soc.* **40**, 855–8.

Vaserstein, L. N. (1969). On the stabilization of the general linear group over a ring. *Math. USSR Sb.* **8**, 383–400.

Voronoi, G. F. (1896). On a generalization of continued fractions (in Russian). Doctoral dissertation, Warsaw.

Wada, H. (1966). On the class number and the unit group of certain algebraic number fields. *J. Fac. Sci. Univ. Tokyo,* Sect. I **13**, 201–9.

—— (1970). A table of fundamental units of purely cubic fields. *Proc. Japan Acad.* **46**, 1135–40.

Wahlin, G. E. (1932). The multiplicative representation of the principal units of a relative cyclic field. *J. Reine Angew. Math.* **167**, 122–8.

Wall, C. T. C. (1974). Norms of units in group rings. *Proc. Lond. Math. Soc.* (3) **29**, 593–632.

Washington, L. C. (1982). *Introduction to cyclotomic fields*. Springer-Verlag, New York–Heidelberg–Berlin.

Watters, J. F. (1968). On the adjoint group of a radical ring, *J. Lond. Math. Soc.* **43**, 725–9.

Wedderburn, J. H. M. (1905). A theorem on finite algebras. *Trans. Am. Math. Soc.* **6**, 349–52.

Weiss, M. J. (1936). Fundamental systems of units in normal fields. *Am. J. Math.* **58**, 249–54.

Whitcomb, A. (1968). The group ring problem. Ph.D. thesis, University of Chicago.

Witt, E. (1931). Über die Kommutativität endlicher Schiefkörper. *Hamb. Abh.* **8**, 413.

—— (1952). Die algebraische Struktur des Gruppenringes einer endlichen Gruppe uber einem Zahlenkörper. *J. Math.* **190**, 231–45.

Yamamoto, Y. (1971). Real quadratic number fields with large fundamental units. *Osaka J. Math.* **8**, 261–70.

Yokoi, H. (1960). On unit groups of absolute abelian number fields of degree pq. *Nagoya Math. J.* **16**, 73–81.

—— (1968). On real quadratic fields containing units with norm -1. *Nagoya Math J.* **33**, 139–52.

—— (1970). On the fundamental unit of real quadratic fields with norm 1. *J. Number Theory* **2**, 106–15.

—— (1974). The diophantine equation $x^2 + dy^3 = 1$ and the fundamental unit of a pure cubic field $\mathbb{Q}(\sqrt[3]{d})$. *J. Reine Angew. Math.* **268/269**, 174–9.

Zalesskii, A. E. (1963). Solvable subgroups of the multiplicative group of a locally finite algebra. *Mat. Sb.* **61,** 408–17.

—— (1965). Solvable groups and crossed products. *Mat. Sb.* **67**(109), 154–60.

—— (1972). On a problem of Kaplansky. *Soviet Math.* **13,** 449–52.

Zariski, O. and Samuel, P. (1958). *Commutative algebra,* Vol. I. Van Nostrand, Princeton, N.J.

—— and —— (1960). *Commutative algebra,* Vol. II. Van Nostrand, Princeton, N.J.

Zassenhaus, H. J. (1936). Über endliche Fastkörper. *Hamb. Abh.* **11,** 187–220.

—— (1972). On units of orders. *J. Algebra* **20,** 368–95.

—— (1974). On the torsion units of finite group rings. In *Studies in mathematics,* pp. 119–26. Instituto de Alta Cultura, Lisboa.

Zhmud, E. M. and Kurennoi, G. C. (1967). On finite groups of units of an integral group ring. *Vesnik Kharkov Gos. Univ.* **26,** 20–26.

Author index

Subject index